Gene Amplification and Analysis

Volume 2
Structural Analysis of Nucleic Acids

Editors:
Jack G. Chirikjian
Georgetown University Medical Center
Washington, D.C. 20007

Takis S. Papas
Laboratory of Tumor Virus Genetics
National Cancer Institute
National Institutes of Health
Bethesda, MD. 20205

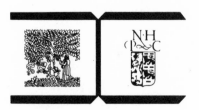

ELSEVIER/NORTH-HOLLAND
NEW YORK   AMSTERDAM   OXFORD

CHEMISTRY

6830-0384

©1981 by Elsevier North Holland, Inc.

Published by:

Elsevier North Holland, Inc.
52 Vanderbilt Avenue, New York, New York 10017

Sole distributors outside of the United States and Canada:

Elsevier Science Publishers B.V.
P.O. Box 211, 1000 AE Amsterdam, The Netherlands

Library of Congress Cataloging in Publication Data

Main entry under title:

Structural analysis of nucleic acids.
    (Gene amplification and analysis, ISSN 0275-2778; 2)

    Bibliography: p.
    Includes index
    1. Nucleic acids. I. Chirikjian, Jack G.   II. Papas, Takis S.   III. Series: Gene
    amplification and analysis; v. 2.
QP620.S76   574.87'328   81-5504
ISBN 0-444-00636-2   AACR2
ISSN: 0275-2778

Manufactured in the United States of America

GENE AMPLIFICATION AND ANALYSIS

ERRATA:
"The Use of Terminal Transferase for Molecular Cloning and Nucleic Acids
Analysis"

Ranajit Roychoudhury

42A

To Be Inserted Between Pages 42 and 43

In our previous studies with synthetic primer and Mg^{+2} ion, we detected
a maximum of two ribo- additions with paper chromatography for the separation
of products[4,12].

The proportion of mono- or di- addition products was influenced by the
primer-to-substrate ratio[12]. When the substrate concentration was 50-100 fold
higher than the primer concentration, more than two ribo- additions with Mg^{+2}
ion were detected only with purine triphosphates (rATP and rGTP) as substrates
(Roychoudhury and Wu, 1974, unpublished results). An improved separation procedure (homochromatography) permitted clear identification of the products.
The pyrimidine nucleotides still yielded a maximum of two ribo- additions, as
observed earlier[12]. Terminal transferase accepts dideoxynucleoside triphosphates as substrates[19,20]. This reaction was used for 3'-terminal labeling
of several synthetic oligonucleotides[21]. With dideoxyadenosine triphosphates
and Mg^{+2} ion, complete conversion of d(pT)6 to a single nucleotide addition
product, d(pT-T-T-T-T)-ddA, was obtained[21]. Thus, the reaction is useful
for single terminal addition without the alkali and phosphatase steps. The
ddNTP or cordycepin triphosphate labeled products can be used for DNA sequence
analysis[26] or for specific joining of oligonucleotides with T4 RNA ligase[27].
However, because of the absence of 3'-OH groups, such products cannot be used
for primer extension[4,13-15,22].

When Mg^{+2} ion was replaced with Co^{+2} ion, we noticed not only a highly
efficient primer utilization[23], but also multiple addition[14] of all four common
ribonucleotides at the 3' end of a given primer[23]. This higher priming efficiency in the presence of Co^{+2} ion prompted us to apply this reaction to the
labeling of duplex DNA fragments generated by different restriction endonucleases. We noticed that all types of duplex DNA ends were labeled with much
higher efficiency in the presence of Co^{+2} ion than in the presence of Mg^{+2}
ion, with either ribonucleoside or deoxyribonucleoside triphosphates as substrates[23].

Recently, $[\alpha-^{32}P]$cordycepin triphosphate (3'-deoxyadenosine-5'-triphos-
phate) has been used to label duplex DNA fragments in the presence of Co^{+2} ion,
and these DNA fragments were used for sequence analysis[26].

Such high labeling efficiency enabled us to use terminal transferase for
the analysis of the primary structure of nucleic acids (oligonucleotides and
specific regions of duplex DNA) and to use this reaction for molecular cloning
of important genes. This paper presents the relevant techniques and other information that may be useful to other workers.

To Be Inserted Between Pages 50 and 51

For preparative isolation, 2.5 ml reaction mixtures with 0.5 mM primer and 1 mM dNTP were incubated in the presence of 1,875 units of terminal transferase. Other conditions were similar to those described above. The products were chromatographed on a column (1 X 15 cm) of DE-52 cellulose. A linear gradient of 1,200 ml of 0.05-0.5 M NaCl in 20 mM sodium acetate (pH, 4.7) was used to separate oligonucleotides. The amount of radioactive nucleotide in each peak was determined by drying 100-μl aliquots on paper disks and counting in a scintillation counter. The total concentration of oligonucleotides was determined with the millimolar absorbance data of Cassani and Bollum[47].

From the position of the peak in the elution profile and the molar ratio of [^{14}C]nucleotide to total nucleotide in each peak, the chain length of each product was unequivocally established[46].

Specific oligonucleotide primer molecules are increasingly used in copying RNA and DNA templates for molecular cloning. We have investigated enzymatic extension of chemically synthesized short oligodeoxynucleotide primer with ribonucleotides. In order to compare the polymerization characteristics of all four common ribonucleotides under identical conditions and measure directly the amount of primer distributed in different addition products, the primer was labeled at the 5' end with a [^{32}P]phosphate residue. When this primer, d($\overset{*}{p}$T-T-T-T-T), was extended with ribonucleotides, the resulting products were expected to migrate slower and thus be well resolved by homochromatography[48].

$$\text{d}(\overset{*}{p}\text{T-T-T-T-T}) \quad \underset{\substack{\text{and terminal}\\ \text{transferase}}}{\overset{\text{Co}^{2+}\text{plus}}{\underline{\text{unlabeled rNTP}}}} \longrightarrow \begin{array}{l} \text{d}(\overset{*}{p}\text{T-T-T-T-T}) \\ \text{d}(\overset{*}{p}\text{T-T-T-T-T})\text{-rN} \\ \text{d}(\overset{*}{p}\text{T-T-T-T-T})\text{-rN...rN}_n \\ \text{(where n = 4 to 6 residues)} \end{array}$$

This technique[48] will show not only the distribution of products, but also the number of residues added in each reaction. By counting the radioactivity in each band, we can determine the exact amount of primer converted into a product. To check the maximal degree of polymerization, we used a 1,000-fold excess of rNTP.

The reaction mixture (15 μl) contained 10 pmol of d($\overset{*}{p}$T-T-T-T-T) and 10,000 pmol of unlabeled rNTPs in 140 mM potassium cacodylate, 30 mM Tris base (final pH, 6.9), 1 mM CoCl$_2$, 100 μM dithiothreitol, and 16 units of terminal transferase. After 22 h at 37°C, aliquot (3 μl) from each

Chapter 6

EXONUCLEASE VII OF E. COLI

John W. Chase and Lynne D. Vales

Chapter 7

THE EXTRACELLULAR NUCLEASE FROM ALTEROMONAS ESPEJIANA: AN ENZYME HIGHLY
SPECIFIC FOR NONDUPLEX STRUCTURE IN NOMINALLY DUPLEX DNAS`

Horace B. Gray, Jr., Thomas P. Winston, James L. Hodnett, Randy J. Legerski,
David W. Nees, Chik-Fong Wei, and Donald L. Robberson

Chapter 8

S_1 NUCLEASE OF ASPERGILLUS ORYZAE

George W. Rushizky

Chapter 12

SEQUENCE AND STRUCTURE ANALYSIS OF END-LABELED RNA WITH NUCLEASES

John N. Vournakis, James Celantano, Margot Finn, Raymond E. Lockard, Tanaji Mitra, George Pavlakis, Anthony Troutt, Margaret van den Berg, and Regina M. Wurst

Chapter 13

PHAGE T4 POLYNUCLEOTIDE KINASE

William R. Folk

Chapter 14

T4 RNA LIGASE AS A NUCLEIC ACID SYNTHESIS AND MODIFICATION REAGENT

Richard I. Gumport and Olke C. Uhlenbeck

Chapter 18

STRUCTURAL GENE IDENTIFICATION UTILIZING EUKARYOTIC CELL-FREE TRANSLATIONAL
SYSTEMS

Bruce M. Paterson and Bryan E. Roberts

Chapter 19

IN VITRO TRANSLATION OF EUKARYOTIC MESSENGER RNA

Don Hendrick

Chapter 20

DNA TOPOISOMERASES

James C. Wang and Karla Kirkegaard

Chapter 24

HYDROPHOBIC CHROMATOGRAPHY OF NUCLEIC ACIDS AND PROTEINS ON TRITYLATED AGAROSE

Peter Cashion, Ali Javed, Dolores Harrison, Jane Seeley, Victor Lentini, and Ganesh Sathe

Chapter 25

HYBRIDIZATION ANALYSIS OF SPECIFIC RNAs BY ELECTROPHORETIC SEPARATION IN AGAROSE GELS AND TRANSFER TO DIAZOBENZYLOXYMETHYL PAPER

J. Claiborne Alwine

Chapter 26

ELECTRON MICROSCOPY OF NUCLEIC ACIDS

Claude F. Garon

Preface

The series <u>Gene Amplification and Analysis</u> was established with the purpose of rapidly communicating information about various aspects of bio-technology. The second volume of the series emphasizes experimental pro-cedures dealing with the analysis of nucleic acid structure and was developed primarily as a technical resource for experimental techniques commonly utilized in molecular biology. Most of the papers in this volume deal with enzymatic procedures, although a number of important chemical methods have also been included. It is noteworthy that several articles discuss the use of previously reported enzymes in new roles as important reagents in the analysis of RNA and DNA. One of our original aims was to make the volume an inclusive one. This proved to be unachievable in view of the continuous developments in molecular biology. We encouraged individual authors to present a detailed presentation of experimental procedures and to include important descriptive information. Several areas were discussed by more than one group to include small but significant differences in these procedures.

We wish to thank the authors of the individual chapters for their contributions. Special appreciation is also extended to Bethesda Research Laboratories, Inc., (BRL, Inc.) for resources that made this effort possible.

Jack G. Chirikjian
Takis S. Papas
May 12, 1981

Contributors

Numbers in brackets indicate the page on which the author's contribution begins.

J. C. ALWINE, The Department of Microbiology, School of Medicine, University of Pennsylvania, Philadelphia, Pennsylvania (565)

C. BRADY, Laboratory of Biochemistry and Laboratory of Molecular Biology, National Cancer Institute, National Institutes of Health, Bethesda, Maryland (383)

P. CASHION, Biology Department, University of New Brunswick, Fredericton, New Brunswick, Canada (551)

J. CELANTANO, Department of Biology, Syracuse University, Syracuse, New York (267)

J. W. CHASE, The Department of Molecular Biology, Albert Einstein College of Medicine, Bronx, New York (147)

J. G. CHIRIKJIAN, Department of Biochemistry, Georgetown University, Washington, D.C. (1)

D. COURT, Laboratory of Biochemistry and Laboratory of Molecular Biology, National Cancer Institute, National Institutes of Health, Bethesda, Maryland (383)

R. J. CROUCH, Laboratory of Molecular Genetics, National Institute of Child Health and Human Development, National Institutes of Health, Bethesda, Maryland (217)

M. FINN, Department of Biology, Syracuse University, Syracuse, New York (267)

W. R. FOLK, Department of Biological Chemistry, University of Michigan Medical School, Ann Arbor, Michigan (299)

C. F. GARON, Laboratory of Biology of Viruses, National Institute of Allergy and Infectious Diseases, National Institutes of Health, Bethesda, Maryland (573)

M. L. GEFTER, Massachusetts Institute of Technology, Cambridge, Massachusetts (369)

H. B. GRAY, Department of Biophysical Sciences, University of Houston, Houston, Texas (169)

R. I. GUMPORT, The Department of Biochemistry, School of Chemical Sciences and School of Basic Medical Sciences, University of Illinois, Urbana, Illinois (313)

D. HARRISON, Biology Department, University of New Brunswick, Fredericton, New Brunswick, Canada (551)

C. P. HARTMAN, Bethesda Research Laboratories, Inc., P.O. Box 577, Gaithersburg, Maryland (17)

D. V. HENDRICK, Bethesda Research Laboratories, Inc., P.O. Box 577, Gaithersburg, Maryland (439)

J. L. HODNETT, Department of Biophysical Sciences, University of Houston, Houston, Texas (169)

K. IATROU, The Biological Laboratories, Harvard University, 16 Divinity Avenue, Cambridge, Massachusetts (537)

A. JAVED, Biology Department, University of New Brunswick, Fredericton, New Brunswick, Canada (551)

C. W. JONES, The Biological Laboratories, Harvard University, 16 Divinity Avenue, Cambridge, Massachusetts (537)

F. C. KAFATOS, Department of Biology, Harvard University, Cambridge, Massachusetts and University of Athens, Athens, Greece (537)

R. L. KARPEL, Department of Chemistry, University of Maryland Baltimore County, Catonsville, Maryland (509)

K. KIRKEGAARD, Department of Biochemistry and Molecular Biology, Harvard University, Cambridge, Massachusetts (455)

W. KONIGSBERG, Department of Molecular Biophysics and Biochemistry, Yale University, P.O. Box 333, New Haven, Connecticut (475)

R. J. LEGERSKI, Department of Biophysical Sciences, University of Houston, Houston, Texas (169)

V. LENTINI, Biology Department, University of New Brunswick, Fredericton, New Brunswick, Canada (551)

J. W. LITTLE, Department of Microbiology, Arizona Health Sciences Center, University of Arizona, Tucson, Arizona (135)

R. E. LOCKARD, Department of Biochemistry, The George Washington University Medical School, Washington, D.C. (229) (267)

J. L. MANLEY, Department of Biological Sciences, Columbia University, New York, New York (369)

K. MCKENNEY, Laboratory of Biochemistry and Laboratory of Molecular Biology, National Cancer Institute, National Institutes of Health, Bethesda, Maryland (383)

T. MITRA, Department of Biology, Syracuse University, Syracuse, New York   (267)

B. MOSS, Laboratory of Biology and Viruses, National Institute of Allergy and Infectious Diseases, National Institute of Health, Bethesda, Maryland (253)

D. W. NEES, Department of Biophysical Sciences, University of Houston, Houston, Texas   (169)

T. S. PAPAS, Laboratory of Tumor Virus Genetics, National Cancer Institute, National Institutes of Health, Bethesda, Maryland   (1)

B. M. PATERSON, National Cancer Institute, Laboratory of Biochemistry, Bethesda, Maryland   (417)

G. PAVLAKIS, National Institutes of Health, Recombinant DNA Unit, Bethesda, Maryland   (267)

D. RABUSSAY, Department of Biological Sciences, Florida State University, Tallahassee, Florida   (17)

L. RIESER, Department of Biology, Syracuse University, Syracuse, New York (229)

D. L. ROBBERSON, Department of Molecular Biology, University of Texas System Cancer Center, M.D. Anderson Hospital and Tumor Institute, Houston, Texas   (169)

B. E. ROBERTS, Department of Biological Chemistry, Harvard Medical School, 25 Shattuck Street, Boston, Massachusetts   (417)

M. ROSENBERG, Laboratory of Biochemistry and Laboratory of Molecular Biology, National Cancer Institute, National Institutes of Health, Bethesda, Maryland   (383)

R. ROYCHOUDHURY, Division of Molecular Biology, International Plant Research Institute, San Carlos, California   (41)

G. W. RUSHIZKY, Laboratory of Nutrition and Endocrinology, NIAMDD, NIH, Bethesda, Maryland   (205)

G. SATHE, Biology Department, University of New Brunswick, Fredericton, New Brunswick, Canada   (551)

U. SCHMEISSNER, Laboratory of Biochemistry and Laboratory of Molecular Biology, National Cancer Institute, National Institutes of Health, Bethesda, Maryland   (383)

R. A. SCHULZ, Department of Biochemistry, Georgetown University, Washington, D.C.   (1)

J. SEELEY, Biology Department, University of New Brunswick, Fredericton, New Brunswick, Canada   (551)

H. SHIMATAKE, Laboratory of Biochemistry and Laboratory of Molecular Biology, National Cancer Institute, National Institutes of Health, Bethesda, Maryland   (383)

R. T. SIMPSON, Developmental Biochemistry Section, National Institute of
   Arthritis, Metabolism and Digestive Diseases, National Institutes of
   Health, Bethesda, Maryland   (347)

G. Thireos, The Biological Laboratories, Harvard University, 16 Divinity Avenue,
   Cambridge, Massachusetts   (537)

A. TROUTT, Department of Biology, Syracuse University, Syracuse, New York (267)

S. G. TSITILOU, Department of Biology, University of Athens, Panepistimiopolis,
   Kouponia, Athens, Greece   (537)

O. C. UHLENBECK, The Department of Biochemistry, School of Chemical Sciences and
   School of Basic Medical Sciences, University of Illinois, Urbana,
   Illinois   (313)

L. D. VALES, The Department of Molecular Biology, Albert Einstein College of
   Medicine, Bronx, New York   (147)

M. VON DEN BERG, Department of Biology, Syracuse University, Syracuse, New
   York   (267)

J. N. VOURNAKIS, Department of Biology, Syracuse University, Syracuse, New
   York   (267) (229)

J. C. WANG, Department of Biochemistry and Molecular Biology, Harvard University,
   Cambridge, Massachusetts   (455)

C. -F. WEI, Department of Biophysical Sciences, University of Houston, Houston,
   Texas   (169)

K. R. WILLIAMS, Department of Molecular Biophysics and Biochemistry, Yale
   University, P.O. Box 3333, New Haven, Connecticut   (475)

T. P. WINSTON, Department of Biophysical Sciences, University of Houston,
   Houston, Texas   (169)

R. M. WURST, Department of Biology, M.I.T., Cambridge, Massachusetts   (267)

Gene Amplification
and Analysis

Volume 2
Structural Analysis of Nucleic Acids

ENZYMATIC SYNTHESIS OF DUPLEX DNA BY
AVIAN MYELOBLASTOSIS VIRAL REVERSE TRANSCRIPTASE

Takis S. Papas,[+] Robert A. Schulz[*] and Jack G. Chirikjian[*]

[+]Laboratory of Tumor Virus Genetics
National Cancer Institute
National Institute of Health
Bethesda, Maryland 20205

[*]Department of Biochemistry
Georgetown University Medical Center
Washington, D.C. 20007

I. INTRODUCTION

The discovery of reverse transcriptase had an immediate and profound impact upon the understanding of RNA tumor viruses and more generally on the field of molecular biology[1,2]. An important application of the AMV DNA polymerase is the reverse transcription of mRNA to enable functional structural genes to be constructed[3-8]. As a result, opportunities have arisen to directly study the chromosomal arrangements of specific genes and the ways in which they are regulated. cDNA transcripts of retroviral genomic RNA have been used to study integrated proviral sequences in normal and transformed cells[9]. Such probes, specific for particular viruses, have been used to screen libraries of normal cellular DNA to identify naturally occurring proviral sequences and to study the organization of these sequences within the genome[10]. Several recent reviews have dealt with the properties and uses of reverse transcriptase[11,12].

This chapter summarizes experimental conditions that we have used to synthesize a complete duplex dNA of the avian myeloblastosis virus (AMV) 35S genomic RNA with the AMV reverse transcriptase. Because the AMV genome is quite large (approximately 7,500 bases), the method presented here should be readily applicable to similar RNAs that contain poly rA.

II. PURIFICATION OF AMV REVERSE TRANSCRIPTASE

In transcribing mRNAs utilizing AMV reverse transcriptase, an important requirements is to obtain concentrated enzyme purified free of contaminating degradative activities. A simple and rapid procedure for preparing AMV reverse transcriptase from viral protein has recently been published by Myers et al.[14]. The enzyme, free of ribonuclease and deoxyribonuclease activities, transcribes high molecular weight RNA into full-size DNA. The procedure for purification of the enzyme uses two chromatography steps and can be completed within 36 hrs[14]. For synthesis of complete cDNA from RNA template, the following criteria have to be met[13,15].

o The RNA template must be intact and free of degraded material. A small amount of fragmented material represents a great excess over intact template when molar quantities are considered. Therefore, the quality of the template RNA must be examined under stringent denaturing conditions.

o The enzyme must be concentrated and free of contaminating degradative activities such as RNase and DNase.

o Optimal enzyme concentration must be established for each RNA template used.

o Addition of pyrophosphate to the reaction[16] enhances the yield of complementary DNA and inhibits anti-complementary DNA synthesis[17]. Pyrophosphate stabilizes the RNA-DNA duplex, which is degraded in the absence of pyrophosphate.

III. EXPERIMENTAL PROCEDURES

A. Preparation of Viral RNA

RNA was extracted by a modification of a procedure previously described[18]. Approximately 0.5 grams (wet weight) of pelleted virus were lysed in 5 ml of lysis buffer containing 50 mM Tris-acetate (pH 7.5), 100 mM LiCl, 5 mM EDTA, 1% (w/v) sodium dodecyl sulfate, 1% (w/v) dithiothreitol, and 10% (v/v) phenol. After disruption of the virus at room temperature, the suspension was transferred to a sterilized 30-ml centrifuge tube. The virus suspension was then extracted with 5 ml of phenol and 5 ml of a chloroform-isoamyl alcohol mixture (24:1). After mixing intermittently for 10 min at room temperature, the mixture was separated by centrifugation. The aqueous layer was removed and reextracted twice with phenol-chloroform-isoamyl alcohol. After the final extraction, the aqueous phase was removed to a 30 ml tube. To this we added 0.1 volume of 20% (w/v) acetate (pH 5.5) and 2.5 volumes of absolute ethanol. The RNA was allowed to precipitate overnight at -20°C and total RNA was pelleted by centrifugation. The pellet was resuspended in 0.5 ml of TLE buffer--50 mM Tris-acetate (pH 7.5), 100 mM LiCl, and 5 mM EDTA--and layered on an 11.2 ml 5-30% linear sucrose gradient in TLE and 0.2% (w/v) sarcosyl. RNA fractionation was carried out at 40,000 rpm for 3 h at 4° in a Bechman SW41 rotor. The fractions were monitored for A_{260}, and the viral 70S RNA peak pooled and precipitated. 70S RNA was pelleted, resuspended in 0.5 ml of TLE, heated at 90°C for 3 min, and then quickly chilled on ice. Centrifugation was carried out as noted above, except that the duration was 4.5 h.

4

B. Preparation of Complementary DNA

All reactions were carried out in 100 µl volumes in siliconized tubes. The standard reaction mixture was prepared in ice and contained 50 mM Tris-HCl (pH 8.3 at 42°C), 10 mM $MgCl_2$, 30 mM β-mercaptoethanol, 70 mM KCl, the four deoxynucleoside triphosphates (including the $[^3H]$-labeled species) at 500 µM, oligo(dT)$_{10}$ at 100 µg/ml, 4 mM sodium pyrophosphate, 10-20 µg of 35S RNA, and reverse transcriptase at 20 units/µg of RNA. Some variations in these conditions during optimization of reaction conditions may be necessary. The template RNA was heated at 60°C for 3 min and then quickly chilled on ice before being added to the mixture. After incubating at 37°C for 1 h, the reaction was terminated by chilling on ice and adding EDTA to 10 mM and sodium lauryl sarcosinate (SLS) to 1% (w/v). The nucleic acids were then extracted with phenol-chloroform-isoamyl alcohol.

cDNA products to be fractionated on alkaline sucrose gradients were processed by diluting the reaction mixture to 250 µl. We then added 250 µl of a 2X alkaline sucrose suspension buffer containing 0.2 M NaOH, 1.4 M NaCl, and 20 mM EDTA. The sample was layered on a 11.2 ml 10-30% linear sucrose gradient containing 0.1 M NaOH, 0.7 M NaCl, and 10 mM EDTA. Centrifugation was carried out at 40,000 rpm for 20 h at 10°C in a Beckman SW41 rotor. Fractions were monitored by radioactivity and the peaks of $[^3H]$-cDNA were pooled, neutralized with Tris-HCl (pH 7.5) to 50 mM and 5 N HCl to 0.1 N, and precipitated with ethanol. For sizing cDNA on denaturing gels, neutralized fractions were chromatographed on a siliconized Sephadex G-50 column (11.5 ml bed volume) and the excluded cDNA was precipitated with ethanol in the presence of tRNA carrier at 50 µg/ml.

C. Preparation of Duplex DNA

The standard reaction was carried out in a 100 µl volume in a siliconized tube and contained 50 mM Tris-acetate (pH 8.3 at 42°C), 10 mM $MgCl_2$, 30 mM β-mercaptoethanol, 70 mM KCl, 500 mM deoxynucleotide triphosphates (including $[^{32}P]$-labeled compounds), 4 µM sodium pyrophosphate, full-length $[^3H]$-cDNA at 0.5-5 µg/ml, and 20-50 units of reverse transcriptase. The reaction was incubated at 42°C for 1 h and terminated by chilling on ice with the addition of EDTA to 10 mM and SDS to 1%. Incorporation was measured by determining TCA-precipitable radioactivity. The double-strandedness of the DNA was measured by the sensitivity to S_1 nuclease. After phenol-chloroform-isoamyl alcohol extraction, the material was either chromatographed on Sephadex G-50 as described above or fractionated on neutral sucrose gradients under the conditions

used for 35S RNA preparation. The peak DNA fractions obtained from these
procedures were precipitated and stored at -20°C until further use.

D. S_1 Nuclease Assay

Products of the reverse transcriptase reaction were treated with S_1 nuclease
in a 50 μl reaction mixture containing 50 mM sodium acetate (pH 4.5), 200 mM
NaCl, 1 mM $ZnCl_2$, heat-denatured calf thymus DNA at 100 μg/ml, double-stranded
AMV DNA, and 20 units of S_1 nuclease. Incubations were carried out at 37°C.
Aliquots were removed from the reaction mixture and S_1 resistance was determined.

E. Agarose Gel Electrophoresis

The horizontal gel system described by McDonnell et al.[19] was used for
methylmercury hydroxide, formaldehyde, and native agarose gels. Electrophoresis of RNA in gels containing methylmercury hydroxide has been previously described[20]. Gels were prepared by dissolving an appropriate weight of agarose
in buffer E containing 25 mM boric acid, 2.5 mM $NaB_2O_4 \cdot 10H_2O$, 5 mM sodium
sulfate, and 0.5 mM Na_2EDTA (pH 8.2). After autoclaving for 3 min and allowing to cool to 60°C, methylmercury hydroxide was added to 6 mM prior to capping
gels. RNA samples were dissolved in 25 μl of sample buffer and electrophoresed
at 15°C at 45 mA for 6 h as described by Bailey and Davidson[21]. The RNA was
made visible by staining the gel in 0.5 M ammonium acetate containing ethidium
bromide at 50 μg/ml for 30 min.

Electrophoresis of DNA in denaturing formaldehyde agarose gels was carried
out as described by Lehrack et al.[22]. Gels were prepared by heating a 2.8% or
3.0% agarose-water suspension in the autoclave for 5 min; it was then diluted
with 1 volume of 4.4 M formaldehyde in a mixture of 36 mM Na_2HPO_4 and 4 mM
NaH_2PO_4. Precipitated DNA was dissolved in 25 μl of sample buffer containing
50% formamide, 2.2 M formaldehyde, 10% glycerol, 0.04% bromophenol blue and
0.4% xylene cyanole FF. The samples were heated at 65°C for 3 min and electrophoresed at a constant voltage of 90 V for 3 h. Unlabeled DNA was stained with
ethidium bromide at 100 μg/ml in 0.5 M ammonium acetate overnight.

Electrophoresis of duplex DNAs on native agarose gels was described by
McDonnell et al.[19]. Precipitated DNA was resuspended in 20 μl of water and
5 μl of 50% glycerol, 1% SDS, and 0.02% bromophenol blue. The samples were
electrophoresed at a constant voltage of 90 V for 3 h in 1X running buffer.
Unlabeled DNA bands were made visible by staining with ethidium bromide. Gels
containing [^{32}P]-labeled DNAs were soaked in 100% methanol for at least 1 h,

Fig. 1. Comparision of AMV and RSV RNAs on denaturing 1.4% agarose-6 mM methylmercury hydroxide gel. 1 µg of AMV (slot A) and RSV (slot B) were suspended in 25 µl of sample buffer and electrophoresed at 15°C at 45 mA for 6 h.

dried under vacuum, and exposed to Kodak XR2 film for various periods at -70°C.

F. Electron Microscopic Analysis of DNA

Grids were prepared and DNA was deposited on them by a modification of the method described by Westphal and Lai[24]. DNA was placed in hyperphase solution (55% formamide, 2.6 M urea, 0.009 M EDTA, 0.09 M Tricine, and 0.09 N NaCl), the mixture was heated at 53°C for 30 s and immediately placed on ice. The mixture was then spread as a Kleinschmidt monolayer on deionized water (hypophase)[24]. Collodion-coated grids treated with cytochrome C were used for picking up DNA by touching the spread hyperphase. After washing with acetone, the grids were dried on filter paper. For contrast enhancement, the grids were rotary-shadowed with platinum-palladium evaporated from an electron wire. Grid were examined with a Siemens electron microscope (Model 101) operating at 40 keV and a magnification of 10,000. Size was measured by comparision with standard DNA molecules from SV40, treated as above.

IV. SYNTHESIS OF LINEAR DUPLEX DNA

A. Analysis of RNA on Denaturing Gels

The isolation of an intact RNA species is a prerequisite for the synthesis of complete DNA. Freshly pelleted virus is essential in obtaining high yields of intact 35S RNA; aged virus preparactions yielded significantly smaller amounts of template. The virus is lysed and RNA is extracted as described in the methods section. A single gradient peak does not ensure homogeneity, as extensive handling of RNA can result in nicking and degardation. A better test for intactness is electrophoresis of the RNA species on denaturing gels. The integrity of the RNA template was checked by using denaturing gels, such as methylmercury hydroxide, acid urea, and formaldehyde. Total RNA from freshly pelleted AMV was extracted and analyzed on denaturing methylmercury hydroxide gels (Fig. 1). Three RNA species from the total AMV RNA preparation were detected with ethidium bromide staining (lane A). A faster-migrating band that comigrated with the cellular 18S RNA was removed on passage through a poly rU-Sepharose column. Measurement of the relative amounts of the poly rA-containing RNA bands by densitometry showed the slower-migrating species to be present in approximately 10-fold excess over the faster-migrating RNA. These two RNA species contain 7,500 and 7,000 ribonucleotides, repectively, on the basis of migration relative to RSV (Rous sarcoma virus, Schmidt-Ruppin strain D) genomic and ribosomal RNA markers.

B. Synthesis of Full-length cDNA from RNA

The work of Myers et al.[25] and Buell et al.[26] provided several critical observations useful in attempting to transcribe complete DNA copies. Particular attention was paid to screening of purified reverse transcriptase for RNase. As previously reported[16], addition of 4 mM pyrophosphate causes a large shift in the size distribution of the transcripts, with 80% of the transcripts sedimenting at the size of the complete DNA copy (Fig. 2). The peak material was neutralized, precipitated with ethanol, and examined on denaturing gels. Two distinct species of DNA were detected: the larger species of 2.6×10^6 daltons, representing a complete copy of the viral genome[26], and a smaller species of 2.0-2.3×10^6 daltons. Reactions carried out in the absence of pyrophosphate result in low yields of complete cDNA transcripts.

The large size and secondary structure of AMV RNA limited the synthesis of long cDNA transcripts until high enzyme-to-RNA ratios were used. We found it necessary to determine the ratio which results in maximal yields of large cDNA transcripts. Increasing the enzyme concentration while keeping the RNA concentration constant resulted in a 20-fold increase in deoxytriphosphate incorporation. Additional enzyme inhibits incorporation, possibly owing to phosphate present in the enzyme storage buffer. Sizing the products of such reactions on a denaturing gel demonstrated that changing the enzyme-to-RNA ratio resulted in significant variation in the size distribution of the cDNA products. Essentially no full-size transcripts are seen at enzyme-to-RNA ratios (units/μg RNA) of 5 or 200 (Fig. 3). However, at ratios of 10, 20, or 50, a significant portion of the RNA was converted into a full-length cDNA transcript (Fig. 3). Careful analysis of large transcripts shows two distinct cDNA species of 2.6×10^6 and 2.3×10^6 daltons.

The cDNA transcript was subjected to S_1 digestion, and shown to contain 11% double-stranded character (Fig 4). This may be due to intermolecular interactions of cDNA molecules; however, a significant proportion of this double-stranded material rapidly regains nuclease resistance after denaturation by heat. This result is most consistent with the possibility that the molecule contains a terminal or internal hairpin structure[27].

C. Synthesis of Duplex DNA

With optimal reaction conditions determined for first-strand synthesis, we examined the cDNA-dependent kinetics of second-strand synthesis in the absence of oligo dT. Aliquots removed at the various times were examined for DNA syn-

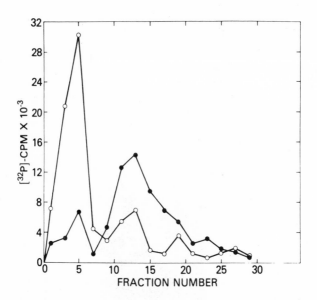

Fig. 2. Analysis of AMV cDNA products by alkaline sucrose gradient centrifugation and denaturing 1.4% agarose-4.4 M formaldehyde gel electrophoresis.
Standard reverse transcriptase reactions containing (o) or lacking (●)
4 mM sodium pyrophosphate were prepared as described in the text. The reaction was terminated after 1 h by adding EDTA to 10 mM and SDS to 1% (w/v).
The aqueous layer was adjusted to 0.1 M NaOH, 0.7 M NaCl, and 10 mM EDTA.
The samples were layered on an 11.2 ml 10-30% linear sucrose gradient containing 0.1 M NaOH, 0.7 M NaCl, and 10 mM EDTA. Centrifugation was for 20 h
at 10°C and 40,000 rpm in a Beckman SW41 rotor.

thesis as well as for resistance to S_1 digestion (Fig 5). An increase in DNA

synthesis up to 30 m in of incubation was obtained; beyond this point, no

further synthesis occurred. It is interesting to note that the increase in S_1

resistance of the first strand parallels the rate of second-stranded DNA (sDNA)

synthesis. After 30 min, almost 100% of the template was protected from

S_1 digestion.

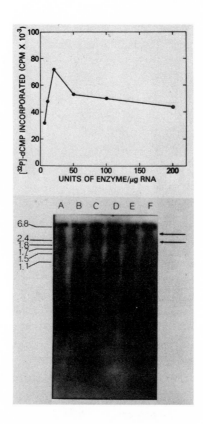

Fig. 3. Titration of reverse transcriptase with AMV 35S RNA. [^{32}P]-cDNA was synthesized under standard conditions in a 100 μl reaction mixture (RNA at 1 μg/ml and [^{32}P]-dCTP at 2.0 Ci/mmol) with various amounts of reverse transcriptase. (A) Incorporation of [^{32}P]-dCMP in cDNA with increasing enzyme units. 10 μl were removed from each reaction after 60 min, and precipitated with TCA. (B) Formaldehyde-agarose gel electrophoresis of [^{32}P]-cDNA products. 25 μl were removed from the terminated reaction and precipitated in ethanol. Samples were prepared and electrophoresed on a 1.5% agarose-formaldehyde gel as noted in the caption of Fig. 2. The different slots represent various units of reverse transcriptase per reaction: a, 5 units; B, 10; C, 20; D, 50; E, 100; and F, 200 units. The gel was dried and subjected to autoradiography.

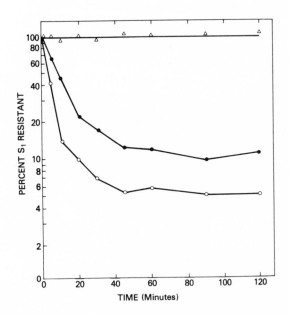

Fig. 4. S_1 nuclease digestion of full-length AMV cDNA. [^{3}H]-cDNA was isolated and neutralized after alkaline sucrose gradient centrifugation. 50 μl reactions containing 50 ng of cDNA were carried out with conditions noted in the text. The cDNA was treated with S_1 nuclease under native (●) and heat-denatured conditions (o). 5 μl aliquots were removed at indicated times and precipitated with TCA. Values represent percent of TCA-precipitable counts remaining compared to a control reaction lacking enzyme (△).

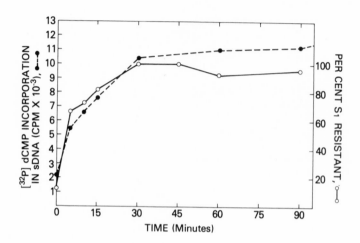

Fig. 5. Time course of AMV sDNA synthesis. 66 ng of [³H]-cDNA was incubated in a 100 µl standard reaction mixture with [³²P]-dCTP as the labeled compound (2.4 Ci/mmol). Three 2 µl aliquots were removed at the times indicated, with one of these precipitated with TCA to monitor incorporation of [³²P]-dCMP into the second strand (●). The other two were used to determine the kinetics of acquisition of S_1 resistance of the first strand (o). One aliquot was added to the standard S_1 reaction mix containing enzyme; the other was used in a control reaction lacking enzyme. Values represent percent of TCA-precipitable counts remaining compared to a control reaction lacking enzyme.

Because sDNA synthesis occurs in the absence of exogenous primer and the size of the second strand after S_1 digestion of the duplex is the same as that of the template first strand, the presence of a hairpin structure at the 3' end of the cDNA is strongly suggested. To verify the existence of the this hairpin structure, cDNA was subjected to S_1 nuclease digestion under native and denaturing conditions. It was found that 11% of the cDNA template behaves as if it were double-stranded, whereas 5.3% of the cDNA double-stranded character is regained after heat denaturation (Table 1). To examine the effect of S_1 nuclease digestion on double-stranded DNA, DNA was synthesized with [³H]-dCTP in the first strand and [³²P]-dGTP in the second strand. Second-strand synthesis increased the S_1 resistance of the [³²H]-cDNA to 92%, and the sDNA was totally resistant to digestion. Heat denaturation increases the susceptibility of the two strands to nuclease attack; however, more than half of the double-stranded material regains resistance during the initial minutes of the reaction. This further suggests the presence of a hairpin structure that facilitates rapid re-annealing of the first and second strands.

TABLE I

S_1 NUCLEASE ANALYSIS OF cDNA AND cDNA-sDNA*

		S_1 Resistance, %	
DNA	Conditions	1st Strand	2nd Strand
$[^3H]$-cDNA	$-S_1$	100	NA
	$+S_1$	11.2	NA
	$+S_1$, 100°C	5.3	NA
$[^3H]$-cDNA- $[^{32}P]$-sDNA	$-S_1$	100	100
	$+S_1$	91.9	100
	$+S_1$, 100°C	53.0	57.8

*63 ng of $[^3H]$-cDNA and 52 ng of $[^3H]$-cDNA-$[^{32}P]$-sDNA were subjected to S_1 nuclease digestion with standard reaction conditions. Samples were denatured by heating at 100°C for 1 min and then quickly-chilling in ice. Values repre-sent the average of triplicate determinations. NA, not applicable.

The double-stranded DNA synthesized by the second reverse transciptase reaction was fractionated on a neutral sucrose gradient to determine the size distribution of the DNA chains. Both the $[^3H]$-labeled first strand and the $[^{32}P]$-labeled second strand cosedimented as a sharp peak (Fig. 6). Peak frac-tions were precipated with ethanol and electrophoresed on native agarose gels. Two distinct major species with weights of 5.2×10^6 and 4.0×10^6 daltons can be detected (Figure 6 insert). Electron microscopic analysis of DNA isolated from the neutral gradient showed linear duplex DNA molecules of 5.2×10^6 daltons (Fig. 7). These species are present in 20 of every 100 molecules, whereas smaller duplexes may have resulted during the preparation for viewing.

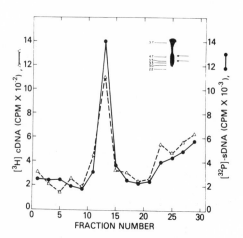

Fig. 6. Analysis of duplex DNA by gradient centrifugation and electrophoresis.
[^{32}P]-sDNA was synthesized in a standard reaction mixture containing [^{32}P]-dGTP
(2.2 Ci/mmol) as the labeled compound and 71 ng of [^{3}H]-cDNA as the template-
primer. The recovered DNA samples were resuspended in 500 μl of gradient
buffer and fractionated on 5-30% linear sucrose gradients. Samples (20 μl)
were precipitated with trichloroacetic acid, filtered and monitored for
radioactivity. Peak fractions 10-15 were pooled, precipitated with ethanol and
the DNA was electrophoresed on a nondenaturing 1.4% agarose gel for 2 h at 100 V.

V. CONCLUSIONS

A product of the avian myeloblastosis virus genome, reverse transcriptase,
can be used to synthesize both negative and positive DNA strands from AMV RNA.
Full-length DNA copies of intact RNA can be transcribed using conditions similar to those of Myers et al.[25]. Important considerations are the use of intact
RNA template, high enzyme-to-RNA ratios, and the addition of pyrophosphate to
reactions. The product of the two-step reaction catalyzed by reverse transcriptase is a linear duplex of 5.2×10^6 daltons that is covalently linked and
90-100% S_1-resistant.

Analysis of the viral RNA template, genomic length cDNA, and linear duplex
on agarose gels demonstrate two distinct species of each. RNAs of 7,600 and
7,000 ribonucleotides, cDNAs of 2.6×10^6 and 2.3×10^6 daltons, and duplexes
of 5.2×10^6 and 4.0×10^6 daltons have been identified. Double-stranded DNAs
synthesized with the method described in this chapter were inserted into bacterial plasmids. Isolated clones were characterized by restriction enzyme

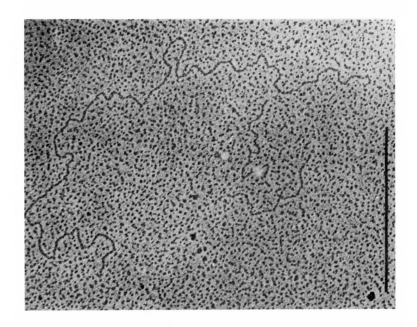

Fig. 7. Analysis of duplex DNA by electron microscopy. The DNA was viewed by using a Siemens electron microscope with a magnification x 10,000. Size measurements were obtained by comparison to SV40 DNA. The inserted bar represents 1 µm, which corresponds to 1×10^6 daltons under the conditions used.

analysis and utilized as probes to study the expression and regulation of AMV genomes in cells transformed by this virus. The procedures outlined with some modifications are applicable for the synthesis of double-stranded DNAs from a variety of mRNAs.

ACKNOWLEDGEMENTS

Those aspects of this work carried out at Georgetown University Medical Center were supported by USPHS grants CA-16914. J.G.C. was supported as a Scholar of the Leukemia Society of America, Inc.

REFERENCES

1. Mizutoni, S., and Temin, H.M. (1970) Nature (London) 226, 1211-1213.
2. Baltimore, D. (1979) Nature (London) 226, 1209-1211.
3. Efstratiadis, A., Kafatos, F.C., Maxam, A.M., and Maniatis, T. (1976) Cell 7, 279-288.
4. Rougeon, F., and Mach, B. (1976) Proc. Natl. Acad. Sci. USA 73, 3418-3422.
5. Rabbitts, T.H. (1976) Nature (London) 260, 221-225.

6. McReynolds, L.A., Monahan, J.J., Bendure, D.W., Woo, S.L.C., Paddock, G.V., Salser, W., Dorson, J., Moses, R.E., and O'Malley, B.W. (1977) J. Biol. Chem. 252, 1840-1843.

7. Gordon, J.I., Burns, A.T.H., Christmann, J.L., and Deely, R.G. (1978) J. Biol. Chem. 253, 8629-8639.

8. Wickens, M.P., Buell, G.N., and Schimke, R.T. (1978) J. Biol. Chem. 253, 2483-2495.

9. Vande Woude, G.F., Oskarsson, M., Enquist, L.W., Nomura, S., Sullivan, M., and Fischinger, P.J. (1979) Proc. Natl. Acad. Sci. USA 76, 4464-4468.

10. Lautenberger, J.A., Schulz, R.A., Garon, C.F., Tsichlis, P.N., and Papas, T.S. (1981) Proc. Natl. Acad. Sci. USA 78, 1518-1522.

11. Efstratiadis, A., and Villa-Kamaroff, L. (1979) in Genetic Engineering (Setlow, J.K. and Hollaender, A., eds.) pp. 15-36, Plenum Press, New York and London.

12. Gerard, G.F., and Grandgenett, D.P. (1980) in Molecular Biology of RNA Tumor Viruses (Stephenson, J., ed.) pp. 345-394, Academic Press, New York, New York.

13. Myers, J.C., Ramirez, F., Kacian, D.L., Flood, M., and Spiegelman, S. (1980) Anal. Biochem. 101, 88-96.

14. Kacian, D.L., Watson, K.F., Burney, A., and Speigelman, S. (1971) Biochim. Biophys. Acta 246, 365-383.

15. Schulz, R.A., Chirikjian, J.G., and Papas, T.S. (1981) Proc. Natl. Acad. Sci. USA 78, 2057-2061.

16. Kacian, D.L., and Myers, J.C. (1976) Proc. Natl. Acad. Sci. USA 73, 2191-2195.

17. Myers, J.C., and Spiegelman, S. (1978) Proc. Natl. Acad. Sci. USA 75, 5329-5333.

18. Stephenson, M.L., Wirthlin, L.R.S., Scott, J.F., and Zamecnik, P.C. (1972) Proc. Natl. Acad. Sci. USA 69, 1176-1180.

19. McDonnell, M.W., Dimon, N., and Studier, F.W. (1977) J. Mol. Biol. 110, 119-146.

20. Deeley, R.G., Gordon, J.I., Burns, A.T.H., Mullinix, K.P., Binastein, M., and Goldberger, R.F. (1977) J. Biol. Chem. 252, 8310-8319.

21. Bailey, J.M., and Davidson, N. (1976) Anal. Biochem. 70, 75-85.

22. Lehrack, H., Diamond, D., Wozney, J.M., and Boedtker, H. (1977) Biochemistry 16, 4743-4751.

23. Westphal, H., and Lai, S.-P. (1977) J. Mol. Biol. 116, 525-548.

24. Kleinschmidt, A.K., and Zahn, R.K. (1959) Z. Naturforsch. 14b, 770-779.

25. Myers, J.C., Speigelman, S., and Kacian, D.L. (1977) Proc. Natl. Acad. Sci. USA 74, 2840-2843.

26. Buell, G.N., Wickens, M.P., Payver, F., and Schimke, R.T. (1978) J. Biol. Chem. 253, 2471-2482.

27. Salser, W. (1978) in Genetic Engineering (Chakrabarty, A.M., ed.) pp. 53-81, CRC Press Inc., West Palm Beach, Fla.

E. COLI DNA POLYMERASE I: ENZYMATIC FUNCTIONS AND THEIR APPLICATION IN
POLYMER FORMATION, NICK TRANSLATION AND DNA SEQUENCING

Christopher P. Hartman[+] and Dietmar Rabussay[‡*]

[+]Bethesda Research Laboratories
P.O. Box 577
Gaithersburg, Maryland 20760

[‡]Department of Biological Science
Florida State University
Tallahassee, Florida 32306

[*]Present address: Bethesda Research Laboratories

I. INTRODUCTION

The elucidation of the double-helical structure of DNA[1] allowed the construction of rather specific models for DNA replication and the prediction of some of the properties of enzymes involved in DNA synthesis. In 1958, the first DNA synthesizing enzyme, now called DNA polymerase I or Pol I, was purified from E. coli[2], and it stimulated great interest in the enzymology of DNA replication. Accumulating knowledge about DNA polymerase I, extensive genetic analysis of E. coli DNA replication, and emerging DNA replication mechanisms of other biological systems made it clear that this enzyme is only one component (although a prominent one) of the complex machinery of DNA metabolism. The isolation of a viable E. coli strain carrying a mutation in the structural gene (pol A) for DNA polymerase I[3] facilitated the search for other, less abundant DNA polymerases in these cells. Thus, more than 10 yr after the isolation of DNA polymerase I, DNA polymerases II[4,5] and III[6] were discovered.

Despite a transient lack of attention in the past, DNA polymerase I is now of high interest: it is one of the most studied enzymes of nucleic acid metabolism, it plays an important role in DNA replication and repair[7-12], and it has become increasingly important as a preparative and analytical tool in nucleic acid chemistry and molecular biology. Excellent synopses on DNA polymerase I and its physiological roles can be found in two extensive reviews[13,14].

In this paper, we present an overview of the properties and enzymatic functions of DNA polymerase I and a summary of some of its applications, including the use of this enzyme in the synthesis of oligodeoxynucleotides and polydeoxynucleotides, in nick translation (labeling of DNA), and in DNA sequencing (Sanger method).

II. PROPERTIES OF E. COLI DNA POLYMERASE I

A. Purification and Physicochemical Properties

DNA polymerase I, the most abundant of the DNA sythesizing enzymes in E. coli, is present at about 350 copies in a rapidly growing wild-type cell[13]. The cellular abundance of this enzyme has been enhanced about 100-fold by lytic infection with a λ phage carrying the polA gene with its own promoter[15]. From these infected cells, the enzyme can be purified with high yields and exceptional purity[16].

The enzyme consists of a single polypeptide chain of about 1,000 amino acid residues and has a molecular weight of 109,000[17]. It contains only one disulfide bridge and one free sulfhydryl group, which can bind a mercury atom without affecting the enzymatic activity[17]. Labeling with radioactive mercury-203 has proved useful in DNA-binding studies and may become important in the elucidation of the three-dimensional structure of the enzyme. Under standard conditions, one enzyme molecule polymerizes approximately 700 nucleotides per minute at 37°C[13].

B. Enzymatic Activities

E. coli DNA polymerase I has five enzymatic functions which include a DNA polymerase[2], a 3' ——>5' exonuclease (E. coli exonuclease II)[18,19], a 5'——>3' exonuclease (E. coli exonuclease VI)[20,21], a pyrophosphorolysis and a pyrophosphate exchange activity[22]. The principal function of the enzyme is the polymerizing activity[21] that catalyzes the condensation of deoxyribonucleoside-5'-triphosphates (dNPPP) into DNA:

$$\text{dNPPP} + (\text{dNP})_n \xrightarrow[]{\text{DNA}} (\text{dNP})_{n+1} + \text{PPi.}$$

There is an absolute requirement for a DNA template and a DNA or RNA primer with a 3'-hydroxyl terminus. As shown in Figure 1, DNA synthesis by DNA polymerase I results in different types of products, the type depending on the kind of template-primer offered. Intact duplexes with no internal 3'-hydroxyl group are not a substrate for DNA polymerase I (Fig. 1A). Duplexes with nicks function as template-primers: starting from a nick, the enzyme proceeds 5'——>3' by either degrading or displacing the lead strand in front of it (Fig. 1B,C). The degradation is catalyzed by the enzyme's intrinsic 5'——>3' exonuclease. Gapped duplexes also serve as template-primers, and that results in filling of the gaps. Even large gaps (greater than 100 nucleotides) in extensively damaged DNAs can be filled, as long as there is a primer with a 3'-hydroxyl group available[26,27] (Fig. 1D). Long single strands can be complemented into double strands if a suitable primer is present. The primer can be a "snap back" 3'-hydroxyl end or a piece of DNA or RNA complementary to the 3' end of the single strand (Fig. 1E,F). Primers that terminate in a 3'-phosphate[23] or a 2', 3'-dideoxyribonucleotide[24] are refractory to DNA synthesis. DNA polymerase I does not show a preference or particular specificity for the template; it will copy any base sequence in the template with equal efficiency. The direction of chain growth is always 5'——>3'; thus, the polarity of the new chain is opposite that of the template. DNA polymerase I acts processively, i.e., the enzyme remains associated with, and travels along, the template for many polymerization steps without release.

The 3'——>5' exonuclease is important for copying the template with high

20

ACTION	TEMPLATE-PRIMER	PRODUCT
A. No Change	Intact Duplexes	
B. Nick Transla-tion	Nicked Duplexes	
C. Strand Displacement	Nicked or Frayed Duplexes	
D. Gap Filling	Gapped Duplexes	
E. Chain Elongation	Single Strand with primer	
F. Chain Elongation	Single Strand (hairpin structure)	

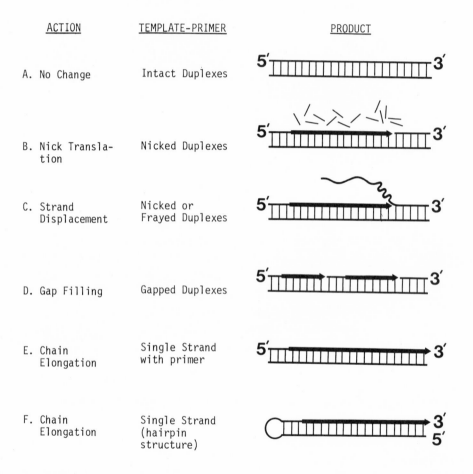

Fig. 1. The use of different template-primers by E. coli DNA polymerase I (after Kornberg[13]). Thin lines represent the template-primer; heavy arrows, newly synthesized DNA. Schemes 1-5 also apply, correspondingly, to circular DNA. A nick is a break in one of the strands of the duplex, usually bounded by a 3'-hydroxyl and a 5'-phosphate group.

fidelity. This intrinsic activity of DNA polymerase I removes non-base-paired (mismatched) nucleotides at the 3' end of the primer on the growing chain. Therefore, this function has also become known as the "proofreader." Single-stranded chains and frayed or non-base-paired ends of duplex DNA can be degraded by this activity (Figure 2). The products of this reaction are exclusively

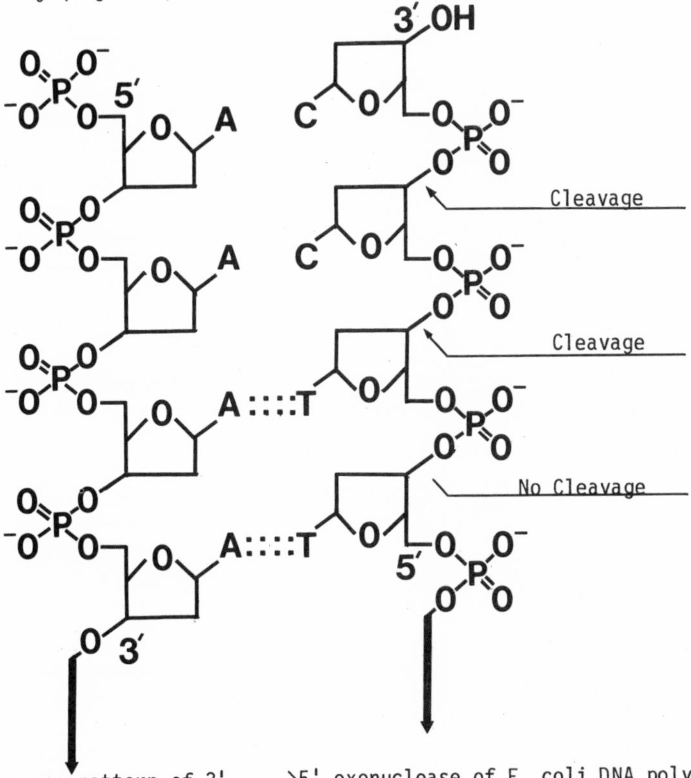

Fig. 2. Cleavage pattern of 3'——>5' exonuclease of E. coli DNA polymerase I. This activity differs from the 5'——>3' exonuclease in the direction of cutting and in the cleavage of only non-base-paired nucleotides. Base pairs that are normally closed but are melted under assay conditions will also be cleaved.

5'-monophosphates. The 3'——>5' exonuclease activity is very specific: it will start to cleave only at 3'-hydroxyl termini in the deoxyribo configuration and will not sever a DNA chain that terminates in a 3'-phosphate or a 2', 3'-dideoxyribonucleotide[23].

The 5'——>3' exonuclease degrades one strand of duplex DNA from a 5' end and releases mono or oligonucleotides (Fig. 3). Non-base-paired nucleotides over 100 in length will not be cleaved[25]. The 5'——>3' exonuclease will attack nucleic acids with either 5'-hydroxyl, 5'-monophosphate, 5'-diphosphate,

or 5'-triphosphate termini[25]. The primary purpose of the 5'——>3' exonuclease
is to remove the "lead strand" ahead of the enzyme: the degraded strand will
subsequently be replaced with newly-synthesized material that is strictly complementary to the template strand. The ability of this intrinsic exonuclease
to remove even aberrant structures, such as pyrimidine dimers or RNA from a
DNA-RNA hybrid duplex, predestines DNA polymerase I for DNA excision repair.

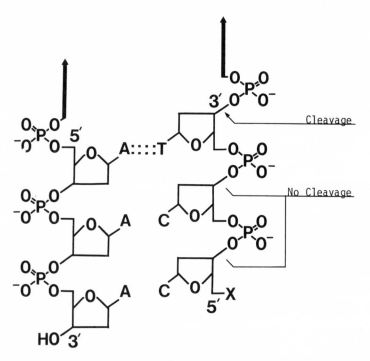

Fig. 3. Cleavage pattern of 5'——>3' exonuclease of E. coli DNA polymerase I.
This activity cleaves only at base-paired nucleotides on the lead strand, thereby freeing the template strand for continued DNA synthesis.

Pyrophosphorolysis and pyrophosphate exchange are proof of the reversibility
of the polymerization reaction. Pyrophosphorolysis results in the degradation
of a DNA chain in the presence of excess pyrophosphate (PPi):

$$\text{DNA} \quad\quad\quad\quad\quad \text{DNA}$$
$$(dNP)_n + {}^{32}PPi \longrightarrow (dNP)_{n-1} + dNP{}^{32}PP \text{ or } dNPP{}^{32}P.$$

This reaction has an absolute requirement for a DNA template-primer and is inhibited by a 2',3'-dideoxyribonucleotide terminus[24].

Pyrophosphate exchange involved the exchange of the β and phosphates of the deoxyribonucleoside-5'-triphosphates with inorganic pyrophosphate:

$$^{32}PPi + dNPPP \xrightarrow{\text{DNA}} PPi + dNP^{32}PP \text{ or } dNPP^{32}P.$$

This reaction can even occur when only one of the four triphosphates is present. DNA is necessary although it is not affected by this reaction. Because the presence of potent pyrophosphatases in all cells prevents the accumulation of the required high concentrations of pyrophosphate, pyrophosphorolysis and pyrophosphate exchange probably do not occur to any great extent in vivo.

C. Substrates and Cofactors

All enzymatic functions of E. coli DNA polymerase I require a template-primer (or more specifically, the base-paired primer terminus with a free 3'-hydroxyl group) as a substrate. If DNA is to be synthesized, the appropriate deoxyribonucleoside-5'-triphosphates are also required as substrates. The Km for all four deoxyribonucleoside-5'-triphosphates is 33 μM. In addition, E. coli DNA polymerase I depends on the presence of cofactors like Zn^{2+} and Mg^{2+} for its polymerase activity[26,28]. Both cations are also essential for the activities of the Klenow fragment (see below) but are dispensable for the 5'——>3' exonuclease activity of the small fragment[26,28].

Zn^{2+} is found in a ratio of one ion per molecule of DNA polymerase I[28,29]. When zinc is removed by dialysis against 1,10-phenanthroline, other ions can be substituted for it; e.g., Co^{2+} restores approximately 80% and Mn^{2+} approximately 45% of the original polymerase activity. The postulated role of zinc, which is probably present in all nucleotide polymerases[30], is the conversion of the 3'-hydroxyl primer terminus to a highly nucleophilic alkoxide ion, which attacks the α-phosphate of the deoxyribonucleoside-5'-triphosphate-magnesium (or manganese) chelate[28].

Magnesium (Mg^{2+}), the other metal cofactor, is not present in DNA polymerase I, but must be added to the reaction mixture, if DNA is to be synthesized. Magnesium is essential for the binding of the deoxyribonucleoside-5'-triphosphates to the enzyme[26] and may also participate in the binding of polymerase to the DNA template-primer[31]. Manganese can substitute for magnesium with almost 100% efficiency[32], but changes the substrate specificity: in the presence of Mn^{2+}, ribonucleoside triphosphates can serve as substrates and ribonucleotides are incorporated[33]. The 3'——>5' exonuclease activity is stimulated by Mg^{2+} if deoxyribonucleoside-5'-triphosphates are absent[31].

D. Cleavage of DNA Polymerase I into Two Polypeptides with Distinct Enzymatic Activities

DNA polymerase I can be cleaved by subtilisin (EC 3.4.32.14) or trypsin (EC 3.4.4.4) into two active enzyme moieties[34,35] with molecular weights of 76,000 daltons[36] and 36,000 daltons[37]. Both fragments retain distinct enzymatic capabilities. This splitting of the molecule demonstrated conclusively that DNA polymerase I had at least two enzymatically active sites. The larger, C-terminal fragment (called the "large fragment of DNA polymerase I" or the "Klenow fragment") contains the polymerase, 3'——>5' exonuclease, pyrophos-phorolysis, and pyrophosphate exchange activities[36]. The only activity associated with the smaller, N-terminal fragment is the 5'——>3' exonuclease activity[35-38]. The two enzymes can be separated by hydroxylapatite[38] or phos-phocellulose[37] chromatography.

Although the large fragment can catalyze the four reactions mentioned above, it cannot synthesize poly dAT from the corresponding deoxyribonucleoside-5'-triphosphates[36], as can the intact DNA polymerase I[39,40]. This difference appears to be due to the absence of the 5'——>3' exonuclease activity, which is thought to be responsible for forming template-primers in de novo polynucleotide synthesis. Addition of the small fragment to the reaction mixture restores the exponential rate of polynucleotide synthesis that is typical for the intact DNA polymerase I[37]. The Klenow fragment does synthesize poly dAT if a primer-template is used, but this reaction differs from that of intact DNA polymerase I in two ways[36]: the deoxyribonucleoside-5'-triphosphates are completely converted to poly dAT, as opposed to only partial conversion by DNA polymerase I; and the rate of incorporation of the deoxyribonucleoside-5'-triphosphates is linear.

III. FUNCTIONAL APPLICATIONS

A. Polymer Formation

Polymer formation by DNA polymerase I without added template-primer came as a surprise when it was discovered about 20 yr ago[39]. Even today, its mechanism is still far from completely understood. The first "artificial" polymer synthe-sized was poly dAT, a copolymer in which deoxyadenylate and thymidylate alter-nate with high precision[39]. A variety of other polymers have also been synthe-sized with or without added short template-primers[40-42], including the copolymers poly dGC and poly dIC and the homopolymers poly dA poly dT, poly dG poly dC and poly dI poly dC. Copolymers consist of two strands in a double-helical conformation, each strand containing two kinds of nucleotides (in "alternating copolymers," e.g., poly dAT, the two kinds of nucleotides alternate with high regularity). Homopolymers are duplexes consisting of two complementary strands, each of which contains only one kind of nucleotide. De novo synthesis of polymers (i.e., synthesis in the absence of added primer) is most likely initiated from DNA fragments as small as 2 or 3 base pairs, which are present as contaminants in some of the DNA polymerase molecules in a reaction mixture[17]. De novo synthesis always has a lag time of several hours before substantial amounts of polymer can be detected, and the rate of synthesis then increases exponentially[43]. The lag period is a function of the size of the template-primer. For example, if poly dAT synthesis is primed by dAT oligomers, priming by $(dAT)_2$ will result in a lag time of over 6 hr whereas $(dAT)_6$ will reduce the lag to less than 2 hr[13]. The lag time also depends on the concentration and base sequence of the priming fragments as well as other reaction conditions.

During the lag period, a process termed "reiterative replication" must be initiated. This process may be at the heart of de novo polymer synthesis. Its existence has been deduced from studies with dAT oligomers and AT-rich natural DNAs[44]. The envisioned mechanism of reiterative replication is the following: one strand of the short, completely base-paired template-primer, such as $(dAT)_6$, is bound to the template site of DNA polymerase I. The complementary strand is free to "slip" along the template strand, thereby exposing unpaired nucleotides. If slippage occurs in the proper direction, the 3'-hydroxyl end of the slipped strand can be elongated until the 3' end of the slipped strand and the 5' end of the template strand are flush again. This sequence of events--slippage of one strand relative to the enzyme-bound template strand and elongation of the slipped strand from its 3'-hydroxyl end--can occur repeatedly until the slipping strand becomes long enough to fold back on itself and form a hairpin. The strongest evidence of such a mechanism comes from the distinct temperature optimum for polymer formation, which depends on the size of the primer: the smaller the priming oligonucleotide, the lower the temperature optimum. Moreover, the temperature optimum does not change with the growth of the slipping strand, but seems to depend only on the original size of the (polymerase-bound) template strand. Reiterative replication proceeds at a linear rate. However, polymer synthesis with DNA polymerase I shows an exponential increase after the lag phase has been overcome. This implies that new template-primers must become available for the vast numbers of unused polymerase molecules that are present in a typical reaction mixture. It has been shown that exponential reaction rates depend on nucleolytic action that originates either from traces of contaminating endonucleases or from the intrinsic 5'———>3' exonuclease of DNA polymerase I[36]. Both activities plausibly can create fragments with template-primer activity.

De novo polymer synthesis goes through a peak of polymer product accumulation if the reaction is left to itself. When the deoxyribonucleoside triphosphate substrates are exhausted, the synthesized polymers are broken down; the degradation seems to result mainly from excisions by the intrinsic 5'———>3' exonuclease.

The formation of a particular polymer requires stringently controlled reaction conditions, which have been established empirically and many of which are not understood. Trace contaminations of the enzyme or a slight change in pH, buffer composition, ionic strength, or temperature can result in a completely different product[36]. Intercalating agents can alter the reaction from copolymer to homopolymer synthesis[45,46].

B. Nick Translation

Nick translation is in reality nick migration. DNA polymerase I can bind at a nick in double-stranded DNA and move this nick linearly along the template primer without a net loss of polymerized nucleotides (Fig. 1). This process requires strict coordination of the 5'———>3' exonuclease and polymerizing activities; all DNA hydrolyzed by the exonuclease has to be rapidly replaced by new DNA. The lead strand is preferentially degraded to 5'-mononucleotides and sometimes to oligonucleotides. The 5'———>3' exonuclease activity, although clearly detectable in the absence of DNA synthesis, is stimulated about 10-fold by simultaneous DNA synthesis[47]. Nick translation has been used extensively for preparing highly radioactive DNA. This DNA has proved useful in nucleic acid hybridization and nucleic acid binding experiments[48-50]. Linear phage λ DNA and circular SV40 DNA have been labeled with (α-^{32}P) triphosphates to 10^8 cpm/µg[51], and even higher specific activities can be achieved. The nick-translation methods now used in creating highly radiolabeled DNA involve the use of bovine pancreatic DNase I (EC 3.1.4.5) in conjunction with E. coli DNA polymerase I[52-54]. Bovine pancreatic DNase I, a readily available endonuclease, is used to create nicks in intact double-stranded DNA. Only an extremely small amount of pancreatic DNase is needed to form nicks; in general, 1-2 ng/µg of DNA is used, but concentrations as low as 0.1 ng/µg of DNA have been used. The smaller the amount of DNase I, the fewer nicks will be created in a given period. Thus, the length of DNA pieces newly synthesized by DNA polymerase I will increase with decreasing amounts of DNase I. After the nick-translation reaction, the

DNA can be separated from the unincorporated nucleotides by Sephadex G-50 chromatography. A detailed protocol and a typical example of the time course of incorporation of radioactive label into DNA over several hours are provided in the Appendix.

Several side reactions can interfere with proper nick translation, including ligation, strand displacement, and template switching (Fig. 4). If a nick is bounded by a 3'-hydroxyl group and a 5'-phosphate group, it can be sealed (covalently closed) by DNA ligase[55]. Therefore, ligase and DNA polymerase I compete for nicks in the structure mentioned above. Nicks bound by other than a 3'-hydroxyl and a 5'-phosphate group cannot be sealed by ligase, but may function as a starting point for nick translation. Strand displacement can occur if the 5'-oligonucleotide escaped degradation because of its length or through inactivation of the 5'——>3' exonuclease. Strand displacement results in a net gain of DNA. Template switching requires strand displacement or a template-primer with a loose 5'-end. DNA polymerase I can adopt the unpaired single strand as a template and produce a branched, double-stranded DNA structure. This results in a net synthesis of DNA.

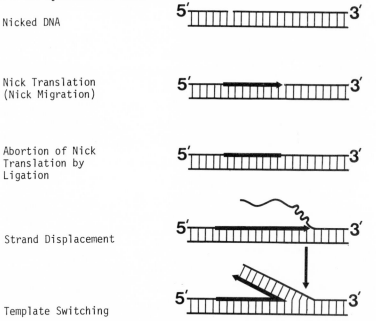

Nicked DNA

Nick Translation (Nick Migration)

Abortion of Nick Translation by Ligation

Strand Displacement

Template Switching

Fig. 4. Different courses of DNA synthesis starting at a nick (after Kornberg[13]) DNA polymerase I starting at a nick can proceed in various ways, depending on the components and reaction conditions involved. See text for description of the different options.

C. DNA Sequencing (Sanger Method)

Essentially three DNA sequencing methods that use DNA polymerases have been developed: the "plus and minus" method[56], partial ribo-substitution[57], and the new Sanger method with chain-terminating inhibitors[58-60]. The first two methods can use the large fragment of E. coli DNA polymerase I or T4 DNA polymerase (both enzymes lack the 5'———>3' exonuclease activity). The chain-terminating method, which is rapidly becoming the method of choice, depends on the availability of high-quality Klenow fragment. The large fragment of DNA polymerase I is important for two reasons: it can incorporate the 2',3'-dideoxyribonucleotides into polymerizing DNA strands[61] (most other polymerases, except Pol I and AMV reverse transcriptase,* cannot), and it will not degrade DNA that has become inactive as a template-primer, such as DNA that has been terminated with a 2',3'-dideoxyribonucleotide at the 3' end[62]. Intact E. coli DNA polymerase I would degrade the DNA from the 5'-end once DNA synthesis has been stopped by the incorporation of a 2',3'-dideoxyribonucleotide[62]. Hence, it is important that the 5'———>3' exonuclease activity be completely removed if the large fragment is to be used in the Sanger sequencing method[58].

The principle of the new Sanger method with chain-terminating inhibitors is as follows: 2',3'-dideoxyribonucleoside triphosphates terminate DNA chain growth, because no 3'-hydroxyl group is available to form a phosphoester bond with the next deoxyribonucleoside triphosphate. Arabinonucleotides have a similar effect. Arabinose, a stereoisomer of ribose, has a 3'-hydroxyl group that is in a trans-position to the 2'-hydroxyl group. From such a conformation, E. coli DNA polymerase I cannot extend a chain. Both the dideoxy- and the arabinose derivatives are incorporated in the place of their corresponding deoxyribonucleotides[24]. When DNA polymerase acts on a template-primer in the presence of the four deoxyribonucleoside triphosphates and one of the chain-terminating analogues, a mixture of extended primers with different lengths will be obtained. The resulting fragments will possess identical 5' ends, but different 3' termini because chain termination by the analogue is a random event (depending mainly on the ratio of terminating triphosphate to normal triphosphate). The polymerized fragments are made radioactive by using one α^{32}P-labeled deoxynucleotide during synthesis. These polymerized fragments are separated from

---

*Avian myeloblastosis viral reverse transcriptase can also incorporate 2', 3'-dideoxyribonucleotides, but is much less stable than the Klenow fragment of Pol I.

the template and then from each other according to their size by denaturing polyacrylamide gel electrophoresis. The pattern of band sizes obtained indicates the distribution of that base in the newly synthesized DNA whose chain-terminating analogue has been used in the incorporation mixture. If parallel DNA syntheses and gel separations are carried out with each of the four chain-terminating analogues, a pattern of bands in four parallel gel tracks allows one to read the entire base sequence of the newly synthesized DNA.

Some technical aspects should be kept in mind if the Sanger sequencing method is to be used. The first step involves the purification of the DNA template in single-stranded form. For small DNA, strand separation can be achieved by gel electrophoresis[63] or by alkaline RPC-5 analog column chromotography[64]. For larger DNA of markedly asymmetric base composition, alkaline cesium chloride density-gradient centrifugation can be used[65]. If the natural density difference between two complementary DNA strands is not sufficient for separation, it may be possible to increase the difference with poly (U,G). The poly (U,G) method of Szybalski et al.[65] involves the preferential binding of guanine-rich or uracil-rich ribopolymers to one of the DNA strands; this results in a difference between the densities of the two strands that is adequate for clear separation on a density gradient. The M13 viral DNA cloning system[74-76] was recently developed for obtaining template-primers and for performing sequencing[77].

The next step is the acquisition of a suitable DNA primer. Several types of primers have been used successfully[66-69]. Synthetic oligonucleotides[56,58] that will bind to complementary regions of the template can be readily purchased; they can also be synthesized with E. coli DNA polymerase I[39,40] or with calf thymus terminal deoxynucleotidyl transferase[70]. These primers are useful with templates known to contain particular regions of oligonucleotides. For an uncharacterized DNA template, the correct primer is not easily identified[66]. In such cases restriction fragments of the template DNA under investigation have been successfully used as primers[56,58]. These are relatively easily obtained by digesting the template with an appropriate restriction enzyme and separating the DNA fragments by gel electrophoresis[63] or by alkaline RPC-5 analog column chromatography[64]. The denatured fragments are isolated[71] and then annealed to the template[58]. A template-primer can also be obtained by appropriate use of E. coli exonuclease III[72,73].

Once the template-primer is formed, it can be extended, in separate reactions, by the Klenow fragment in the presence of all four deoxyribonucleoside triphosphates and one of the four chain-terminating analogues. To achieve

the best pattern of bands on the autoradiogram, the exact ratio of dideoxyribo-
nucleotides to deoxyribonucleotides must be varied according to the DNA under
investigaton. Under optimal conditions, a stretch of 200-300 bases can usually
be sequenced in one experiment. More detailed information and the literature
on DNA sequencing with chain-terminating inhibitors are summarized in the
Appendix.

IV. DISCUSSION AND CONCLUSIONS

E. coli DNA polymerase I fulfills several functions in the cell: it is involved in DNA replication, recombination, and repair[12]. The first function is
essential for the survival of the cell; the other two are dispensable under
favorable growth conditions. Proof of the essentiality of DNA polymerase I
comes from the existence of conditionally lethal mutants in the structural gene
of this enzyme[79-81]. Other evidence is consistent with this notion; despite
considerable efforts, no viable mutant lacking DNA polymerase I activity has
been found, and the introduction of polA mutations into recombination-deficient
and repair-deficient cells (e.g., recA, recBC, uvrB) has not been possible.

The physiological role of DNA polymerase I can best be deduced from the
properties of different mutants deficient in this enzyme. PolA1[3], the first
mutant isolated, has a normal growth rate. However, it displays increased sensitivity to UV irradiation and methylmethane sulfonate. Other properties con-
ferred by this suppressible nonsense mutation (summarized by Kornberg[13]) are
also consistent with a general deficiency in gap filling capacity. PolA1 cell
extracts have about 1% of the DNA polymerase activity of wild-type cells, but
almost normal levels of 5'———>3' exonuclease (as measured in the absence of
DNA synthesis). Surprisingly, the residual DNA polymerizing activity was found
in molecules the size of normal DNA polymerase I (5.4S), whereas most of the
exonuclease activity resided in a protein that was the size of this enzyme's
small fragment (2.8S). Thus, a small amount of fully functional DNA polymerase
I is probably synthesized by infrequent read-through of the amber codon, whereas
almost normal amounts of a (trimmed) amber fragment displaying the 5'———>3'
exonuclease activity are produced[82]. If the low DNA polymerase activity measured in cell extracts reflects the intracellular activity, it must take very
few functional molecules to support DNA replication at a normal rate. There
must be a hierarchy of functions that directs DNA polymerase I primarily to
its role in DNA replication. Only if the needs for DNA replication have been

satisfied will "excess" DNA polymerase I molecules engage in repair and recombination. Similar considerations have been made regarding DNA ligase[13].

Two conditionally lethal (temperature-sensitive) mutations demonstrate the importance of both the polymerase and the 5'———>3' exonuclease activities for cell survival. PolAexl affects only the 5'———>3' exonuclease activity that is abolished at 43°C. Even at the permissive temperature (30°C) the 5'———>3' activity is depressed[79,80]. On nicked double-stranded DNA, the deficient DNA polymerase advances while displacing, rather than degrading, the lead strand in front of it. As opposed to the wild-type enzyme, concurrent polymerization inhibits, rather than stimulates, 5'———>3' exonuclease. The mutant BT4113 is defective in both the polymerase and the 5'———>3' exonuclease activities at the nonpermissive temperature. However, if the mutant polymerase is cleaved into its large and small fragments, the polymerase activity is normal, but the exonuclease remains defective. Hence, the defect in the small-fragment region also affects other parts of the molecule.

From the properties of the mutants mentioned above and the mutant polA12, one can conclude that the gap-filling and excision functions of DNA polymerase I are essential for chromosome replication, whereas repair and recombination can probably be maintained by other pathways. Excision and gap filling must be coordinated to excise and replace RNA primers at the 5' ends of nascent DNA fragments. The lack of proper coordination of these functions and the resulting inability to join nascent DNA fragments probably explain the lethality of some mutations in DNA polymerase I.

DNA polymerase I has become one of the most useful enzymes in molecular biology. In Section III, the use of Pol I was described for synthesizing deoxyribonucleotide homopolymers and copolymers, for nick translation, and for DNA sequencing. Other applications include the use of its various nucleolytic activities and gap filling ability. Some aspects relevant to the applications of DNA polymerase I are discussed below.

The ability of DNA polymerase I to synthesize homopolymers and copolymers in vitro has been invaluable. Repetitive DNAs (copolymers or homopolymers) of highly defined composition and structure can be used for many preparative and analytical purposes in molecular biology: for isolation and characterization of mRNAs, DNA fragments, nucleases, DNA and RNA polymerases, and for physicochemical studies on DNA structure and function. The ability of DNA polymerase I to synthesize such polymers de novo (in the absence of added template-primer) is due to its strong affinity for small DNA fragments that copurify with the enzyme. Even the best enzyme preparations contain on the average 0.5 phosphate residues per enzyme molecule[17]. A small fraction of polymerase molecules probably contains oligonucleotides of repetitive sequence (i.e., oligo dAT) which can serve as small template-primers. However, it is puzzling that the same enzyme preparation under different reaction conditions (pH, salt concentration) will yield different highly homogeneous polymer products. The reaction conditions for de novo synthesis of different polymers have been found empirically, and it is not known why particular reaction conditions lead to a particular product.

De novo synthesis always starts with a considerable lag phase during which the ongoing reactions have not been detectable by available methods. Nevertheless, several stages of the reaction can be conceived from experiments with short synthetic template-primers. The first step is probably the growth of one of the strands of a small template-primer by slippage and reiteration. This reaction proceeds at a constant rate. It has not been excluded that in rare instances true de novo synthesis may occur, i.e., synthesis starting exclusively from mononucleotides without any template-primer. Slippage-reiteration events continue until the synthesized strand can be converted into a new template-primer for the many idle DNA polymerases in the reaction mixture. All conversions depend on nucleolytic action, either by a contaminating nuclease or by the intrinsic 5'——>3' exonuclease activity of DNA polymerase I. Once new template-primers become available the reaction will proceed exponentially. It is possible that similar mechanisms of polymer formation play a role in vivo. In E. coli, a transposon derived from the insertion sequence IS2 is likely to be formed by slippage-reiteration[83]. Homopolymeric and copolymeric regions rich in A and T have been found in yeast mitochondrial DNA[84], and satellite DNAs consisting mainly of alternating dAT copolymers exist in several animal species[85-89].

Nick translation has been used extensively for radiolabeling DNA to high specific activities. Nicking of a double-stranded DNA substrate is usually achieved with small amounts of an endonuclease, such as pancreatic DNase I. It is important that nicks be introduced in optimal numbers and at proper distances from each other. One nick per 800-1,500 nucleotides is desirable. The use of too many or too few nicks will result in suboptimal labeling and may change the properties of the original substrate to an unacceptable degree. Side reactions, such as strand displacement and strand switching, are in many cases undesirable and can be minimized by choosing the correct reaction conditions. (See Appendix A)

DNA polymerase I has the unique ability to initiate DNA synthesis at a nick
(E. coli DNA polymerases II and III lack this ability) and to move the nick
along the DNA by degrading the 5' end of the strand ahead of it while extending
the 3' end behind it. This nick translation requires a high degree of coordination of the polymerizing and 5'————>3' exonuclease functions of DNA polymerase
I. A nonlethal mutation of polA that greatly affects this coordination has been
identified. The mutant polA12 shows drastically increased sensitivity to DNA
damage by UV irradiation or methylmethane sulfonate at 43°C, compared to the
wild-type[90]. When DNA polymerase I isolated from polA12 cells is assayed for
nick translation ability at 30°C or 43°C, this function cannot be detected.
However, both the polymerase and the 5'————>3' exonuclease activities are found
at almost normal rates if tested separately. The mutant enzyme can also fill
gaps, but cannot continue its action into nick translation. Physicochemical
studies have shown structural differences between the mutant and the wild-type
enzyme. It is plausible that the spatial arrangement of the mutant enzyme has
been altered in such a way that polymerization and 5'————>3' exonucleolytic
activity are no longer coordinated[91]. Because semi-conservative DNA replication can be sustained in polA12 cells to an extent sufficient for cell growth,
a low level of coordinated exonuclease and polymerase activity directed to act
in DNA replication must exist in vivo.

Until recently, DNA sequencing has been difficult and slow, compared with
protein and RNA sequencing. But during the last few years several methods for
DNA sequencing have been developed in quick succession. The first extensively
exploited procedure was the "plus and minus" method of Sanger and Coulson[56].
It is based on the use of DNA polymerase I to complement defined regions of
single-stranded DNA under controlled conditions. The method is relatively
rapid and simple, but the distribution of the fragments of different sizes is
not always sufficiently uniform, and consecutive runs of nucleotides are not
easily determined. These disadvantages are largely overcome by the ribo-
substitution method[57], which makes use of the ability of DNA polymerase I to
incorporate ribonucleotides into DNA in the presence of Mn^{2+}. A third method,
described by Maxam and Gilbert[92], applies specific chemical degradation of DNA.
The advantage of this method is that any DNA, single- or double-stranded, can be
used, but it is more laborious and hazardous. A fourth procedure, based on the
"plus and minus" method but using chain-terminating nucleotides has been
developed by Sanger et al.[58]. It is becoming the method of choice for reasonable large single-stranded DNA fragments for which a primer is readily available. The obvious advantages of this method over others include high efficiency

and accuracy, lower exposure of the investigator to radioactivity and avoidance of toxic chemicals. Until recently, these advantages could not be fully exploited because of difficulties in obtaining the DNA that is to be sequenced in a suitable single-stranded form with an appropriate primer. This problem has been largely solved by the development of a cloning vehicle derived from the single-stranded DNA phage M13, which allows one to obtain high yields of single-stranded DNA restriction fragments, and to use a "universal" primer[74-76]. It can be expected that this technique and the availability of compact kits of materials[78] will further stimulate DNA sequencing. With the surge of the use of the dideoxynucleotide method the large fragment (Klenow fragment) of DNA polymerase I will become increasingly important.

V. APPENDIX

A. Nick Translation Procedure

A proven detailed procedure for radiolabeling DNA is given below:

Part 1: Place 1.8 µl of each of the unlabeled deoxyribonucleoside 5'-triphosphates (1 mM) in a vial. Add a sufficient amount of $[\alpha\text{-}^{32}P]$deoxyribonucleoside-5'-triphosphates to give a specific activity of 200 Ci/mmol. Add DNA (1-2 µg) and then 10 µl of a 10-fold concentrated reaction buffer containing 500 mM Tris-HCl (pH 7.8), 67 mM $MgCl_2$, 50 mM 2-mercaptoethanol, and nuclease-free bovine serum albumin at 0.5 mg/ml. Adjust the volume of the reaction mixture with distilled water (total volume after addition of DNA polymerase I should be 100 µl.

Part 2: Add 1 µl of pancreatic DNase I (100 ng/ml) to the reaction mixture and incubate at 15°C for 10 min. The addition of bovine pancreatic DNase I is required for extensive labeling of the substrate when E. coli DNA polymerase I from Bethesda Research Laboratories, Inc. is used, because this enzyme preparation contains no detectable endonucleolytic activity. When used without bovine pancreatic DNase I, this DNA polymerase I yields labeled DNA of greater than 1,000 base pairs with Herpes Simplex I DNA that contains one or two nicks per molecule[93].

Part 4: Add 10 µl of 0.3 M ethylenediaminetetraacetic acid (sodium salt, pH 8.0) to stop the reaction. A typical graph of incorporation kinetics is shown in Fig. 5.

B. DNA Sequencing with Chain-Terminating Nucleotides

The sequencing method has been described in detail[58-60]. Recent improvements in methods for obtaining template-primers and for performing sequencing[74-76] have been summarized in a manual that is part of the M13 cloning-sequencing kit of Bethesda Research Laboratories, Inc.[78]. An example of a DNA-sequencing gel pattern is shown in Fig. 6.

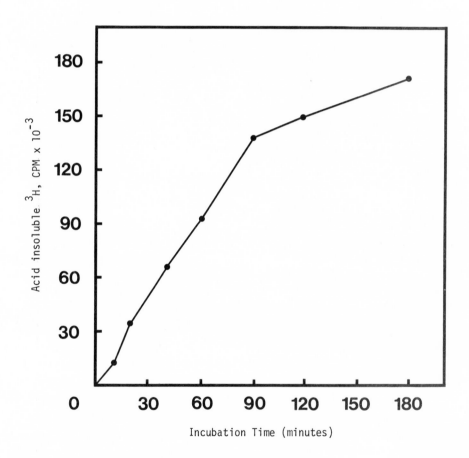

Fig. 5. Incorporation of ^{3}H deoxyadenosine-5'-triphosphate into nicked adenovirus-2 DNA. Reaction conditions are described in <u>Appendix</u> A.

36

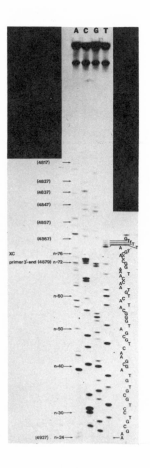

Fig. 6. Autoradiograph of a DNA-sequencing gel (sequencing was performed by
Bethesda Research Laboratories, Inc.). Single-stranded ØX174 was used as a
template with the 72 base-pair Hae III fragment of ØX174 DNA as the primer.
The annealed template-primer was excised with exonuclease III. The rest of
the procedure was performed according to the method of Sanger et al.[58] with
[α-^{32}P]dATP as the labeled nucleotide. The inhibitors used were (left to
right) the dideoxyribonucleotide derivatives of adenine, cytosine, guanine, and
thymine. Electrophoresis (8% polyacrylamide gel) was carried out on an apparatus available from Bethesda Research Laboratories, Inc.; this apparatus yields
a gel which can be used with standard medical X-ray film to produce an autoradiogram. The pattern is read from bottom to top to determine the DNA sequence,
starting from the 3' end of the exonuclease III-digested restriction fragment
primer. The sequence starts with base 4,927 and continues to base 4,827 of the
the ØX174 genome. The original primer 3' end is indicated at base 4,879. XC,
indicates the position of the dye marker xylene cyanol FF, which corresponds to
that of an oligonucleotide of approximately 76 bases.

REFERENCES

1. Watson, J.D. and Crick, F.H.C. (1953) Nature, 171, 737-741.
2. Lehman, I.R., Bessman, M.J., Simms, E.S. and Kornberg, A. (1958) J. Biol. Chem., $\underline{233}$, 163-170.
3. DeLucia, P. and Cairns, J. (1969) Nature, $\underline{244}$, 1164-1166.
4. Moses, R.E. and Richardson, C.C. (1970) Biochem. Biophys. Res. Comm., $\underline{41}$, 1557-1564.
5. Moses, R.E. and Richardson, C.C. (1970) Biochem. Biophys. Res. Comm., $\underline{41}$, 1565-1571.
6. Kornberg, T. and Gefter, M.L. (1971) Proc. Nat. Acad. Sci., U.S.A., $\underline{68}$, 761-764.
7. Gross, J. and Gross, M. (1969) Nature, 224, 1166-1168.
8. Kato, T. and Kondo, S. (1970) J. Bact., $\underline{104}$, 871-881.
9. Town, C.D., Smith, K.C. and Kaplan, H.S. (1971), Science, $\underline{172}$, 851-854.
10. Cooper, P.K. and Hanawalt, P.C. (1972) Proc. Nat. Acad. Sci. U.S.A., $\underline{69}$, 1156-1160.
11. Coukell, M.B. and Yanofsky, C. (1970) Nature, $\underline{228}$, 633-635.
12. Lehman, I.R. and Uyemura, D.G. (1976) Science, $\underline{193}$, 963-969.
13. Kornberg, A. (1980) DNA Replication, W. H. Freeman and Company, San Francisco.
14. Wickner, S.H. (1978) in Ann. Rev. Biochem, Snell, E.E., ed.; Annual Reviews, Inc., Palo Alto, CA, pp. 1163-1192.
15. Kelley, W.S., Chalmers, K. and Murray, N.E. (1977) Proc. Nat. Acad. Sci. U.S.A., $\underline{74}$, 5632-5636.
16. Kelley, W.S. and Stump, K. (1979) J. Biol. Chem., $\underline{254}$, 3206-3210.
17. Jovin, T.M., Englund, P.T. and Bertsch, L.L. (1969) J. Biol. Chem. $\underline{244}$, 2996-3008.
18. Brutlag, D. and Kornberg, A. (1972) J. Biol. Chem., $\underline{247}$, 241-248.
19. Lehman, I.R. and Richardson, C.C. (1964) J. Biol. Chem. 239, 233-241.
20. Klett, R.P., Cerami, A. and Reich, E. (1968) Proc. Nat. Acad. Sci. U.S.A., $\underline{60}$, 943-950.
21. Cozzarelli, N.R., Kelly, R.B. and Kornberg, A. (1969) J. Mol. Biol., $\underline{45}$, 513-531.
22. Deutscher, M.P. and Kornberg, A. (1969) J. Biol. Chem., $\underline{244}$, 3019-3028.
23. Deutscher, M.P. and Kornberg, A. (1969) J. Biol. Chem., $\underline{244}$, 3029-3037.
24. Atkinson, M.R., Deutscher, M.P., Kornberg, A., Russel, A.F. and Moffatt, J.G. (1969) Biochemistry, $\underline{8}$, 4897-4904.
25. Kelly, R.B., Atkinson, M.R., Huberman, J.A. and Kornberg, A. (1969) Nature, $\underline{224}$, 495-501.
26. Englund, P.T., Huberman, J.A., Jovin, T.M. and Kornberg, A. (1969) J. Biol. Chem., $\underline{244}$, 3038-3044.
27. Kelly, R.B., Cozzarelli, N.R., Deutscher, M.P., Lehman, I.R. and Kornberg, A. (1970) J. Biol. Chem., $\underline{245}$, 39-45.
28. Springgate, C.F., Moldvan, A.S., Abramson, R., Engle, J.L. and Loeb, L.A. (1973) J. Biol. Chem., $\underline{248}$, 5987-5993.
29. D'Aurora, V., Stern, A.M., and Sigman, D.S. (1977) Biochem. Biophys. Res. Comm. $\underline{78}$, 170-176.
30. Vallee, B.L. (1977) in Biological Aspects in Inorganic Chemistry, Dolphin, D., ed., John Wiley, New York, p. 38.
31. Englund, P.T., Kelly, R.B. and Kornberg, A. (1969) J. Biol. Chem., $\underline{244}$, 3045-3062.
32. Slater, J.P., Tamir, I., Loeb, L.A. and Mildvan, A.S. (1972) J. Biol. Chem., $\underline{247}$, 6748-6794.
33. Van De Sande, J.H., Loewen, P.C., and Khorana, H.G. (1972) J. Biol. Chem., $\underline{247}$, 6140-6150.

34. Brutlag, D., Atkinson, M.R. Setlow, P. and Kornberg, A. (1969) Biochem. Biophys. Res. Comm., 37, 982-989.
35. Klenow, H. and Henningsen, I. (1970) Proc. Nat. Acad. Sci. U.S.A., 65, 168-175.
36. Setlow, P., Brutlag, D. and Kornberg, A. (1972) J. Biol. Chem., 247, 224-231.
37. Setlow, P. and Kornberg, A. (1972) J. Biol. Chem., 247, 232-240.
38. Jacobsen, H., Klenow, H. and Overgaard-Hansen, K. (1974) Eur. J. Biochem. 45, 623-627.
39. Schachman, H.K., Adler, J. Radding, C.M., Lehman, I.R. and Kornberg, A. (1960) J. Biol. Chem., 235, 3242-3249.
40. Burd, J.F. and Wells, R.D. (1970) J. Mol. Biol., 53, 435-459.
41. Kornberg, A. (1965) in Evolving Genes, and Proteins, Bryson, V. and Vogel, H.J., eds., Academic Press, New York, pp. 403-417.
42. Radding, C.M., Josse, J. and Kornberg, A. (1962) J. Biol. Chem., 237, 2869-2882.
43. Radding, C.M., Josse, J. and Kornberg, A. (1962) J. Biol. Chem., 237, 2877-2882.
44. Kornberg, A., Bertsch, L.L., Jackson, J.F. and Khorana, H.G. (1964) Proc. Nat. Acad. Sci. U.S.A., 51, 315-323.
45. McCarter, J.A., Kadohama, N. and Tsiapalis, C. (1969) Can. J. Biochem., 47, 391-399.
46. Olson, K., Luk, C. and Harvey, C.L. (1972) Biochem. Biophys. Acta., 277, 269-275.
47. Lehman, I.R. (1967) Ann. Rev. Biochem., 36, 645-668.
48. Dunn, A.R., Matthews, M.B., Chow, L.T., Sambrook, J. and Keller, W. (1978) Cell, 15, 511-526.
49. Dunn, A.R. and Hassell, J.A. (1977) Cell, 12, 23-36.
50. Casey, J. and Davidson, N. (1977) Nucleic Acids Res., 4, 1539-1552.
51. Rigby, P.W.J., Dieckmann, M., Rhodes, C. and Berg, P. (1977) J. Mol. Biol. 113, 237-251.
52. Maniatis, T., Jeffrey, A. and Kleid, D.G. (1975) Proc. Nat. Acad. Sci. U.S.A., 72, 1184-1188.
53. Weinstock, R., Sweet, R., Weiss, M., Cedar, H. and Axel, R. (1978) Proc. Nat. Acad. Sci. U.S.A., 75, 1229-1303.
54. Botchan, M., Topp, W. and Sambrook, J. (1976) Cell, 9, 269-287.
55. Lehman, I.R. (1974) Science, 186, 790-797.
56. Sanger, F. and Coulson, A.R. (1975) J. Mol. Biol., 94, 441-448.
57. Barnes, W.M. (1978) J. Mol. Biol., 119, 83-99.
58. Sanger, F., Nicklen, S. and Coulson, A.R. (1977) Proc. Nat. Acad. Sci. U.S.A., 74, 5463-5467.
59. Sanger, F. and Coulson, A.R. (1978) FEBS Letters, 87, 107-110.
60. Smith, A.J.H. (1980) in Methods of Enzymology, Academic Press, New York, Vol. 65, part I, pp. 560-580.
61. Helfman, W.B., Hendler, S.S., Schannahoff, D.H. and Smith, D.W. (1978) Biochemistry, 17, 1607-1611.
62. Agarwal, S.S., Dube, D.K. and Loeb, L.A. (1979) J. Biol. Chem., 254, 101-106.
63. Jeppesen, P.G.N. (1974) Anal. Biochem., 58 195-207.
64. Eshaghpour, H. and Crothers, D.M. (1978) Nucleic Acid Res., 5, 13-21.
65. Szybalski, W., Kubinski, M. and Summers, W.C. (1971) in Methods in Enzymology, Grossman, L. and Moldave, K., eds., Academic Press, N.W., 21, 383-415.

66. Goulian, M., Goulian, S.H., Codd, E.E. and Blumenfield, A.Z. (1973) Biochemistry, 12, 2893-2901.
67. Oertel, W. and Schaller, H. (1972) FEBS Letters, 27, 316-320.
68. Oertel, W. and Schaller, H. (1973) Eur. J. Biochem., 35, 106-113.
69. Kaptein, J.S. and Spencer, J.H. (1978) Biochemistry, 17, 841-850.
70. Chang, L.M.S. and Bollum, F.J. (1971) J. Biol. Chem., 246, 909-916.
71. Galbert, F., Sedat, J. and Ziff, E.G. (1974) J. Mol. Biol., 87, 377-407.
72. Smith, A.J.H. (1979) Nucleic Acids Res., 6, 831-848.
73. Schreier, P.H. and Cortese, R. (1979) J. Mol. Biol., 129, 169-172.
74. Heidecker, G., Messing, J. and Gronenborn, B. (1980), Gene, 10, 68-73.
75. Anderson, S., Gait, M.J., Mayol, L. and Young, I.G. (1980), Gene, 10, 68-73.
76. Messing, J., Crea, R. and Seeburg, P.H. (1981) Nucleic Acid Res., 9, 309-322.
77. Sanger, F. Coulson, A.R., Barrell, B.G., Smith, A.J.H. and Roe, B. (1980) J. Mol. Biol., 143, 168-178.
78. M13 DNA Cloning/"Dideoxy" Sequencing Manual (1980), Bethesda Research Laboratories, Inc., Rockville, MD.
79. Konrad, E.B. and Lehman, I.R. (1974) Proc. Nat. Acad. Sci. U.S.A., 71, 2048-2051.
80. Uyemura, D., Eichler, D.C. and Lehman, I.R. (1976) J. Biol. Chem., 251, 4058-4089.
81. Olivera, B.M. and Bonhoeffer, F. (1974) Nature, 250, 513-514.
82. Lehman, I.R. and Chien, J.R. (1973) J. Biol. Chem., 248, 7717-7723.
83. Ghosal, D. and Saedler, H. (1978) Nature, 275, 611-617.
84. Borst, P. (1972) Ann. Rev. Biochem., 41, 333-376.
85. Sueoka, N. (1961) J. Mol. Biol., 3, 31-40.
86. Swartz, M.N., Trautner, T.A. and Kornberg, A. (1962) J. Biol. Chem., 237, 1961-1967.
87. Rosenberg, H., Singer, M. and Rosenberg, M. (1978) Science, 200, 394-410.
88. Carlson, M. and Brutlag, D. (1977) Cell, 11, 371-379.
89. Botchan, M.R. (1974) Nature, 251, 288-295.
90. Monk, M. and Kinross, J. (1972) J. Bact., 109, 971-983.
91. Uyemura, D. and Lehman, I.R. (1976) J. Biol. Chem., 251, 4078-4086.
92. Maxam, A.M. and Gilbert, W. (1977) Proc. Nat. Acad. Sci. U.S.A., 74, 560-564.
93. Berninger, M., Bethesda Research Labs., Personal Communication.

THE USE OF TERMINAL TRANSFERASE FOR MOLECULAR CLONING AND NUCLEIC ACIDS ANALYSIS

Ranajit Roychoudhury

Division of Molecular Biology
International Plant Research Institute
San Carlos, California 94070, U.S.A.

I. INTRODUCTION

Terminal transferase (E.C.2.7.7.3.1.) from calf thymus is a small protein (32,000 daltons) with two subunits of 24,000 and 8,000 daltons[1]. A high-molecular-weight protein (79,000 daltons) consisting of a single polypeptide with very similar enzymatic activity has also been detected in calf thymus[2]. The enzyme catalyzes the primer-dependent but template-independent polymerization of deoxynucleotides at the 3'-OH ends of single-stranded DNA or oligonucleotides[3]. It also catalyzes a limited polymerization of ribonucleotides[4]. Such ribonucleotide addition has been used to develop a new method for specific labeling of 3' ends with a single ribonucleotide or a $[^{32}P]$phosphate[5]. This technique permits not only the identification of the original 3'-terminal deoxynucleotide by nearest-neighbor analysis[5-7], but also the sequence analysis of oligonucleotides[7-11].

All four common ribonucleotides can be added at the 3' end of a given primer[12]. The reaction may be driven to completion with a 10- to 50-fold excess of any of the four common ribonucleotides[12]. The addition of a single specific ribonucleotide in a given primer ensures the insertion of an alkali-sensitive linkage after the primer extension by DNA polymerase. Thus, this procedure is a powerful tool for DNA sequence analysis[13-15].

An oligoribonucleotide, r(A-A-A-A-A-A), can be made to act as a primer if one or two deoxynucleotides are added at the 3' end by using the enzyme polynucleotide phosphorylase to yield the products r(A-A-A-A-A-A)-dT and r(A-A-A-A-A-A) dT-dT[16]. r(A-A-A-A-A-A) also acts as a primer for polymerization of $[\alpha-^{32}P]$dCTP in the presence of terminal transferase[17]. The incorporation, however, is about 250 fold less than that obtained with d(pT-T-T-T) as a primer. A covalent linkage between r(A-A-A-A-A-A) and the product poly(dC) has been demonstrated by nearest-neighbor analysis[17].

II. END-GROUP LABELING OF OLIGONUCLEOTIDES

To determine the specificity of the 3'-terminal labeling reaction, we have used synthetic oligonucleotides of defined chain length and nucleotide sequence[28-30]. The oligonucleotides d(A-C-C-A), d(A-C-C-A-T-C), and d(A-C-C-A-T-C-C-A) were prepared by chemical synthesis[31] with protected dinucleotide blocks as intermediates[32].

A. Reaction Steps for Transferase Labeling and Analysis of the End Group

1. Depending on the primer-to-substate ratio, the reaction in the presence of Mg^{2+} ion yields either a mono- or a di-addition product:

$$d(A-C-C-A-T-C) + ppp\overset{*}{p}A_r \xrightarrow[\text{transferase}]{Mg^{2+},} d(A-C-C-A-T-C)\overset{*}{p}A_r + PPi$$

$$d(A-C-C-A-T-C) + 2\ ppp\overset{*}{p}A_r \longrightarrow d(A-C-C-A-T-C)\overset{*}{p}A_r\overset{*}{p}A_r + 2\ PPi$$

(The subscript r indicates a ribonucleotide, and $\overset{*}{p}$ denotes a $[^{32}P]$ phosphate.)

2. Treatment of such products with alkali yields two species of oligonucleotides:

$$d(A-C-C-A-T-C)\overset{*}{p}A_r$$
$$d(A-C-C-A-T-C)\overset{*}{p}A_r\ \overset{*}{p}A_r \xrightarrow{\text{alkali}}$$

$$d(A-C-C-A-T-C)\overset{*}{p}A_r$$
$$d(A-C-C-A-T-C)\overset{*}{p}A_r\overset{*}{p} + A_r$$

3. Incubation of these products with alkaline phosphatase yields a homogeneous mono-addition product:

$$d(A-C-C-A-T-C)\overset{*}{p}A_r \xrightarrow{\text{Phosphatase}} d(A-C-C-A-T-C)\overset{*}{p}A_r$$
$$d(A-C-C-A-T-C)\overset{*}{p}A_r\overset{*}{p} \qquad d(A-C-C-A-T-C)\overset{*}{p}A_r + {}^{32}P_i$$

4. To label the 3'end with a single $[^{32}P]$ phosphate, the mono-addition product is treated with periodate and cyclohexylamine.

$$d(A-C-C-A-T-C)\overset{*}{p}A_r \xrightarrow[\text{cyclohexylamine}]{IO_4^-\ \text{and}} d(A-C-C-A-T-C)\overset{*}{p}$$

5. Thus, the labeling of the 3' terminus of an unknown oligodeoxynucleotide enables one to identify the original 3'-terminal nucleotide by nearest-neighbor analysis[6].

$$d(A-C-C-A-T-C)\overset{*}{p}A_r \xrightarrow[\text{phosphodiesterase}]{\text{spleen}} dAp/dCp/dCp/dAp/dTp/\underline{d\overset{*}{C}p}/A_r$$
$$d(A-C-C-A-T-C)\overset{*}{p} \qquad\qquad dAp/dCp/dCp/dAp/dTp/\underline{d\overset{*}{C}p}/$$

nearest
neighbors

When all four nucleotides are separated from one another with paper
chromatography or electrophoresis, only the nearest neighbor is detected
as the radioactive nucleotide[6].

B. Procedures

1. Enzymes:

Terminal transferase was isolated by the method of Chang and Bollum[1],
with slight modification of the final step[12]. To remove the traces of exonuclease and reduce substantially the amount of endonuclease contamination, the
final enzyme preparation was subjected to rechromatography on a column of
hydroxylaptite. This step resulted in a 50% loss in the yield of the final
product (Feix and Roychoudhury, 1972-unpublished results), but it improves
the quality of the enzyme remarkably. Only an enzyme preparation completely
free of exonuclease activity and very low in endonuclease activity is suitable
for gene manipulation and sequence analysis. The final specific activity of
the enzyme was 25,000 units/mg of protein, as tested with d(pT)6 as primer in
the presence of Mg^{2+} ion.

E. coli alkaline phosphatase and calf spleen phosphodiesterase were products of Worthington. The spleen phosphodiesterase may be further purified to
remove the traces of contaminating nucleotidase activity with unsuspected base
specificity[33]. Such nucleotidase activity may cause gross error in the yield
of radioactive nucleotide in nearest-neighbor analysis. The contaminating
traces of nucleases in the alkaline phosphatase preparation may be inhibited
by the procedure of Shinagawa and Padmanabhan[34].

2. Preparation of Concentrated Buffer for Transferase Reaction:

Cacodylic acid (13.8 g, Fisher Scientific Company, pure grade) is suspended in glass-distilled water (35 ml) in a narrow tall beaker (150-ml capacity,
Corning) and neutralized to a pH of 7.0 with constant stirring over a magnetic
stirrer by slow addition of solid pellets of KOH (analytic grade). The solution is then cooled to 0°C and placed over an ice bucket. The ice bucket, with
the beaker containing the magnetic spin bar, is then placed over the magnetic
stirrer, and a gentle stirring is resumed. The electrode of a pH meter is
immersed in such a way that the magnetic spin bar does not come into contact
with the electrode. Solid powder of Tris base (3.0 g, TRIZMA base from Sigma)
is then added to the solution. The pH should be close to 7.6. If it exceeds
7.6, it may be lowered by adding small amounts of the powder of cacodylic acid.

Glass-distilled water is then added to a final volume of 88 ml (the solution
may be poured into a 100-ml measuring cylinder to determine the exact volume).
To this solution is then added 2 ml of 0.1 M dithiothreitol with constant stirring over the magnetic stirrer. Less dithiothreitol (1 ml) may be added, but
the addition of more dithiothreitol (to more than 2 mM in the final solution)
would be disastrous. The last addition is that of the metal ions. The addition of 10 ml of 0.1 M $CoCl_2$ (or 1 M $MgCl_2$) is carried out very slowly, dropwise
from a pipette with constant stirring by a magnetic stirrer. This clear solution (1 M potassium cacodylate, 250 mM Tris base, 10 mM $CoCl_2$ or 100 mM $MgCl_2$,
and 2 mM dithiothreitol) may be kept at 4°C for more than a year. One stock
was used for more than 2 years.

3. Labeling Reaction:

The reaction is carried out in a siliconized tube (6 X 50 mm, Scientific
Products). To the tube are added 1 µl of 10 X cacodylate $CoCl_2$ buffer, 1 µl
oligonucleotide primer (1-5 pmol of 3'-OH ends), 100-500 pmol of $[\alpha^{32}P]rNTP$
A, G, C, or U, specific activity 400-600 Ci/mmol from Amersham or NEN, previously evaporated in the same tube under reduced pressure in a desiccator, (because
it is supplied as 50% ethanol solution), and 20 units (1-5 µl) of terminal
transferase. The final volume is made up to 10 µl. The tube is sealed with
Parafilm and incubated at 35°C for 20 min. The reaction is terminated by
addition of 10 µl of 0.6 M KOH. The content of the tube is then carefully
drawn into a siliconized capillary tube which is then sealed at both ends with
a flame from a bunsen burner. The capillary tube should be long enough so that
the reaction mixture remains in the middle of it. This will permit the cutting of the two ends of the tube with a diamond pencil (or a graphite file)
without disturbing the solution in the middle. The sealed tube is then placed
horizontally on a styrofoam block and incubated at room temperature for 20-24 h.

The ends of the capillary tube are cut and the contents are transferred to
the original 6 X 50⁻ mm tube by gently blowing on the end with a pipettor
(screw type). The solution is then neutralized by adding a few grains of HEPES
(Calbiochem) powder and mixing the contents in a vortex mixer.

The pH of the solution may be tested with extremely thin strips of
"ColorpHast" pH indicator strips (E. Merck). When the pH is between 7.0 and
9.0, the alkaline phosphatase treatment may be carried out in the same tube.
To the tube are added 5 µl of 2% SDS and 0.2 unit (in 5-10 µl volume) of E.
coli alkaline phosphatase (Worthington Product BAPF). The SDS will completely

inhibit any nonspecific nucleases that may be present as contaminants in the alkaline phosphatase preparation[34].

The capillary tube is rinsed with 30 μl of water and transferred to the phosphatase reaction mixture. The tube is sealed with Parafilm and incubated at 37°C for 1 h. Because the volume of the solution is more than 50 μl, it will survive the 37°C incubation without being dry at the bottom. This procedure has been designed to minimize both the loss of material and the number of operations and to obviate the removal of either the salts or the triphosphates from the reaction mixture.

If the chain length of the oligonucleotide is small (six or lower), paper chromatography is the method of choice for isolation of the oligonucleotide. If it is longer than six residues, the material may be isolated by gel filtration on Sephadex G-25 (superfine).

For incubation with greater amounts of oligonucleotide, the reaction may be scaled up proportionately. However, the concentration of oligonucleotide should not exceed 1 A_{260}/ml. With an incubation volume greater than 100 μl, a sealed capillary tube is not necessary. All operations may be carried out in the 6 X 50-mm tube or a larger tube.

4. Paper Chromatography:[5]

Schleicher and Schuell 2043b (slow) or 2040b (fast) paper may be used. At the origin, 50 μl of 0.1 M EDTA is applied as a streak. After drying, 20 μl of a solution of 50 mM (750 A_{260}/ml)rATP is applied over the streak of EDTA. The drying may be accomplished with a hot air dryer. When drying is complete, an aliquot (10 or 20 μl) of the reaction mixture is applied as a streak over the streak of ATP. Care must be taken to ensure that the boundary of streaking does not exceed the boundary of ATP. This operation will substantially reduce the nonspecific binding of labeled oligonucleotide at the origin of the chromatogram. The chromatogram is then developed in n-propanol, concentrated NH_4 OH, and water (in a volume ratio of 55:10:35) by the descending technique. When the solvent front reaches the bottom, the chromatogram is air dried. Paper strips, 1 cm apart, are cut and counted by Cerenkov radiation to indicate the position of the labeled oligonucleotide. The oligonucleotide peak may be eluted with 0.1 N NH_4OH, evaporated to dryness, and dissolved in a suitable volume of water or buffer.

5. Gel Filtration:

The gel filtration (for larger oligonucleotides) may be carried out by using 0.1 M triethylammonium bicarbonate (TEAB) as the solvent. The peak fractions detected by Cerenkov radiation may be evaporated and used for further experiments.

6. Nearest-neighbor Analysis:[6]

The original 3'-terminal nucleotide, now next to the incorporated ribonucleotide (nearest neighbor), can be identified by complete digestion of the oligonucleotide with spleen phosphodiesterase[33]. The separation of the four nucleoside 3'-phosphates by paper chromatography[35] or paper electrophoresis[5] will reveal the identity of the original 3'-terminal nucleotide as the only radioactive nucleotide.

7. Preparation of Oligonucleotides with a [^{32}P]phosphate at the 3'End:[5]

To 100 µl of oligonucleotide such as d(A-C-C-A-T-C)$\overset{*}{p}$A$_r$ are added 50 µl of cyclohexylamine glutamate solution and 50 µl of 0.1 M NaIO$_4$. The cyclohexylamine glutamate solution is prepared as follows: 3.7 g of glutamic acid (25 mmol) is dissolved in 30 ml of water, and the pH is adjusted to 8.0 by addition of about 14 ml of 1 M NaOH. The final volume is made up to 50 ml after addition of 5.75 ml (50 mmol) of cyclohexylamine. The resulting pH is close to 11.0.

After incubation at 45°C for 90-120 min, 50 µl of 0.4 M ribose is added to destroy the excess of periodate. After further incubation at room temperature for 30 min, 1 ml of water is added, and then 20-30 µl of 1 N HCl to adjust the pH to 8.0. The total reaction mixture may then be applied to a small DEAE column in a pasteurpipette, washed, and eluted with TEAB as described earlier[25], or it may be isolated by gel filtration.

8. Microscale Procedure:[10]

The labeled oligonucleotide is oxidized in a solution containing 4 µl of water, 1 µl of 2 M TEAB, and 2 µl of 0.1 M NaIO$_4$ at room temperature for 1 h. Then, 5 µl of 1 M n-propylammonium bicarbonate (pH, 7.5) is added and the solution kept at 45°C for 3 h. Excess periodate is then destroyed by adding 1 µl of 0.2 M glycerol. The TEAB is removed by evaporation in a desiccator. The product is oligonucleotide 3'-phosphate.

9. Analysis of the 3'-terminal Nucleotide Composition of DNA Fragments Produced by DNase Digestion:[36,37]

To 15 ml of calf thymus DNA (8 A$_{260}$/ml) in 50 mM ammonium acetate and 1 mM EDTA (pH, 5.5) is added 0.8 unit of spleen DNase. When the hyperchromicity reaches 2% after 15 min at 22°C, the reaction is terminated by shaking the contents with 1 ml of chloroform:isoamyl alcohol (24:1 v/v). The

digest is dialyzed against running water and then adjusted to 0.05 M NaOH
for 2 min to ensure complete denaturation of DNA fragments. The pH is then
adjusted to 5.5 by adding 1 M acetic acid. The oligonucleotides are dephosphorylated with 0.2 unit/ml of acid phosphomonoesterase B at 37°C for 14 h.
The average chain length is 54 nucleotide residues[36].

The reaction mixture (50 µl) contains 1-10 µM oligonucleotides (0.025-
0.25 A_{260} units), 150-500 µM [α-^{32}P]rATP of specific activity 0.3-1 Ci/mmole,
200 mM potassium cacodylate (pH, 7.0), 8 mM $MgCl_2$, 1 mM 2-mercaptoethanol,
and 45 units of terminal transferase. After incubation at 37°C for 12-14 h
in a sealed capillary tube, the contents are mixed and shaken with 0.2 volume
of chloroform:isoamyl alcohol (24:1 v/v). The unreacted ATP is separated by
gel filtration on Sephadex G25.

The excluded peak fractions are pooled, mixed with 1.2 A_{260} units of
calf thymus DNA, lyophilized, dissolved in 100 µl of 20 mM ammonium acetate
(pH, 5.5) and digested at room temperature with 0.6 unit of spleen DNase and
0.6 unit of spleen phosphodiesterase. After 1 h the enzymes are inactivated,
and the pH is adjusted to 8-9 by adding 1 M NH_4OH.

A portion (50 µl) of this solution (20,000 cpm) is applied to a column of
DEAE cellulose (0.5 X 15 cm) preequilibrated with 50 mM ammonium acetate
(pH, 9.3) and washed with the same buffer until the peaks of Pi, dCMP, dTMP,
and dAMP were eluted. A step of 150 mM ammonium acetate (pH, 8.9) is then
applied. During this elution, two more peaks (rAMP and dGMP) are eluted.
Thus, this procedure isolates all of the components of the digestion as separate
peaks. From the radioactivity in each peak, the composition of 3'-terminal
nucleotides liberated by spleen DNase is determined.[36] In this way, the determination of the original 3'-terminal nucleotides is achieved without the alkali
and phosphatase steps[36].

10. Labeling of Oligonucleotides with Dideoxyadenosine Monophosphate
     Residues:[19,20]

In 1969, Cozzarelli et al.[19] and Atkinson et al.[20] demonstrated that terminal transferase accepts ddNTP as a substrate. Several chemically-synthesized
oligonucleotides of defined chain length and nucleotide sequence have been used
as primers for single-terminal addition with dideoxyadenosine triphosphate
(ddATP).[21] The reaction mixture (200 µl) contains pmol of primer oligonucleotide, 200 mM potassium cacodylate (pH, 7.2), 5 mM 2-mercaptoethanol, 8 mM $MgCl_2$,
7.6 nmol of [α-^{32}P]ddATP (7 X 10^5 cpm per nmol), and 13 units of terminal transferase. After 16 h at 37°C, the reaction mixture is applied to a DEAE paper

strip and chromatographed (descending) in 0.5 M TEAB (pH, 7.6) for 4 h. Paper strips are cut out and counted. The labeled oligonucleotide located in this way is eluted with 1 M TEAB and subjected to nearest-neighbor analysis[5,6]. In all cases, almost 100% of the label is detected in the original 3'-terminal nucleotide of a given primer. When tested with d(pT)$_8$ as primer, the amount of ddAMP incorporated was identical with that of a given primer; this indicates a high degree of purity of the [α-^{32}P]ddATP used for single-terminal addition[21].

11. Labeling of the 3' End with Cordycepin Monophosphate:

Because of the absence of the 3'-OH group in cordycepin triphosphate (in this respect similar to ddATP), the reaction conditions described[26] for labeling of duplex DNA fragments can be used for single-terminal labeling of oligonucleotides[19,27].

12. Merits of the Methods:

[α-^{32}P]ddNTP or [α-^{32}P]cordycepin triphosphate used for labeling the 3' end enables one to obtain strictly single-terminal addition. Thus, the homogeneous population of oligonucleotide essential for DNA sequence analysis[38] is obtained.

When purine rNTPs are used for 3'-terminal labeling at a concentration 10-50 times higher than the 3'-OH ends, several labeled residues are added[14,23]. Hydrolysis of these products with alkali produces a homogeneous population of labeled oligonucleotides containing two [^{32}P]phosphate moieties on either side of the first ribonucleotide[25]. Thus, the phosphatase step is not necessary. Maxam and Gilbert have described a modification of this procedure that uses the volatile base piperidine for the same objective[38].

Although the method with [α-^{32}P]rNTP requires one more step than that with ddNTP[19-21] or cordycepin triphosphate, it has very useful advantages. First, because of the absence of 3'-OH groups, the oligonucleotides labeled with chain-terminating residues cannot be used for further chain extension[4,12-15]; this is a powerful tool for DNA sequence analysis[13-15,18]. Second, the use of purine triphosphates (rATP and rGTP) enables one to introduce two ^{32}P labels instead of one[25,38]. Third, the insertion of a single ribonucleotide at the 3' end of an oligonucleotide followed by primer extension, ensures an alkali-sensitive linkage between the primer and the product sequences. This property allows the separation of the primer from the product after alkaline hydrolysis[4,13,14,18] and enables one to do further analysis of the isolated product[13,14,18,39-42].

Fourth, the addition of one or two specific ribonucleotides complementary to a given template enables one to increase the hybridization efficiency of short oligonucleotides and thereby obtain more defined chain extension at a known location in a DNA molecule. Thus, the primer extension by DNA polymerase I can be accomplished with greater fidelity. This is especially useful when oligo(dT) with one or two specific nucleotides is used as primer[43].

III. SYNTHESIS OF SPECIFIC PRIMERS

F.J. Bollum first discovered[44,45] an enzymatic method for specific limited addition of nucleotides to a given primer in the absence of a template DNA. (The polymerase from calf thymus had not been well characterized, and the reaction of limited addition discovered by Bollum was actually catalyzed by terminal transferase[3].) He also described the procedure for separation, isolation, and analysis of the products[45]. By using a known nucleotide sequence, specific primers corresponding to a known location in the template DNA or template mRNA may be enzymatically synthesized with terminal transferase. At a triphosphate-to-primer ratio of 1:3, most combinations of primer and triphosphate produce distributions containing one to five monomer additions[46]. These products can be easily separated and isolated[45,46]. When the product of single-terminal addition has been isolated, the extended product may be used as primer for a second addition. In this way, additions of two to ten specific nucleotide residues may be achieved. Starting with a trinucleotide primer, oligonucleotides 10-12 residues long may be synthesized and isolated by paper chromatography[45], column chromatography[46], or thin-layer sheet chromatography[30]. As an example of such synthesis, d(pA-A-A) was used as primer for limited addition of dCMP, dTMP, or dAMP residues[46].

The reaction mixture (100 µl) contained 200 mM potassium cacodylate, (pH, 7.2), 4 mM $MgCl_2$, 1 mM 2-mercaptoethanol, and 13-122 µM d(pA-A-A) with 1.43-fold [^{14}C]dCTP (in 1 mM $CoCl_2$), 1.48-fold each of [^{14}C] dTTP and [^{14}C]dATP, and 75 units of terminal transferase. After incubation overnight at 35°C, the reaction was terminated by adding 10 µl of 0.2 M EDTA and subjected to chromatography. Strips (1.5 in. wide) of DE-81 paper were attached to glass frames and developed for 16 h with 0.5 M ammonium bicarbonate. The locations of the products were detected by cutting paper strips and counting the radioactivity[46].

The reaction was applied as a streak at the origin of a DEAE TLC plate and developed in homomixture 6 of Jay et al.[10].   After drying in a 65°C oven, the plate was subjected to autoradiography.

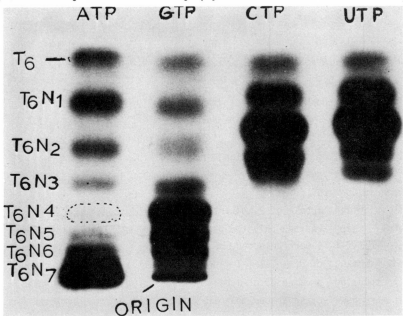

Fig. 1. Multiple addition of ribonucleotides with d(pT)$_6$ as primer.

As shown in Fig. 1, the purine ribonucleotides yielded more than six ribo-addition products. Owing to poor resolution near the origin, the products with more than six additions were not clearly separated. However, the addition of six nucleotides is clearly distinguishable. With rCTP and rUTP as substrates, a maximum of four residues were added with di-addition product as the predominant species. In the rCTP lane, the third and fourth products were not well separated, because of some imperfection in the thin-layer plate; this will be shown more clearly later.

A quantitative anslysis of radioactivity in different products indicated that more than 90% of the labeled primer was converted to ribo-addition products. It is, however, quite possible that the higher oligomers were produced by polymerization of contaminating deoxytriphosphates used in these experiments. If that is true, the products will be unaffected by alkali treatment. However, if ribonucleotides had been added, one would expect the following products after the alkali treatment, as shown with a single-addition and multiple-addition product (pN = ribonucleotide, rN = ribonucleoside):

$$d(\overset{*}{p}T-T-T-T-T) \xrightarrow[\text{transferase}]{\text{rNTP, terminal}} \begin{array}{l} d(\overset{*}{p}T-T-T-T-T)pN \\ d(\overset{*}{p}T-T-T-T-T)pNpNpNpNpNpN \end{array}$$

$$\begin{array}{l} d(\overset{*}{p}T-T-T-T-T)pN \\ d(\overset{*}{p}T-T-T-T-T)pNpNpNpNpNpN \end{array} \xrightarrow{\text{alkali}} \begin{array}{l} d(\overset{*}{p}T-T-T-T-T)pN \\ d(\overset{*}{p}T-T-T-T-T)pNp + 4Np + rN \end{array}$$

Thus, whatever the length, only two radioactive products migrating slower than d($\overset{*}{p}$T-T-T-T-T) would be detected after alkali treatment.

The remaining aliquots (12 µl) of the reaction mixture shown in Fig. 1 were adjusted to 0.3 M KOH and incubated at room temperature for 20 h. After alkaline hydrolysis, 5-µl aliquots were subjected to analysis by homochromatography.

It is evident from Fig. 2 that all of the higher oligomers shown in Fig. 1 completely disappeared and only two radioactive products migrating slower than d($\overset{*}{p}$T-T-T-T-T) were detected. Because of the extra phosphate present at the 3' terminus of d($\overset{*}{p}$T-T-T-T-T)pNp, two negative charges were added; the corresponding d-value[10] for mobility thus increased significantly.

It may, however, be argued that inasmuch as the ribonucleotides were unlabeled, a contaminating deoxynucleotide added after the first ribonucleotide will also produce d($\overset{*}{p}$T-T-T-T-T)prNp, and smaller products of the type dNprNp in the middle of the chain liberated after the alkali treatment will remain undetected because of the absence of radioactivity in such a product. That this is not the case was conclusively resolved by using [α-^{32}P]rNTPs.

The reaction mixture (50 µl) contained 50 pmol of d(pT-T-T-T-T), 5,000 pmol of [α-^{32}P]rCTP, and 16 units of terminal transferase. Other conditions were similar to those described in Fig. 1. After 24 h at 37°C, the products were isolated by gel filtration on Sephadex G-25. The products were pooled and evaporated in a desiccator. The material was dissolved in 50 µl of water. An aliquot (2 µl) was applied as a spot on a DEAE TLC plate. A small amount of labeled d($\overset{*}{p}$T-T-T-T-T) was applied on the same spot as an internal marker. As shown in Fig. 3A, four products were formed by the additon of [^{32}P]rCMP residues. The remaining aliquot (48 µl) was adjusted to 0.3 M KOH and in-cubated at room temperature for 20 h. The reaction mixture was then passed through a small column (bed volume, 100 µl) of pyridinium Dowex-50 and evaporated to dryness. The content of the tube was dissolved in 50 µl of water, and a 1-µl aliquot was subjected to analysis by homochromatography.

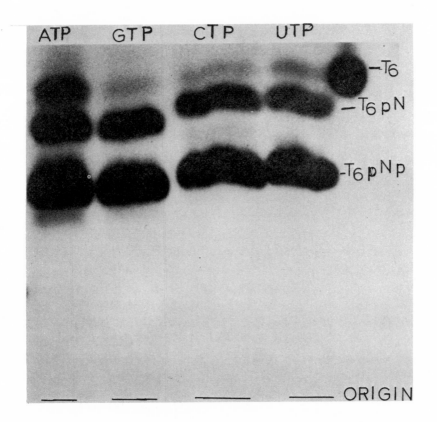

Fig. 2. Disappearance of multiple products after alkaline hydrolysis.

The marker d($\overset{*}{p}$T-T-T-T-T) was also run on an adjacent lane, as shown in Fig. 3B.

If the chain extension was due exclusively to the addition of ribonucleotides, we would expect the following products after the alkali treatment:

d(pT-T-T-T-T)$\overset{*}{p}$rC

d(pT-T-T-T-T)$\overset{*}{p}$rC$\overset{*}{p}$rC

d(pT-T-T-T-T)$\overset{*}{p}$rC$\overset{*}{p}$rC$\overset{*}{p}$rC

d(pT-T-T-T-T)$\overset{*}{p}$rC$\overset{*}{p}$rC$\overset{*}{p}$rC$\overset{*}{p}$rC

$\xrightarrow{\text{alkali}}$

d(pT-T-T-T-T)$\overset{*}{p}$rC

d(pT-T-T-T-T)$\overset{*}{p}$rC$\overset{*}{p}$ + rC

d(pT-T-T-T-T)$\overset{*}{p}$rC$\overset{*}{p}$ + r$\overset{*}{C}$p + rc

d(pT-T-T-T-T)$\overset{*}{p}$rC$\overset{*}{p}$ + 2r$\overset{*}{C}$p + rC

It is evident from the above equations that only three radioactive products - d(pT-T-T-T-T)$\overset{*}{p}$rC$\overset{*}{p}$, d(d(pT-T-T-T-T)$\overset{*}{p}$rC, and r$\overset{*}{C}$p - would be detected after the alkali treatment. If there were any deoxynucleotides in between, we would

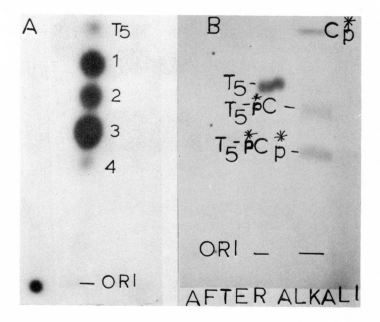

Fig. 3.   Evidence for exclusive polymerization of ribonucleotides.

expect a product of the type dNprC* from the end of the chain and a product of
the type dNprCp* from the middle of the chain. No such products were detected.
As expected from the above equations, only three radioactive products were
detected after the alkali treatment (Fig. 3B). Therefore, all four newly added
residues were composed of ribonucleotides only.

By using d(T-T-T-T-T) as primer and rGTP as substrate at a 1:1 ratio,
we isolated the single addition product d(T-T-T-T-T)-rG in pure form with
one chromatographic step. When this was incubated with rCTP as substrate
at a 1:1 ratio, the product d(T-T-T-T-T)-rG-rC was obtained in 50% yield.
The details of such synthesis will be reported elsewhere[49].

IV.   APPLICATION OF RIBONUCLEOTIDE-TERMINATED PRIMER EXTENSION FOR SYNTHETIC
AND ANALYTIC PURPOSES

Enzyme-catalyzed reactions for polynucleotide synthesis depend on primers
that become covalently linked to the polymerized products. Therefore, the ideal

primer for synthesis of specific polynucleotides (normal or modified bases)
would be one that could be conveniently separated from the products after the
desired stretch of polynucleotide has been synthesized. With this objective
in mind, we have explored the possibility of adding one ribonucleotide at the
3' end of a short oligodeoxynucleotide primer and its later use as a primer for
further synthesis of polynucleotides. The advantage of using a small oligonucleotide as the starting material is that, following a limited synthesis, the
newly synthesized product could be separated from the starting material by
alkaline hydrolysis and then checked for correctness of additions.

As an example of this synthetic approach, we have used d(pT-T-T-T-T)-rA
as primer for the synthesis of poly(dA), poly(dG), poly(dC), or poly(dT)[12].
In all cases, the desired products of expected chain length were synthesized.
These products could be separated from the primer (or unutilized primer) either
by gel filtration or by sucrose density-gradient centrifugation after the alkaline hydrolysis of the product. This approach permits the separation of the
primer sequences from the product sequences[4].

The reaction mixture (50 µl) contained 200 pmol of d(T-T-T-T-T-T)p̄rA,
200 nmol of [^{3}H]dATP, 40 mM potassium cacodylate (pH, 6.8), 8 mM MgCl$_2$, 1 mM
dithiothreitol, and 400 units of terminal transferase. After 90 min at 37°C,
the reaction was terminated by adjusting to 50 mM EDTA and 1% formaldehyde.
The tube was heated to 90°C for 1 min, quick-chilled, and layered on a 5-20%
neutral sucrose gradient. Centrifugation was carried out in an SW56 rotor
at 50,000 rpm for 16 h at 5°C. A total of 36 fractions (5 drops in each
approximately 100 µl) were collected. The [^{32}P]primer covalently linked to
the [^{3}H]poly(dA) product banded in the middle of the gradient. The unused
primer remained near the top of the gradient. Of the total of 200 pmol of
[^{32}P]primer applied to the gradient, 150 pmol banded together with the poly(dA)
product. During the 70-min incubation, 66,400 pmol of [^{3}H]dAMP residues were
incorporated. Therefore, the average chain length of the product was 442.
From the sharpness of the peak, it appeared that the size distribution was
narrow[4].

The peak fractions were pooled, dialyzed, and evaporated to dryness. The
material was dissolved in 200 µl of 0.5 N NaOH; after 20 h at room temperature,
this was layered onto a 5-20% alkaline sucrose gradient. After centrifugation,
each fraction (100 µl) was dried on a paper disk and counted. All the [^{32}P]pri-
mer, d(T-T-T-T-T-T)p̄rAp, was detected at the top of the gradient[4].

A. Enzymatic Synthesis of Polynucleotides on a Water-Soluble Polymeric Support

Another synthetic approach is to use oligothymidylate primers covalently linked to water-soluble polymer support[50]. An oligonucleotide primer is attached to polyvinyl alcohol with a 5'-phosphodiester linkage by chemical methods[51]. This material is then used as primer for addition of one or two ribonucleotides. The ribonucleotide-terminated primer attached to polyvinyl alcohol is then used for polynucleotide synthesis. In this way, homopolymers of different chain lengths attached to polyvinyl alcohol with an alkali-sensitive linkage may be synthesized. Such homopolymers may be used as templates for DNA polymerase to produce duplexes of defined chain lengths. Because of the attachment to polyvinyl alcohol, the 5' end of the template would remain unaffected by the 5'——>3' exonuclease activity of DNA polymerase I during the synthetic period. After the complete duplex has been formed, such products may be removed from polyvinyl alcohol by alkaline hydrolysis.

Natural DNA templates of known sequence and defined chain lengths may also be attached to the ribonucleotide-terminated primer with T4 RNA ligase. Such polyvinyl alcohol-linked templates may be used to study finer aspects of DNA replication and transcription. The situation would be analogous to the use of membrane-bound templates for DNA replication.

Owing to attachment of the polynucleotide to polyvinyl alcohol, its density will be different from that of polynucleotides unattached to the polymer support. This will allow the separation of the two by density-gradient centrifugation[52]. Because polyvinyl alcohol is water-soluble, all enzymatic reactions are expected to proceed in a manner similar to that with normal incubation conditions. This is illustrated in the following experiment.

The reaction mixture (200 µl) contained 40 mM potassium cacodylate (pH, 6.8), 8 mM $MgCl_2$, 200 µM dithiothreitol, 1 mM [α-^{32}P]rNTP, 1.6 A_{260} unit of PV-(pT)$_x$, and 800 units of terminal transferase. PV-(pT)$_x$ represents oligothymidylate covalently linked to polyvinyl alcohol where x = 3-12 thymidylate residues. At intervals of 10, 20, 30, 60, and 180 min, 5-µl aliquots were monitored for acid-insoluble radioactivity according to the method of Bollum[45a]. After the plateau had been reached (1-2 h) the remaining 170 µl was subjected to gel filtration. The excluded material contained ribonucleotide-terminated oligothymidylates covalently linked to polyvinyl alcohol.

To obtain a single addition product, the material was treated with alkali and phosphatase[5]. The reaction mixture was then treated with 1% SDS and 10 mM

EDTA and passed through a combination column with Sephadex G-50 (0.7 X 27 cm) at the top and Bio-Gel A 1.5 m (0.7 X 86 cm) at the bottom. Under these conditions, phosphatase was separated from PV-(pT)$_x$-rN.

In our original experiment, we prepared PV-(pT)$_x$-rU. This was then used as primer for the synthesis of poly(dA). For comparison, we used three other primers, d(pT)$_7$, PV-(pT)$_x$, and PV-rU (this represents a single rUMP residue covalently attached to polyvinyl alcohol).

The reaction mixtures (200 µl) contained 40 mM potassium cacodylate (pH, 6.8), 10 mM MgCl$_2$ 200 µM dithiothreitol, 1.6 mM [^{3}H]dATP, and 400 units of terminal transferase. The respective primer concentrations were 2.5 A$_{260}$/ml of PV-rU, 3 A$_{260}$/ml of PV-(pT)$_x$, 2.5 A$_{260}$/ml of PV-(pT)$_x$-rU, and 3 A$_{260}$/ml of d(pT)$_7$. At the intervals indicated in Fig. 4, aliquots (10 µl) were monitored for acid-insoluble radioactivity. The remaining material from the PV-(pT)$_x$-rU reaction (carried out in a 500 µl volume) after 20 h of incubation was subjected to gel filtration on the combination column, as describe above. It is evident from the Fig. 4 inset that PV-(pT)$_x$ and PV-(pT)$_x$-rU both show a lag period compared with the standard primer, d(pT)$_7$. Between PV-(pT)$_x$ and PV-(pT)$_x$-rU, the ribonucleotide-terminated primer shows less activity. As expected, a single ribonucleotide attached to polyvinyl alcohol (PV-rU) was not accepted as primer by terminal transferase. This incubation also showed that there was no unprimed synthesis of poly(dA) during the incubation period, even after 20 h (data not shown). The gel-filtration profile in Fig. 4 indicates that about half the dATP was utilized for poly(dA) synthesis[50].

B. Ribonucleotide-Terminated Primer Extension for Analytic Purposes

The DNA chain growth from the primer terminus involves a mechanism that was shown to contain a proofreading function[53]. According to this mechanism, mismatched bases incorporated by erroneous polymerization reactions at the 3' terminus of DNA chains are first removed by the 3' ⟶ 5' exonuclease activity of DNA polymerase before the correct base-paired nucleotides are successively added in the 5' ⟶ 3' direction. In view of this observation, the question arose as to whether the proofreading function also extends to the removal of single ribonucleotides (matched or mismatched to the template), which may be mistakenly incorporated instead of deoxynucleotides. To find an answer to this question, we have prepared two types of ribonucleotide-terminated primers for further elucidation of the proofreading function of DNA polymerase I[22].

We prepared a primer, (dT)$_{30}$, and extended it by one or two ribonucleotides

58

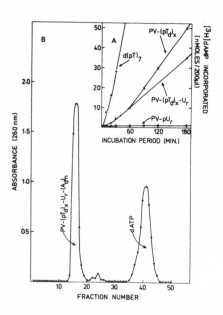

Fig. 4. Use of ribonucleotide-terminated primer for polymer synthesis on a water-soluble polymeric support. Reprinted from FEBS letters 50, 140 (1975).

catalyzed by terminal transferase[22]. The incorporated ribonucleotides were labeled with phosphorus-32, so that we could follow the fate of the nucleotides at the primer terminus (matched or mismatched) during the polymerization reaction with [^{3}H]dTTP as substrate, according to the following scheme:

1. (a) $(dT)_{30}$ + $[\alpha-^{32}P]rATP$ $\xrightarrow[\text{transferase}]{Mg^{+2}, \text{ terminal}}$ $(dT)_{\overline{30}}\overset{*}{p}rA$

    $(dT)_{\overline{30}}\overset{*}{p}rA\overset{*}{p}rA$

    (b) $(dT)_{30}$ + $[\alpha-^{32}P]rUTP$ $(dT)_{\overline{30}}\overset{*}{p}rU$

    $(dT)_{\overline{30}}\overset{*}{p}rU\overset{*}{p}rU$

2. These primers were then annealed to poly(dA), so that $(dT)_{\overline{30}}\overset{*}{p}rA$ was a mismatched primer terminus and $(dT)_{30}\overset{*}{p}rU$ was a base-paired primer terminus. When [^{3}H]dTTP was added to the reaction mixture with DNA polymerase I, the DNA chain growth proceeded by either preserving or removing the 3'-terminal ribonucleotide(s). This could be studied simultaneously by counting the acid-insoluble radioactivity in the hydrogen-3 and phosphorus-32 channels.

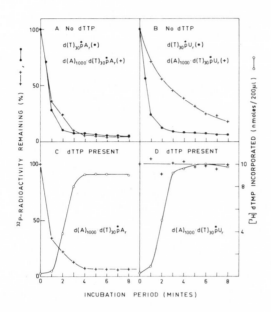

Fig. 5. Analysis of the proof-reading function of DNA polymerase I in relation to the ribonucleotide-terminated primer extension. Reprinted with permission from the Journal of Biological Chemistry.

$\overset{*}{p}$ denotes a $[^{32}P]$phosphate and $\overset{\circ}{T}$ represents $[^3H]$dTMP. The primer and template were incubated in 180 mM glycylglyclglycine-KOH buffer (pH, 7.5) and 18 mM $MgCl_2$ at 90°C for 10 min and allowed to cool to room temperature slowly over a period of 1 h. The reaction mixture was then transferred to an ice bath and adjusted to DNA synthesis conditions. The final reaction mixture (200 µl) contained (in Fig. 5A, B) 2.8 nmol of $(dT)\overline{_{30}}\overset{*}{p}rA$ or 4 nmol of $(dT)\overline{_{30}}\overset{*}{p}rU$ with or without a large excess (40 nmol) of $(dA)\overline{_{1000}}$, 100 mM glycylglycine-KOH buffer (pH, 7.5), 10 mM $MgCl_2$, and 12 units of the large fragment[53,54] of DNA polymerase I. For DNA synthesis (Fig. 5C,D), the conditions were the same, except that 3 nmol of $(dT)\overline{_{30}}\overset{*}{p}rU$ and 10 nmol of $[^3H]$dTTP were present. At indicated intervals, 20-µl aliquots were removed and assayed for acid-insoluble radioactivity on glass-fiber filters.

The use of ribonucleotide-terminated primer thus allows analytic study of the specificity of the 3'——>5' exonuclease activity of DNA polymerase I. In the absence of dTTP and $d(A)\overline{_{1000}}$, the 3'——>5' exonuclease led to a rapid degradation of the single $[^{32}P]$rAMP or $[^{32}P]$rUMP at the 3' termini of primers (Fig. 5A,B). In the case of $(dT)\overline{_{30}}\overset{*}{p}rA$ where the terminal nucleotide was mismatched, a comparable rate of degradation was observed even when the primer was annealed to the template (Fig. 5A). Under similar conditions, the excision of the terminal ribonucleotide residue from $(dT)\overline{_{30}}$ $\overset{*}{p}rU$ was markedly reduced in the presence of a template to which it formed a fully base-paired primer terminus. The influence of this secondary structure on 3'——>5' exonuclease activity is similar to that observed by Brutlag and Kornberg with all deoxy- substrates[53].

The above results show that a ribonucleotide at the primer terminus is excised by the 3'——>5' exonuclease in the absence of DNA synthesis. Under conditions for chain extension (dTTP present), analysis of the ribonucleotide-terminated primer extension revealed that the 3'——>5' exonuclease completely removes a mismatched ribonucleotide in $(dT)\overline{_{30}}\overset{*}{p}rA$ and completely preserves a base-paired ribonucleotide in $(dT)\overline{_{30}}\overset{*}{p}rU$ for further chain growth from the ribonucleotide-3' terminus of the primer (Fig. 5C,D).

Similar results were obtained when two ribonucleotides were present at the 3' end. Whole molecules of polymerase gave the same results (not shown). Therefore, the proofreading function of DNA polymerase I appears to screen the products for mismatched base pairs only, whereas a nucleotide residue containing an incorrect sugar moiety is not recognized as a mistake if it is erroneously incorporated under DNA-synthesis conditions[22].

The ribonucleotide-terminated primer extension by DNA polymerase I provides a powerful tool for analysis of the nucleic acid structure in the unknown region of the template[55]. The synthetic octanucleotide d(A-C-C-A-T-C-C-A) was annealed to single-stranded fl DNA template and a ribonucleotide was incorporated by repair synthesis. Extension of this ribonucleotide-3' terminus with DNA polymerase followed by isolation of the extended product allowed the sequence analysis of 50 nucleotides into the internal region of the template[56]. Sekiya and Khorana[13] used this technique to analyze the extended products in the promoter region of the tyrosine tRNA gene.

The transducing bacteriophage Φ80 DNA carrying the E. coli tyrosine tRNA gene was used as a template. The sequence beyond the C-C-A end of the tRNA gene was previously determined[39]. An oligonucleotide d(G-C-T-C-C-C-T-T-A-T-C-G) complementary to the Φ80 psu III[+] 1 strand was chemically synthesized, and a single rGMP residue was added with terminal transferase. This product was then joined enzymatically to the oligonucleotide d(T-A-C-T-G-G-C-C-T) to yield the oligonucleotide d(T-A-C-T-G-G-C-C-T-G-C-T-C-C-C-T-T-A-T-C-G)rG.

The reaction mixture contained 100 mM NaCl, 100 mM piperazine-N-N'-bis (3-ethanesulfonic acid) buffer (pH, 6.9), Φ80 psuIII[+] DNA at 3 pmol/ml, and ribonucleotide-terminated primer at 3 pmol/ml. This was heated to 100°C for 2 min and kept at 58°C for 18 h. After cooling to 5°C, the reaction mixture was adjusted to 10 mM $MgCl_2$, 10 mM dithiothreitol, 1 μM [α-^{32}P]dNTP, 5-10 μM unlabeled dNTP, and DNA polymerase I at 30 units/ml. The incubation was carried out at 5°C for 6 h.

This 22-nucleotide sequence was annealed to the template and subjected to partial repair synthesis[57] according to the following plan:

```
3'———//———A-A-A-T-G-A-C-C-G-G-A-C-G-A-G-G-G-A-A-T-A-G-C-C-C-T-T-C-G| Promoter
                5'- T-A-C-T-G-G-C-C-T-G-C-T-C-C-C-T-T-A-T-C-GrG   plus dNTP, pol 1
Extended
Product ———→ 5'T-A-C-T-G-G-C-C-T-G-C-T-C-C-C-T-T-A-T-C-GrG |GAAGCGGGGCGCA|
```

One of the products of partial repair synthesis, liberated after alkali treatment, is shown in the boxed sequence.

The reaction mixture was heated to 100°C for 2 min and immediately applied to an agarose 1.5-m column (0.9 x 24 cm) preequilibrated with 50 mM TEAB (pH, 7.6).

The peak corresponding to the elongated primer (eluted after the 80 DNA) was evaporated to dryness, dissolved in 15 µl of 60% glycerol containing loading dye, and subjected to electrophoresis on a 15% polyacrylamide gel in 7 M urea. The radioactive bands were eluted from gel, evaporated, and desalted by gel filtration through a column (0.9 x 24 cm) of Sephadex G-50.

The elongated primer was then hydrolyzed with alkali, neutralized with acetic acid, and subjected to homochromatography[48]. Several radioactive bands were detected by autoradiography. The bands were scraped off the DEAE thin-layer plates[33], eluted[33], hydrolyzed with alkali to destroy the RNA eluted from the plates, and purified by gel filtration. The purified oligonucleotide was then subjected to two-dimensional mapping[11]. One of the maps revealed the sequence shown within the box in the promoter region. In this way, several other bands were analyzed. The overlapping sequences in each band confirmed the sequence of a total of 29 nucleotides in the promoter region with elements of a twofold rotational symmetry[14].

V. SEQUENCE ANALYSIS OF OLIGONUCLEOTIDES

The labeling of the 3' end of a given oligonucleotide with a labeled ribonucleotide or a [^{32}P]phosphate[5] permitted the development of a new method for sequence analysis of oligonucleotides[8].

An oligonucleotide is first digested partially with venom phosphodiesterase to yield a population of oligonucleotides of successively decreasing chain lengths. These oligonucleotides are then labeled with a single ribonucleotide or a [^{32}P]phosphate[5], and labeled oligonucleotides are separated according to chain length. Nearest-neighbor analysis of each chain reveals the identity of the 3'-terminal deoxynucleotide. From these results, the sequence is deduced[8] according to the following scheme:

Oligonucleotide population after partial digestion	Products of one ribonucleotide addition	Products of nearest-neighbor analysis
d(A-C-C-A-T-C-C-A) ...	d(A-C-C-A-T-C-C-A)p̄rN ...	dAp̄
d(A-C-C-A-T-C-C) ...	d(A-C-C-A-T-C-C)p̄rN ...	dCp̄
d(A-C-C-A-T-C) ...	d(A-C-C-A-T-C)p̄rN ...	dCp̄
d(A-C-C-A-T) ...	d(A-C-C-A-T)p̄rN ...	dTp̄
d(A-C-C-A) ...	d(A-C-C-A)p̄rN ...	dAp̄
d(A-C-C) ...	d(A-C-C)p̄rN ...	dCp̄

The nucleotide sequence is unambiguously established by nearest-neighbor analysis. Our original method[8] was developed before the technique of homochromatography[48] became popular. We have since simplified the procedure according to that of Jay et al.[10] whereby all the nucleotides in a given oligodeoxynucleotide can be deduced by two-dimensional mapping[48]. This is illustrated by using the chemically synthesized oligonucleotide d(A-G-T-C-C-A-T-C-A-C-T-T-A-A), corresponding to amino acid residues 36-40 of the lysozyme gene of bacteriophage T4[58]. The plan of the simplified procedure is summarized in the following scheme:

T4 lysozyme
tetradecamer, d(A-G-T-C-C-A-T-C-A-C-T-T-A-A)

$[\alpha\text{-}^{32}P]$rNTP, Co^{+2}
terminal transferase

d(A-G-T-C-C-A-T-C-A-C-T-T-A-A)-(prN*)$_n$

alkali

d(A-G-T-C-C-A-T-C-A-C-T-T-A-A)prN*p* + (n-2)rN*p + rN

phosphatase

d(A-G-T-C-C-A-T-C-A-C-T-T-A-A)prN*

periodate, propylammonium
bicarbonate, pH 7.5

d(A-G-T-C-C-A-T-C-A-C-T-T-A-A)p*

partial spleen phosphodiesterase

G-T-C-C-A-T-C-A-C-T-T-A-Ap*
T-C-C-A-T-C-A-C-T-T-A-Ap*
C-C-A-T-C-A-C-T-T-A-Ap*
C-A-T-C-A-C-T-T-A-Ap*
A-T-C-A-C-T-T-A-Ap*
T-C-A-C-T-T-A-Ap*
C-A-C-T-T-A-Ap*
A-C-T-T-A-Ap*
C-T-T-A-Ap*
T-T-A-Ap*
T-A-Ap*
A-Ap*
Ap*

rN = ribonucleoside
N*p = ribonucleoside 3' phosphate
*p = $[^{32}P]$phosphate

The oligonucleotide was labeled with a [^{32}P]phosphate in a microscale procedure as described earlier. The labeled oligonucleotide (100 pmol) was then incubated in a reaction mixture (10 µl) containing 30 µg of carrier E. coli tRNA (Boehringer), 3 mM potassium phosphate (pH, 6), 0.3 mM EDTA, 0.1% Tween 80, and 0.25 µg of spleen phosphodiesterase. A siliconized 6 x 50-mm tube was maintained at 85°C in a temperature block. Aliquots (2 µl) were withdrawn at 5, 10, 20, 40, and 60 min and transferred to the heated tube. The mixtures of dried oligonucleotides were dissolved in 5 µl of water. From this solution a 1-µl sample was applied to a cellulose acetate strip, subjected to electrophoresis at a pH of 3.5, transferred to a DEAE thin-layer plate, subjected to homochromatography with homo-mix V of Jay et al.[10], and auto-radiographed. The nucleotide sequence was deduced from the mobility-shift analysis, by the method of Tu et al.[11].

VI. LABELING OF DUPLEX DNA FRAGMENTS

A. Labeling with Ribonucleotides:

The different factors affecting the ribonucleotide additions at the 3' termini of DNA fragments have already been discussed[24,25]. Only the optimal conditions are described here.

Most commonly used DNA fragments are produced by restriction endonuclease cleavage, either for ordering of the fragments or for DNA sequence analyis. The reaction mixture (volume based on DNA) contains 20 A_{260}/ml of DNA, 10 mM Tris-HCl (pH, 7.8), 10 mM $MgCl_2$, 1 mM dithiothreitol, and the appropriate restriction endonuclease at 1,000 units/ml. Where essential, 50-150 mM NaCl or KCl is also included. After 2 h at 37°C, a 1-µl aliquot is monitored by agarose gel electrophoresis to check the completeness of digestion. When the digestion is complete, the concentration of DNA termini (picomoles of 3'-OH ends) is determined from the number of fragments produced by restriction endonuclease cleavages.

The final labeling reaction mixture (200 µl) contains 20 µl of concentrated cacodylate buffer system, 10 µl of restriction endonuclease-digested DNA (used directly without any further treatment), 10 units of terminal transferase per picomole of 3'-OH ends, and a 10- to 20-fold excess of [α-^{32}P]rATP or rCTP. If the reaction mixture contains 10 pmol of 3'-OH ends, then 100-200 pmol of [^{32}P] rNTP is dried in a siliconized tube (10 x 75 mm) just before incubation, and all other components are added to this tube. The final volume is made up to 200 µl with distilled water. After incubation at 37°C for 30 min, the reaction is

terminated by adding 20 µl of 500 mM EDTA, 10 µl of 1% SDS, and 20 µl of neutralized phenol. This treatment will instantly destroy all enzymatic activity,
including the activity of the contaminating nucleases. After gentle mixing
in a vortex mixer at low speed, the reaction mixture is warmed to 37°C and
applied to a column of Sephadex G-50 (0.7 x 80 cm) preequilibrated with 50 mM
Tris-HCl (pH, 8.0), 100 mM NaCl, and 1 mM EDTA. Fractions of 0.5-1 ml are
collected at room temperature by using 13 x 50-mm siliconized tubes in a
Gilson F80 fraction collector. The tubes are directly counted for Cerenkov
radiation. The excluded peak is pooled and ethanol-precipitated. Fig. 6
shows <u>Hinc</u> II digested (7 cuts) SV40 DNA labeled with [α-^{32}P]rCTP. The differences in the amount of radioactivity in different fragments were explained
earlier[24].

B. Labeling with Cordycepin Monophosphate

Plasmid pTU4 DNA[59] digested with <u>Hae</u> II endonuclease[60] was used for incorporation of [α-^{32}P]cordycepin monophosphate. The reaction mixture (200 µl) contained a buffer system as described earlier, 18 pmol of 3'-OH ends, 164 pmol of
[α-^{32}P]cordycepin 5' triphosphate (570 Ci/mmol from NEN) and 15 units of terminal transferase (P.L. Biochemicals). At intervals of 0, 0.5, 1, 2, 5, 10, 20,
30, 45, and 150 min, 5-µl aliquots were monitored for acid-insoluble
radioactivity.

The incorporation reached a plateau after 30 min of incubation (Fig. 7).
The reaction was terminated as described earlier and a 40-µl sample was frac-
tioned by gel filtration on a column (0.9 x 45 cm) of Sephadex G-50 (fine).
The elution profile is shown in Fig. 7. From the radioactivity in the excluded
peak, the amount of [^{32}P]cordycepin monophosphate incorporated into DNA was
calculated. Under these conditions, about 18% of the DNA ends were labeled.

In a separate experiment with <u>Sst</u>-II cut pTU4 DNA and [α-^{32}P]cordycepin
triphosphate of lower specific activity (250 Ci/mmol), approximately 40% of
the DNA ends were labeled with an enzyme-to-DNA 3' end ratio of 1:3. In a
parallel experiment in which the cordycepin triphosphate (4 µM) was replaced
by [α-^{32}P]rCTP at a lower concentration (2.5 µM), almost 90% of the 3' ends
were labeled. The lower degree of incorporation may be due to the lower
affinity of the substrate for terminal transferase, to the presence of traces
of impurities, or both. The lower estimation may also result from using an
incorrect value for the specific activity of [α-^{32}P]cordycepin triphosphate
specified in the shipment of this particular batch of triphosphate. By using

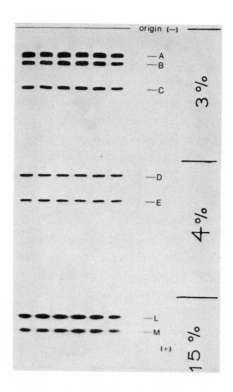

Fig. 6. Labeling of duplex <u>Hind</u> II fragments of SV40 DNA with [^{32}P]rCMP residues. A 20 x 40-cm step gel with 15% gel at the bottom was used to trap the small bands L and M.

a higher level of enzyme (10 units/pmol of DNA ends) and a higher concentration of cordycepin triphosphate (10 μM), it is possible to obtain 100% incorporation.

When the DNA 3' ends are labeled with a ribonucleotide or a [^{32}P]phosphate, such labels are susceptible to attack by contaminating exonuclease or phosphatase activity. More often, traces of 3'——>5' exonuclease activity are associated with some commercial preparations of restriction endonucleases. When end-labeled DNA fragments are digested with a second restriction

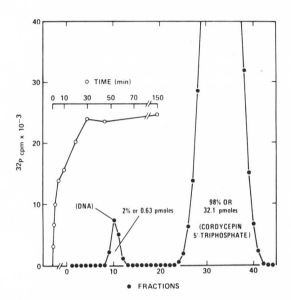

Fig. 7. Kinetics of labeling with [α-^{32}P]cordycepin triphosphate and isolation of product. Reprinted from Gene 10, 177 (1980).

endonuclease and then subjected to polyacrylamide-gel electrophoresis and auto-radiography, there is often a decrease in or removal of the radioactivity from the separated fragments. In contrast, DNA labeled with [^{32}P]cordycepin triphosphate remains completely insusceptible to degradation by either exonuclease or phosphatases (Fig. 8).

The reaction mixtures (200 μl) contained Sst II-digested pTU4 DNA (48 pmol of 3'-OH ends), 4 μM (α-^{32}P]cordycepin triphosphate or 2.5 μM [α-^{32}P]rCTP, and 15 units of terminal transferase. The labeled DNAs from both samples were digested with 25 units of Hind III at 37° for 2 h followed by 25 units of Eco RI for 1 h. The samples were subjected to electrophoresis in 3.5% polyacrylamide gel and then autoradiography. It is clearly evident from Fig. 8 that the fragments labeled with [^{32}P]cordycepin monophosphate completely retained the label, whereas label from the [^{32}P]rCMP-labeled fragments was almost completely removed by the contaminating exonucleases present in either the Hind III, Eco RI, or both, used in this experiment.

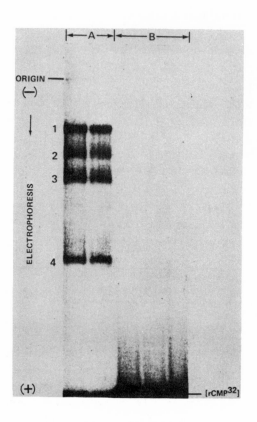

Fig. 8. Comparison of the recovery of radioactivity in labeled DNA fragments after recutting with <u>Hind</u> II plus <u>Eco</u> RI containing traces of exonuclease contamination. Reprinted from Gene <u>10</u>, 177 (1980).

VII. SEQUENCE ANALYSIS OF DNA

The discovery of restriction endonucleases[61] heralded a new era in molecular biology and provided the most powerful approach to be used so far for the nucleotide sequence analysis of DNA. Specific restriction fragments generated from a defined location in a DNA molecule can now be labeled at their ends with either a radioactive nucleotide[5,23,25,66,67] or a [32P]phosphate.[5,68,69] Such

DNA ends may also be labeled with a radioactive nucleoside 3',5'-biphos-phate[70-73]. The end-labeled DNA fragments may then be used for sequence analy-sis either by two-dimensional mobility-shift analysis[48,10,11] (short sequences, 1-20 nucleotides) or by one-dimensional gel electrophoresis (long sequences, 20-200 nucleotides) with chemical-degradation[38,62] or interrupted-synthesis techniques[63-65].

Kelly and Smith[74] isolated and characterized the first type II site-speci-fic restriction endonuclease, Hind II. This enzyme recongnized the general structure, 5'G-T-Y-R-A-C 3', where Y stands for a pyrimidine and R a purine
            3'C-A-R-Y-T-G 5'
nucleotide. We have used terminal transferase to label the seven DNA fragments produced after digestion of SV40 DNA with Hind II (Hinc II). The single ribo-nucleotide terminated fragments were then digested with a second restriction endonuclease to obtain a homogeneous population of labeled DNA fragments.

The labeled DNA fragments were isolated and digested partially with pan-creatic DNase to obtain labeled oligonucleotides of successively decreasing chain lengths. These were then subjected to two-dimensional mobility-shift analysis[11]. In this way, we have determined about 20 nucleotides at all seven Hind II cuts in SV40 DNA[24]. The details of these procedures have been describ-ed elsewhere[25]. Similarly, we have determined the sequences adjacent to the Hind III cleavage sites (six cuts) of SV40 DNA (reference 75 and unpublished results).

In rare cases, the restriction endonucleases produce staggered ends with protruding single-stranded 3' ends like those generated by Pst I, Hae II, and Kpn 1. Such DNA ends cannot be labeled with repair synthesis (Pol 1 or reverse transcriptase) or phosphorylation (polynucleotide kinase). These protruding ends, however, are ideal substrates for terminal transferase. In a circular DNA molecule with multiple cleavage sites for a particular restriction enzyme, the protruding 3' ends consist of the same nucleotide sequence. Therefore, a direct two-dimensional map of the partial pancreatic DNase digest of a mixture of DNA fragments labeled with a single ribonucleotide residue catalyzed by terminal transferase yields the recognition sequence of such restriction endonucleases. In this way, we have determined the recognition sequences of the restriction enzyme Hae II[60] and Kpn I[76]. The analysis of the Kpn I recognition sequence is presented as an example.

Endonuclease Kpn I[77] makes two cuts in plasmid pCRl DNA[78], so that four Kpn I-generated protruding single-stranded 3' ends are produced after diges-tion. These four ends were labeled with a single [^{32}P]rCMP residue.

The reaction mixture (100 µl) for Kpn-I digestion contained 60 µg of pCRI DNA, 10 mM Tris-HCl (pH, 7.5), 5 mM NaCl, 10 mM $MgCl_2$, 5 mM dithiothreitol, and 20 units of Kpn-I. After 6 h at 37°C, a 5-µl aliquot was monitored by agarose-gel electrophoresis. Two fragments were produced. One third of this reaction mixture was used directly for terminal-transferase labeling.

The labeling reaction mixture (500 µl) contained 33 µl of Kpn-I-digested pCR1 DNA (20 µg = 9.6 pmol of 3'-OH ends), 50 µl of concentrated cacodylate buffer mixture, 640 pmol of $[\alpha\text{-}^{32}P]rCTP$, and 32 units of terminal transferase. After 4 h at 37°C, the reaction was terminated by adding 20 µl of 10 M KOH. After 20 h of alkaline hydrolysis, the reaction mixture was neutralized by adding 15 µl of 17 M acetic acid. The pH was adjusted to 8 with 100 mM Tris-HCl buffer. The solution was then treated with 0.6 unit (2 µl) of E. coli alkaline phosphatase and incubated at 45°C for 1 h. The reaction mixture was then treated with 50 µl of 0.5 M EDTA and 50 µl of neutralized phenol. After thorough mixing, the tube was heated at 45°C for 10 min. This treatment inactivates the phosphatase, owing to denaturation by the phenol. This whole reaction mixture was then applied to a column (0.6 x 80 cm) of Sephadex G-50 (fine), and 20-drop fractions were collected. The excluded peak was ethanol-precipitated.

An aliquot of this labeled DNA (840,000 cpm) was then partially digested with pancreatic DNase. The reaction mixture (20 µl) contained 50 mM TEAB (pH, 8.0), calf thymus DNA, 500 µg/ml, 5 mM $MnCl_2$, 3 mM $CoCl_2$, 1 mM $CaCl_2$, and 2 µg (1 µl) of pancreatic DNase (Worthington). This combination of metal ions reduces the specificity of cleavage by DNase I so that oligonucleotides of all sizes are produced (Tu, Roychoudhury, and Wu, 1976, unpublished results). At intervals of 2, 5, and 20 min, 3-µl aliquots were transferred to a small siliconized tube (6 x 50 mm) maintained at 90°C on a temperature block. The remaining solution was digested for 2 h and combined in the tube maintained at 90°C. The dried oligonucleotides were dissolved in 5 µl of water containing $[^{14}C]dpT$ marker (about 10,000 cpm). A 1-µl aliqout of this sample was applied to a strip of cellulose acetate. Further processing for obtaining a two-dimensional map was carried out according to the method of Tu et al.[11].

Fig. 9 shows the two-dimensional map of $[^{32}P]rCMP$-labeled oligonucleotides obtained after partial digestion of a mixture of Kpn-I-generated ends of pCR1 DNA. The sequence from the 3' end can be read as 3'rC-C-A-T-G-G-T/C, with heterogeneity appearing in the seventh nucleotide. Therefore, all the 3' terminal Kpn-I-generated ends consist of the common residues 3'-C-A-T-G-G

The direction of electrophoresis is indicated as "pH 3.5", and the

mobility by homochromatography is indicated as "HOMO" (Fig. 9). By arrang-
ing this common sequence in a duplex with antiparallel orientation, we obtain:

$$5'...G-G-T-A-C$$
$$C-A-T-G-G...\ 5'$$

Therefore, the complete duplex recognition sequence for the <u>Kpn</u>-I endonuclease
should be:

$$5'...\overline{G-G-T-A-C-C}...3'$$
$$3'...C\ C\ A\ \underline{T-G\ G}...5'$$

Thus, <u>Kpn</u>-I recognizes a hexanucleotide sequence with elements of a two-fold
rotational symmetry and generates protruding single-stranded 3' ends four
nucleotides long[76]. The recognition sequence for endonuclease <u>Hae</u> II, which
also generates protruding single-stranded 3' ends, was determined in a similar
manner[60].

For determination of longer sequences, DNA fragments labeled with terminal
transferase are subjected to chemical degradation by the method of Maxam and
Gilbert[38,62]. The partial sequence of the <u>cro</u> gene of phage λ, determined by
Schwarz <u>et al</u>[79], was shown by Roychoudhury and Wu[25]. A clean readable pattern
was also shown by Chang <u>et al</u>.[80]. Fig. 10 shows the nucleotide sequences spanning the replication region of pTU4 DNA labeled with cordycepin monophosphate,
as determined by Tu and Cohen[26].

VIII. USE IN MOLECULAR CLONING

An excellent review of different conditions for addition of homopolymer
tracts to duplex DNA ends was recently published by Nelson and Brutlag[81].
Other general conditions have also been described[25]. Therefore, only some
additional information that may be helpful to other investigators is presented here.

It is often desirable to add very short homopolymer tracts (10-20
nucleotides long). This may require 1-2 min of incubation, according to
the previously determined kinetics of incorporation. However, short incubations may well result in a very small number of DNA molecules tailed at
the 3' end. Unless the enzyme is in very large excess over the number of
3' ends, such an attempt may be unsuccessful. (For further details, see
Nelson and Brutlag.[81]) We therefore incubate our preparations for 5-10 min,
depending on the nature of incorporation previously determined with an aliqout
of the main preparation. This testing should be done with a radioactive

```
—— Gp 5'                    dGTP,          —— Gp 5'
—— C-T-G-C-A-OH 3'      ————————→     —— C-T-G-C-A-G-G-G...G_n
                           transferase
                                              Pst I recognition sequence
```

Other protruding 3' ends are those generated by restriction endonucleases Kpn I[76], Hae II[60], Hha I[87], Sst I[88,90], (Sac I)[89], Sst II[90], (Sac II)[89], and Hph I[91]. Because the double-strand breaks by Hph I occur several nucleotides away from the recognition sequence, there is no need for incorporation of any specific nucleotide at the 3' end. Thus, Hph I-generated 3' ends may be tailed with any of the four common nucleotide residues. The protruding 3' ends of other endonucleases listed above may be reconstructed according to the following scheme:

```
2. Kpn I,    —— Cp 5'                dCTP          —— Cp 5'
             —— G-G-T-A-C-OH 3'   ——————————→   —— G-G-T-A-C-C-C...C_n
                                    transferase
                                                  Kpn I recognition sequence
```

```
3. Hae II,   Cp/Tp 5'               dTTP or dCTP    Cp/Tp 5'
             G/A-G-C-G-C-OH 3'   ——————————→     G/A-G-C-G-C-C...C_n
                                    transferase
                                                     G/A-G-C-G-C-T...T_n
                                                  Hae II recognition sequence
```

```
4. Hha I,    —— Cp 5               dCTP           —— Cp 5'
             —— G-C-G-OH 3'     ——————————→     —— G-C-G-C-C-C...C_n
                                   transferase
                                                  Hha I recognition sequence
```

```
5. Sst I,    —— Cp 5'              dCTP,           —— Cp 5'
             —— G-A-G-C-T-OH 3'  ——————————→    —— G-A-G-C-T-C-C-C...C_n
                                   transferase
                                                  Sst I recognition sequence
```

```
6. Sst II,   —— G-Gp 5'            dGTP,           —— G-Gp 5'
             —— C-C-G-C-OH 3'    ——————————→    —— C-C-G-C-G-G...G_n
                                   transferase
                                                  Sst II recognition sequence
```

When the restriction endonuclease produces a staggered break with protruding 5' ends, such ends may be repaired with reverse transcriptase to generate a duplex blunt end. The duplex blunt end is then tailed with terminal transferase to reconstruct the original restriction endonuclease recognition sequence. We have carried out this procedure and successfully reconstructed the Eco RI and Hind III sites of plasmid pMB9 DNA. Molecular cloning of a specific DNA fragment in such a reconstructed site with pBR322 as the cloning vehicle indicated that the inserted gene was expressed in the recombinant plasmid. After isolation of the plasmid DNA, the inserted segment could be recovered by digestion with Eco RI (Ratzkin and Roychoudhury, unpublished results).

D. Procedure for Reconstruction of Eco RI and Hind III Sites of pMB9 DNA

Hind III Sites:

pMB9 DNA was digested with Hind III under conditions of the reverse transcriptase repair reaction according to the method of Bahl et al.[92].

The reaction mixture (200 μl) contained 50 mM Tris-HCl, (pH, 8.3), 50 mM KCl, 10 mM MgCl$_2$, 10 mM dithiothreitol, 20 μg of pMB9 DNA, and 20 units of Hind III.   After 2 h of 37°C, a 5-μl aliquot was monitored by agarose-gel electrophoresis. All the DNA was converted to the linear form. To the reaction mixture was then added 25 μl of a solution containing dATP, dGTP, dCTP, and dTTP (all at 10 mM) and 2 μl (40 units) of AMV reverse transcriptase. The final volume was made up to 250 μl, so that the dNTP concentrations were all 1 mM.

The repair of the staggered end thus produced a blunt end, according to the following scheme:

pMB9 Hind III-cut staggered end		Duplex blunt end at Hind III site
—— T-T-C-G-Ap$^{5'}$	dNTPs, → Reverse Transcriptase	—— T-T-C-G-Ap$^{5'}$
—— A-OH		—— A-A-G-C-T

When these blunt ends were extended with dTMP residues, the product of such a reaction resulted in the reconstitution of the Hind III recognition sequence:

—— T-T-C-G-Ap$^{5'}$	dTTP, → transferase	—— T-T-C-G-Ap$^{5'}$
—— A-A-G-C-T		—— A-A-G-C-T-T-T...T$_n$
		Hind III recongition sequence

After 30 min of reverse transcriptase reaction at 37°C, the reaction was terminated by adding 20 μl of 0.5 M EDTA and 200 μl of neutralized phenol. After phenol extraction, the aqueous layer was applied to a column (0.5 x 80 cm) of Sephacryl S-200, and the excluded peak was recovered and ethanol precipitated. The recovery of DNA was 80% (16 μg). This DNA was then tailed with poly(dT) tracts (Fig. 11A).

1. Eco RI site: pMB9 DNA (20 μg) was digested under similar conditions to

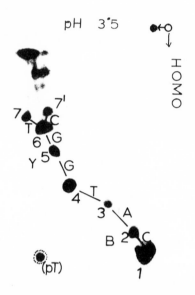

Fig. 9. Two-dimensional map of the recognition sequence for <u>Kpn</u>-I endonuclease. Reprinted with permission from Infromation Retrieval Limited.

triphosphate of high specific activity (400-600 Ci/mmol), so that a very small amount of sample is needed in testing. Nearest-neighbor analysis of the test sample carrying a 3' terminal nucleotide, different from that of the radioactive nucleoside triphosphate will indicate the average number of residues added[23].

When a structural gene is inserted into a cloning vehicle by homopolymer tailing, the inserted gene cannot be recovered intact at the beginning of homopolymer tract, because of the absence of any restriction site at this point. It is often desirable to isolate the inserted gene for sequence analysis, as well as for reengineering of the cloned DNA segment, for a better expression of the gene products. To achieve such an objective terminal transferase may be used for a novel reaction at the 3' end - a reaction that leads to the reconstruction of the original recogniton sequence of the desired restriction endonuclease.

Restriction endonucleases often make staggered breaks in a DNA molecule. Such breaks may contain either a protruding single-stranded 5' end or a pro-

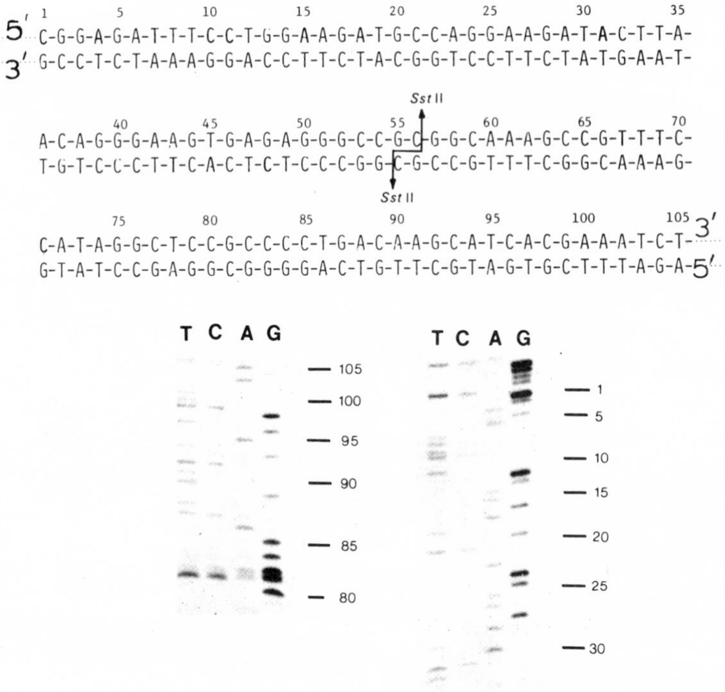

Fig. 10. Nucleotide sequence in the replication region of plasmid pTU₄ DNA. Reprinted from Gene 10, 177 (1980).

truding single-stranded 3' end. When the 3' end is protruding, it may be used directly for the reconstruction of the recognition sequence. The most widely used example is the regeneration of the recognition sequence of endo-nuclease Pst 1[82-84].

A. Reconstruction of the Recognition Sequence:

Endonuclease Pst 1 recognizes a hexanucleotide sequence[86], 5' C-T-G-C-A-G 3', and produces a staggered break after the A residue to gene-rate the protruding 3' end, —— Gp 5'

—— C-T-G-C-A-OH 3'

1. The addition of poly(dG) tracts to this end reconstitutes the recognition sequence, (where n > 10 residues).

obtain the staggered duplex ends. These ends were then repaired with reverse transcriptase under conditions indentical with those described for <u>Hind</u> III ends, except that only dATP and dTTP were used.

$$\begin{array}{ll} \text{--- C-T-T-A-Ap}^5 \\ \text{--- G-OH} \end{array} \quad \xrightarrow[\substack{\text{Reverse} \\ \text{T transcriptase}}]{\text{dATP, dTTP}} \quad \begin{array}{ll} \text{--- C-T-T-A-Ap}^{5'} \\ \text{--- G-A-A-T-T} \end{array}$$

When these repaired blunt ends were tailed with poly(dC) tracts we achieved the reconstruction of the <u>Eco</u> RI recognition sequence (Fig. 11C).

$$\begin{array}{ll} \text{--- C-T-T-A-Ap}^{5'} \\ \text{--- G-A-A-T-T} \end{array} \quad \xrightarrow[\text{Transferase}]{\text{dCTP}} \quad \begin{array}{ll} \text{--- C-T-T-A-Ap}^{5'} \\ \underline{\text{--- G-A-A-T-T-C-C-C}} \ldots \text{ C}_n \end{array}$$

<u>Eco</u> RI recognition sequence

2. <u>Addition of homopolymer tracts</u>: For comparison of nucleotide additions at different duplex-DNA termini (natural or modified for the reconstruction of the recognition sequence of the desired restriction endonuclease), four different reactions were carried out with the same amount of enzyme and the same number of 3'-OH ends. Thus, the residues added by using dATP, dGTP, dCTP, or dTTP could be compared with each other.

3. <u>Poly(dT) tracts</u> - The reaction mixture (500 µl) contained 16 µg of <u>Hind</u> III end repaired pMB9 DNA (about 10 pmol of 3'-OH ends), 50 µl of 10 x buffer 100 µM $[\alpha$-^{32}P]dTTP, and 200 units of terminal transferase (a gift of Dr. John Wilson.[93]) At intervals of 0, 2, 4, 6, 8, and 10 min at 37°C aliquots (10-µl) were monitored for acid-insoluble radioactivity (Fig 11A). This step restored the <u>Hind</u> III recognition sequence.

4. <u>Poly(dA) tracts</u> - The reaction mixture (500 µl) contained 10.9 µg of <u>Sma</u>-I digested pCR1 DNA (four fragments = 10 pmol of 3'-OH ends) and 100 µM $[\alpha$-^{32}P]dATP. Other conditions were similar to those described above. At intervals of 0, 10, 20, 30, and 40 min at 37°C, aliquots were monitored for acid-insoluble radioactivity. A much longer period is necessary for the addition of poly(dA) tracts than that for poly(dT) tracts (Fig. 11B)

5. <u>Poly(dC) tracts</u> - The reaction mixture (500 µl) contained 16.6 µg of <u>Eco</u> RI site-repaired pMB9 DNA (about 10 pmol of 3'-OH ends) and 100 µM $[\alpha$-^{32}P]dCTP. Other conditions were similar to those decscribed above. To reduce the rate of dCMP addition, the incubation was carried out at 23°C. At intervals of 0, 2, 4, 6, 8, and 10 min, 10-µl aliquots were monitored for acid-insoluble radioactivity (Fig. 11C). This procedure restored the <u>Eco</u> RI recognition sequence at the two ends of the linear pMB9 DNA.

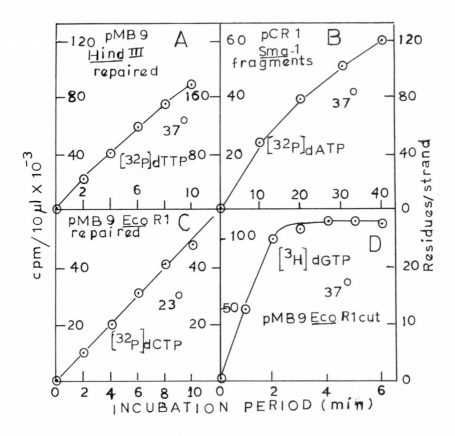

Fig. 11. Comparison of the addition of homopolymer tracts to duplex DNA ends with all four common deoxynucleotides in the presence of Co^{+2} ion.

6. <u>Poly(dG) tracts</u> - The reaction mixture (500 µl) contained 16.6 µg of <u>Eco</u> RI-cut pMB9 DNA (10 pmol of 3'-OH ends) and 100 µM [^{3}H]dGTP. <u>Eco</u> RI digestion was carried out in a final volume of 25 µl. The whole reaction mixture was then used for the transferase reaction in the same tube. Thus, no manipulation was done on the DNA preparation, which was then diluted 20-fold in the transferase buffer condition. At intervals of 0, 1, 2, 3, 4, 5, and 6 min, 10-µl aliquots were monitored for acid-insoluble radioactivity with

glass-fiber filters (for counting hydrogen-3). This procedure destroyed the Eco RI recognition sequence. A DNA segment cloned with such dG-tailed DNA could not be recovered by Eco RI digestion. After 2 min of incubation, the reaction reached a plateau, and no further addition of dG residues was noticed (Fig. 11D).

It is clear from the above description that the addition of homopolyer tracts directly to the duplex-DNA termini is an efficient approach to a novel reaction at the 3' end for in vitro construction of recombinant DNA. The use of Co^{+2} ion enables one to carry out these steps. When such steps are not required (e.g. in the restoration of recognition sequence), one may use λ exonuclease to convert the duplex DNA to an efficient primer for addition of homopolymer tracts[94]. The destruction of the Eco RI recognition sequence, as described above, is sometimes very useful for identifying the location of the promoter of the inserted gene, in that it allows the determination of the orientation of a small Eco RI fragment carrying the promoter DNA (Roychoudhury and Ratzkin, unpublished results).

It is also evident from Fig. 11 that, in the presence of Co^{+2} ion, the affinity of the enzyme for the poly(dT) tract is highest. In 2 min of incubation, about 40 dT residues were added (Fig 11A). Under similar conditions, it required 10 min of incubation for addition of about 40 dA residues (Fig 11B). When the temperature was lowered by 14°C (to 23°C), there was a reduction in the rate of addition of dC residues. Thus, after 2 min of incubation, about 10 dC residues were added, compared with 40 dT residues in 2 min at 37°C (Fig. 11A and 11C). Incubation at 37°C (data not shown) resulted in an addition of 30 dC residues after 2 min of incubation. This result clearly indicates that the affinity of the enzyme for the poly(dT) tract is higher than that for the poly(dC) tract in the presence of Co^{+2} ion. This is consistent with the observation[85] that the initiator Km for $d(pC)_5$ is much higher than that for $d(pT)_5$. A higher Km value indicates a lower affinity.

It is clear from Fig. 11D that the addition of dG residues to poly(dG) tract was automatically terminated after the addition of about 27 dG residues. The use of lower temperatures[43] may enable one to achieve further reduction in the number of dG residues, owing to the stacking of the bases. However, lower temperature drastically reduces the priming activity of duplex-DNA termini, and there is thus a danger of obtaining very few DNA molecules tailed at the 3' end. This problem is not encountered when the exposed end of the duplex contains a protruding 3' end, as is found at the Pst I site.

In that case, an incubation temperature of 15°C may be used[43]. Under this condition, only 15 dG residues were added at the Pst I site of pBR322 DNA[43].

IX. CONCLUSION

Terminal-transferase activity is strongly influenced by the choice of buffer system[95]. Other factors affecting the activity in cacodylate buffer were studied in considerable detail[96]. Some of the details of assay conditions were described by Coleman[97]. Katcoff et al.[98] used calcium chloride for limited addition of dC residues at the 3' end of cDNA. The reaction mixture was used directly for annealing with dG-tailed pBR322 DNA in the presence of 150 mM RbCl[98].

There are some useful modifications of the final step for isolation of terminal transferase. The material isolated according to the method of Chang and Bollum[1] may be subjected to further hydroxylapatite chromatography for removal of contaminating nucleases. The preparation may also be purified by oligo(dT) cellulose chromatography[99].

An interesting application of terminal transferase for analysis of nucleic acid structure is the isolation of 3' terminal region of unique DNA molecules[100]. The method involves limited digestion of large DNA with spleen DNase to produce fragments of desired chain lengths, addition of poly(dA) tracts to the original 3'-terminal fragments (other fragments contain 3'-phosphoryl groups and are thus not substrates for transferase), and isolation of poly(dA)-tailed fragments by oligo(dT) cellulose chromatography[100].

In addition to the methods outlined here, the restriction-endonuclease recognition sequences may also be added by ligating synthetic adapter molecules ("linker" oligonucleotides)[101-104]. When such a material is not available, the application of terminal transferase for reconstruction of the restriction site provides a useful and efficient method for molecular cloning of important genes. We have also noticed that a higher number of DNA molecules with reconstructed restriction sites are obtained by this approach than by blunt-end ligation. However, the synthetic adapter provides more versatility, because it allows the creation of any restriction site at the end of an inserted DNA segment[104].

ACKNOWLEDGMENTS

This work was supported by grants from the Deutsche Forschungsgemeinshaft (Scwerpunkt Synthese Makromolekularer Naturstoffe, and SFB46, Molekulare Grundlagen der Entwicklung) in West Germany and by Research grants GM24904 from

80

the National Institutes of Health and 77-20313 from the National Science
Foundation. I am grateful to Dr. Hans Kossel of the University of Freiburg,
West Germany, and to Dr. Ray Wu of Cornell University, for their encouragement
and support. I am especially thankful to Dr. Ernest Jay and Dr. David Tu for
their advice and help in two-dimensional mapping and gel electorphoresis. I am
grateful to Dr. Jeffrey Labovitz for critical reading of the manuscript and to
Dr. Stanley N. Cohen of Stanford University for providing me with the results
of pTU4 DNA sequence.

REFERENCES

1. Chang, L. M. S. and Bollum, F. J. (1971). J. Biol. Chem. 246, 909-916.
2. Johnson, D. and Morgan, R. A. (1976). Biochem. Biophys. Res. Commun. 72, 840-849.
3. Bollum F. J. in "The Enzymes" (P. D. Boyer ed.) 3rd ed. Vol. 10 p. 145 Academic Press, New York, 1974.
4. Roychoudhury, R. and Kossel, H. (1971). Eur. J. Biochem. 22, 310-320.
5. Kossel, H. and Roychoudhury, R. (1971). Eur. J. Biochem. 22, 271-276.
6. Josse, J., Kaiser, A. D. and Kornberg, A. (1961). J. Biol. Chem 236, 864-875.
7. Kossel, H., Roychoudhury, R., Fischer, D. and Otto, A. in "Methods in Enzymology "(L. Grossman and K. Moldave ed.). Vol. 29 part E, p 322, Academic Press, New York 1974.
8. Roychoudhury, R., Fischer, D. and Kossel, H. (1971). Biochem. Biophys. Res. Commun. 45, 430-435.
9. Wu, R., Tu, C. D. and Padmanabhan, R. (1973). Biochem. Biophys. Res. Commun. 55, 1092-1099.
10. Jay, E., Bambara, R., Padmanabhan, R. and Wu, R. (1974). Nucleic Acids Res. 1, 331-354.
11. Tu, C. P. D., Jay, E., Bahl, C. P. and Wu, R. (1976). Anal. Biochem. 74, 73-93.
12. Roychoudhury, R. (1972). J. Biol. Chem. 247, 3910-3917.
13. Sekiya, T. and Khorana, H. G. (1974). Proc. Natl. Acad. Sci. U.S.A. 71, 2978-2982.
14. Sekiya, T., Van Ormond, H. and Khorana, H. G. (1975). J. Biol. Chem. 250, 1087-1098.
15. Podmanabhan, R., Wu, R. and Calender, R. (1974). J. Biol. Chem. 249, 6197-6207.
16. Feix, G. (1972). Biochem. Biophys. Res. Commun. 46, 2141-2147.
17. Feix G. (1971). Fed. Eur. Biochem. Soc. Lett. 18, 280-282.
18. Padmanabhan, R. (1977) Biochemistry 16, 1996-2003.
19. Cozzarelli, N. R., Kelly, R. B. and Kornberg, A. (1969). J. Mol. Biol. 45, 513-531.
20. Atkinson, M. R., Deutcher, M. P., Kornberg, A., Russel, A. F. and Moffat, J. G. (1969). Biochemistry, 8, 4897-4904.
21. Olson, K. and Harvey, C. (1975). Nucleic Acids Res. 2, 319-325.
22. Kossel, H. and Roychoudhury, R. (1974). J. Biol. Chem. 249, 4094-4099.
23. Roychoudhury, R., Jay, E. and Wu, R. (1976) Nucleic Acids Res. 3, 863-877.
24. Roychoudhury, R., Tu, C. P. and Wu, R. (1979). Nucleic Acids Res. 6, 1323-1333.

25. Roychoudhury, R. and Wu, R. in "Methods in Enzymology" (L. Grossman and K. Moldave, ed.) Vol. 65, part I, p. 43, Academic Press, New York 1980.
26. Tu, C. P. D. and Cohen, S. N. (1980) Gene 10, 177-183.
27. Sugino, A., Snopek, T. J. and Cozzarelli, N. R. (1977). J. Biol. Chem. 252, 1732-1738.
28. Kossel, H. and Seliger, H. in "Progress in the Chemistry of Organic Natural Products" (W. Herz, H. Grisebach and G. W. Kirby, ed.) Vol. 32, article, Recent Advances in Polynucleotide Synthesis pp 297-508, Springer-Verlag/Wien. New York. 1975.
29. Khorana, H. G., et al. (1972). J. Mol. Biol. 72, 209. Khorana, H. G. et al. (1976). J. Biol. Chem. 251, 565., Khorana et al. in "Methods in Enzymology" (Ray Wu, ed.) Vol 68 p 109, Academic Press. New York, 1979. Khorana, H. G., (1979) Science, 203, 614.
30. Narang et al. in "Methods in Enzymology" (Ray Wu, ed.) Vol. 68, p 90 Academic Press. New York 1979. Narang et al. in "Methods in Enzymology" (L. Grossman and K. Moldave, ed.) Vol 65, p 610. Academic Press. New York 1980.
31. Schott, H., Fischer, D. and Kossel, H. (1973) Biochemistry 12, 3447-3453.
32. Kossel, H., Buchi, H. and Khorana, H. G. (1967). J. Amer. Chem. Soc. 89, 2185.
33. Wu, R., Jay, E. and Roychoudhury, R. (1976). Methods Cancer Res. 12, 87-176.
34. Shinagawa, M. and Padmanabhan, R. (1979) Anal. Biochem. 95, 458-464.
35. Wu, R. (1970). J. Mol. Biol. 51, 501-521.
36. Bertazzoni, U., Ehrlich, S. D. and Bernerdi, G. (1973). Biochim. Biophys. Acta 312, 192-201.
37. Bernerdi, G., Ehrlich, S. D. and Thiery, J. in "Methods in Enzymology" (L. Grossman and K. Moldave, Ed.) Vol 29, part E, p 341. Academic Press, New York. 1974.
38. Maxam, A. and Gilbert, W. in "Methods in Enzymology" (L. Grossman and K. Moldave, ed.) Vol 65, part I p 499. Academic Press. New York. 1980.
39. Loewen, P. C., Sekiya, T. and Khorana, H. G. (1974) J. Biol. Chem. 249, 217-226.
40. Lowen, P. C. and Khorana, H. G. (1973) J. Biol. Chem. 248, 3489-3499.
41. Jackson, J. F., Kornberg, R. D., Berg, P., Rajbhandary, U. L., Stuart, A., Khorana, H. G. and Kornberg, A. (1965) Biochim. Biophys. Acta. 108, 243-248.
42. Wu, R., Donelson, J. E., Padmanabhan, R. and Hamilton, R. (1972) Bull. Inst. Pasteur 70, 203-233.
43. Villa-Komaroff, L. Efstratiadis, A., Broome, S., Lomedico, P., Tizard, R., Naber, S. P., Chcik, W. L. and Gilbert, W. (1978) Proc. Nat. Acad. Sci. U.S.A. 75, 3727-3731.
44. Bollum, F. J. (1960). J. Biol. Chem. 235, PC 18-20.
45. Bollum, F. J. (1962). J. Biol. Chem. 237, 1945-1949.
45a. Bollum, F. J. (1959). J. Biol. Chem 234, 2733-2734.
46. Chang, L. M. S. and Bollum, F. J. (1971) Biochemistry, 10, 536-542.
47. Cassani, G. and Bollum, F. J. (1969). Biochemistry 8, 3928-3936.
48. Brownlee, G. G. and Sanger, F. (1969). Eur. J. Biochem. 11, 395-399.
49. Roychoudhury, R., Kossel, H. and Wu, R. (unpublished results).
50. Roychoudhury, R., Kuhn, S., Schott, H. and Kossel, H. (1975). Fed. Eur. Biochemistry Soc. Lett. 50, 140-143.
51. Schott, H. (1973). Angew. Chem. 85, 263-264, Angew, Chem. Internat. Edit 12, 246.
52. Scheffler, I. E. and Richardson, C. C. (1972). J. Biol. Chem. 247, 5736-5745.
53. Brutlag, D. and Kornberg, A. (1972) J. Biol. Chem. 247, 241-248.

54. Klenow, H., Overgaard-Hansen, K. and Patkar, S. A. (1971). Eur. J. Biochem. 22, 371-381.
55. Wu, R., Bahl, C. P. and Narang, S. A. (1978). Prog. Nuc. Acid. Res. Mol. Biol. 21, 101-141.
56. Sanger, F., Donelson, J. E., Coulson, A. R., Kossel, H. and Fischer, D. (1973). Proc. Natl. Acad. Sci. U.S.A. 70, 1209-1213.
57. Bahl. C. P., Wu, R., Itakura, K., Katagiri, N. and Narang, S. A. (1976). Proc. Natl. Acad. Sci. U.S.A. 73, 91-94.
58. Padmanabhan, R., Jay, E. and Wu, R. (1974). Proc. Natl. Acad. Sci. U.S.A. 71, 2510-2514.
59. Tu, C. P. D. and Cohen, S. N. (1980) Cell 19, 151-160.
60. Tu, C. P. D., Roychoudhury, R. and Wu, R. (1976). Biochem. Biophys. Res. Commun. 72, 355-362..
61. Roberts, R. J. in "Methods in Enzymology" (Ray Wu, ed.) vol 68, p 27, Academic Press, New York 1979.
62. Maxam, A. M. and Gilbert, W. (1977). Proc. Natl. Acad. Sci. U.S.A. 74, 560-564.
63. Sanger, F. and Coulson, A. R. (1975). J. Mol. Biol. 94, 441-448.
64. Sanger, F., Nicklen, S., and Coulson, A. R. (1977). Proc. Natl. Acad. Sci. U.S.A. 74, 5463-5467.
65. Smith, A. J. H. in "Methods in Enzymology" (L. Grossman and K. Moldave, ed.) Vol 65, p 560 Academic Press. New York 1980.
66. Donelson, J. E. and Wu, R. (1972). J. Biol. Chem. 247, 4661-4668.
67. Englund, P. T. (1971). J. Biol. Chem. 246, 3269-3276.
68. Novogrodsky, A. and Hurwitz, J. (1966). J. Biol. Chem. 241, 2923-2932.
69. Richardson, C. C. (1965) Proc. Natl. Acad. Sci. U.S.A. 54, 158-165.
70. Hinton, D. M., Baez, J. A. and Gumport, R. I. (1978) Biochemistry 17, 5091-5097.
70a. Hinton, D. M. and Gumport, R. I. (1979). Nucleic Acids Res. 7, 453-464.
71. England, T. E., Gumport, R. I. and Uhlenbeck, O. C. (1977). Proc. Natl. Acad. Sci. U.S.A. 74, 4839-4842.
72. England, T. E. and Uhlenbeck, O. C. (1978). Nature 275, 560-561.
72a. England, T. E. and Uhlenbeck, O. C. (1978). Biochemistry, 17, 2069-2076.
73. Kikuchi, Y., Hishinuma, F. and Sakaguchi, K. (1978). Proc. Natl. Acad. Sci. U.S.A. 75, 1270-1273.
74. Kelly, T. J. and Smith, H. O. (1970). J. Mol. Biol. 51, 393-409.
75. Jay, E., Roychoudhury, R. and Wu, R. (1976). Biochem. Biophys. Res. Commun. 69, 678-686.
76. Tomassini, J., Roychoudhury, R., Wu, R. and Roberts, R. J. (1978). Nucleic Acids Res. 5, 4055-4064.
77. Smith, D. L., Blattner, F. R. and Davies, J. (1976). Nucleic Acids Res. 3, 343-353.
78. Armstrong, K. A., Hershfield, V. and Helinski, D. R. (1977). Science 196, 172-174.
79. Schwarz, E., Scherer, G., Hobom, G. and Kossel, H. (1978). Nature, 272, 410-414.
80. Chang, J. C., Temple, G. F., Poon, R., Neumann, K. H. and Kan, Y. W. (1977). Proc. Natl. Acad. Sci. U.S.A. 74, 5145-5149.
81. Nelson, T. and Brutlag, D. in "Methods in Enzymology" (Ray Wu, ed.) Vol. 68, p 41, Academic Press. New York 1979.
82. Dugaiczyk, A. (1976), cited in Boliver, F., Rodriguez, R. L., Greene, P. J., Betlach, M. C., Hynecker, H. L., Boyer, H. W., Crosa, J. H. and Falkow, S. (1977). Gene 2, 95.
83. Mann, M. B., Rao, R. N. and Smith, H. O. (1978) Gene, 3, 97-112.
84. Boliver, F. (1978) Gene 4, 121-136.

85. Kato, K., Goncalves, J. M., Houts, G. E. and Bollum, F. J. (1967).
     J. Biol. Chem. 242, 2780-2789.
86. Brown, N. L. and Smith, M. (1976). Fed. Eur. Biochem. Sci. Lett. 65,
     284-287.
87. Roberts, R. J. Myers, P. A., Morrison, A. and Murray, K. (1976).
     J. Mol. Biol. 103, 199-208.
88. Muller, F., Stoffel, S. and Clarkson, S. G. (1978) cited in Roberts, R. J.
     (1978) Gene 4, 183-193.
89. Arrand, J. R., Myers, P. A. and Roberts, R. J., unpublished results,
     cited in Roberts, R. J. (1978) Gene 4, 183-193.
90. Goff, S. P. and Rambach, A. (1978) Gene 3, 347-352, and unpublished reults.
91. Kleid, D., Humayun, Z., Jeffrey, A. and Ptashne, A. (1976). Proc. Natl.
     Acad. Sci. U.S.A. 73, 293-297.
92. Bahl, C. P., Wu, R., Stawinsky, J. and Narang, S. A. (1977). Proc. Natl.
     Acad. Sci. U.S.A. 74, 966-970.
93. Wilson, J. T., Wilson, L. B., deRiel, J. K., Villa-Komaroff, L.,
     Efstratiadis, A. E., Forget, B. G. and Weissman, S. M. (1978). Nucleic
     Acids Res. 5, 563-581.
94. Lobban, P. E. and Kaiser, A. D. (1973). J. Mol. Biol. 78, 453-471.
95. Chirpich, T. P. (1978). Biochim. Biophys. Acta 518, 535-538.
96. Chirpich, T. P. (1977). Biochem. Biophys. Res. Commun. 78, 1219-1226.
97. Coleman, M. S. (1977). Arch. Biochem. Biophys. 182, 525-532.
98. Katcoff, D., Nudel, U., Zevin-Sonkin, D., Carmon, Y., Shani, M.,
     Lehrach, H., Frischauf, A. M. and Yaffe, D. (1980). Proc. Natl. Acad.
     Sci. U.S.A. 77, 960-964.
99. Okamura, S., Crane, F., Messner, H. A. and Mak, T. W. (1978). J. Biol.
     Chem. 253, 3765-3767.
100. Cassuto, E., Rosenberg, M. and Yot, P. (1973). Nature New Biol. 246,
     196-199.
101. Bahl, C. P., Marians, K. J., Wu, R., Stawinsky, J. and Narang, S. A.
     (1976). Gene 1, 81-92.
102. Heynecker, H. L., Shine, J., Goodman, H. M., Boyer, H. W., Rodenberg,
     J., Dickerson, R. E., Narang, S. A, Itakura, K., Lin, S. and Riggs, A. D.
     (1976). Nature 263, 748-752.
103. Scheller, R. H., Dickerson, R. E., Boyer, H. W., Riggs, A. D. and
     Itakura, K. (1977). Science 196, 177.
104. Bahl, C. P., Wu, R., Brousseau, R., Sood, A., Hsiung, H. M. and Narang,
     S. A. (1978). Biochem. Biophys. Res. Commun. 81, 695-703.

RESTRICTION ENDONUCLEASES:

SPECIFICITIES, DIVERSITIES AND COMPUTER ANALYSIS

ROBERT W. BLAKESLEY

Bethesda Research Laboratories, Inc.
P.O. Box 577
Gaithersburg, Maryland 20760

I. INTRODUCTION

Restriction endonucleases are site-specific enzymes which degrade double
stranded DNAs. The fidelity to a particular nucleotide sequence, and the diversity of sequences represented within this class of enzymes, makes restriction
endonucleases an extremely important research tool in molecular biology. Much
of recent technology, e.g., DNA sequencing, mapping, genome characterization/

identification and cloning is dependent on the action of restriction endonucleases. The importance of these enzymes to science is affirmed by the recent
awarding of Nobel prizes to Drs. Arber,[194] Nathans[195] and Smith[196] for their
initial discoveries and work with restriction enzymes. Several review articles
on restriction endonucleases are published, the most recent of which is that of
Roberts.[193]

Interestingly, though, very little is known about these enzymes. Usually,
biochemists begin a careful study after the discovery of an enzyme, detailing
its kinetic and physical properties. This knowledge establishes the basis upon
which a number of clinical, industrial or research applications are frequently
made. The restriction endonucleases (Type II) on the other hand, were almost
immediately recognized as important research tools from which an entirely new
area of technology with far-reaching consequences developed. Since the potential
of these enzymes was so significant, the biochemistry of restriction endonucleases was for the most part ignored. However, the few examples of basic
enzyme data that have been collected so far indicate that restriction enzymes
are much more complex than were originally perceived. As more detail is obtained,
these enzymes will become more versatile as tools, and undoubtedly provide model
systems for studying protein/nucleic acid interactions.

This article is composed of several tables summarizing much of the current
(March 1981) information on restriction endonucleases. Since the primary use
of the enzymes is as research tools, the tables are formatted toward that purpose.
Hopefully, the next issuance of this type of article will include significant
amounts of basic data on the enzymology of restriction endonucleases.

Table I lists the different sequences recognized by restriction endonucleases,
and the number of cleavages for five of the most commonly used DNAs. In the case
of the sequenced DNAs (pBR322,[197] ØX174[198] and SV40[199]) a computer program designed to search for the recognition sequence within a DNA sequence was used to
determine the frequencies. In some instances, the computer-generated frequencies
may not be observable experimentally due to the size (< 50 base pairs) or dupli-
cate lengths of some of the fragments.

In Table II are found those enzymes whose activities were observed, but for
which no recognition sequence is known. Where data is available, the number of
cleavages on the various DNAs is listed.

Restriction endonucleases have been found primarily in bacteria. The large
number of genuses represented in Table III indicate the rather wide distribution
of these enzymes in these microorganisms. Bacillus, Haemophilus and Streptomyces
are particularly fruitful genuses for restriction endonucleases.

Collated in Table IV are typical reaction conditions used for some restriction endonucleases. This data was taken from the literature or, in some cases, was determined in our laboratories.[192] Certain organic solvents, e.g. glycerol, can affect specificity as evidenced in the case for Bam HI.[191] Unfortunately, for nearly all of the parameters listed, optimal conditions as would be determined by kinetic analysis are not represented. However, this empirical data will allow the enzyme to produce the expected cleavage pattern on typical standard DNAs such as that from phage lambda or Adenovirus-2. For other DNAs, especially those that are superhelical or partially methylated, restriction enzyme digestion may produce unexpected results. This is also true for reactions performed under non-standard conditions, e.g., in unsiliconized glassware or high DNA concentrations. Thus, a pilot reaction is recommended before critical experiments or cleavages of large amounts of DNA are executed.

Table V is a computer-determined listing of cleavage frequencies in three sequenced DNAs for 256 possible 4-, 5-, or 6-base palindromic sequences. This table was kindly contributed by Drs. C. Fuchs and W. Szybalski of the University of Wisconsin (see also 201). This table is especially useful when one discovers a putative novel restriction endonuclease. By determining the cleavage frequencies on these readily obtainable DNAs, the new enzyme's recognition sequence can be quickly narrowed to a few if not to a unique sequence. A computer program recently was developed to perform the same type of analysis for nearly 1,500 possible recognition sequences.[218] Further experimental data, e.g. from the "5th Channel Method"[202] would verify a sequence so determined.

The final table, Table VI, lists the hybrid site resulting from ligating two DNA fragment ends generated by two restriction endonucleases (Enzyme 1 and Enzyme 2). The enzyme combinations are grouped depending upon the type of end (blunt, or extended) generated during cleavage of the DNA. The "general hybrid" site is predicted from combining the two parental enzymes' sequences. In certain instances, the sequence is ambiguous due to sequence ambiguities (e.g., purine or pyrimidine wobbles) in the parental sequences. If, however, the true sequence at the restriction site is known (e.g., using a site in sequenced pBR322 DNA) then the ambiguous site can be resolved to the "specific hybrid" site. In all cases, the last column (Enzyme 3) lists all known enzymes that could cleave the hybrid site. This table should prove a valuable aid to cloners when predicting specific restriction enzymes for use in constructing hybrid DNA molecules and what enzymes would excise the inserted DNA at that site.

88

ACKNOWLEDGMENTS

Assistance in preparation and supplying of data for these tables by the many scientific colleagues outside of and within Bethesda Research Laboratories is gratefully acknowledged.

II. TABLE I

DNA CLEAVAGE FREQUENCY FOR RESTRICTION ENDONUCLEASES WITH KNOWN RECOGNITION SEQUENCES

SEQUENCE[j]	Number of Cleavages					ENZYMES
	pBR322	ØX174	SV40	λ	Ad-2	
↓AATT	9	25	37	-	-	Eco RI*
AG↓CT	16	24	35	>50	>50	Alu I, Oxa I[a]
C↓CGG	26	5	1	>50	>50	Hap II, Hpa II[b] Mno I, Msp I[c], SfaGU I[d]
CCTC (GAGG)	26	34	51	>50	>50	Mnl I[g]
CG↓CG	23	14	0	>50	>50	AccII[a], BceR[a], FnuD II, Hin 1056 I[a], Tha I
↓GATC	22	0	8	>50	>50	Bst I*, Dpn II[a], FnuA II[a], FnuC I, FnuE I[d], Mbo I, Mno III[a] Mos I[a], Pfa I[a], Sau 3A[d]
G^{Me}A↓TC	>10	0	0	>50	-	Dpn I
GCG↓C	31	18	2	>50	>50	Cfo I, FnuD III, Hha I, HinGU I[a], Mnn IV[a]
GG↓CC	22	11	19	>50	>50	Blu II[a], BspR I[a], BsuR I, Bsu 1076[a], Bsu 1114[a], FnuD I, Hae III[b], Hhg I[a], Mnn II[a], Ngo II[a], Pal I[a], Sfa I, Sfa 9116 I[a], Szo I[a]
GT↓AC	3	11	11	>50	>50	Rsa I
T↓CGA	7	10	1	>50	>50	Taq I, Tth HB8 I[a], Tth 2172 I[a]
NGCN						Bsu RI*[a]
AGACC (GGTCT)	4	7	4	-	-	EcoP I[h]
↓CC(A/T)GG	6	2	16	>35	>35	Atu I[a], AtuD I[a] Ecl II[a], Eco R II
CC↓(A/T)GG	6	2	16	>35	>35	Apy I, Atu II, Rru II
CC(C/G)GG	10	1	0	>15	>15	Nci I[a]
C↓TNAG	8	14	19	>30	-	Dde I
GAAGA(TCTTC)	11	11	16	>50	>50	Mbo II[e]

TABLE I (Continued)

SEQUENCE	pBR322	ØX174	SV40	λ	Ad-2	ENZYMES
GACGC(GCGTC)	11	14	0	>50	>50	Hga I[b]
GATGC(GCATC)	22	12	6	>50	>50	SfaN I[a]
G↓ANTC	10	21	10	>50	>50	FnuA I, Hha II[a], Hinf I
GC$\binom{A}{T}$GC	21	14	23	>30	>30	Bbv I[a]
GC↑NGC	42	31	25	>50	>50	Fnu4HI
GGTGA(TCACC)	12	9	4	>50	>50	Hph I[e]
G↓G$\binom{A}{T}$CC	8	1	6	>30	>30	Ava II, Bam Nx, Cau I[a], HgiBI, HgiC II[a], HgiE I[a]
GGATG	12	8	11	>50	>50	Hin Gu II[i]
G↓GNCC	15	2	11	>30	>30	Asu I, Sau 96 I
A↓AGCTT	1	0	6	6	11	Bbr I[a], Chu I[a], Hin 91 R[a], Hin 173[a], Hinb III[a], Hinb(1076) III[a], Hind III, Hinf II[a], Hsu I
A↓GATCT	0	0	0	6	12	Bgl II[d]
AGG↓CCT	0	1	7	5	12	Stu I
$\binom{A}{T}$GG↓CC$\binom{A}{T}$	7	6	11	-	-	Hae I
AGT↓ACT	1	0	0	4	5	Rru I
AT↓CGAT	1	0	0	12	-	Cla I
ATGCAT	0	0	3	-	-	Ava III[a]
CAAPuCA	5	11	12	>25	-	Tth III II[i]
CAGCAG	6	5	12	-	-	EcoP 15[h]
CAG↓CTG	1	0	3	15	22	Pvu II
C↓CCGGG	0	0	0	3	12	Xma I
CCC↓GGG	0	0	0	3	12	Sma I
CCGC↓GG	0	1	0	4	>25	Ecc I[a], Sac II, Sbo I[a], Sbr I[a], Shy I[a], Sst II, Tgl I[a]
CCTAGG	0	0	2	2	2	Avr II
CGAT↓CG	1	0	0	3	7	Pvu I, Rsh I, Rsp I[a], Xni I[a], Xor II
C↓GGCCG	1	0	0	2	10	Xma III
C↓TCGAG	0	1	0	1	5	Blu I, Msi I[a], Scu I[a], Sex I[a], Sgo I[a], Sla I, Slu I[a], Spa I[a], Xho I, Xpa I

TABLE I (Continued)

SEQUENCE	pBR322	ØX174	SV40	λ	Ad-2	ENZYMES
CTGCA↓G	1	1	2	18	25	Bce 170[a], Bsu 1247[a], Pst I, Sal PI[a], Xma II[a], Xor I[a]
C↓PyCGPuG	1	1	0	8	15	Ava I, Avr I[a]
G↓AATTC	1	0	1	5	5	Eco RI, Rsh 630 I[a]
G$\binom{A}{T}$ GC$\binom{A}{T}$↓C	8	3	0	20	–	HgiA I
GAGCT↓C	0	0	0	2	7	Sac I, Sst I
GCATG↓C	1	0	2	4	–	Sph I
G↓GATCC	1	0	1	5	3	Bam FI[a], Bam HI, Bam KI[a], Bam NI[a], Bst I
GGTAC↓C	0	0	1	2	8	Kpn I
G↓GPyPuCC						HgiC I
G↓TCGAC	1	0	0	2	3	HgiC III, HgiD II, Sal I, Xam I[a]
GT$\binom{C\,Y\,T}{A\,A\,G}$↓AC	2	2	1	7	8	Acc I
GTT↓AAC	0	3	4	13	6	Hpa I
GPu↓CGPyC	6	7	0	>14	>14	Acy I, Aos II, Asu III[a], Hgi DI
GTPy↓PuAC	2	13	7	34	>20	Chu II[a], Hinc(1160) II[a], Hinc(1161) II[a], Hinc II[a], Hind II, Mnn I[a]
T↓CTAGA	0	0	0	1	4	Xba I
T↓GATCA	0	0	1	7	5	AtuC I[a], Bcl I, Cpe I[a], Sst IV[a]
TGC↓GCA	4	1	0	>10	>15	Aos I
TGG↓CCA	1	0	0	15	17	Bal I
TT↓CGAA	0	0	0	5	1	Asu II, Mla I
Pu↓GATCPy	8	0	3	>20	>20	Xho II
PuGCGC↓Py	11	8	1	>30	>30	Hae II, HinH I[a], Ngo I[a]
PuPuA↓TPyPy	14	16	24	>10	>10	Eco RI'
G↓GTNACC	0	0	0	11	8	BstE II, BstP I, Eca I, Fsp AI
GACN↓NNGTC	–	–	–	3	–	Tth 23 I, Tth 110 I, Tth 3 I
GAA(N)$_4$ TTC	2	3	0			Xmn I[a]
GCC(N)$_4$↓NGGC	3	0	1	22	12	Bgl I
ACC(N)$_6$ GGT						HgiE II

TABLE I (Continued)

SEQUENCE	pBR322	ØX174	SV40	λ	Ad-2	ENZYMES
AAC(N)$_6$GTGC	2	0	0	-	-	EcoK[a]
TGA(N)$_8$TGCT	0	0	1	-	-	EcoB[a]

[a]This is the recognition sequence, but the cleavage location at or near this sequence is unknown.
[b]This sequence is cleaved regardless of whether the external cytosine is methylated or not.
[c]This sequence is cleaved regardless of whether the internal cytosine is methylated or not.
[d]This sequence is cleaved regardless of whether the internal adenine is methylated or not.
[e]The cleavage occurs 8 bases 3' from this recognition sequence.
[f]The cleavage occurs 5 bases 3' from this recognition sequence.
[g]The cleavage occurs 5-10 bases 3' from this recognition sequence.
[h]The cleavage occurs 24-26 bases 3' from this recognition sequence.
[i]The cleavage occurs 9-11 bases 3' from this recognition sequence.
[j]Each sequence is written 5'——>3' with an N representing A, C, G or T, Pu, an A or G and Py, a C or T. The arrow indicates the point of cleavage within the sequence.

III. TABLE II
DNA CLEAVAGE FREQUENCY FOR RESTRICTION ENDONUCLEASES WITH
UNDETERMINED RECOGNITION SEQUENCES

ENZYME	DNA: NUMBER OF CLEAVAGES
Aca I	
Aim I	
Ani I	
Ani II	
Ani III	
Asu IV	
Atu BVI	ØX:0, SV40:1, λ:>14
Bam HI.1	pBR:>7, ØX:2, SV40:9, λ:>20
Bce 1229	λ:>10
Bce 14579	λ:>10
Bme 205	λ:>10
Bme 899	λ:>5
Bme I	SV40:4, λ:>10, Ad-2:>20
Bpu I	SV40:2, λ:6, Ad-2:>30
Bsp 1286	
Bst AI	
Bst EI	
Bst EIII	λ:>7
Bsu M	λ:>10
Bsu 1145	λ:>20
Bsu 1192	λ:>10
Bsu 1193	λ:>30
Bsu 1231	λ:>20
Bsu 1259	λ:>8
Bsu 6633	λ:>20
●Cau II	SV40:0, λ:>30, Ad-2:>30
Cvi I	
Ecl I	SV40:4, λ:10
Fnu 84I	ØX:>10, λ:>50

TABLE II (Continued)

ENZYME	DNA: NUMBER OF CLEAVAGES
Hap I	λ:>30
Hin 1056 II	ØX:5, λ:>30, Ad-2:>30
Hin GLU	
Mgl I	
Mgl II	
Mnn III	λ:>50, Ad-2:>50
Mno II	SV40:3, λ:>10, Ad-2:>6
Mvi I	λ:1
Mvi II	
Nde I	pBR:1, ØX:3, SV40:0
Oxa II	
PaeR7	
Rrb I	ØX:1, SV40:5, Ad-2:4
Sac III	λ:>30, Ad-2:>30
Sag I	
Sal II	λ:>30
Sau I	λ:2
Sgr I	SV40:0, λ:0, Ad-2:7
Sna I	λ:2
Ssp I	
Sst III	pBR:6, ØX:5, SV40:1, λ:>10
Taq II	ØX:6, SV40:4, λ:>30, Ad-2:>30

IV. TABLE III
MICROORGANISMS AS SOURCES FOR RESTRICTION ENDONUCLEASES

GENUS	SPECIES	STRAIN	SOURCE	ENZYME	REF
Achromobacter	immobilis		ATCC 15934	Aim I	54
Acinetobacter	calcoaceticus		R. J. Roberts	Acc I, Acc II	18
Agrobacterium	tumefaciens	B6		Atu I	41
		C58	E. Nester	Atu CI	117
		ID 1138	C. Kado	Atu DI	127
		ID 135	C. Kado	Atu II	40, 143
		IIBV7	G. Roizes	Atu BVI	161
Anabaena	catanula		CCAP 1403/1	Aca I	53
	cylindrica		CCAP 1403/2a	Acy I	168, 179
	oscillarioides		CCAP 1403/11	Aos I, Aos II	168, 180
	subcylindrica		ATCC 29211	Asu I, Asu II	53, 168
				Asu III, Asu IV	203

TABLE III (Continued)

GENUS	SPECIES	STRAIN	SOURCE	ENZYME	REF
	variabilis		K. Kurray	Ava I, Ava II,	51, 52, 135
				Ava III	136, 137
		UW	W. Szybalski	Avr I, Avr II	154
Anacystis	nidulans		A. deWaard	Ani I, Ani II,	168
				Ani III	168
Arthrobacter	luteus		ATCC 21606	Alu I	3, 4
	pyridinolis		R. DiLauro	Apy I	126
Bacillus	amyloliquefaciens	F	ATCC 23350	Bam FI	17
		H	F. E. Young	Bam HI, Bam HI.1	79-83, 191
		K	T. Kaneko	Bam KI	17
		N	T. Ando	Bam NI, Bam Nx	84, 100, 183

TABLE III (Continued)

GENUS	SPECIES	STRAIN	SOURCE	ENZYME	REF
	brevis		ATCC 9999	Bbv I	63
	caldolyticus		A. Atkinson	Bcl I	92
	cereus	Rf sm st	T. Ando	Bce R	17
			T. Ando	Bce 170	17
			IAM 1229	Bce 1229	17
			ATCC 14579	Bce 14579	17
	globigii	RUB 550	G. A. Wilson	Bgl I, Bgl II	58-60, 156, 165
		RUB 561	G. A. Wilson	Bgl I	140
		RUB 562	G. A. Wilson	Bgl II	140
Bacillus	megaterium	899	B899	Bme 899	17
		B205-3	T. Kaneko	Bme 205	17
			J. Upcroft	Bme I	37
	pumilus	AHU 1387A	T. Ando	Bpu I	101
	sphaericus		IAM 1286	Bsp 1286	17
		R	P. Venetianer	Bsp RI	31, 190

TABLE III (Continued)

GENUS	SPECIES	STRAIN	SOURCE	ENZYME	REF
	stearothermophilus	1503-4R	NCA 1503	Bst I, Bst I*	85, 185
		240	A. Atkinson	Bst AI	92
		ET	N. Welker	Bst EI, Bst EII,	102, 174
				Bst EIII	102
			ATCC 12980	Bst PI	189, 213
	subtilis	Marburg 168	T. Ando	Bsu M	17
		X5	T. Trautner	Bsu RI, Bsu RI*	32, 33, 123
			ATCC 6633	Bsu 6633	17
			ATCC 14593	Bsu 1145	17
			IAM 1076	Bsu 1076	17
			IAM 1114	Bsu 1114	17
			IAM 1192	Bsu 1192	17
			IAM 1193	Bus 1193	17
			IAM 1231	Bsu 1231	17
			IAM 124α	Bsu 1247	17
			IAM 1259	Bsu 1259	17

TABLE III (Continued)

GENUS	SPECIES	STRAIN	SOURCE	ENZYME	REF
Bordetella					
	bronchiseptica		ATCC 19395	Bbr I	15
Brevibacterium					
	albidum		ATCC 15831	Bal I	91
	luteum		ATCC 15830	Blu I, Blu II	30, 66
Caryophanon					
	latum	L	H. Mayer	Cla I	145
Chloroflexus					
	aurantiacus		A. Bingham	Cau I, Cau II	142
Chromobacterium					
	violaceum		ATCC 12472	Cvi I	54
Clostridium					
	formicoaceticum		ATCC 23439	Cfo I	162, 182
Corynebacterium					
	humiferum		ATCC 21108	Chu I, Chu II	54
	petrophilum		ATCC 19080	Cpe I	93
Desulfovibrio					
	desulfuricans	Norway	H. Peck	Dde I	162

TABLE III (Continued)

GENUS	SPECIES	STRAIN	SOURCE	ENZYME	REF
Diplococcus	pneumoniae		S. Lacks	Dpn I, Dpn II	22, 23, 177
Enterobacter	cloacae		H. Hartmann	Ecl I, Ecl II	103
		DSM 30056		Eca I	151
		DSM 30060		Ecc I	15
Escherichia	coli	B	W. Arber	Eco B	104-107, 138, 152
		K	M. Meselson	Eco K	108, 109, 139
		(PI)	K. Murray	Eco PI	109-113, 171
		P15	W. Arber	Eco P15	114, 206
		RY13	R. N. Yoshimori	Eco RI, Eco RI*, Eco RI'	44, 71-78, 120, 173, 175; 99

TABLE III (Continued)

GENUS	SPECIES	STRAIN	SOURCE	ENZYME	REF
Flavobacterium		R245	R. N. Yoshimori	Eco RII	35, 42-45, 186
Fusobacterium	nucleatum	HP 1039	N. Brown	Fsp AI	166
		A	M. Smith	Fnu AI, Fnu AII	19
		C	M. Smith	Fnu CI	19
		D	M. Smith	Fnu DI, Fnu DII, Fnu DIII	19
		E	M. Smith	Fnu EI	19
		84	M. Smith	Fnu 84I	19
		4H	M. Smith	Fnu 4HI	155
Haemophilus	aegyptius		ATCC 11116	Hae I, Hae II, Hae III	33-36, 61, 94-96, 121, 144
	aphrophilus		ATCC 19415	Hap I, Hap II	5-7, 15, 131
	gallinarum		NCTC 3438	Hga I	5, 6, 119, 130

TABLE III (Continued)

GENUS	SPECIES	STRAIN	SOURCE	ENZYME	REF
	haemoglobinophilus		ATCC 19416	Hhg I	15
	haemolyticus		ATCC 10014	Hha I, Hha II	27, 28, 47
	influenzae	b, 1076	J. Stuy	Hinb (1076) III	20
		c, 1160	J. Stuy	Hinc (1160) II	20, 159
		c, 1161	J. Stuy	Hinc (1161) II	20, 159
		H-1	M. Takanami	Hin HI	5, 6, 97
		GU	J. Chirikjian	Hin GU I, Hin GU II	29, 122, 212
		Rb	C. A. Hutchison	Hinb III	15, 49
		Rc	A. Landy, G. Leidy	Hinc II	86
		Rd	S. H. Goodgal	Hind II, Hind III	9, 55-57, 86-88, 131
		Rd 123		Hin GLU	133

TABLE III (Continued)

GENUS	SPECIES	STRAIN	SOURCE	ENZYME	REF
		Rf	C. A. Hutchison	Hinf I, Hinf II	12, 28, 47-49
		91R	J. Chirikjian	Hin 91R	29
		173	J. Chirikjian	Hin 173	29
Haemophilus	influenzae				
		1056	J. Stuy	Hin 1056 I,	20, 159
				Hin 1056 II	20
	parahaemolyticus		C. A. Hutchison	Hph I	49, 50
	parainfluenzae		J. Setlow	Hpa I, Hpa II	8-14, 131, 144
	suis		ATCC 19417	Hsu I	15
Herpetosiphon	giganteus	HP 1023	N. Brown	Hgi AI	149, 166
		Hpg 5	H. Reichenbach	Hgi BI	216
		Hpg 9	H. Reichenbach	Hgi CI, Hgi CII Hgi CIII, 216	216
		Hpa 2	H. Reichenbach	Hgi DI, Hgi DII	216

TABLE III (Continued)

GENUS	SPECIES	STRAIN	SOURCE	ENZYME	REF
		Hpg 24	H. Reichenbach	Hgi EI, Hgi EII	216
Klebsiella	pneumoniae	OK8	J. Davies	Kpn I	69, 90
Mastigocladus	laminosus		CCAP 1447/1	Mla I	168, 214
Moraxella	bovis		ATCC 10900	Mbo I, Mbo II	24, 46, 129
	glueidi	LG1	J. Davies	Mgl I	69
		LG2	J. Davies	Mgl II	69
	nonliquefaciens		ATCC 17953	Mnl I	16
			ATCC 17954	Mnn I, Mnn II,	89
				Mnn III, Mnn IV	89
	osloensis		ATCC 19975	Mno I, Mno II	
				Mno III	14, 15
			ATCC 19976	Mos I	24

TABLE III (Continued)

GENUS	SPECIES	STRAIN	SOURCE	ENZYME	REF
Myxococcus	-		R. J. Roberts	Msp I	62, 176
	stipitatus	Mxs 2-H	H. Reichenbach	Msi I	
	virescens	V-2	H. Reichenbach	Mvi I, Mvi II	115
Neisseria	cinerea		NRCC 31006	Nci I	204
	denitrificans		NRCC 31009	Nde I	204
	gonorrhoea	MUTK1		Ngo I	98
		MUTK1	CDC 66	Ngo II	124, 158
Oerskovia	xanthineolytica		R. Shekman	Oxa I, Oxa II	172
Proteus	vulgaris		ATCC 13315	Pvu I, Pvu II	63, 207
Providencia	alcalifaciens		ATCC 9866	Pal I	178
	stuartii	164	J. Davies	Pst I	68, 69

TABLE III (Continued)

GENUS	SPECIES	STRAIN	SOURCE	ENZYME	REF
Pseudomonas	aeruginosa		G. A. Jacoby	Pae R7	205
	facilis		M. VanMontagu	Pfa I	20, 30
Rhodopseudomonas	sphaeroides	28/5	S. Kaplan	Rsa I	169
		630	S. Kaplan	Rsh 630 I	169
			R. Lascelles	Rsp I	25
		2.4.1	S. Kaplan	Rsh I	141
Rhodospirillum	rubrum		A. deWaard	Rru I, Rru II	211
			J. Chirikjian	Rrb I	116
Serratia	marcescens	Sb	C. Mulder	Sma I	64, 65
Spirabillus	-	SAI	B. Torheim	Ssp I	118
	nathans		S. Kaplan	Sna I	169
Staphylococcus	aureus	3A	E. E. Stobberingh	Sau 3A	19, 26, 157

TABLE III (Continued)

GENUS	SPECIES	STRAIN	SOURCE	ENZYME	REF
Streptococcus					
	agalatiae	PS 96	E. E. Stobberingh	Sau 96I	134
	faecalis			Sag I	38
	ssp. liquefaciens	GU	J. Chirikjian	Sfa GUI	128
		ND 547	D. Clewell	Sfa NI	117
		9116		Sfa 9116 I	38
	ssp. zymogenes	TR	R. Wu	Sfa I	38
	zooepidemicus			Szo I	38
Streptomyces					
	achromogenes		ATCC 12767	Sac I, Sac II, Sac III	1, 132, 1
	albus		CMI 52766	Sal PI	70
	ssp. pathocidicus		KCC S0166	Spa I	147
		G	J. M. Ghuysen	Sal I, Sal II	125
	aureofaciens	IKA 18/4		Sau I	
	bobiliae		ATCC 3310	Sbo I	148

TABLE III (Continued)

GENUS	SPECIES	STRAIN	SOURCE	ENZYME	REF
	bradiae		ATCC 3535	Sbr I	148
	cuspidoras		KCC S0316	Scu I	160
	exfoliatus		KCC S0030	Sex I	147
	goshikiensis		KCC S0294	Sgo I	147
	griseus		ATCC 23345	Sgr I	1
	hygroscopicus		F. Walter	Shy I	153
	lavendulae		ATCC 8664	Sla I	148
	luteoreticuli		KCC S0788	Slu I	147
	phaeochromogenes		NRRL B-3559	Sph I	181
	tubercidicus		KCC S-0054	Stu I	215
	—	Stanford	A. Goff,	Sst I, Sst II,	2, 132
			A. Rambach	Sst III,	2, 163
				Sst IV	163, 164
Thermoplasma	acidophilum	122-1B3	ATCC 25905	Tha I	21
Thermopolyspora	glauca		ATCC 15345	Tgl I	63

TABLE III (Continued)

GENUS	SPECIES	STRAIN	SOURCE	ENZYME	REF
Thermus	aquaticus	YTI	J. I. Harris	Taq I, Taq II	15, 39
	thermophilus				
		23		Tth 23 I	209
		110		Tth 110 I	208
		111		Tth 111 I	208
		HB8	ATCC 27634	Tth HB8 I	210
			NRCC 2172	Tth 2172I	209
Xanthomonas					
	amaranthicola		ATCC 11645	Xam I	125
	badrii		ATCC 10014	Xba I	60
	holcicola		ATCC 13461	Xho I, Xho II	20, 63, 66, 67
	malvacearum		ATCC 9924	Xma I, Xma II,	64
				Xma III	170
	manihotis	7AS1		Xmn I	217
	nigromaculans		ATCC 23390	Xni I	89
	oryzae	507	M. Ehrlich	Xor I, Xor II	146
	papavericola		ATCC 14180	Xpa I	66

V. TABLE IV
REACTION CONDITIONS FOR CERTAIN RESTRICTION ENDONUCLEASES

ENZYME	TEMP °C	pH	TRIS-HCl mM	Mg^{++} mM	Me$^+$ mM	RSH mM	OTHER	REF.
Alu I	37	7.6-7.9	6	6		6E		3
Asu I	37	7.5	20	10	100S			
Atu II	37	7.5	20	7				40
Bam HI	37	8.5	20	10	(100S)[b]			83
Bam HI.1	37	8.5	10	7			36% (v/v) glycerol	191
Bcl I	50	7.4	10	9		0.5D		92
Bgl I	37	7.5	20	15	50P	NR		156
Bgl II	37	7.4	20	7		7E		192
Bsp RI	37	8.2	25	20[c]	(50-100S)[d]	10E		31, 190
Bst I	37	7-9.5	100	0.5-2[e]	f			185
Bst I*	37	9.1	100	<10			Active to 70°C	185
Bst EII	60	7.9	6	6	50S	6E	10% glycerol	102
Bst PI	37	7.5	90	10	100S			189
Bsu RI	37	7.4	10	10	100S	5E		123
Bsu RI*	37	8.5	25	10		5E	25% glycerol	123
Cfo I	37	7.5	100	5		6E		162
Dde I	37	7.5	100	5	100S	6E		162
Dpn I	37	7.6	50	5	40S			177
Ecl I	37	7.4	25	10				103

TABLE IV (Continued)

ENZYME	TEMP °C	pH	TRIS-HCl mM	Mg++ mM	Me+ mM	RSH mM	OTHER	REF.
Eco RI	37	7-7.5	100	5	50S		Inactivated above 42°C	78
Eco RI*	37	8.5	25	2 Mn++			Inhibited by PCMB	78, 173
Eco RI*	37	8-8.5	250	20			40% glycerol or 58% ethylene glycol	173, 175
Eco RII	37	7.5	25	5				186
FnuDI	37	7.9	6	6		6E		19
FnuDII, III	37	7.9	6	6	50S	6E		19
Fnu 4HI	37	7.9	6	6		6E		155
Hae II	37	7.5	50	5		0.5D	Inactivated above 42°C	121
Hae III	37	7.5	50	5^g	h	NR	Active to 70°C	121
Hap I	37	7.6	10	7		7E		6
Hga I	37	7.6	10	7		7E		6
Hha I	37	8	50	5				121
Hin HI	37	7.6	10	7		7E		6
Hinc II	37	7.9	10	6.6	60S	1D		191
Hind III	37	7.4	40	10	50S			55, 187
Hinf I	37	7.5	6	6	100S	6E		191
Hpa I	37	7.4	20	10	20S	1D		188, 192
Hpa II	37	7.4	20	10		1D	Inhibited at 100 mM KCl	188

TABLE IV (Continued)

ENZYME	TEMP °C	pH	TRIS-HCl mM	Mg++ mM	Me+ mM	RSH mM	OTHER	REF.
Hph I	37	7.4	10	10	6P	10E		50
Kpn I	37	7.5	6	6	6S	6E		90
Mla I	40	7.4	6.7	6.7	60P	6.7E	Active 30-50°C	214
Mno I	37	7.9	6	6		6E		14
Nci I	37	7.5	6	6	65	6E	Inhibited above 150 mM KCl	204
Ngo II	37	8.5	100	1	20S		Active up to 45°C	158
Pst I	37	7.5	20	10	50 NH_4^+		Complete inhibition- 0.3 M NaCl	69, 192
Rsa I	34	7.9	10	6		0.5D	Active to 200 mM NaCl	203
Rsh I	37	7.9	10	6		0.5D	Inhibited by Na+ or K+ >100 mM	141
Sau 3A	37	7.5	6	6	50S	6E		26, 157
Sal I	37	7.6	8	6	150S			125, 192
Sal PI	37			10	50			70
Sau 96I	30	7.4	6	15	100S	6E		134
Sfa I	37	7.5	6.6	6.6		6.6E		38
Sla I	37	7.9	10	10	100S	10E		148
Sma I	37	8.0	15	6				191
Sph I	37	7.5	6	6	50S	6E	Inhibited at 100 mM NaCl	
Sst I	37	7.5	14	6	90P	6E		192

TABLE IV (Continued)

ENZYME	TEMP °C	TRIS-HCl mM	pH	Mg^{++} mM	Me^+ mM	RSH mM	OTHER	REF.
Sst II	37	14	7.5	6	90P	6E		192
Stu I	37	10	7.9	10	100S			
Taq I	65	10	8.4	6		6E		39
Tha I	37	6	7.4	10				21, 192
Tth HB8I	60	20	8	5	50S	10E		210
Tth 111I	65	8	7.4	8	50S	8E	10% active at 37°C; inactive > 200 µM NaCl	208
Xba I	37	6	7.9	6		6E		60
Xho I	37	8	7.4	6	150S	6E		66, 192
Xor II	37	6	7.4	12		6E		146

[a]Abbreviations: D, dithiothreitol; E, 2-mercaptoethanol; S, sodium chloride; P, potassium chloride; RSH, mercaptan; NR, not required.
[b]Addition of 100mM NaCl supresses Bam HI.1 activity without significant effect on Bam HI activity.
[c]Mn^{++} partially substitutes for Mg^{++}; the presence of Zn^{++} is inhibitory.
[d]Up to 100mM NaCl does not alter activity; whereas 400mM NaCl is completely inhibitory.
[e]Only 10% activity is obtained when Mn^{++} replaces Mg^{++}.
[f]Activity is inhibited 50% at 50mM NaCl.
[g]Only 50% activity is obtained when Mn^{++} replaces Mg^{++}.
[h]Activity is inhibited 50% at 100mM NaCl.

VI.   TABLE V
SEQUENCE FREQUENCIES IN ØX174, SV40 and pBR322 DNAs

FREQUENCY			
ØX174	SV40	pBR322	SEQUENCE
0	0	0	ACCGGT
			ACTAGT
			AGATCT
			CACGTG
			CCCGGG
			GAGCTC
			TACGTA
			TCTAGA
			TTCGAA
0	0	1	ACATGT
			AGTACT
			ATATAT
			ATCGAT
			CGATCG
			CGCGCG
			CGGCCG
			GCTAGC
			GTATAC
			GTCGAC
			TATATA
			TCCGGA
			TGGCCA
			UCTAGY
0	0	2	YCCGGU
0	0	4	UTCGAY
0	1	0	GGGCCC
			GGTACC
			TGATCA
0	1	1	GAATTC
			GATATC
			GGATCC
0	1	4	AGCGCT
			GCCGGC
0	1	7	UCCGGY

TABLE V (Continued)

ØX174	SV40	pBR322	SEQUENCE
	FREQUENCY		
			YGATCU
0	2	0	CCTAGG
			TAGCTA
0	2	0	TGTACA
0	2	1	GCATGA
0	2	2	GUGCYC
0	2	4	UCATGY
0	3	0	ATGCAT
			CCATGG
0	3	1	CAGCTG
0	3	3	GGUYCC
			GUATYC
0	3	8	UGATCY
0	5	0	CCYUGG
0	6	1	AAGCTT
0	7	2	TUTAYA
0	8	22	GATC
1	0	0	CCGCGG
			CTCGAG
			GCGCGC
1	0	1	CYCGUG
			GACGTC
1	0	2	CUCGYG
1	0	3	GTGCAC
			TYCGUA
			UTATAY
1	0	4	AYATUT
			TGCGCA
1	0	7	CGUYCG
1	0	10	CCSGG
1	1	4	GAYUTC
1	2	0	CTATAG
			TAATTA

TABLE V (Continued)

ØX174	SV40	FREQUENCY pBR322	SEQUENCE
1	2	1	CATATG
			CTGCAG
1	2	4	CAYUTG
1	3	0	CAATTG
			TTATAA
1	4	2	GAUYTC
1	5	7	GGSCC
1	6	4	TAUYTA
1	6	8	GGRCC
1	7	0	AGGCCT
1	7	1	AATATT
1	9	1	YAGCTU
1	9	3	UGGCCY
2	0	0	ACGCGT
			CGTACG
			YTCGAU
2	0	1	ACUYGT
			TCGCGA
			YACGTU
2	0	3	TAYUTA
2	0	4	GGCGCC
2	0	6	YGGCCU
2	0	7	GYGCUC
2	1	0	CTTAAG
2	2	2	UGTACY
2	2	4	GCUYGC
			GTUYAC
2	3	1	TTGCAA
2	3	3	AYGCUT
2	3	5	UTGCAY
2	4	1	ATTAAT
			CUATYG
2	4	5	CUGCYG

TABLE V (Continued)

FREQUENCY			
ØX174	SV40	pBR322	SEQUENCE
2	4	9	AGYUCT
2	5	1	YCTAGU
2	6	3	ATUYAT
2	6	4	TCYUGA
2	6	6	TGYUCA
2	8	3	CYATUG
2	11	3	TTTAAA
2	11	15	GGNCC
2	16	6	CCRGG
3	0	4	AACGTT
			YCGCGU
3	0	5	AYCGUT
			TUCGYA
3	0	7	YGCGCU
3	1	9	GGYUCC
3	2	4	CYGCUG
			TCATGA
			YATATU
3	2	8	GUTAYC
3	3	0	YGTACU
3	3	2	CCUYGG
3	4	0	GTTAAC
3	7	1	AYTAUT
3	7	2	TUGCYA
3	8	4	ATYUAT
3	9	0	AGUYCT
3	9	3	CAUYTG
3	11	1	TTUTAA
3	12	1	YAATTU
3	12	5	CTAG
3	16	16	CCNGG
4	0	3	CGYUCG
4	0	4	UCGCGY

TABLE V   (Continued)

ØX174	SV40	pBR322	SEQUENCE
		FREQUENCY	
4	0	7	TCSGA
4	3	0	CYTAUG
4	4	4	ACYUGT
4	5	1	CTRAG
4	5	3	GYATUC
4	6	5	GYTAUC
4	7	2	YTATAU
4	10	4	AUTAYT
5	1	7	GYCGUC
5	1	26	CCGG
5	4	0	AAATTT
5	4	1	TASTA
5	4	6	CTYUAG
6	3	2	AGRCT
6	10	4	CTUYAG
6	11	6	TYATUA
6	17	5	TYTAUA
6	22	11	CASTG
7	0	6	GUCGYC
7	1	11	AUCGYT
7	1	13	GCYUGC
7	2	1	CUTAYG
7	7	2	UAATTY
7	13	4	TTYUAA
7	16	3	YTTAAU
7	17	3	AUGCYT
7	25	8	ACRGT
8	0	6	ACSGT
			UACGTY
8	0	9	CGRCG
8	1	11	UGCGCY
8	5	1	TUATYA
8	6	9	GTSAC

TABLE V   (Continued)

	FREQUENCY		
ØX174	SV40	pBR322	SEQUENCE
8	11	3	UAGCTY
8	12	6	YTGCAU
8	18	3	AAUYTT
9	2	8	TCUYGA
9	3	2	TYGCUA
9	3	3	TGUYCA
9	8	8	GTRAC
9	11	2	AUATYT
9	12	6	AAYUTT
9	14	5	UTTAAY
10	1	7	TCGA
10	5	4	GASTC
10	9	7	CARTG
10	10	7	TGRCA
10	14	7	CTSAG
11	5	6	GARTC
11	8	8	YCATGU
11	11	3	GTAC
11	14	2	UATATY
11	18	8	TATA
11	19	22	GGCC
13	1	21	CGSCG
13	7	2	GTYUAC
13	8	9	TGSCA
14	0	23	CGCG
14	9	8	TCRGA
14	15	6	AASTT
14	15	12	ATSAT
14	16	7	AGSCT
14	19	8	CTNAG
14	23	21	GCRGC
15	11	7	TARTA
15	16	9	ATRAT

TABLE V (Continued)

∅X174	SV40	pBR322	SEQUENCE
		FREQUENCY	
15	25	14	ACNGT
16	9	8	TTSAA
16	31	18	CANTG
17	2	21	GCSGC
17	14	17	GTNAC
18	2	31	GCGC
18	9	15	TCNGA
18	36	21	TGCA
18	39	9	TTRAA
19	0	10	ACGT
20	15	8	TANTA
20	19	9	AGNCT
21	1	30	CGNCG
21	10	10	GANTC
21	22	9	ATAT
22	17	26	CATG
22	22	4	AARTT
23	18	16	TGNCA
24	35	16	AGCT
25	37	9	AATT
29	31	21	ATNAT
31	25	42	GCNGC
34	48	17	TTNAA
35	47	15	TTAA
36	37	10	AANTT

VII. TABLE VI

RESTRICTION ENDONUCLEASE SITE COMBINATIONS FOR LIGATION[1]

ENZYME 1	ENZYME 2	HYBRID SITE GENERAL	SPECIFIC	ENZYME 3
1 base 5'-extension				
Apy I	Apy I	CCRGG[2]		Apy I, Eco RII
	Fnu 4HI	CCRGC[3]		-
	Tth 3I	CCRNGTC[3]		-
			CCRGGTC	Apy I, Eco R II
Fnu 4HI	Apy I	GCRGG[3]		-
	Fnu 4HI	CGNGC[3]		Fnu 4HI
			GCRGC	Fnu 4HI, Bbv I
Tth 3I	Apy I	GACNRGG[3]		-
			GACCRGG	Apy I, Eco R II
	Fnu 4HI	GACNNGC[3]		-
	Tth 3I	GACNNNGTC[3]		Tth 3I
2 base 5'-extension				
Acc I	Acc I	GT$\left\{{C \atop A}\right\}\left\{{T \atop G}\right\}$AC[3]		Acc I
			GTCTAC	Acc I
			GTCGAC	Acc I, Hinc II, Sal I, Taq I
			GTATAC	Acc I
			GTAGAC	Acc I
	Acy I	GTCGYC[3]		-
	Asu II	GTCGAA[3]		Taq I
	Cla I	GTCGAT[3]		Taq I
	Hpa II	GTCGG[3]		-
	Taq I	GTCGA[3]		Taq I
Acy I	Acc I	GUCGAC[3]		-
	Acy I	GUCGYC		Acy I
			GACGCC	Acy I, Hga I
			GACGTC	Acy I
			GGCGTC	Acy I
			GGCGCC	Acy I, Hae II, Hha I
	Asu II	GUCGAA		-
	Cla I	GUCGAT		-
	Hpa II	GUCGG		-

TABLE VI (Continued)

ENZYME 1	ENZYME 2	HYBRID SITE GENERAL	SPECIFIC	ENZYME 3
	Taq I	GUCGA		–
Asu II	Acc I	TTCGAC[3]		Taq I
	Acy I	TTCGYC		–
	Asu II	TTCGAA		Asu II, Taq I
	Cla I	TTCGAT		Taq I
	Hpa II	TTCGG		–
	Taq I	TTCGA		Taq I
Cla I	Acc I	ATCGAC[3]		Taq I
	Acy I	ATCGYC		–
	Asu II	ATCGAA		Taq I
	Cla I	ATCGAT		Cla I, Taq I
	Hpa II	ATCGG		–
	Taq I	ATCGA		Taq I
Hpa II	Acc I	CCGAC[3]		–
	Acy I	CCGYC		–
	Asu II	CCGAA		–
	Cla I	CCGAT		–
	Hpa II	CCGG		Hpa II
	Taq I	CCGA		–
Taq I	Acc I	TCGAC[3]		Taq I
	Acy I	TCGYC		–
	Asu II	TCGAA		Taq I
	Cla I	TCGAT		Taq I
	Hpa II	TCGG		–
	Taq I	TCGA		Taq I
3 base 5'-extension				
Ava II	Ava II	GGRCC[2]		Ava II, Sau 96I
	Sau 96I	GGRCC[3]		Ava II, Sau 96I
Dde I	Dde I	CTNAG[3]		Dde I
Hinf I	Hinf I	GANTC[3]		Hinf I
Sau 96I	Ava II	GGRCC[3]		Ava II, Sau 96I
	Sau 96I	GGNCC[3]		Sau 96I

TABLE VI (Continued)

ENZYME 1	ENZYME 2	HYBRID SITE GENERAL	SPECIFIC	ENZYME 3
			GGRCC	Ava II, Sau 96I
			GGSCC	Sau 96I
4 base 5'-extension				
Ava I	Ava I	CYCGUG[3]		Ava I
			CCCGAG	Ava I
			CCCGGG	Ava I, Nci I, Hpa II, Sma I, Xma I
			CTCGAG	Ava I, Taq I, Xho I
			CTCGGG	Ava I
	Sal I	CTCGAC[3]		Taq I
	Xho I	CTCGAG[3]		Ava I, Taq I, Xho I
	Xma I	CCCGGG[3]		Ava I, Nci I, Hap II, Sma I, Xma I
Bam HI	Bam HI	GGATCC		Bam HI, Sau 3A, Xho II
	Bcl I	GGATCA		Sau 3A
	Bgl II	GGATCT		Sau 3A, Xho II
	Sau 3A	GGATCN		Sau 3A
			GGATCU	Sau 3A
			GGATCU	Bam HI, Sau 3A, Xho II
			GGATCT	Sau 3A, Xho II
	Xho II	GGATCY		Sau 3A, Xho II
			GGATCC	Bam HI, Sau 3A, Xho II
			GGATCT	Sau 3A, Xho II
Bcl I	Bam HI	TGATCC		Sau 3A
	Bcl I	TGATCA		Bcl I, Sau 3A
	Bgl II	TGATCT		Sau 3A
	Sau 3A	TGATCN		Sau 3A
			TGATCA	Bcl I, Sau 3A
			TGATCG	Sau 3A
			TGATCY	Sau 3A
	Xho II	TGATCY		Sau 3A
Bgl II	Bam HI	AGATCC		Sau 3A, Xho II
	Bcl I	AGATCA		Sau 3A
	Bgl II	AGATCT		Bgl II, Sau 3A, Xho II

TABLE VI (Continued)

| ENZYME 1 | ENZYME 2 | HYBRID SITE | | ENZYME 3 |
		GENERAL	SPECIFIC	
	Sau 3A	AGATGN		Sau 3A
			AGATCU	Sau 3A
			AGATCC	Sau 3A, Xho II
			AGATCT	Bgl II, Sau 3A, Xho II
	Xho II	AGATCY		Sau 3A, Xho II
			AGATCC	Sau 3A, Xho II
			AGATCT	Bgl II, Sau 3A, Xho II
Eco RI	Eco RI	GAATTC		Eco RI
Hind III	Hind III	AAGCTT		Alu I, Hind III
Sal I	Ava I	GTCGAG[3]		Taq I
	Sal I	GTCGAC		Hind II, Sal I, Taq I
	Xho I	GTCGAG		Taq I
Sau 3A	Bam HI	NGATCC		Sau 3A
			AGATCC	Sau 3A, Xho II
			GGATCC	Bam HI, Sau 3A, Xho II
			YGATCC	Sau 3A
	Bcl I	NGATCA		Sau 3A
			UGATCA	Sau 3A
			CGATCA	Sau 3A
			TGATCA	Bcl I, Sau 3A
	Bgl II	NGATCT		Sau 3A
			AGATCT	Bgl II, Sau 3A, Xho II
			GGATCT	Sau 3A, Xho II
			YGATCT	Sau 3A
	Sau 3A	GATC		Sau 3A
	Xho II	NGATCY		Sau 3A
			AGATCT	Bgl II, Sau 3A, Xho II
			AGATCC	Sau 3A, Xho II
			YGATCY	Sau 3A
			GGATCT	Sau 3A, Xho II
			GGATCC	Bam HI, Sau 3A, Xho II
Xba I	Xba I	TCTAGA		Xba I
Xho I	Ava I	CTCGAG[3]		Ava I, Taq I, Xho I

TABLE VI (Continued)

ENZYME 1	ENZYME 2	HYBRID SITE GENERAL	SPECIFIC	ENZYME 3
	Sal I	CTCGAC		Taq I
	Xho I	CTCGAG		Ava I, Taq I, Xho I
Xho II	Bam HI	UGATCC		Sau 3A, Xho II
			AGATCC	Sau 3A, Xho II
			GGATCC	Bam HI, Sau 3A, Xho II
	Bcl I	UGATCA		Sau 3A
	Bgl II	UGATCT		Sau 3A, Xho II
			AGATCT	Bgl II, Sau 3A, Xho II
			GGATCT	Sau 3A, Xho II
	Sau 3A	UGATCN		Sau 3A
			AGATCU	Sau 3A
			AGATCC	Sau 3A, Xho II
			AGATCT	Bgl II, Sau 3A, Xho II
			GGATCU	Sau 3A
			GGATCC	Bam HI, Sau 3A, Xho II
			GGATCT	Sau 3A, Xho II
	Xho II	UGATCY		Sau 3A, Xho II
			AGATCC	Sau 3A, Xho II
			AGATCT	Bgl II, Sau 3A, Xho II
			GGATCC	Bam HI, Sau 3A, Xho II
			GGATCT	Sau 3A, Xho II
Xma I	Ava I	CCCGGG[3]		Ava I, Hpa II, Nci I, Sma I, Xma I
	Xma I	CCCGGG		Ava I, Hpa II, Nci I, Sma I, Xma I
Xma III	Xma III	CGGCCG		Hae III, Xma III
5 base 5'-extension				
Bst EII	Bst EII	GGTNACC[3]		Bst EII
Eco RII	Eco RII	CCRGG		Apy I, Eco RII
Blunt-no extension				
Alu I	Alu I	AGCT		Alu I
	Pvu II	AGCTG		Alu I
Aos I	Aos I	TGCGCA		Aos I, Hha I
Bal I	Bal I	TGGCCA		Bal I, Hae III

TABLE VI (Continued)

ENZYME 1	ENZYME 2	HYBRID SITE GENERAL	SPECIFIC	ENZYME 3
	Hae III	TGGCC		Hae III
Dpn I	Dpn I	G^{Me}ATC		Dpn I, Sau 3A
Hae III	Bal I	GGCCA		Hae III
	Hae III	GGCC		Hae III
	Stu I	GGCCT		Hae III
Hind II	Hind II	GTYUAC		Hind II
			GTTAAC	Hind II, Hpa I
			GTCGAC	Acc I, Hind II, Sal I
	Hpa I	GTYAAC		Hind II
			GTTAAC	Hind II, Hpa I
Hpa I	Hind II	GTTUAC		Hind II
			GTTAAC	Hind II, Hpa I
	Hpa I	GTTAAC		Hind II, Hpa I
Pvu II	Alu I	CAGCT		Alu I
	Pvu II	CAGCTG		Alu I, Pvu II
Rru I	Rru I	AGTACT		Rru I, Rsa I
	Rsa I	AGTAC		Rsa I
Rsa I	Rru I	GTACT		Rsa I
	Rsa I	GTAC		Rsa I
Sma I	Sma I	CCCGGG		Ava I, Hpa II, Nci I, Sma I, Xma I
Stu I	Bal I	AGGCCA		Hae III
	Hae III	AGGCC		Hae III
	Stu I	AGGCCT		Hae III, Stu I
Tha I	Tha I	CGCG		Tha I
2 base 3'-extension				
Hha I	Hha I	GCGC		Hha I
			UGCGCY	Hae II, Hha I
Pvu I	Pvu I	CGATCG		Pvu I, Sau 3A
Sst II	Sst II	CCGCGG		Sst II, Tha I
3 base 3'-extension				
Bgl I	Bgl I	GCC(N)$_4$NGGC[4]		Bgl I
4 bp 3'-extension				
Hae II	Hae II	UGCGCY		Hae II, Hha I

TABLE VI (Continued)

ENZYME 1	ENZYME 2	HYBRID SITE GENERAL	SPECIFIC	ENZYME 3
Hgi AI	Hgi AI	GRGCRC[3]		Hgi Ai
			GAGCTC	Alu I, Hgi AI, Sst I
	Sst I	GAGCTC[3]		Alu I, Hgi AI, Sst I
Kpn I	Kpn I	GGTACC		Kpn I, Rsa I
Pst I	Pst I	CTGCAG		Pst I
Sph I	Sph I	GCATGC		Sph I
Sst I	Hgi AI	GAGCTC[3]		Alu I, Hgi AI, Sst I
	Sst I	GAGCTC		Alu I, Hgi AI, Sst I

[1]The enzyme combinations listed are those expected to generate ends which will ligate readily without additional modifications. Only those combinations of blunt-end producing enzymes which will provide a cleavable hybrid site are listed in this table. The "general hybrid" site is predicted from combining the two parental sequences. In some cases, the sequence is ambiguous due to sequence ambiguities (e.g., purine/pyrimidine "wobbles") in the parental sequence. If, however, the actual sequence at the restriction site is known (e.g., using a site in sequenced pBR322 DNA), then the ambiguous site can be resolved to the "specific hybrid" site. The predicted sequence after ligation of the 3'-hydroxyl acceptor (generated by Enzyme 1) with the 5'-phosphate donor (generated by Enzyme 2) is written left to right, 5'——>3'. Sequences complimentary to these are not listed. For some enzymes, more than one site would be possible. Generated sequences can be cleaved by Enzyme 3. The following characters and their substitutions are used in the sequences: (A, C, G, and T represent the four common deoxynucleotides) N = A, C, G or T; R = A or T; U = A or G; Y = C or T; and S = C or G.
[2]Only 50% of the possible combination of ends will form the hybrids; however, all formed will be cleaved.
[3]Only 25% of the possible combination of ends will form the hybrid; however, all formed will be cleaved.
[4]Only 1/64 of the possible combination of ends will form the hybrid; however, all formed will be cleaved.

REFERENCES

1. Arrand, J.R., Myers, P.A. and Roberts, R.J., unpublished observations.
2. Goff, S. and Rambach, A. (1978) Gene 3, 347-352, and unpublished observations.
3. Roberts, R.J., Myers, P.A., Morrison, A. and Murray, K. (1976) J. Mol. Biol. 102, 157-165.
4. Yang, R.C.A., Van DeVoorde, A. and Fiers, W. (1976) Eur. J. Biochem. 61, 119-138.
5. Takanami, M. (1973) FEBS Letters 34, 318-322.
6. Takanami, M. (1974) Methods in Molecular Biology 7, 113-133.
7. Sugisaki, H. and Takanami, K. (1973) Nature New Biol. 246, 138-140.
8. Allet, B. (1973) Biochemistry 12, 3972.
9. Danna, K. J., Sack, G.H. and Nathans, P. (1973) J. Mol. Biol. 78, 363-376.
10. Garfin, D.E. and Goodman, H.M. (1974) Biochem. Biophys. Res. Comm. 59, 108-116.
11. Gromkova, R. and Goodgal, S.H. (1972) J. Bact. 109, 987-992.
12. Murray, K. and Morrison, A., unpublished observations.
13. Sharp, P.A., Sugden, B. and Sambrook, J. (1973) Biochemistry 12, 3055-3063.
14. RajBhandary, U.L. and Baumstark, B., unpublished observations.
15. Roberts, R.J. and Myers, P.A., unpublished observations.
16. Zabeau, M., Green, R., Myers, P.A. and Roberts, R.J., unpublished observations.
17. Shibata, T., Ikawa, S., Kim, C. and Ando, T. (1976) J. Bact. 128, 473, 476.
18. Zabeau, M. and Roberts, R.J., unpublished observations.
19. Lui, A.C.P., McBride, B.C., Vovis, G.F. and Smith, M. (1979) Nucleic Acids Res. 6, 1-15
20. Olson, J.A., Myers, P.A. and Roberts, R.J., unpublished observations.
21. McConnell, D., Searcy, D. and Sutcliffe, G. (1978) Nucleic Acids Res. 5, 1729-1739.
22. Lacks, S. and Greenberg, W. (1975) J. Biol. Chem. 250, 4060-4072.
23. Lacks, S. and Greenberg, W. (1977) J. Mol. Biol. 114, 153-168.
24. Gelinas, R.E., Myers, P.A. and Roberts, R.J. (1977) J. Mol. Biol. 114, 169-180.
25. Bingham, A., Atkinson, A. and Darbyshire, J., unpublished observations.
26. Sussenbach, J.S., Monfoort, C.H., Schiphof, R. and Stubberingh, E.E. (1976) Nucleic Acids Res. 3, 3193-3202.
27. Roberts, R.J., Myers, P.A., Morrison, A. and Murray, K. (1976) J. Mol. Biol. 103, 199-208.
28. Subramanian, K.N., Weissman, S.M., Zain, B.S. and Roberts, R.J. (1977) J. Mol. Biol. 110, 297-317.
29. Smith, L., Blakesley, R.W. and Chirikjian, J.G., unpublished observations.
30. Van Montagu, M., unpublished observations.
31. Kiss, A., Sain, B., Csordas-Toth, E. and Venetianer, P. (1977) Gene 1, 323-329.
32. Bron, S., Murray, K. and Trautner, T.A. (1975) Mol. Gen. Genet. 143, 13-23.
33. Bron, S. and Murray, K. (1975) Mol. Gen. Genet. 143, 25-33.
34. Middleton, J.H., Edgell, M.H. and Hutchinson, C.A., III (1972) J. Virol. 10, 42-50.
35. Subramanian, K.N., Pan, J., Zain, B.S. and Weissman, S.M. (1974) Nucleic Acids Res. 1, 727-752.
36. Yang, R.C.A., Van DeVoorde, A. and Fiers, W. (1976) Eur. J. Biochem. 61, 101-117.
37. Gelinas, R.E., Myers, P.A. and Roberts, R.J., unpublished observations.
38. Wu, R., King, C. and Jay, E. (1978) Gene 4, 329-336.

39. Sato, S., Hutchison, C.A. and Harris, J.I. (1977) Proc. Nat. Acad. Sci. USA 74., 542-546.
40. LeBon, J., Kado, C., Rosenthal, L. and Chirikjian, J.G. (1978) Proc. Nat. Acad. Sci. USA 75, 4097-4101.
41. Roizes, G., Patillon, M. and Kovoor, A. (1977) FEBS Letters 82, 69-70.
42. Bigger, C.H., Murray, K. and Murray, N.E. (1973) Nature New Biology 244, 7-10.
43. Boyer, H.W., Chow, L.T., Dugaiczyk, A., Hedgpeth, J. and Goodman, H.M. (1973) Nature New Biology 244, 40-43.
44. Yoshimori, R.N., Ph.D., (1971) Thesis.
45. Yoshimori, R.N., Roulland-Dussoix, D., Goodman, H.M. and Boyer, H., unpublished observations.
46. Brown, N.L., Hutchison, C.A., III and Smith, M., J. Mol. Biol., in press.
47. Mann, M. and Smith, H.O. (1978) Gene 3, 97-112, and unpublished observations.
48. Hutchison, C.A. and Barrell, B.G., unpublished observations.
49. Middleton, J.H., Stankus, P.V., Edgell, M.H. and Hutchison, C.A., III, unpublished observations.
50. Kleid, D., Humayun, Z., Jeffrey, A. and Ptashne, A. (1976) Proc. Nat. Acad. Sci USA 73, 293-297.
51. Murray, K., Hughes, S.G., Brown, J.S. and Bruce, S. (1976) Biochem. J. 159, 317-322.
52. Fuchs, C., Rosenvold, E.C., Honigman, A. and Szybalski, W. (1978) Gene 4, 1-23.
53. Hughes, S.G., Bruce, T. and Murray, K. (1980) Biochem. J. 185, 59-63.
54. Endow, S.A. and Roberts, R.J., unpublished observations.
55. Old, R., Murray, K. and Roizes, G. (1975) J. Mol. Biol. 92, 331-339.
56. Roy, P.H. and Smith, H.O. (1973) J. Mol. Biol. 81, 427-444.
57. Roy, P.H. and Smith, H.O. (1973) J. Mol. Biol. 81, 445-459.
58. Pirrotta, V. (1976) Nucleic Acids Res. 3, 1747-1760.
59. Wilson, G.A. and Young, F.E. in Microbiology (1976), ed. D. Schlessinger, Amer. Soc. Microbiol., Washington, pp. 350-357.
60. Zain, B.S. and Roberts, R.J. (1977) J. Mol. Biol. 115, 248-255.
61. Murray, K., Morrison, A., Cooke, H.W. and Roberts, R.J., unpublished observations.
62. Van Montagu, M., Myers, P. and Roberts, R.J., unpublished observation.
63. Gingeras, T.R., Milazzo, J.P. and Roberts, R.J. (1978) Nucleic Acids Res. 5, 4105-4127.
64. Endown, S.A. and Roberts, R.J. (1977) J. Mol. Biol. 112, 521-529.
65. Greene, R. and Mulder, C., unpublished observations.
66. Gingeras, T.R., Myers, P.A., Olson, J.A., Hanberg, F.A. and Roberts, R.J. (1978) J. Mol. Biol. 118, 113-122.
67. Sims, J., unpublished observations.
68. Brown, N.L. and Smith, M. (1976) FEBS Letters 65, 284-287.
69. Smith, D.L., Blattern, F.R. and Davies, J. (1976) Nuc. Acid Res. 3, 343-353.
70. Chater, K. (1977) Nucleic Acids. Res. 4, 1989-1988.
71. Allet, B., Jeppesen, P.G.N., Katagiri, K.J. and Delius, H. (1973) Nature 241, 120-123.
72. Dugaiczyk, A., Hedgpeth, J., Boyer, H.W. and Goodman, H.M. (1974) Biochemistry 13, 503-512.
73. Greene, P.J., Betlach, M.C., Goodman, H.M. and Boyer, H.W. (1974) Methods Mol. Biol. 7, 87-111.
74. Hedgpeth, J., Goodman, H.M. and Boyer, H.W. (1972) Proc. Nat. Acad. Sci. USA 69, 3448-3452.
75. Morrow, J.F. and Berg, P. (1973) Proc. Nat. Acad. Sci. USA 69, 3365-3369.
76. Mulder, C. and Delius, H. (1973) Proc. Nat. Acad. Sci. USA 69, 3215-3219.

77. Petterson, U., Mulder, C., Delius, H. and Sharp, P.A. (1973) Proc. Nat. Acad. Sci. USA 70, 200-204.
78. Polisky, B., Greene, P., Garfin, D.E., McCarthy, B.J., Goodman, H.M. and Boyer, H.W. (1975) Proc. Nat. Acad. Sci. USA 72, 3310-3314.
79. Haggery, D.M. and Schleif, R.E. (1976) J. Virol. 18, 659-663.
80. Perricaudet, M. and Tiollais, P. (1975) FEBS Letters 56, 7-11.
81. Roberts, R.J., Wilson, G.A. and Young, F.E. (1977) Nature 265, 82-84.
82. Wilson, G.A. and Young, F.E. (1975) J. Mol. Biol. 97, 123-126.
83. Smith, L. and Chirikjian, J.G. (1979) J. Biol. Chem. 254, 1003-1006.
84. Shibata, T. and Ando, T. (1976) Biochem. Biophys. Acta 442, 184-196.
85. Catterall, J. and Welker, N. (1977) J. Bact. 129, 1110-1120.
86. Landy, A., Ruedisueli, E., Robinson, L., Foeller, C. and Ross, W. (1974) Biochemistry 13, 2134-2142.
87. Kelly, T.J., Jr. and Smith, H.O. (1970) J. Mol. Biol. 51, 393-409.
88. Smith, H.O. and Wilcox, K.W. (1970) J. Mol. Biol. 51, 379-391.
89. Hanberg, F., Myers, P.A. and Roberts, R.J., unpublished observations.
90. Tomassini, J., Roychoudhury, R., Wu, R. and Roberts, R.J. (1978), Nucleic.
91. Gelinas, R.E., Myers, P.A., Weiss, G.A., Roberts, R.J. and Murray, K. (1977) J. Mol. Biol. 114, 433-440.
92. Bingham, A.H.A., Atkinson, T., Sciady, D. and Roberts, R.J. (1978) Nucleic Acids Res. 5, 3457-3467.
93. Fishermen, J., Gingeras, T.R. and Roberts, R.J., unpublished observations.
94. Barrell, B.G. and Slocombe, R., unpublished observations.
95. Roberts, R.J., Breitmeyer, J.B., Tabachnik, N.F. and Myers, P.A. (1975) J. Mol. Biol. 91, 121-123.
96. Tu, C.P.D., Roychoudhury, R. and Wu, R. (1976) Biochem. Biophys. Res. Comm. 72, 355-362.
97. Takanami, M. and Kojo, H. (1973) FEBS Letters 29, 267-270.
98. Wilson, G.A. and Young, F.E., unpublished observations.
99. Murray, K., Brown, J.S. and Bruce, S.A., unpublished observations.
100. Shibata, T. and Ando, T. (1975) Mol. Gen. Genet. 138, 269-380.
101. Ikawa, S., Shibata, T. and Ando, T. (1976) J. Biochem. 80, 1457-1460.
102. Meagher, R.B., unpublished observations.
103. Hartmann, H. and Goebel, W. (1977) FEBS Letters 80, 285-287.
104. Eskin, B. and Linn, S. (1972) J. Biol. Chem. 247, 6176-6182.
105. Lautenberger, J.A. and Linn, S. (1972) J. Biol. Chem. 247, 6176-6182.
106. Smith, J.D., Arber, W. and Kuhnlein, V. (1972) J. Mol. Biol. 63, 1-8.
107. Van Ormondt, H., Lautenberger, J.A., Linn, S. and deWaard, A. (1973) FEBS Letters 33, 177-180.
108. Haberman, A., Heywood, J. and Meselson, M. (1972) Proc. Nat. Acad. Sci. USA 69, 3138-3141.
109. Meselson, M. and Yuan, R. (1968) Nature 217, 1110-1114.
110. Brockes, J.P. (1973) Biochem. J. 133, 629-633.
111. Brockes, J.P., Brown, P.R. and Murray, K. (1972) Biochem. J. 127, 1-10.
112. Brockes, J.P., Brown, P.R. and Murray, K. (1974) J. Mol. Biol. 88, 437-443.
113. Haberman, A. (1974) J. Mol. Biol. 89, 545-563.
114. Reiser, J. and Yuan, R. (1977) J. Biol Chem. 252, 451-456.
115. Morris, D.W. and Parish, J.H. (1976) Arch. Microbiol. 108, 227-230.
116. LeBon, J., LeBon, T., Blakesley, R.W. and Chirikjian, J.G., unpublished observations.
117. Sciaky, D. and Roberts, R.J., unpublished observations.
118. Torheim, B., unpublished observations.
119. Brown, N.L. and Smith, M. (1977) Proc. Nat. Acad. Sci. USA 74, 3213-3216.
120. Modrich, P. and Zabel, D. (1976) J. Biol. Chem. 251, 5866-5874.
121. Blakesley, R. W., Dodgson, J., Nes, I. and Wells, R. (1977) J. Biol. Chem. 252, 7300-7306.

122. Chirikjian, J., George, A. and Smith, L.A. (1978) Fred. Proc. 37, 1415.
123. Heininger, K., Horz, W. and Zachau, H.G. (1977) Gene 1, 291-303.
124. Clanton, D.J., Woodward, J.M. and Miller, R.V. (1978) J. Bact. 135, 270-273.
125. Arrand, J.R., Myers, P.A. and Roberts, R.J. (1978) J. Mol. Biol. 118, 127-135.
126. DiLauro, R., unpublished observations.
127. LeBon, J. and Chirikjian, J., unpublished observations.
128. Coll, E. and Chirikjian, J., unpublished observations.
129. Endow, S.A. (1977) J. Mol. Biol. 114, 441-450.
130. Sugisaki, H. (1978) Gene 3, 17-28.
131. Yang, R.C.A., Danna, K., Van De Voorde, A. and Fiers, W. (1975) Virology 68, 260-265.
132. Yang, R.C.A., Szostak, J. and Wu, R., unpublished observations.
133. Tanyashin, I., Li, L.I. Muizhnieks, I.O. and Baev, A.A. (1976) Dokl. Akad. Nauk. SSSR 231, 226-228.
134. Sussenbach, J.S., Steenbergh, P.H., Rost, J.A., van Leeuwen, W.J. and van Embden, J.D.A. (1978) Nucleic Acids Res. 5, 1153-1163.
135. Sutcliffe, J.G. and Church, G.M. (1978) Nucleic Acid Res. 5, 2313-2319.
136. Roizes, G., Nardeus, T.C. and Monier, R. (1979) FEBS Letters 104, 39-44.
137. Shimatake, H. and Rosenberg, M., unpublished observations.
138. Ravetch, J.V., Horiuchi, K. and Zinder, N.D. (1978) Proc. Natl. Acad. Sci. USA 75, 2266-2270.
139. Kan, N.C., Lautenberger, J.A., Edgell, M.H. and Hutchinson, C.A., III (1978) Fed. Proc. 37, 1499.
140. Duncan, C.H., Wilson, G.A. and Young, F.E. (1978) J. Bact. 134, 338-344.
141. Lynn, S.P., Cohen, L.K., Gardner, J.F. and Kaplan, S. (1979) J. Bact. 138, 505-509.
142. Bingham, A.H.A. and Darbyshire, J., unpublished observations.
143. LeBon, J.M. (1978) Fed. Proc. 37, 1413.
144. Mann, M.B. and Smith, H.O. (1977) Nucleic Acids Res. 4, 4211-4221.
145. Mayer, H., Grosschedi, R., Schutte, H. and Hobom, G., unpublished observations.
146. Wang, R.Y.H., Shedlarski, J.G., Farber, M.B., Kuebbing, D. and Ehrlich, M. (1980) Biochem. Biophys. Acta. 606, 371-385.
147. Takahashi, H., unpublished observations.
148. Takahashi, H., Shimizu, M., Saito, H., Ikeda, Y. and Sugisaki, H. (1979) Gene 5, 9-18.
149. Brown, N.L., McClelland, M. and Whitehead, P.R., (1980) Gene 9, 49-68.
150. Comb, D., Schildkraut, I. and Roberts, R.J., unpublished observations.
152. Lautenberger, J.A., Kan, N.C., Lackey, D., Linn, S., Edgell, M.H. and Hutchison, C.A., III (1978) Proc. Natl. Sci., USA 75, 2271-2275.
153. Walter, F., Hartman, M. and Roth, M. (1978) Abstracts of 12th FEBS Symposium, Dresden.
154. Rosenvold, E. and Szybalski, W., unpublished observations.
155. Leung, D.W., Lui, A.C.P., Meriless, A., McBride, B.C. and Smith, M. (1979) Nucleic Acids. Res. 6, 17-25.
156. Lee, Y.H. and Chirikjian, J.G. (1979) J. Biol. Chem. 254, 6838-6841.
157. Stubberingh, E.E., Schiphof, R. and Sussenbach, J.S. (1977) J. Bact. 131, 645-649.
158. Clanton, D.J., Riggsby, W.S. and Miller, R.V. (1979) J. Bact. 137, 1299-1307.
159. Stuy, J.H. (1978) Antonie van Leeuwenhoek 44, 367-376.
160. Takahashi, H., Shimotzu, H. and Saito, H., unpublished observations.
161. Roizes, G., Pages, M., Lecou, C., Patillon, M., and Kovoor, A. (1979) Gene 6, 43-50.
162. Makula, R.A. and Meagher, R. B. (1980) Nucleic Acids Res. 8, 3125.

163. Hu, A.W., Kuebbing, D. and Blakesley, R., unpublished observations.
164. Kuebbing, D. and Blakesley, R. (1979) Fed. Proc. 38, 780.
165. Lautenberger, J.A., White, C.T., Edgell, M.H. and Hutchison, C.A., III (1979) Fed. Proc. 38, 293.
166. Brown, N.L. (1979) XIth Int. Cong. Biochem. Abst., p. 44.
167. Timko, J., Zelinka, J. and Zelinkova, E. (1979) XIth Int. Cong. Biochem. Abst., p. 44.
168. deWaard, A., van Ormondt, H., Maat, J., van Beveren, C.P., Dijkema, R., Duyvesteyn, M., Koomey, J.M. and Mulder, C. (1979) XIth Int. Cong. Biochem. Abst. p. 45.
169. Lynn, S.P., Cohen, L.K., and Kaplan, S. Gardner, J.F., (1980) J. Bacti. 142, 380-383.
170. Kunkel, L.M. Silberklang, M., and McCarthy, B.J. (1979) J. Mol. Biol. 132, 133-139.
171. Bachi, B., Reiser, J. and Pirotta, V., (1979) J. Mol. Biol. 128, 143-163.
172. Stotz, A. and Phillippson, P., unpublished observations.
173. Tikchonenko, T.I., Karamov, E.V., Zavizion, B.A., and Naroditsky, B.S. (1978) Gene 4, 195-21.
174. Lautenberger, J.A., Edgell, M.H. and Hutchison, C.A., III, personal communications.
175. Mayer, H. (1978) FEBS Letters 90, 341-344.
176. Waalwijk, C. and Flavell, R.A. (1978) Nucleic Acids. Res. 5, 3231-3236.
177. Geier, G. and Modrich, P. (1979) J. Biol. Chem. 254, 1408-1413.
178. Baski, K. and Rushizky, G.W.V. (1979) Anal. Biochem. 99, 207-212.
179. deWaard, A., Korsuize, J., van Berern, C.P. and Maat, J. (1978) FEBS Lett. 96, 106-110.
180. deWaard, A., van Bevern, C.P., Duyvesteyn, M. and van Ormandt, H. (1979) FEBS Lett. 101, 71-76.
181. Fuchs, L.Y., Covarrubias, L., Escalante, L., Sanchez, S. and Bolivar, F. (1980) Gene 10 39-46.
182. Cizewski, V. and Kuebbing, D., unpublished results.
183. Ikawa, S., Shibata, T. and Ando, T. (1979) Agric. Biol. Chem. 43, 873-875.
184. Makino, O., Kawamura, F., Saito, H. and Ikeda, Y. (1979) Nature 277, 64-66.
185. Clarke, C.M. and Hartley, B.S. (1979) Biochem. 177, 49-62.
186. Hughes, S.G. and Hattman, S. (1975) J. Mol. Biol. 98, 645-647.
187. Smith, H.O. (1974) Methods in Molec. Biol. 7, 71-85.
188. Garfin, D.E. and Goodman, H.M. (1974) Biochem. Biophys. Res. Commun. 59, 108-116.
189. Pugatsch, T. and Weber, H. (1979) Nucleic Acids Res. 7, 1429-1444.
190. Koncz, C., Kiss, A. and Venetianer, P. (1978) Eur. J. Biochem. 89, 523-529.
191. George, J., Blakesley, R.W. and Chirikjian, J.G. (1980) J. Biol. Chem. 255, 6521-6524.
192. Blakesley, R.W., et al., unpublished observations.
193. Roberts, R.J. (1980) Nucleic Acids Res. 8.
194. Arber, W. (1979) Science 205, 361-365.
195. Nathans, D. (1979) Science 206, 903-909.
196. Smith, H.O. (1979) Science 205, 455-461.
197. Sutcliffe, J.G. (1979) Cold Spring Harbor Symp Quant. Biol. 43, 77-90.
198. Sanger, F., Coulson, A.R., Friedmann, T., Air, G.M., Barrell, B.G., Brown, N.L., Fiddes, J.C., Hutchison, C.A., III, Slocombe, P.M. and Smith, M. (1978) J. Mol. Biol. 125, 225-246.
199. Reddy, V.B., Thimmappaya, B., Dhar, R., Subramanian, K.N., Zain, B.S. Pan, J., Ghosh, P.K., Celma, M.L. and Weissman, S.M. (1978) Science 200, 494-502.
200. Blakesley, R.W., unpublished observations.
201. Fuchs, C., Rosenvold, E.C., Honigman, A. and Szybalski, W. (1978) Gene 4, 1-23.

202. McConnel, D.J., Searcy, D.G. and Sutcliffe, J.G. (1978) Nucleic Acids Res. 5, 1729-1739.
203. deWaard, A., personal communication.
204. Watson, R.J., Zuker, M., Martin, S.M. and Visentin, L.P. (1980) FEBS Lett. 118, 47-50 and unpulished observations.
205. Hinkle, N.F. and Miller, R.V. (1979) Plasmid 2, 387-393.
206. Hadi, S.M., Bachi, B., Shepard, J.C.W., Yuan, R.J., Neichen, K. and Bickle, T.A. (1979) J. Mol. Biol. 134, 655-666.
207. Rosenberg, M., unpublished observations.
208. Shinomiya, T. and Sato, S. (1980) Nucleic Acids Res. 8, 43-56.
209. Visentin, L.P. and Clanton, D., unpublished observations.
210. Venegas, A., Vicuna, R., Alonso, A., Valdes, F. and Yudelevich, A. (1980) FEBS Lett. 109, 156-158.
211. Duyvesteyn, M.G.C., deWaard, A. and vanOrmondt, H. (1980) FEBS Lett. 117, 241-246.
212. Blakesley, R.W., Tolstoshev, C.M., Nardone, G. and Chirikjian, J.G., unpublished observations.
213. Ferry, S. and Blakesley, R.W., upublished observations.
214. Duyvesteyn, M. and deWaard, A. (1980) FEBS Lett. 111, 423-426.
215. Shimotsu, H., Takahashi, H. and Saito, H. (1980) Gene 11, 219-225.
216. Mayer, H. and Hobom, G., personal communication.
217. Lin, B-C., Chien, M-C., and Lou, S-Y. (1980) Nucleic Acids Res. 8, 6189-6198.
218. Tolstoshev, C. and Blakesley, R., manuscript in preparation.

LAMBDA EXONUCLEASE

John W. Little

Department of Microbiology, Arizona Health Sciences Center
University of Arizona, Tucson, Arizona 85724

I. INTRODUCTION

Lambda exonuclease is a protein made early in the life cycle of bacteriophage lambda. It is encoded by the redA gene of the phage. As discussed below,
its biological function is to support generalized recombination. This review
focuses, however, on its use in vitro in the analysis and modification of DNA
structure.

Lambda exonuclease was discovered in 1962 by Korn and Weissbach,[1,2] who
showed that a new deoxyribonuclease activity appeared after phage infection or
induction. It was soon found that a defective lambda mutant, Tll, massively
overproduces lambda exonuclease,[3] and this strain was subsequently used as
a plentiful source of enzyme for large-scale purification.[4,5,6] The molecular
basis of the Tll defect has since been determined, and our understanding of
lambda gene regulation now allows the use of more favorable strains for enzyme
purification.

Two kinds of evidence suggest that lambda exonuclease is involved in genetic
recombination in vivo. First, phage mutants called "red" have been isolated;
they cannot support phage recombination in the absence of the host recombination
system controlled by the recA gene.[7] Two types of red mutations, redA and redB,
have been isolated; several redA mutations encode nonfunctional or thermosensi-
tive forms of lambda exonuclease.[8,9] Second, lambda exonuclease can catalyze
in vitro a reaction, strand assimilation, which is believed to be involved in
genetic recombination.[10]

It is interesting that the product of the redB gene, known simply as "beta",
copurifies with lambda exonuclease in at least one procedure,[11] suggesting that
the two proteins interact strongly in vitro. Beta protein by itself has not been
shown to have any in vitro function; moreover, preparations of exonuclease with
and without beta have only subtle differences in catalytic properties that do not
suggest an obvious role for beta in genetic recombination.[12,13] Conceivably it
could act to stabilize the exonuclease. In any case, the question as to the role
of beta in genetic recombination remains open and interesting, and its involvement suggests that our understanding of phage recombination is incomplete.

II. MODIFICATIONS OF PUBLISHED PURIFICATION PROCEDURES

Two similar purification procedures were developed independently and have
been described in considerable detail.[6,14] These procedures used the defective
lambda strain Tll. Tll has since been shown to be a mutant in the rightward
early promoter pR (see reference 15 for review). In consequence, a prophage

carrying a T11 mutation will be unable to express any genes under pR control. Among these genes is the cro gene, whose product partially represses expression of the leftward promoter pL. Exonuclease is expressed under pL control; in a wild-type infection, cro protein will largely repress pL expression, and the level of exonuclease will be low; late gene expression will also take place, leading within 60 min. to cell lysis. In contrast, in a host in which T11 is growing, pL expression is at a high rate, and the late genes are never turned on; in consequence, exonuclease accumulates, and the cells can be grown for several hours without lysing. In extracts from such cells, about 1% of the soluble protein is lambda exonuclease.

Experimentally, it is much more convenient to maintain the T11 strain as a prophage than to make phage stocks and infect cells with them. Formerly, it was necessary to induce the prophage by treatments such as UV irradiation or mitomycin C treatment. These treatments are now known to lead to destruction of prophage repressor by proteolytic cleavage.[16,17] It is easier, however, to use strains that make a thermolabile repressor, such as cI857; in this case, induction is achieved simply by raising the temperature.

The purification procedure previously described[5,6] had certain drawbacks, many of which can perhaps be avoided by using newer strains of lambda as a source of enzyme. Primary among the drawbacks was that the growth and induction of cells was a long procedure. This came about basically because the cells were induced by UV irradiation; for large-scale cultures, it was most convenient to use a medium transparent to UV light, and the cells grew slowly in this minimal medium. Moreover, because the UV source was within the fermentor chamber, the optimal dose had to be determined empirically. More recently, J. Carbon (personal communication) has used mitomycin C at 1 µg/ml to induce the prophage, although he followed the same regimen of growth in minimal medium, inducing treatment, and addition of bactotryptone at 1%. Yields of enzyme from this modified procedure were excellent. It would be of interest to know whether the procedure would work the same way if cells were grown in rich medium before mitomycin treatment as well.

An alternative to refining this procedure for growth of cells containing a T11 prophage would be to use our detailed understanding of lambda biology to modify the virus. Two changes can be made to implement this approach. The first is in the means of inactivating repressor. The standard method of inducing a lambda prophage has become thermal induction of a lysogen encoding a temperature-sensitive repressor; the allele in common use is cI857ts. Use of thermal induction would allow growth in a rich medium and a simple, direct, reliable, and inexpensive means of inactivating repressor. The second change stems from the nature of the defect in Tll strains. Tll is a mutation that blocks expression of cro from the lambda promoter pR[15]. Available mutations in cro would serve the same purpose. pR mutations also block phage replication, which requires the products of genes O and P. The ideal combination of genes would be a cro mutation with mutations in two other genes: a mutation in gene Q to block expression of late phage genes, and a mutation in gene S to block any residual tendency of the induced cells to lyse[18]. This cI857 cro Q S phage would be thermoinducible and, because it could replicate, it would make several hundred copies of the phage DNA after induction; the high gene dosage should yield high levels of exonuclease.

A multiply mutant strain of lambda that was developed for other purposes by Gottesman and Gottesman[19] has been used as a source of lambda exonuclease. This strain, SG5519, contains the thermoinducible cI allele cI857ts, the pR mutation x13, and the int29 allele, which blocks prophage excision. This last feature is useful, because it avoids a frequent problem observed with the Tll strain, namely, that induced cultures were eventually overgrown with cells cured of the prophage (J. Little, unpublished observations). Induced cultures of SG5519 were used as a source of exonuclease, and the following modifications were made in the published procedures (R.J. Roberts and T. Maniatis, personal communication):

1. Cells were grown in rich medium (4X YT, containing per liter 32 g of bactotryptone, 20 g of yeast extract, and 5 g of NaCl) at 32°C to a Klett reading of 115 (red filter); the temperature was raised to 44°C and after 20 min was lowered to 38°C. After 4 h more, cells were harvested. The crude extract was prepared immediately[6].

2. Upon streptomycin sulfate precipitation[6], the enzyme remained in the supernatant, in contrast to the result obtained with Tll lysogens induced with UV light[6,14] or mitomycin C (Carbon, personal communication) — namely, that the enzyme precipitated with the nucleic acids. Perhaps damaged DNA in the latter cases has an increased affinity for the enzyme.

3. Enzyme was then precipitated from the supernatant by the addition of an equal volume of saturated ammonium sulfate over a 30-min period followed by overnight stirring. The precipitate was recovered by centrifugation for 30 min at 12,000 g, resuspended in 0.05 M glycylglycine-NaOH (pH, 7.0), and dialyzed against a combination of 0.02 M Tris-HCl (pH, 7.4) and 0.15 M NaCl.

This material was further fractionated by chromatography on DEAE-cellulose and phosphocellulose by minor modifications of the published procedures[6,14]. Overall yields of enzyme from this procedure were comparable with those obtained by induction of Tll lysogens.

Two other aspects of my published procedure[6] deserve modification. First, cells were lysed soon after harvest, and the extract was treated promptly with streptomycin sulfate; although this was done to prevent possible changes in the nucleic acid substrate to which the enzyme bound during streptomycin precipitation, it made a long day indeed of the first day. It would be desirable to explore the possibility of freezing the cells at -70°C to counteract possible nuclease action. Such frozen cell pastes would probably be active and stable for long periods of time.

The second modification deals with the dialysis of the PEG phase after phase partition. This step was originally worked out on a small scale; when the preparation was scaled up, the dialysis was done in a large number of dialysis sacs whose geometry reproduced that of the small-scale prodcedure. The use of fewer sacs should be feasible when the dialysis conditions have been worked out[14]. If the purification procedure is altered at earlier stages, as described above, this step can be avoided entirely.*

III. CATALYTIC PROPERTIES

Lambda exonuclease has several distinctive catalytic properties that make it useful as a reagent for analyzing and modifying the structure of DNA.

A. Preference for double-stranded DNA The enzyme degrades native DNA at a rate at least 100 times faster than long single-stranded DNA[20]. It attacks oligonucleotides at a low rate[20]. Attack on short single-stranded DNAs has been observed in some studies to proceed at a rate up to 10% of that seen with native DNA[21]. The differences in rate of attack on single-stranded DNA as a function of chain length has been attributed[21] to the relative inaccessibility of the ends of a randomly-coiled macromolecule when the ends are in the center of the coil; the fraction of time when the ends are inaccessible increases with increasing chain length.

---

*I would appreciate receiving information relevant to purification of lambda exonuclease and will make such information available to interested investigators.

A dramatic example of the preference of the enzyme for double-stranded DNA is the comparison between its attack on mature lambda DNA, which has protruding 5' single-stranded ends, and on the same substrate with these ends filled in by treatment with DNA polymerase I. Attack on lambda DNA was initally blocked and then proceeded at about one-third the rate observed with the filled-in termini[20]. This finding suggests that attack on the 12-base-long cohesive ends took at least as much time as did digestion of the remaining portion, which is 50,000 base pairs long. Competition experiments[21] in which flush-ended molecules were added shortly after enzyme and lambda DNA had been mixed suggested that attack on the single-stranded lambda DNA cohesive ends is not processive[21], in contrast with the case of double-stranded DNA.

The short 5' protruding termini produced by many restriction endonucleases (for example, EcoRI and HpaII) are also attacked poorly, in comparison with termini with flush ends (such as HaeIII ends) or recessed 5' termini (such as PstI ends) (R.J. Roberts, personal communication); see Roberts[22] for specificities of restriction enzymes.

It is unclear at present whether the differences in rates of attack on substrates with various secondary structures reflect differences in K_m for binding to double- versus single-stranded DNA, or in the V_{max} of the rate of hydrolysis, or in the degree of processivity.

B. Attack from the 5' terminus   Lambda exonuclease attacks duplex DNA and single-stranded DNA exclusively from the 5' terminus[20,21].

C. Preference for a 5'-phosphoryl group over a 5'-hydroxyl group   The relative rates of these reactions have not been determined precisely, but the difference is at least tenfold[20].

D. Processivity   The processivity of the enzyme is very high--i.e., once bound to a particular DNA molecule, an enzyme molecule continues to digest that molecule in preference to falling off and attacking another substrate molecule[12,23]. Thus, in reactions containing an excess of 5' termini over active enzyme species, the enzyme will completely degrade some molecules of the population before attacking others at all. In contrast, under conditions of enzyme excess, attack on all molecules can be made approximately synchronous.

E. Inability to initiate hydrolysis at a nick or a gap   Unlike exonuclease III, for example, lambda exonuclease cannot start digesting duplex DNA at a nick (single-strand break)[12] or a gap[24].

F. Single-strand assimilation   Lambda exonuclease can degrade the 5'-terminated strand at a single-strand branch, converting this structure to a simple nick[10]. Such reactions are thought to be important in genetic recombination and are presumed to bear on the <u>in vivo</u> role of the enzyme[25].

IV. USE AS A REAGENT

A. <u>Determination of secondary structure at termini</u> The termini generated
by restriction-enzyme cleavage of duplex DNA have different susceptibilities to
lambda exonuclease (Roberts, personal communication). This property of the
enzyme should allow a rapid screening method for distinguishing protruding 5'
termini from the other types when new restriction enzymes are being characterized. It should also be useful in preparing templates for DNA sequencing.

B. <u>Discrimination between 5'-phosphoryl, 5'-hydroxyl, and blocked 5'termini</u>
Lambda exonuclease has been used to show that the 5' termini of some viral DNA
molecules, such as adenovirus 2 and <u>B</u>. <u>subtilis</u> phage phi29, are blocked and
inaccessible to enzyme attack[26-29]. Other types of studies show that these termini bear a covalently attached protein molecule. A necessary control in these
studies was to show that the 5' termini could not be phosphorylated by treatment
with polynucleotide kinase, and that kinase treatment did not convert the molecules into substrates for lambda exonuclease.

C. <u>Restriction mapping</u> The processivity of the enzyme and the fact that
attack can be made relatively synchronous under enzyme excess have permitted its
use in restriction mapping. If a long restriction fragment contains a number of
sites for another enzyme, treatment of that fragment with the second enzyme will
generate a set of fragments that form bands in a gel. Which of these fragments
lie closest to the termini can be determined by digesting the large fragment
with lambda exonuclease for increasing periods of time, followed by cleavage with
the second restriction enzyme and gel electrophoresis. Bands arising from terminal regions disappear first, followed progressively by internal fragments.
The inability of lambda exonuclease to attack at nicks helps to make this procedure useful. A recently described enzyme, BAL31, may be more suitable for this
purpose[30].

D. <u>Preparation of substrate for deoxynucleotidyl terminal transferase (TdT)</u>
TdT adds nucleotidyl residues to the 3' termini of DNA. Under standard reaction
conditions, this enzyme works more efficiently if the 3' terminus is single-
stranded. The first techniques developed for formation of recombinant DNA molecules[31,32] used limited digestion of initially duplex DNA by lambda exonuclease
to create protruding 3' termini. It was then found that, in the presence of Co^{2+}
ion, TdT adds nucleotides to the 3' terminus of flush-ended molecules[33]. Conditions for carrying out this reaction efficiently are given by Chang <u>et al</u>.[34].

Consequently it is no longer necessary to pretreat the DNA with lambda exonuclease. However, it has recently been found[35] that only a small proportion of the termini are tailed in the presence of Co^{2+} ion when lambda exonuclease is not used. Reference 35 describes conditions which use lambda exonuclease followed by TdT which result in greater than 90% of the molecules being tailed.

   E. Determination of direction of transcription Lambda exonuclease has been used (e.g. 36-38) to determine the direction of transcription of various cloned eukaryotic genes for which purified RNA products, such as ribosomal RNAs and histone mRNAs, are available. In general, restriction fragments thought to carry the gene in question are partially digested with lambda exonuclease, and the DNA is then hybridized with the purified RNA. Hybrids are detected either by electron microscopy or, more usually, by filter hybridization with labeled RNA species. If exonuclease treatment has digested all or part of the noncoding strand in the coding region, the DNA will hybridize with the RNA, even without prior denaturation of the DNA. If, in contrast, the exonuclease has digested the coding strand, the RNA will not hybridize, even after denaturation of the DNA. Proper interpretation of these experiments depends on knowing the locations of restriction sites relative to the gene in question.

   Most published experiments of this type have involved the use of both lambda exonuclease and, in separate reactions, E. coli exonuclease III, which digests duplex DNA from its 3' termini[39]. The two treatments should give complementary results, and in general this has been observed.

   The use of restriction enzymes that leave protruding 5' termini might lead to complications in the case of lambda exonuclease, because such ends are attacked relatively poorly. For example, if the protrusion were removed from one end before the other, that end would become a good substrate, and the strand whose 5' terminus was at that end might be degraded completely, or at least to an extent greater than 50% of its length. It is unclear from published reports whether such complications were encountered. Until detailed quantitative data are available for comparing attack on various termini at various temperatures, it will be difficult to assess this problem. One simple solution would be to fill in the ends with DNA polymerase before exonuclease digestion.

   F. Assessment of terminal redundancy in viral DNA molecules Many viral DNAs bear terminal redundancies at their two ends; that is, a short sequence at one end is present as a direct repeat at the other in the fashion abcd....xyzab. If such DNA is treated with lambda exonuclease or exonuclease III under conditions of limited digestion, the terminal redundancies are made single-stranded. Because the repeat is direct, the two terminal single-stranded regions have complementary sequences (e.g.,ab on one end and b'a' on the other) and can hybridize to one another, forming a circular molecule. This technique was devised by MacHattie et al. with exonuclease III[40] and has since been used with one or both enzymes by many workers, (e.g., references 41 and 42). Lambda exonuclease offers the advantage that it cannot attack at internal interruptions and so can be used on DNA that bears naturally-occurring nicks[41].

G. Preparation of template for DNA sequencing by chain-terminator methods
For chain-terminator sequencing[43], one wishes to obtain a template DNA that represents only one strand of the two originally present for a particular segment of DNA. Smith[44] has used digestion of restriction fragments with exonuclease III to prepare such templates. Lambda exonuclease has also been used in an analogous fashion (R.J. Roberts, personal communication). In both of these approaches, a fragment is degraded from both ends, so that each strand is reduced to 50% of its original length; consequently, half of the resulting template is derived from one of the strands, and half from the other.

I wish to propose a modification of this approach which should result in preparation of a full-length template strand. Two properties of lambda exonuclease should prove very useful in development of this method: first, the differential rates of attack on different types of termini, and second, the differential rates of attack on 5'-phosphoryl relative to 5'-hydroxyl termini. One strand can be preferentially degraded, if the strand one wishes to save has either a protruding 5' terminus or a 5'-hydroxyl group, whereas the strand to be degraded begins with a flush or recessed 5' terminus and a 5'-phosphoryl group. For example, if the restriction fragment has EcoRI and AluI ends, one would digest first with EcoRI, treat with alkaline phosphatase (if this proved necessary), digest with AluI, and isolate the fragment. Exonuclease digestion should then result in attack predominantly or exclusively from the AluI end.

Single-stranded DNA molecules prepared by this method should also be useful in preparing specific hybridization probes in studies of mRNAs with modifications of techniques described earlier.

H. Preparation of substrate for other enzymes DNA partially digested with lambda exonuclease or exonuclease III has been used to study the mechanism of unwinding of DNA by DNA helicases, or DNA unwinding enzymes[45]. It was found that DNA partially digested with exonuclease III, but not by lambda exonuclease, could be unwound by these enzymes. It was concluded that the enzymes bind to a single-stranded segment of DNA and then translocated in the 5' to 3' direction on that strand, displacing the complementary strand.

ACKNOWLEDGMENTS

I am grateful to R.J. Roberts, T. Maniatis, and J. Carbon for communication of unpublished information and for permission to cite their findings.

V. REFERENCES

1. Weissbach, A., and Korn, D. (1962) J. Biol. Chem. 237, PC3312-3314.
2. Korn, D., and Weissbach, A. (1963) J. Biol. Chem. 238, 3390-3394.
3. Radding, C.M. (1964) Proc. Natl. Acad. Sci. U.S. 52, 965-973.
4. Radding, C.M. (1966) J. Mol. Biol. 18, 235-250.
5. Little, J.W., Lehman, I.R., and Kaiser, A.D. (1967) J. Biol. Chem. 242, 672-678.
6. Little, J.W. (1967) In Methods in Enzymology, Vol. 12, eds. Grossman, L., and Moldave, K., Academic Press, New York, pp. 263-269.
7. Signer, E., and Weil, J. (1968) J. Mol. Biol. 34, 261-271.
8. Shulman, M.J., Hallick, L.M., Echols, H., and Signer, E.R. (1970) J. Mol. Biol. 52, 501-520.
9. Radding, C.M. (1970) J. Mol. Biol. 52, 491-499.
10. Cassuto, E., and Radding, C.M. (1971) Nature New Biol. 229, 13-16.
11. Radding, C.M., and Schreffler, D.C. (1966) J. Mol. Biol. 18, 251-261.
12. Carter, D.M., and Radding, C.M. (1971) J. Biol. Chem. 246, 2502-2512.
13. Radding, C.M. and Carter, D.M. (1971) J. Biol. Chem. 246, 2513-2518.
14. Radding, C.M. (1971) In Methods in Enzymology, Vol. 21, eds. Grossman, L., and Moldave, K., Academic Press, New York, pp. 273-280.
15. Herskowitz, I. (1973) Ann. Rev. Genetics 7, 289-324.
16. Roberts, J.W., and Roberts, C.W. (1975) Proc. Natl. Acad. Sci. U.S. 72, 147-151.
17. Roberts, J.W., Roberts, C.W., Craig, N.L., and Phizicky, E.M. (1978) Cold Spring Harbor Symp. Quant. Biol. 43, 917-920.
18. Moir, A., and Brammar, W.J. (1977) Mol. Gen. Genet. 149, 87-99.
19. Gottesman, S., and Gottesman, M. (1975) Proc. Natl. Acad. Sci. U.S. 72, 2188-2192.
20. Little, J.W. (1967) J. Biol. Chem. 242, 679-686.
21. Sriprakash, K.S., Lundh, N., Huh, M.M., and Radding, C.M. (1975) J. Biol. Chem. 250, 5438-5445.
22. Roberts, R.J. (1980) Nucleic Acids Res. 8, r63-r80.
23. Thomas, K.R., and Olivera, B.M. (1978) J. Biol. Chem. 253, 424-429.
24. Masamune, Y., Fleishman, R.A., and Richardson, C.C. (1971) J. Biol. Chem. 246, 2502-2510.
25. Meselson, M., and Radding, C.M. (1975) Proc. Natl. Acad. Sci. U.S. 72, 358-361.
26. Ito, J. (1978) J. Virol. 28, 895-904.
27. Yehle, C.O. (1978) J. Virol. 27, 776-783.
28. Carusi, E.A. (1977) Virology 76, 380-394.
29. Arrand, J.R., and Roberts, R.J. (1979) J. Mol. Biol. 128, 577-594.
30. Legerski, R.J., Hodnett, J.L., and Gray, H.J. Jr. (1978) Nucl. Acids Res. 5, 1445-1464.
31. Lobban, P.E., and Kaiser, A.D. (1973) J. Mol. Biol. 78, 453-471.
32. Jackson, D.A., Symons, R.H., and Berg, P. (1973) Proc. Natl. Acad. Sci. U.S. 69, 2904-2909.
33. Roychoudhury, R., Jay, E., and Wu, R. (1976) Nucl. Acids Res. 3, 863-877.
34. Chang, A.C.Y., Nunberg, J.H., Kaufman, R.J., Erlich, H.A., Schimke, R.T., and Cohen, S.N. (1978) Nature 275, 617-624.

35. Boseley, P.G., Moss, T., and Birnstiel, M.L. (1980) In Methods in Enzymology, Volume 65, eds. Grossman, L., and Moldave, K., Academic Press, New York, pp. 478-494.
36. Dawid, I.B., and Wellauer, P.K. (1976) Cell 8, 443-448.
37. Gross, K., Schaffner, W., Telford, J., and Birnstiel, M. (1976) Cell 8, 479-484.
38. Kulkas, C., and Dawid, I.B. (1976) Cell 9, 615-625.
39. Richardson, C.C., Lehman, I.R., and Kornberg, A. (1964) J. Biol. Chem. 239, 251-258.
40. MacHattie, L.A., Ritchie, D.A., Thomas, C.A. Jr., and Richardson, C.C. (1967) J. Mol. Biol. 23, 355-363.
41. Rhoades, M., and Rhoades, E.A. (1972) J. Mol. Biol. 69, 187-200.
42. Wadsworth, S., Hayward, G.S., and Roizman, B. (1976) J. Virol. 17, 503-512.
43. Sanger, F., Nicklen, S., and Coulson, A.R. (1977) Proc. Natl. Acad. Sci. U.S. 74, 5463-5467.
44. Smith, A.J.H. (1979) Nucl. Acids Res. 6, 831-848.
45. Kuhn, B., Abdel-Monem, M., Krell, H., and Hoffman-Berling, H. (1979) J. Biol. Chem. 254, 11343-11350.

EXONUCLEASE VII OF E. COLI

JOHN W. CHASE AND LYNNE D. VALES

The Department of Molecular Biology
Albert Einstein College of Medicine
Bronx, New York   10461

I. INTRODUCTION

E. coli exonuclease VII is a single-strand-specific deoxyribonuclease
which can hydrolyze denatured DNA, single-stranded regions extending from the
termini of duplex DNA, or displaced single-stranded regions[1,2]. Hydrolysis is
initiated at both 3' and 5' termini. The enzyme has no detectable activity on
RNA or DNA-RNA hybrid molecules. Exonuclease VII can also excise thymine dimers
from duplex DNA following incision near the damaged region by a UV endonuclease
(corendonuclease) specific for thymine dimers. The purified enzyme is active in
the presence of EDTA. Routine procedures have now been developed which utilize
exonuclease VII as an enzymatic reagent for a variety of applications requiring
the specific removal of single-stranded DNA termini from duplex molecules. The
in vitro biochemical properties of exonuclease VII suggest pathways in DNA replication, recombination and repair where the enzyme might function. Both genetic
and biochemical approaches are being used to investigate the role of the enzyme
in DNA metabolism.

II. ENZYMOLOGY OF EXONUCLEASE VII

A. Substrate Specificity

Studies with both natural and synthetic DNA and RNA molecules have demonstrated the absolute specificity of exonuclease VII for single-stranded DNA[1,2].
In Figure 1 are shown the substrates for exonuclease VII which have been well
characterized.

1. Single-stranded DNA (Fig. 1A). Exonuclease VII is able to degrade
sonically irradiated DNA to a limit product consisting of acid-soluble oligonucleotides. DNA structure has a striking effect on exonuclease VII activity.
For example, DNA which has been broken by sonic irradiation and then denatured

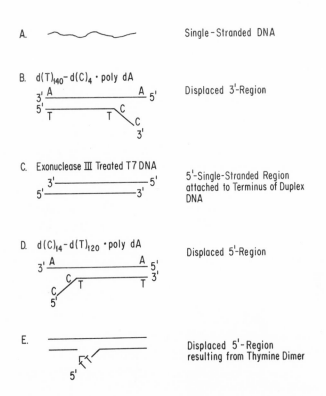

A.   Single-Stranded DNA

B. $d(T)_{140} - d(C)_4 \cdot$ poly dA   Displaced 3'-Region

C. Exonuclease III Treated T7 DNA   5'-Single-Stranded Region
attached to Terminus of Duplex
DNA

D. $d(C)_{14} - d(T)_{120} \cdot$ poly dA   Displaced 5'-Region

E.   Displaced 5'-Region
resulting from Thymine Dimer

<u>Fig. 1.</u> Substrates for exonuclease VII. Reprinted from Chase and Richardson[2].

is a more active substrate than DNA which has only been denatured. This suggests
that the enzyme is sensitive to the presence of any secondary structures or is
stimulated by the higher concentrations of termini present in the preparation of
broken DNA. Competition experiments have shown that duplex T7 DNA and poly(A)
do not compete with single-stranded DNA in the exonuclease reaction, suggesting
not only that duplex DNA and RNA are not substrates for exonuclease VII, but
that it does not even bind efficiently to these molecules. Homopolymer pairs
(including ribo-deoxyribo hybrids) have been found not to serve as substrates
for the enzyme. Exonuclease VII can degrade the duplex poly d(A-T) at a slow
rate, but the activity is dependent upon temperature, suggesting that "breathing" and "slipping" of the DNA produce single-stranded regions which can be
attacked by the enzyme. It has been shown that a 5'-phosphoryl group is not

150

required for exonuclease VII activity and that 5'-hydroxyl termini can be attacked as readily as 5'-phosphoryl termini. The question of whether a 3'-phos-
phoryl group would inhibit the 3' activity of exonuclease VII has not been
studied.

2. 3'-Exonuclease activity (Fig. 1B). A substrate containing a displaced
3'-single-stranded region was prepared by annealing $[^3H]d(T)_{140}-[^{32}P]d(C)_4$ to
poly(dA). Exonuclease VII was able to render 52% of the ^{32}P acid-soluble under
conditions where less than 0.5% of the ^3H was made acid-soluble indicating that
the enzyme is able to remove nucleotides from 3'-single-stranded termini.

3. 5'-Exonuclease activity (Fig. 1C,D). Exonuclease III-treated [^3H, 5'-
^{32}P] DNA (C) was prepared by shearing T7 [^3H] DNA by sonic irradiation and labeling the 5'-termini of these molecules using $[\gamma-^{32}P]ATP$ and polynucleotide
kinase. Limited treatment of these molecules with E. coli exonuclease III produces a substrate for exonuclease VII containing short 5'-single-stranded regions of DNA (10-30 nucleotides in length) extending from a duplex region. Exonuclease VII can remove these single-stranded regions leaving the duplex region
intact. The substrate shown in "D" is synthetic and was prepared in a manner
similar to that shown in "B" except it contains a displaced 5'-single-stranded
region. Exonuclease VII can also remove the displaced single-stranded region
from this molecule. These results demonstrate than exonuclease VII can hydrolyze DNA in a 5' ——> 3' direction.

The exonuclease III-treated T7 DNA substrate has proven extremely valuable
for assay of exonuclease VII. Since the enzyme acts processively (see below)
and does not appear to bind to duplex DNA it is possible to specifically examine the hydrolysis of the short single-stranded regions of the substrate. The
5' termini of the molecule can be labeled with $[\gamma-^{32}P]ATP$ of high specific activity. This, together with the ability to assay the enzyme in the presence of
EDTA, allow a sensitive and specific assay for exonuclease VII directly in crude
cell extracts.

4. 5'-Single-stranded region containing a thymine dimer displaced from a
duplex region (Fig. 1E). The introduction of a thymine dimer in a DNA molecule
causes distortion of that molecule in the region of the dimer. This distortion
is recognized by so-called UV endonucleases which incise or nick the DNA on the
5' side of the dimer. The distortion in the molecule is then relieved by the
displacement of a short single-stranded region containing the dimer. The resulting structure is thus similar to than of the displaced 5'-single-stranded
region in the synthetic substrate shown in "D". Exonuclease VII can recognize
this displaced region containing the dimer and catalyze its excision.

The ability of exonuclease VII to excise thymine dimers _in vitro_ (Fig. 1E) is an important property of the enzyme, since it suggests that it could function in UV repair. This possibility will be considered in detail in a later section.

B. Reaction Products

The initial products of the exonuclease VII reaction are large acid-insoluble oligonucleotides possibly 100 or more nucleotides in length. As the reaction progresses there is a shift in product size distribution. This observation and alkaline sucrose gradient sedimentation studies discussed below imply that the initial products of exonuclease VII hydrolysis are further degraded enzymatically to acid-soluble oligonucleotides. The limit products of exonuclease VII action on single-stranded DNA substrates and from substrates with long single-stranded regions extending from duplex DNA are oligonucleotides varying in length from 2 to greater than 25 nucleotides; most of these products, however, are in the range of tetramers to dodecamers. No mononucleotides have been detected during hydrolysis of any substrates tested to date. More extensive product characterization, including end-group analysis, has been published[2].

C. Physical Properties

Analysis by a combination of sucrose gradient sedimentation and gel filtration indicates that exonuclease VII has a sedimentation coefficient of 6.3S and a Stokes radius of 89Å[2]. From these parameters the native molecular weight of the enzyme is calculated to be 88,000 and the frictional coefficient 3.07. The magnitude of this frictional coefficient suggests that the molecule is extremely asymmetric and may therefore be represented as an ellipsoid with a large major semi-axis _i.e._, either a rigid rod (prolate ellipsoid) at the one extreme or a saucer or donut (oblate ellipsoid) at the other extreme. In either case the major semi-axis would be 50 to 100 times the length of the minor semi-axis. If a prolate ellipsoid model is assumed, it can be estimated that exonuclease VII could cover a piece of DNA more than 100 bases in length. If exonuclease VII attaches to the DNA and progresses along it releasing oligonucleotides (see below), the size of the molecule is consistent with the production of large oligonucleotide products.

D. Mechanism of Action

1. _Exonucleolytic mechanism of hydrolysis._ Although exonuclease VII hydrolyzes single-stranded DNA to yield products which are exclusively oligonucleo-

tides, the mechanism of hydrolysis is clearly exonucleolytic, since a DNA terminus is required[2]. In order to show this, 2 preparations of single-stranded ØX174 DNA were made, each labeled with different radioactive isotopes. One of these preparations was nicked with pancreatic DNase to produce linear molecules and then, after inactivation of the DNase, it was combined with the intact circular molecules and treated with exonuclease VII. The extent of the reaction was monitored by determining the amount of acid-soluble radioactivity at various times during the reaction. The integrity of the DNA molecules in the reaction mixture was monitored by sucrose gradient analysis. The circular DNA molecules remained intact, while, the linear molecules were totally degraded, proving that a DNA terminus is required in order for the enzyme to initiate hydrolysis, and therefore that the mechanism of hydrolysis of the enzyme is exonucleolytic. It has also been independently estimated that the total endonuclease activity on covalently-closed single-stranded DNA is 0.005% of the activity on linear single-stranded DNA.

2. _Processive mechanism of hydrolysis._ Analysis of the reaction products of denatured T7 DNA treated with exonuclease VII to Varying extents has suggested that the enzyme acts by a processive rather than a random mechanism[2]. Sedimentation analysis demonstrated that when exonuclease VII had degraded one-third of the T7 DNA, two-thirds remained intact and when two-thirds had been degraded, one-third remained high molecular weight. An exonuclease acting randomly should have partially degraded all of the molecules rather than selectively attacking a portion of the population. The observation that a significant fraction of the population apparently remains intact under these conditions is consistent with processive exonuclease action.

3. _Model for exonuclease VII hydrolysis._ The following model[2] for the action of exonuclease VII on single-stranded DNA (Fig. 2) has been suggested by the results of the _in vitro_ analysis of the enzyme: (a) Processive hydrolysis begins from either terminus of a single-stranded piece of DNA and progresses at approximately equal rates in both directions (Fig. 2, 1). (b) The initial products of the reaction are mainly large, acid-insoluble oligonucleotides (Fig. 2, 2). When enzyme molecules become free, many of the initial products can be further degraded (Fig. 2, 3). This process may be repeated many times until finally a limit digest containing mostly acid-soluble products is obtained (Fig. 2, 4).

1. Hydrolysis initiated at 5' and 3' termini.

2. Initial products are acid-soluble oligonucleotides (a) and acid insoluble oligonucleotides (b).

3. Acid-insoluble oligonucleotides further degraded.

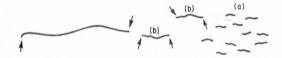

4. *Limit digest contains > 95 % acid-soluble products, $\bar{n}=6$.*

Fig. 2. Model for the hydrolysis of DNA by exonuclease VII. Reprinted from Chase and Richardson[2].

III. GENETICS OF EXONUCLEASE VII

A. Isolation of Mutant Strains

Strains deficient in exonuclease VII activity have been isolated by two procedures. In both cases use was made of the ability of exonuclease VII to be assayed in crude cell extracts. The activity of the enzyme in the presence of EDTA combined with its absolute specificity for single-stranded DNA allows the direct and specific determination of exonuclease VII in crude fractions. In addition, substrates labeled with ^{32}P can be employed so that acid-soluble radio-activity can be determined by autoradiography. In this way semi-automated mass screening can be performed allowing the assay of several hundred-to-a thousand extracts in a single day. Assays performed on approximately 5000 extracts of nitrosoguanidine mutagenized cells yielded 9 strains with reduced levels of exonuclease VII activity[3]. Several of these strains had no detectable exonuclease VII activity and one contained an activity which was temperature-sensitive. These strains have allowed the mapping of an exonuclease VII structural gene (xseA) near the gua operon.

A second approach to the isolation of exonuclease VII-deficient strains involved the isolation of deletion mutants[4]. Temperature-resistant derivatives were isolated at 43°C from a strain containing phage λc1857 inserted into the guaB gene. These strains were then tested for their ability to grow without guanine and all of those that required guanine were assayed for exonuclease VII activity. Three such strains contained no detectable exonuclease VII activity and were sensitive to phage λ infection, suggesting that excision of the prophage genome resulted in at least a partial deletion of the gene(s) involved in the production of exonuclease VII. Since excision of the prophage from the genome can result in chromosomal deletions extending in either direction from guaB, strains were also isolated exhibiting 6-azauracil and temperature resistance. Resistance to 6-azauracil suggests a defect in the upp gene. These strains were found to be defective in both the guaA and guaB genes as well as upp and to have wild-type levels of exonuclease VII activity. These results have made it possible to establish the position of the xseA gene with respect to nearby genetic loci (Fig. 3).

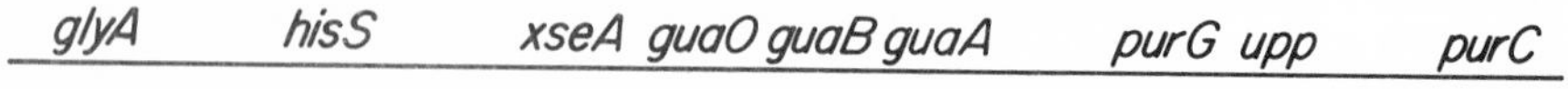

| glyA | hisS | xseA guaO guaB guaA | purG upp | purC |

Fig. 3. Genetic map of E. coli K12 in the region between glyA and purC. Reprinted from Vales, Chase and Murphy[4].

B. Properties of Mutant Strains

A variety of studies have now been performed in order to analyze the effects of a deficiency in exonuclease VII on the cell[3]. In general, the effect of a deficiency in exonuclease VII alone is not striking. The xseA⁻ strains are slightly more sensitive to ultraviolet irradiation than wild-type strains; bacteriophages T7, fd and λred appear to grow normally in these strains and temperature survival is similar to wild-type. The most significant defects so far observed due to a deficiency in exonuclease VII activity are the increased sensitivity to nalidixic acid and the hyper-rec character of the strains. The latter effect is similar to that observed in strains deficient in the 5 ──→ 3' exonuclease activity of DNA polymerase I (polAex⁻) and interestingly the effect is additive in strains with defects in both exonuclease VII and the 5' activity of DNA polymerase I. These double mutant strains have also been found to be more temperature-sensitive than the polAex⁻ mutant strains alone. Several enhanced effects of deficiencies in exonuclease VII and other 5' ──→ 3' exonuclease activities have now been observed. The results of these studies pertaining to DNA excision repair will be discussed in the next section. Although the hyper-rec character of xseA⁻ strains suggests some involvement of exonuclease VII in recombination we have observed no effect in Hfr mating with xseA⁻ recipient strains. The donor ability of xseA⁻ strains has not been investigated.

IV. In vivo FUNCTION OF EXONUCLEASE VII

A. DNA Repair

The ability of exonuclease VII to excise pyrimidine dimers in vitro suggests a role for the enzyme in DNA repair. A variety of studies have now been performed to evaluate the in vivo role of exonuclease VII in DNA excision repair[5,6]. It has been shown that mutant strains deficient only in exonuclease VII are as efficient as wild-type strains at excising pyrimidine dimers. However, the cell contains other exonuclease activities which are able to excise pyrimidine dimers in vitro and the possibility exists that due to the overlapping specificities of these nucleases a deficiency in one may not be detrimental to the cell. In order to evaluate this possibility the excision repair properties of strains deficient in exonuclease VII (xseA⁻), the 5' ──→ 3' exonuclease activity of DNA polymerase I (polAex⁻) and exonuclease V (recB⁻ recC⁻) have been examined. Although exonuclease V is not known to play a direct role in excision repair, it does contribute to post-irradiation DNA degradation which complicates the interpretation of dimer excision data. The results (Fig. 4) of these studies demonstrate that a recB⁻ recC⁻ polAex⁻ strain is only slightly reduced in its ability to remove thymine dimers compared to the wild-type. An additional deficiency in exonuclease VII, however, causes a significant deficiency in dimer excision compared to wild-type. Although excision is reduced in strains deficient in both exonuclease VII and the 5' ──→ 3' exonuclease of DNA polymerase I, it still occurs at a measurable rate. This may be due to residual enzyme levels of either or both of these activities (exonuclease VII deletion mutants have not yet been examined) or to other nucleases capable of dimer excision. The only other exonuclease activity of E. coli known to be capable of pyrimidine dimer excision in vitro, the 5' ——> 3' exonuclease of DNA polymerase III, has not yet been evaluated in vivo. The possibility also exists that as yet unidentified nucleases participate in the excision step.

This data must be interpreted in light of the known in vitro properties of exonucleases VI and VII and current models of DNA excision repair. It is known from in vitro studies that exonuclease VI (the 5' ——> 3' hydrolytic activity of DNA polymerase I) will only hydrolyze short, displaced single-stranded regions of DNA attached to a duplex region. It will not degrade such regions if they are long and it will not attack single-stranded DNA. Exonuclease VII, on the other hand, will only attack single-stranded DNA and appears to prefer to hydrolyze long, single-stranded regions of DNA displaced from a duplex region, although it will attack such regions if they are short but at a greatly reduced rate.

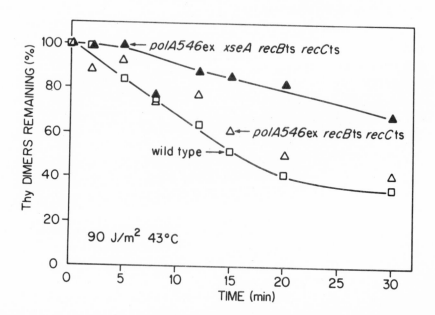

Fig. 4. Rate of dimer excision in exonuclease-deficient mutants. Cultures of E. coli strains KLC124: trpA33 rha⁻ thyA⁻; KLC333: polA546ex recB270 recC271 trpA33 thyA⁻; and KLC341: polA546ex recB270 recC271 xseA7 trpA33 thyA⁻ were grown at 32°C to mid-log phase in M9 medium supplemented with required nutrients and [³H] thymidine. The cells were chilled, harvested, and washed with M9 salts (without glucose or other additives) before irradiation with 90 J/m² UV light. The cells were warmed to 43°C for 5 min, nutrients were added (at t=0) and samples were withdrawn at the indicated times. Content of thymine dimers was determined as previously described[6]. At zero time 0.29% of the thymine was found as dimers. □= KLC124; △= KLC333; ▲=KLC341. Reprinted from Masker and Chase[7].

The three-dimensional structure of the damaged DNA region may well be more important to the mechanism of DNA repair than the exact type of damage. It seems reasonable to imagine that nucleases with a variety of specificities may be necessary to efficiently remove various types of damage from DNA. In the cases of damage resulting from UV-irradiation, low UV doses will introduce relatively few pyrimidine dimers into DNA even in pyrimidine-rich regions. The damaged regions of DNA to be excised will, therefore, be relatively short. High UV doses introduce large numbers of dimers into pyrimidine-rich regions of DNA, resulting in relatively long damaged regions that may be excised in a single step. Nucleases with different specificities may be called into play, depending simply on the length of the damaged region. In the case of damage resulting from different agents, other structural features may determine which nucleases are best suited to function most efficiently. Exonucleases VI and VII may therefore function in a coordinated fashion in removing damage produced by one type of agent and independently in the removal of other types of damage. Either enzyme may be capable of at least partially substituting for the other in a back-up capacity in the event one or the other becomes defective or inoperative.

B. DNA Replication and Recombination

The role of exonuclease VII in DNA replication and recombination has not yet been analyzed in detail, although the hyper-rec character of exonuclease VII-deficient strains and the increased effect in xseA⁻ polAex⁻ mutant strains suggests some involvement in recombination. As has proven to be the case in DNA repair, it may well be necessary to analyze multiple mutant strains in order to evaluate the pathways in which exonuclease VII may function. Now that deletion mutants of exonuclease VII are available this analysis may be more straightforward since it will be possible to utilize strains totally lacking exonuclease VII activity, thus eliminating effects of residual amounts of the enzyme.

V. ASSAY AND PURIFICATION OF EXONUCLEASE VII

The procedures presented below describe the most recent improvements developed in our laboratory for the assay and expecially for the purification of exonuclease VII since the initial publications[1,2]. Note particularly the use of a combination of Tris and phosphate buffers in the assay mixture. Care should also be taken to adjust the stock solution of EDTA to pH 7.9, since the pH optimum of exonuclease VII is rather sharp at pH 7.9.

A. Assay of Exonuclease VII

The standard reaction mixture (0.15 ml) contains 50 m$\underline{M}$ potassium phosphate buffer (pH 7.9), 50 m$\underline{M}$ Tris-HCl buffer (pH 7.9), 8.3 m$\underline{M}$ EDTA, 10 m$\underline{M}$ 2-mercaptoethanol, 1 nmole of sonically irradiated, denatured T7 [^{3}H] DNA, and enzyme. Incubations are for 30 min at 37°C and are stopped by the addition of 0.05 ml of 2.5 mg per ml of salmon sperm DNA and 0.1 ml of 20% (w/v) trichloroacetic acid. After 5 min at 0°C the solution is centrifuged for 10 min at 9000 rpm in the Sorvall SE-12 rotor. All of the supernatant fluid is removed and the radioactivity is determined by counting in 10 ml of hydrofluor scintillation fluid. The enzyme is routinely diluted with a solution containing 0.5 mg per ml of bovine serum albumin, 10 m$\underline{M}$ 2-mercaptoethanol, and 50 m$\underline{M}$ Tris-HCl buffer (pH 7.9). One unit of exonuclease VII activity is defined as the amount causing the production of 1.0 nmole of acid-soluble nucleotide in 30 min at 37°C. The activity is proportional to enzyme concentration at levels of 0.05 to 0.5 unit of enzyme.

B. Purification of Exonuclease VII

A sample purification of exonuclease VII from 500 gm of cells is described below and summarized in Table 1. The preparation can be proportionately increased or decreased. Unless otherwise indicated, all operations are performed at 4°C. The pH measurements of Tris buffers are made at a concentration of 0.05 $\underline{M}$ at room temperature. Imidazole buffer solution is adjusted to pH 6.5 at a concentration of 2 $\underline{M}$ at room termperature so that a 50 m$\underline{M}$ solution will measure pH 6.8 at 4°C.

Growth of cells. Exonuclease VII is routinely isolated from Escherichia coli HMS137, an sbcB$^-$ derivative of JG138 thy$^-$ rha$^-$ polAl lacZ $_{am}$Str^r. This strain contains a deletion of the sbcB region of the genome rendering the cells exonuclease I-negative and thereby eliminating a major contaminant of exonuclease VII preparations. HMS137 is grown at 37°C under forced aeration in a Fermocell (New Brunswick Scientific) in 100 liter of L-broth supplemented with 0.1% glucose, 20 µg per ml of thymine and 0.01 $\underline{M}$ potassium phosphate buffer (pH 7.4). The pH is maintained by the addition of 50% NaOH. Cells are harvested at OD$_{590}$=6 and stored at -80°C. Approximately 600 to 700 gm cell paste is obtained from each fermentation.

TABLE 1

PURIFICATION OF EXONUCLEASE VII

	Fraction	Volume	Activity	Protein	Specific activity	Yield
		ml	units x 10^{-3}	mg/ml	units/mg protein	%
I.	Extract	2030	557	20	13.7	100
II.	Streptomycin pellet	2000	506	4.4	57.5	91
III.	Polyethylene glycol phase	2600	449	-	-	81
IV.	Precipitate after dialysis	480	446	2.4	387	80
V.	Acetone fractionation	65	240	1.5	3,066	50
VI.	DEAE-cellulose	140	179	0.225	6,311	32
VII.	Acetone concentration	40	181	0.5	9,072	32
VIII.	Bio-Rex 70	82	155	0.086	21,965	28

Preparation of extract. Frozen cells (500 g) are suspended in 2 liters of 50 mM Tris-HCl buffer (pH 8.0), 10 mM 2-mercaptoethanol, and 0.1 mM EDTA (buffer A). Cells are broken by sonic irradiation in a Heat Systems Model W-375 sonic oscillator at an output setting of 8 with a 50% duty cycle for a total of 40 min per 500 ml of suspension. The extract is centrifuged for 25 min at 9000 rpm in the Sorvall GS3 rotor to remove debris. The supernatant is removed and recentrifuged for 15 min at 9000 rpm. The supernatant is Fraction 1.

Streptomycin precipitation. To 2030 ml of Fraction I, 570 ml of 5% (w/v) streptomycin sulfate is added slowly with stirring over a 30 min period (final concentration 1.1% (w/v)). The suspension is allowed to stir for 30 min after the last addition. The solution is centrifuged for 30 min at 9000 rpm in the Sorvall GS3 rotor. The pellet is redissolved in 2000 ml of buffer A containing 2 M NaCl by stirring for 2.5 hr. The suspension is centrifuged for 10 min at 9000 rpm in the Sorvall GS3 rotor to remove undissolved material. The supernatant is Fraction II.

Polyethylene glycol-Dextran 500 phase partition. To 2000 ml of Fraction II, 576 g of solid NaCl, 645 ml of 30% (w/w) polyethylene glycol (carbowax 6000) solution in water, and 230 ml of 20% (w/w) Dextran 500 (Pharmacia) solution in water are added with stirring (per ml of Fraction II, 0.288 g of NaCl, 0.322 ml of 30% (w/w) polyethylene glycol solution and 0.115 ml of 20% (w/w) Dextran 500 solution are added). The suspension is allowed to stir for 2 hr and is then centrifuged for 20 min at 9000 rpm in the Sorvall GS3 rotor. The upper clear phase (polyethylene glycol + protein) is Fraction III. The lower turbid phase (Dextran + nucleic acids) is discarded.

Dialysis of the polyethylene glycol phase and precipitate formation. Fraction III is dialyzed in approximately 70 ml portions in dialysis tubing with a diameter of 1.6 cm against 110 liters of buffer A overnight or approximately 10 hr. Under these conditions dialysis results in the formation of a heavy white precipitate containing the enzyme. The dialyzed fractions are pooled and the dialysis bags are washed with about 5 to 10 ml of buffer A per bag to maximize recovery of the precipitate. The pooled fractions are centrifuged for 30 min at 9000 rpm in the Sorvall GS3 rotor. The pellet is redissolved in 500 ml of 20 mM Tris-HCl buffer (pH 7.9), 10 mM 2-mercaptoethanol, 0.1 mM EDTA, 10% (w/v) glycerol and 0.2 M NaCl. The pellet is resuspended with the aid of a Dounce homogenizer and the suspension is allowed to stir for 10 min. The suspension is clarified without significant loss of exonuclease VII activity by centrifugation for 10 min at 9000 rpm in the Sorvall GS3 rotor. The supernatant is Fraction IV.

Acetone fractionation. To 480 ml of Fraction IV, 10 ml of 1 M sodium acetate buffer (pH 5.9) is added at +4°C, followed by the addition of 160 ml of acetone at -20°C with stirring (final concentration 25% (v/v)). Immediately after the acetone addition the suspension is centrifuged for 25 min at 9000 rpm in the Sorvall GS3 rotor. To the supernatant 72 ml of acetone (-20°C) is added (final concentration 32% (v/v)) with stirring. Immediately after the acetone addition the suspension is centrifuged for 25 min at 9000 rpm in the Sorvall GS3 rotor. Exonuclease VII activity precipitates between 25 and 32% (v/v) acetone. The residual acetone is removed from the pellet under vacuum in a desiccator for approximately 3 to 5 min or until acetone is no longer detected. The pellet is then redissolved with the aid of an homogenizer in 65 ml of 20 mM Tris-HCl buffer (pH 8.0), 10 mM 2-mercaptoethanol, 0.1 mM EDTA, 10% (w/v) glycerol (buffer B) containing 0.2 M NaCl. The suspension is clarified by centrifugation for 10 min at 10,000 rpm in the Sorvall SS34 rotor. The supernatant is Fraction V.

DEAE-cellulose chromatography. A column of Whatman DE-52 (2 cm^2 x 27.5 cm) is prepared and washed with 600 ml of buffer B. Fraction V (98 mg of protein) is diluted 1:1.5 with 100 ml of buffer B (final concentration of NaCl is approximately 0.1 M) and is applied to the column. The resin is washed with 110 ml of buffer B containing 0.075 $\underline{M}$ NaCl. The enzyme activity is eluted with a 500 ml linear gradient from 0.075 to 0.3 $\underline{M}$ NaCl in buffer B. The activity elutes at about 0.15 $\underline{M}$ NaCl. Fractions of 10 ml are collected and those containing the major portion of the enzyme activity are pooled (Fraction VI).

Acetone precipitation. To 140 ml of Fraction VI, 2.9 ml of 1 $\underline{M}$ sodium acetate buffer (pH 5.9) is added followed by 94 ml of acetone (-20°C) with stirring (final concentration 40% (v/v)). Immediately after the acetone addition the suspension is centrifuged for 25 min at 9000 rpm in the Sorvall GSA rotor. The pellet is briefly desiccated and then resuspended with the aid of an homogenizer in 40 ml of 50 m$\underline{M}$ imidazole buffer (pH 6.8), 10 m$\underline{M}$ 2-mercaptoethanol, 0.1 m$\underline{M}$ EDTA, 10% (w/v) glycerol (buffer C) containing 0.2 $\underline{M}$ NaCl. The suspension is clarified by centrifugation for 10 min at 10,000 rpm in the Sorvall SS34 rotor. The supernatant is Fraction VII.

Bio-Rex 70 chromatography. A column of Bio-Rad Bio-Rex 70 (2 cm^2 x 12.5 cm) is prepared and washed with 300 ml of buffer C. Fraction VII (20 mg of protein) is applied to the column and the resin is washed with 50 ml of buffer C containing 0.2 $\underline{M}$ NaCl. The enzyme activity is eluted with a 350 ml linear gradient from 0.2 to 1.0 $\underline{M}$ NaCl in buffer C. The major portion of the enzyme activity elutes at about 0.35 $\underline{M}$ NaCl. Fractions of 5 ml are collected and those containing the major portion of the enzyme activity are pooled (Fraction VIII).

Concentration and storage of exonuclease VII. Of a variety of methods tested for the concentration of exonuclease VII from fractions of low protein concentration (including ammonium sulfate and acetone) only absorption and elution from a small DEAE-cellulose column has proven to be reliable. Fraction VIII is dialyzed against 4 liters of 0.075 $\underline{M}$ NaCl in buffer B for 4 hr with one change. The enzyme is applied to a small DEAE-cellulose column (0.8 cm^2 x 2.5 cm) that is prepared and washed with buffer B. The column is then washed with 5 ml of 0.075 $\underline{M}$ NaCl in buffer B and the enzyme is eluted with 0.3 $\underline{M}$ NaCl in the same buffer. Fractions of 1 ml are collected and those containing the major portion of the enzyme activity are pooled. The enzyme is routinely stored at -20°C in this elution buffer made 50% with glycerol. Under these storage conditions no loss in activity (less than 10%) has been detected over a period of 18 months. Preparations of exonuclease VII should not be frozen since freezing and thawing may result in loss of enzymatic activity.

VI. APPLICATIONS OF EXONUCLEASE VII TO THE ANALYSIS OF NUCLEIC ACID STRUCTURE

As predicted after the initial characterization of the _in vitro_ properties of exonuclease VII, the enzyme has proven to be an extremely useful enzymatic reagent. The ability of the enzyme to act exonucleolytically and bidirectionally on single-stranded DNA alone, leaving circular DNA including "hairpins" or "snapbacks" intact, has allowed many applications of exonuclease VII to the analysis and modification of various DNA structures. Current applications of exonuclease VII can be divided into at least 3 categories: (1) analysis of the products obtained after digestion of DNA and DNA-RNA hybrid molecules with exonuclease VII alone or with a combination of nucleases including exonuclease VII has been used to structurally analyze the original molecules. (2) Cloned DNA segments have been excised and purified from their vectors with exonuclease VII exploiting the inability of the enzyme to attack "snapback" molecules. (3) Exonuclease VII alone or in combination with other single-stranded specific nucleases has been used to remove free single-stranded DNA from preparations of various molecules.

Since exonuclease VII is not yet readily available, it is worth noting that the apparent activity of the enzyme measured as a function of acid-soluble radioactivity is often an overestimate of the actual nucleolytic activity required for many applications. This is apparent when the processive hydrolytic mechanism of the enzyme is considered along with the size of the initial products of hydrolysis. Thus the required modification of the substrate may have been accomplished but not be directly assayable by means of acid-soluble radioactivity. Another factor that should be considered in determining the quantity of exonuclease VII needed for a particular application is the structure of the molecule to be modified (particularly the length of the single-stranded regions that must be removed), since DNA structure markedly affects the activity of the enzyme (see Section II above).

Specific examples of the applications of exonuclease VII will now be described.

A. Structural Analysis of DNA

A variety of DNA exo- and endonucleases are known with such a variety of specificities that the analysis of the products obtained after treatment of a DNA molecule with one or more of them can yield a great deal of information about the structure of the original molecule. The reactions may be sequential or in combination and the molecules to be analyzed may themselves be the products of other reactions. Exonuclease VII has now been added to this arsenal of nucleolytic reagents. Its exonucleolytic and bidirectional mechanism of action

makes it ideally suited for many applications. Two examples will serve to illustrate some of the many possibilities.

MacKay and Linn[8] were able to prepose a mechanism of degradation of duplex
DNA by the recBC enzyme (exonuclease V) partially as a result of structural
studies of its reaction intermediates. Reaction intermediates formed during
digestion of T7 DNA were isolated from a preparative neutral sucrose gradient
and banded in $Hg(II)$-Cs_2SO_4 density gradients. The presence of single-stranded
regions in the intermediates was confirmed by the partial susceptibility of the
material to endonuclease S1 which is specific for single-stranded DNA. The intermediates were next treated with a combination of exonuclease VII and exonuclease I, a 3' ⟶ 5' single-stranded specific exonuclease. The extent of
degradation using exonucleases I and VII concomitantly was found to be similar
to that using endonuclease S1 alone, demonstrating that all of the single-
stranded regions were external. Finally, it was shown that the extent of degradation by exonuclease I alone was approximately one-half that of exonucleases I
and VII used concomitantly, thus demonstrating not only that both 3' and 5'
single-stranded regions were present but that they were nearly equal in length.
Based on these observations MacKay and Linn were able to conclude that the recBC
DNase reaction intermediates are duplex molecules containing nearly equal 3' and
5' single-stranded termini. While a complete discussion of the evidence incorporated in the model proposed by these investigators for the mechanism of degradation of duplex DNA by the recBC nuclease is beyond the scope of this review,
the structural analysis of the reaction intermediates was instrumental to the
model suggested.

Exonuclease VII has also been used to characterize DNA used in cloning.
In a study by Nisen et al.[9] involving the cloning of C. crescentus inverted repeat (IR) DNA, exonuclease VII was able to digest all but 7.4% of denatured and
rapidly cooled chromosomal DNA. This undigested DNA represents both the double-
stranded IR DNA stem and the intervening single-stranded loops. It was also
found that approximately 3% of this same DNA was resistant to hydrolysis by
endonuclease S1. It was therefore possible to conclude that approximately 4.4%
of the C. crescentus chromosome is contained between IR DNA sequences. The
inverted repeat DNA prepared after endonuclease S1 hydrolysis was then cloned
in a phage λ vector.

B. Analysis of Spliced RNAs

A technique developed by Berk and Sharp[10] utilizes the different substrate
specificities exhibited by exonuclease VII and endonuclease S1 in order to ana-

lyze the structure of spliced RNA molecules. Such RNA molecules are composed
of at least 2 transcripts joined together; however, the genome sequence between
the transcripts is not represented. The study by Berk and Sharp utilized early
SV40 mRNAs and restriction endonuclease-treated SV40 DNA in order to determine
the structure of the mRNAs coding for the large and small T antigens. In principle the technique is generally applicable to the analysis of any RNA molecule.
In the example outlined in Fig. 5, RNA is first hybridized to ^{32}P-labeled DNA.

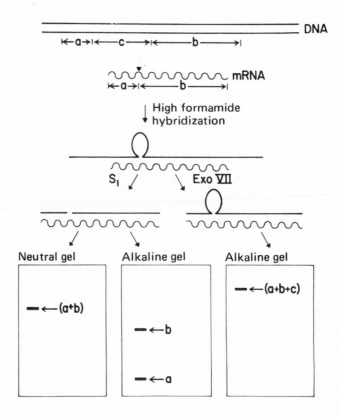

Fig. 5. Strategy for the analysis of spliced mRNA structure by gel electrophoresis of endonuclease S1 and exonuclease VII-digested RNA-DNA hybrids. Reprinted from Berk and Sharp,[10] with permission.

If RNAs are spliced, hybrid structures result containing "loops" of non-hybridized single-stranded DNA at splice points which result from DNA sequences not
represented in the RNA molecule. In addition, the hybrid duplex is flanked by
single-stranded DNA. Treatment of these hybrid molecules with the single-
strand-specific endonuclease S1 results in hydrolysis of all single-stranded
regions - both internal and external - leaving a fully duplex structure but containing interruptions in the DNA at the splice points. Electrophoretic analysis
under neutral conditions reveals a single band, while analysis under alkaline
conditions reveals 2 bands representing the single-stranded DNA fragments.

After analysis with endonuclease S1 the original hybrid molecules are next
analyzed by digestion with exonuclease VII. This treatment removes the flanking
single-stranded regions which extend beyond the 5' and 3' termini of the RNA but
leaves the single-stranded DNA loops at the splice points intact. Analysis of
the product of this reaction by alkaline gel electrophoresis reveals a single
DNA band which represents the total length of the genome between the 5' and 3'
sequences of the RNA. Comparison of the lengths of the sequence (a+b) determined after endonuclease S1 digestion with the length of the sequence (a+b+c)
determined after exonuclease VII digestion allows an estimation of the genome
sequence (c) between RNA transcripts of a and b. A similar application of this
technique to the mapping of adenovirus-associated virus type 2-specific RNAs
allowed Green and Roeder[11,12] to determine that one of the RNAs is spliced containing a 5' terminal "leader" sequence encoded by a non-contiguous region of
the viral genome.

This technique can be used as a means of determining whether an RNA molecule has been produced by splicing as well as the structure of a molecule already known to be spliced. If a hybrid DNA-RNA molecule does not yield identical products upon differential treatment with exonuclease VII and endonuclease
S1, it is likely that the RNA molecule is the product of intramolecular splicing.

C. Excision of DNA Segments from Cloning Vectors

Exonuclease VII has also been applied to recombinant DNA technology in
order to selectively excise specific DNA sequences introduced into the cloning
vector by the poly(dA dT) joining method. It is often desirable to reisolate a
DNA insert from a cloning vector in order to further study the DNA fragment or
to place it into another vector. Recombinant DNA molecules that are formed by
ligation of restriction endonuclease-derived cohesive termini can be utilized to
regain the DNA insert from the vector DNA by cleavage with the same restriction
endonuclease. However, DNA recombinants that have been annealed at synthetic

166

complementary 3'-polynucleotide termini, such as poly(dA) and poly(dT), cannot easily provide donors from which the original DNA insert can be reisolated. Goff and Berg[13] have overcome this limitation by altering the structure of the recombinant DNA molecules such that the insert DNA segment to be retrieved is resistant to exonuclease VII attack, while the vector DNA becomes a suitable substrate for the enzyme. In the example shown (Fig. 6), plasmid pTK1 DNA was cleaved with restriction endonuclease SmaI to produce linear duplex molecules.

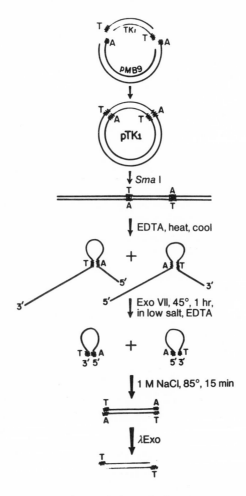

Fig. 6. Overall scheme for the excision of cloned DNA segments from recombinant DNA molecules constructed by the poly(dA·dT) joining method. Reprinted from Goff and Berg,[13] with permission.

In the presence of EDTA to chelate divalent cations and thereby promote formation of the subsequent structure, these molecules were heat denatured and then rapidly cooled, forming single-stranded "snap-backs" with pairing at the poly(dA) and poly(dT) sequences. In this case the insert DNA comprises a single-stranded hairpin loop with vector DNA as single-stranded tails. Exposure of these molecules to exonuclease VII specifically removes the vector tails leaving behind circular DNA structures containing only the insert DNA. At this point the circular single-stranded insert DNA molecules are denatured and complementary strands are annealed thereby forming linear duplex fragments. Upon treatment with λ exonuclease, poly(dT) single-stranded termini are exposed allowing the insertion of the DNA molecule into another vector.

D. Removal of Free Single-Stranded DNA

Since exonuclease VII specifically cleaves single-stranded DNA, it can be used to remove free single-stranded DNA from preparations of any type containing molecules which cannot themselves serve as substrates for the enzyme. In this regard exonuclease VII has been used in the isolation of DNA rings by Hutton and Thomas[14,15]. Any single-strand-specific exonuclease could of course be used for this purpose as well as single-strand-specific endonucleases, assuming the molecules being purified do not contain internal single-stranded DNA. However, the ability of exonuclease VII to function in the absence of added divalent cations makes it ideally suited for this use since the activity of other DNA nucleases is virtually eliminated. This particular application of exonuclease VII has, however, been rather limited so far primarily due to the lack of a convenient and economical source of the enzyme since large quantities are often required. As a result, less convenient methods have been used, such as separation on agarose gels followed by extraction of the desired molecules.

E. Degradation of Adenovirus-Associated Virus (AAV) DNA by Exonuclease VII

Green, Straus and Roeder[12] have recently reported the somewhat surprising observation that exonuclease VII can hydrolyze AAV DNA. The ability of AAV DNA to serve as a substrate for exonuclease VII is unexpected since the terminal 125 nucleotides at each end can form "hairpins" with extensive double-stranded structure. All previous characterization of the enzyme argues against hydrolysis of this duplex region. In view of the known properties of exonuclease VII, including its apparent size and shape and its production of large initial products, the most likely mechanistic explanation would seem to be that the enzyme can bypass either or both of these "hairpins" and initiate hydrolysis at the nearest

168

single-stranded regions as suggested by Green, Straus and Roeder. From the point of view of nucleolytic reaction mechanisms, the mechanism operating here will be extremely interesting and exciting to elucidate. At the same time the observation suggests that results obtained with exonuclease VII from certain applications should be interpreted with caution until this reaction has been more carefully studied. If exonuclease VII can in fact bypass a small terminal duplex region, it may be able to remove very short terminal DNA-RNA hybrid regions, for example. Work relating to this new finding is currently underway in our own laboratory.

ACKNOWLEDGMENTS

Research from the authors' laboratory was supported by Public Health Service grants GM 11301-17 and GM 25451-02 from the National Institute of General Medical Sciences, and CA 13330-08 from the National Cancer Institute. L. D. V. is a Predoctoral Fellow of the National Institute of General Medical Sciences, grant no. GM 07491-03. J. W. C. is an Established Investigator of the American Heart Association, grant no. 78-129.

REFERENCES

1. Chase, J.W. and Richardson, C.C. (1974) J. Biol. Chem. 249, 4545-4552.
2. Chase, J.W. and Richardson, C.C. (1974) J. Biol. Chem. 249, 4553-4561.
3. Chase, J.W. and Richardson, C.C. (1977) J. Bacteriol. 129, 934-947.
4. Vales, L.D., Chase, J.W. and Murphy, J.B. (1979) J. Bacteriol. 139, 320-322.
5. Chase, J.W., and Masker, W.E. (1977) J. Bacteriol. 130, 667-675.
6. Chase, J.W., Masker, W.E. and Murphy, J.B. (1979) J. Bacteriol. 137, 234-242.
7. Masker, W.E. and Chase, J.W. (1978) in DNA Repair Mechanisms, Hanawalt, P.C., Friedberg, E.C. and Fox, C.F., eds., Academic Press, New York, pp. 261-265.
8. MacKay, V. and Linn, S. (1974) J. Biol. Chem. 249, 4286-4294.
9. Nisen, P., Medford, R., Mansour, J., Purucker, M., Skalka, A. and Shapiro, L. (1979) Proc. Nat. Acad. Sci. 76, 6240-6244.
10. Berk, A.J. and Sharp, P.A. (1978) Proc. Nat. Acad. Sci. 75, 1274-1278.
11. Green, M.R. and Roeder, R.G. (1980) J. Virol., in press.
12. Green, M.R., Straus, S.E. and Roeder, R.G. (1980) J. Virol., in press.
13. Goff, S.P. and Berg, P. (1978) Proc. Nat. Acad. Sci. 75, 1763-1767.
14. Hutton, J.R. and Thomas, C.A., Jr. (1975) J. Mol. Biol. 98, 425-438.
15. Hutton, J.R. and Thomas, C.A., Jr. (1975) Biochemistry 14, 1432-1436.

THE EXTRACELLULAR NUCLEASE FROM <u>ALTEROMONAS ESPEJIANA</u>: AN ENZYME HIGHLY

SPECIFIC FOR NONDUPLEX STRUCTURE IN NOMINALLY DUPLEX DNAS

HORACE B. GRAY, JR.,[+] THOMAS P. WINSTON,[+] JAMES L. HODNETT,[+*] RANDY J.
LEGERSKI,[+*] DAVID W. NEES,[+**] CHIK-FONG WEI[+], AND DONALD L. ROBBERSON[++]

[+]Department of Biophysical Sciences, University of Houston, Houston, Texas
77004; [++]Department of Molecular Biology, University of Texas System Cancer
Center, M. D. Anderson Hospital and Tumor Institute, Houston, Texas 77030

---

* Present address: Department of Molecular Biology, University of Texas System
Cancer Center, M.D. Anderson Hospital and Tumor Institute, Houston, Texas 77030.
**Present address: Department of Biochemistry, Rice University, Houston, Texas
77001.

I. INTRODUCTION

The existence of the extracellular nuclease from the marine bacterium
Alteromonas espejiana was first noted incidentally in preparations of bacteriophage PM2 banded in CsCl density gradients. It had been observed that a substantial fraction (up to 50%) of the phage undergoes lysis during this procedure
and that the DNA released is pelleted (D. A. Ostrander and H. B. Gray, Jr.,
unpublished observations). Examination of the released DNA revealed a mixture
of supercoiled closed circular duplex DNA (form I DNA) which is the form found
in the mature phage particle[1,2], circular duplex PM2 DNA containing one or more
single-strand scissions (form II DNA), and the linear duplex form of PM2 phage
DNA (form III DNA). Because such a mixture could not arise from the random
introduction of single-strand breaks (nicks) into form I DNA, the existence of
an enzyme active in CsCl solutions whose starting concentration was near 1.8 M
was inferred. The activities proved to be present in culture fluid in which
noninfected cells of Alteromonas espejiana, the host bacterium of phage PM2,
have been grown. The nuclease activities persist very strongly in preparations
of PM2 phage, but can be removed by repeated banding of the phage in CsCl density

gradients. Laval[3] reported nuclease activities ostensibly associated with phage PM2 that are similar to those of the Alteromonas nuclease[4]. It is very likely that Laval's observations were made on Alteromonas nuclease that contaminated these phage preparations.

The organism that produces the nuclease was originally assigned to the genus Pseudomonas[5], but has been reassigned to the genus Alteromonas (species espejiana) on the basis of more extensive microbiologic tests[6]. The strain originally isolated, given the designation BAL 31 by Espejo and Canelo[5], is available from the senior author and from the American Type Culture Collection (ATCC 29659). This strain also exhibits extracellular alginase, gelatinase, and lipase activities[6] and thus appears well equipped to use marine biologic detritus as a source of nutrients. If the nuclease serves this purpose, a nucleotidase activity should be present, because mononucleotides are not expected to be transported into the cell. A 5'-nucleotidase activity has indeed been found in concentrated culture supernatant (C.-F. Wei and H. B. Gray Jr., unpublished data).

It should be noted that there are apparently two forms of the Alteromonas nuclease, with different kinetic properties, whose presence depends on the initial purification step. Most of the experiments to date have been done with the "slow" (S) form. It should be assumed that the discussion here concerns the S form unless it is indicated that the "fast" (F) form is involved. The preparation of S-form enzyme used in the bulk of the studies displayed two bands (one representing a contaminant) after denaturation and electrophoresis in polyacrylamide gels containing sodium dodecyl sulfate.

II. NUCLEIC ACIDS CONTAINING SITES OR REGIONS OF NONDUPLEX OR ALTERED HELICAL STRUCTURE AS SUBSTRATES FOR NUCLEASE

A. Single-Stranded DNA

The single-stranded circular DNA of coliphage ØX174 is readily degraded to nonsedimenting, dialyzable material by both concentrated crude supernatant[4] and a highly purified preparation; this indicates that an endonucleolytic activity against single-stranded DNA is present. The products of exhaustive digestion of single-stranded DNA are mononucleotides, as indicated by chromatography of the desalted digestion products on columns of DEAE-cellulose as described by Tomlinson and Tener[7] (data not shown; see legend to Figure 6). Coliphage T7 DNA labeled with [^{3}H]thymidine was denatured in alkali and digested with Alteromonas nuclease, as described in the footnote to Table 1, until well after the full hyperchromicity (260 nm) was achieved. The chromatographic position of the digestion products corresponded with that of mononucleotide, as confirmed by chromatography of [^{3}H]thymidine-5'-monophosphate. The elution pattern was very similar to that of Figure 6b, except that the high molecular weight material was not present.

The digestion products were further characterized as 5'-mononucleotides (Table 1). Nonlabeled denatured T7 DNA was exhaustively degraded, and the concentration of nucleotides in the final mixture was calculated from the value of A^{260}, the extinction coefficients of the DNA nucleotides at 260 nm, and the guanine + cytosine content of T7 DNA. After treatment of the reaction mixtures (see footnote to Table 1), digestion with 5'- and 3'-nucleotidases gave the indicated results. Allowing for the activity of the 3'-nucleotidase against the 5'-mononucleotide standards it may be concluded that the Alteromonas nuclease degrades denatured DNA to yield 5'-mononucleotides.

B. Linear Duplex DNA

Linear duplex DNA is degraded from both the 5' and the 3' ends by the Alteromonas nuclease without the introduction of detectable scissions away from the termini[4]. PM2 form I DNA is rapidly converted to form III DNA by the action of the nuclease, and the linear duplex DNA is then degraded from the ends[4]. Samples of linear duplex PM2 DNA progressively shortened by crude concentrated culture supernatants in the reaction buffer described in Table 1 were subjected to analytical band sedimentation in alkaline solution[9]. The alkali-denatured DNA sedimented as a single band, with no observable trailing material, at each extent of degradation examined, an indication that internal scissions are introduced infrequently compared to the number of exonucleolytic events[4]. With a highly purified nuclease sample, the linear duplex DNA of phage T7 (molecular weight near 25 X 10^6 daltons) was shortened by up to 75%. The remaining strands of DNA, averaging only 25% of the full genome length, sedimented as a single broad band in alkali. Because only 5'-mononucleotides and remaining high molecular weight material are found in intermediate digests of linear duplex DNA (below), so that there is hydrolysis of a phosphodiester bond for every nucleotide removed, it may be calculated that fewer than three breaks are introduced at random into a single molecule during the period over which approximately 57,000 phosphodiester bonds are broken in connection with the terminally directed hydrolysis.

The products of exonucleolytic attack may, however, contain substantially long protruding single-stranded ends, but probably not as long as would be produced by degradation of linear duplex DNA by, e.g., E. coli exonuclease III.

TABLE 1

RELEASE OF INORGANIC PHOSPHATE (P_i) FROM PRODUCTS OF <u>ALTEROMONAS</u> NUCLEASE
DIGESTION BY 5'-NUCLEOTIDASE AND 3'-NUCLEOTIDASE

Substrate	Total P_i, nmol	5'-nucleotidase		3'-nucleotidase	
		P_i Released, nmol	% of Total	P_i Released, nmol	% of Total
Nuclease degradation products of denatured T7 DNA	12.6	9.0	71.3	2.4	19.0
Nuclease degradation products of duplex T7 DNA	42.0	33.5	79.8	3.6	8.6
5'-dGMP	64.5	58.4	91.3	5.6	8.8
5'-dTMP	77.5	70.6	91.1	4.0	5.2
3'-dAMP	71.0	0.6	0.8	34.6	48.7

The 5'-nucleotidase from the venom of <u>Crotalus atrox</u> (E.C. 3.1.3.5) (Sigma
Chemical Co.) was used in a reaction mixture (0.5 ml) containing 0.1 M glycine-
NaOH buffer (pH, 8.5), 10 mM $MgCl_2$, the indicated amount of DNA degradation products or nucleotide standard, and 0.25 unit (as defined by Sigma Chemical Co.)
of enzyme. The 3'-nucleotidase from rye grass (E.C. 3.1.3.6) (Sigma) was used
in 0.5 ml of a mixture containing 0.1 M Tris-HCl (pH, 7.8), the indicated amount
of DNA degradation products or nucleotide standard, and 0.25 unit of enzyme.
Reactions were for 2 h at 37°C. Released inorganic phosphate was determined by
the method of Chen <u>et al.</u>[8] with NaH_2PO_4 solutions as standards. <u>Alteromonas</u>
nuclease digests of denatured or duplex T7 DNA were prepared at 30°C in 0.60 M
NaCl, 12.5 mM $MgSO_4$, 12.5 mM $CaCl_2$, 20 mM Tris-HCl (pH, 8.1), and 1 mM EDTA, with
DNA present at concentrations near 50 µg/ml. Appropriately diluted samples of
<u>Alteromonas</u> nuclease in CAM buffer[4] comprised 0.1 volume of the reaction mixture.
Two volumes of 95% ethanol were added to the reaction mixture after the reaction
was stopped with excess EDTA. The supernatants, after centrifugation to remove
precipitates, were evaporated to dryness under a stream of dry nitrogen and resuspended in water for desalting on a column of Bio-Gel P-2 (Bio-Rad Laboratories)
before use in the nucleotidase reaction mixtures.

The endonuclease activity against single-stranded DNA would be fully expected to attack very long single-stranded chains. Electron microscopic observation of partially shortened linear duplex PM2 DNA did not yield evidence of collapsed single-stranded "bush" structures expected for single-stranded terminal chains over about 500 bases long in DNA mounted for microscopy using the aqueous Kleinschmidt technique[4]. The average length of the partially degraded duplexes measured from electron micrographs agreed with the strand length obtained from the sedimentation coefficients in alkali and calculation with Studier's[10] empirical relationship between sedimentation coefficient in alkaline solution and molecular weight. The errors associated with these methods would not have detected single-stranded termini shorter than about 500 bases.

Linear duplex DNA shortened by the action of the Alteromonas nuclease can contain single-stranded ends long enough to promote the joining of the molecules by E. coli polynucleotide ligase. Supercoiled PM2 DNA and ØX174 replicating form DNA, converted to their linear duplex forms by crude preparations of the nuclease and then shortened to various extents, are oligomerized in the presence of E. coli DNA ligase[4], which does not join fully base-paired termini[11]. Both PM2 DNA shortened to an extent undetectable by sedimentation and DNA shortened to an average of 74% of the full genome length showed a substantial fraction (17-25% by mass) of dimeric species after ligase treatment. Moreover, the DNA does not have to be converted from a circular to a linear form by the Alteromonas nuclease for oligomerization to occur. PM2 DNA rendered linear by the action of the single-strand-specific nuclease S_1 from Aspergillus oryzae on the viral form I DNA, a reaction originally noted for the S_1 nuclease in the case of SV40 form I DNA[12], displayed near 70% (by mass) of material sedimenting as dimer- and trimer-length molecules after reduction to 90% of the full genome length by exposure to purified Alteromonas nuclease and subsequent treatment with E. coli ligase[13]. A similar result was obtained when PM2 form I DNA was converted to form II DNA by the limited action of bovine pancreatic DNAse I and then to form III DNA by the Alteromonas nuclease, after which the DNA was shortened to the extent mentioned above and treated with ligase. PM2 form III DNA made by the action of the Hpa II restriction endonuclease, which cleaves the circular PM2 genome at a single unique site[14,15], also showed about the same extent of ligation to dimeric and oligomeric species after exposure to the Alteromonas nuclease[13]. A similar extent of ligation and the presence of trimeric species were noted in these experiments when the DNA was converted from form I to form III by Alteromonas nuclease[13]; this indicated that the lower extent of ligation and the absence of trimeric species in the earlier experiments with crude enzyme preparations[4] were

due either to an impurity in crude preparations that damages single-stranded
ends or to the much greater activity of the ligase preparation used in the
experiments involving the purified nuclease[13].

The behavior of the dimeric species obtained on ligase treatment of nearly
full-length PM2 DNA led to the conclusion that this DNA has been joined in a
"head-to-head" fashion, in which joined strands are self-complementary and can
form a "hairpin" structure with the length of PM2 form III DNA[4]. This type of
joining requires that the single-stranded ends contain palindromic nucleotide
sequences, so that the protruding end is self-complementary (see, e.g., Mertz
and Davis[16]) and can, if long enough, form a small "hairpin" structure.

It is suggested that the exonuclease activity is slowed when a palindromic
sequence long enough to form a "hairpin" structure is exposed as a protruding
single strand, owing to the formation of the self-complementary "hairpin" structure. The probability of molecules in the population that have single-stranded
palindromic sequences at one terminus or both is thus increased. Later treatment with ligase can result in joining of the identical palindromic protruding
single strands produced in corresponding sections of the genomes of different
molecules. This can occur even though the various linear duplex molecules in
the population arise from initial cleavages at a large number of different sites
in the circular PM2 genome, such as would be produced by, e.g., nicking with
pancreatic DNAse I followed by cleavage by the Alteromonas enzyme at the sites
of the nicks.

These combined observations have suggested that short single-stranded termini are produced as a result of exonuclease activity on linear duplex DNA. We
have therefore recently examined the products of Alteromonas exonuclease-treated
PM2 DNA by electron microscopy using the ethidium bromide spreading technique of
Koller et al[17]. The micrographs in Figure 1 illustrate linear duplex strands
of PM2 DNA heavily coated with Pt-Pd, which are approximately 4-6 nm in diameter.
The termini of the duplexes shown in the top two panels have been partially denatured and fixed with glutaraldehyde to reveal single strands with attached protein, presumably Alteromonas nuclease. In most cases, the lengths of the two
single strands at a terminus are quite different. The enzyme with a diameter in
the range of 23 nm is typically found to bind to only one of the two arms at a
site that approximately corresponds to the duplex terminus before partial denaturation with glutaraldehyde. A protruding single strand several hundred bases
long is often seen to extend beyond the site of attachment of the enzyme (Figure
1). Some thinner strands are also detected, which probably correspond to single-
stranded DNA and in which enzyme is bound at or near one terminus (bottom panels

in Figure 1). These structures may represent single strands that are released from the termini of duplexes and then degraded exonucleolytically to 5'-mononucleotides. Examination of similarly exonuclease-treated samples with the formamide modification of the basic protein technique without glutaraldehyde treatment and with EDTA to stop the reaction has revealed a short single-stranded terminus on many of the molecules, as well as free single strands some 500 bases long.

With such large single strands, a predominantly exonucleolytic mode of degradation is expected, if only 5'-mononucleotides, rather than oligonucleotides, are major products of incomplete exonuclease digestion (Figure 6). Finally, it should be noted that most duplexes contain one, but only one, enzyme molecule bound at a terminus and enzyme is occasionally observed bound at internal positions of the duplex. The apparent size of the enzyme is comparable with that observed by Koller et al.[17] for E. coli RNA polymerase and suggests that the activity may require association of multiple monomeric units.

The products of exhaustive digestion of linear duplex DNA with the Alteromonas nuclease are 5'-mononucleotides, as determined from DEAE-cellulose chromatograph and exposure of the products to 5'-and 3'-nucleotidases (Table 1).

The Alteromonas nuclease seems very well suited for the purpose of shortening both strands of linear duplex DNAs from both ends, because it is very stable in storage, is denaturation-resistant, is active over a wide range of ionic strengths and temperatures, is optimally active near neutral pH, has predictable and reproducible kinetic properties, and is easily obtainable free of interfering nuclease activities[4,15,18] (these properties are discussed later). None of the alternative methods for accomplishing this reaction[19-21] has all the advantages of the use of the Alteromonas enzyme for this purpose.

The activity of the Alteromonas nuclease against linear duplex DNA has been exploited as a method for the rapid determination of the order of the fragments produced by digestion of small DNA genomes with restriction endonucleases[15]. The nuclease is used to produce a series of samples of progressively shorter linear duplex DNA, starting from a full-length genome. Digestion of these samples with a restriction endonuclease and then analysis of the resulting fragments with gel electrophoresis yields information about the order of the fragments produced by the restriction enzyme from the order in which the various fragments disappear from, or change their rate of migration in, the electrophoretic patterns.

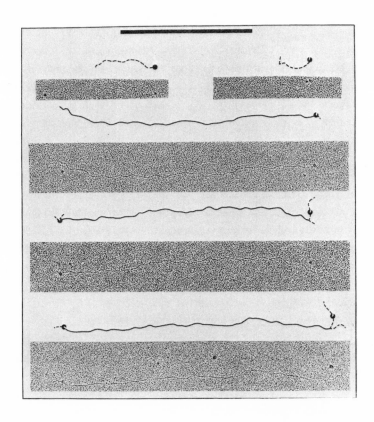

Fig. 1. Electron micrographs illustrating products of Alteromonas exonuclease digestion of PM2 DNA. Samples were prepared for electron microscopy by the procedure of Koller et al.[17] using ethidium bromide at 100 μg/ml. A sample of form I PM2 DNA at a concentration of 50 μg/ml in CAM buffer[4] was incubated in a total volume of 100 μl with highly purified S form Alteromonas nuclease[4] at 15 units/ml for 10 min at 30°C. Twenty-five microliters of 0.4% glutaraldehyde in 0.1 X CAM buffer and 0.01 M Tris-HCl (pH, 7.5) was then added, and the reaction continued for 10 min at 30°C. The sample was diluted 1:10 in 0.1 X CAM buffer and 0.01 M Tris-HCl (pH, 7.5) containing 0.1% glutaraldehyde. Ethidium bromide was added to a final concentration of 100 μg/ml, and droplets were placed on a clean Teflon surface. Freshly cleaved mica was touched to the surface, dehydrated in ethanol, and then shadowed with Pt-Pd (80:20). After being backed with carbon, films were stripped at an air-water interface, mounted on 300 mesh copper screens, and examined in a Philips 300 electron microscope. Single-stranded segments are indicated by dashed lines and duplex segments by solid lines in drawings below each micrograph. Enzyme molecules are indicated by solid circles with or without cleft. Magnification is indicated by bar length of 1 μm.

A major use of the exonuclease activity is to remove unwanted sequences from DNA fragments to be cloned or from cloning vectors themselves, and reports of this use are beginning to appear[22]. It is important for this purpose that an appreciable fraction of the shortened fragments be ligatable to other DNA molecules with fully base-paired termini in the reaction catalyzed by the T4 phage-induced DNA ligase[11,23]. Talmadge et al.[22] have shown that a fully duplex oligonucleotide linker can be ligated to plasmid DNA that has been shortened by the Alteromonas nuclease and that the resulting DNA can be cyclized. This implies that at least a portion of the molecules in their S form nuclease-treated preparations had fully base-paired termini. In experiments carried out at Bethesda Research Laboratories, Inc. (which markets the Alteromonas nuclease), a high percentage of nuclease-shortened duplex fragments undergo T4 DNA ligase-catalyzed joining (Bethesda Research Laboratories Product Profile 8019-2382, 1979), although it cannot be ascertained what fraction of these molecules were joined through protruding single strands, as opposed to the joining of fully base-paired ends. These last experiments involved the F form nuclease.

C. Negatively and Positively Supercoiled Circular DNA

Negatively supercoiled closed circular duplex DNA is converted to the linear duplex form by the Alteromonas nuclease-catalyzed hydrolysis of the form I DNAs of PM2, ØX174 replicating form, and SV40 virus[4]. A stepwise introduction of the two strand breaks required to render linear a covalently closed circular molecule is indicated by the demonstration of form II DNA as an intermediate[4]. The analogous reaction has been well documented in the case of other single-strand-specific nucleases[12,21,24-28].

The dependence of the initial rate of introduction of the first chain scission (initial nicking rate) into covalently closed circular PM2 DNA by single-strand-specific nucleases on the superhelix density (σ, the number of titratable superhelical turns per 10 base-pairs[29,30]) has been examined for the nucleases from Neurospora crassa[14,31], mung bean[14], and Alteromonas espejiana[32]. It is necessary to correct the values of σ of the earlier reports[14,31] to reflect the redetermined value of the amount of unwinding of duplex DNA per intercalated ethidium moiety[33-35]. With this correction, the earlier reports show that the initial nicking rate with the nucleases from mung bean and N. crassa was relatively low and depended only slightly on σ at values of σ between zero and 0.06 - 0.08. For more negatively supercoiled DNAs, the initial nicking rate increased rapidly with increasing values of σ. In contrast, the initial nicking rates of

closed circular PM2 DNA mediated by highly purified samples of the Alteromonas
nuclease have been shown to be readily measurable at values of $-\sigma$ as low as 0.02.
However, even at the increased concentrations of enzyme and extended digestion
periods required to cause nicking at an appreciable rate at near-zero values of
$-\sigma$, nonsupercoiled covalently closed (form I°) PM2 DNA is not cleaved at a detectable rate. This behavior was noted with PM2 DNAs that were made artifi-
cially[36-38] so as to produce a series of DNAs of different superhelix densities
and with the naturally occurring negatively supercoiled viral PM2 DNA in the
presence of various concentrations of ethidium bromide to yield various degrees
of supercoiling[32].

The virtual absence of activity of the Alteromonas nuclease against form I°
DNA in CAM buffer[4] at 20°C permitted the use of higher enzyme concentrations and
extended times of incubation to demonstrate the presence of nuclease-susceptible
regions in negatively supercoiled DNA of near-zero superhelix density[32]. Such
exhaustive digestions are not feasible with the other nucleases studied in this
regard, because these have a low, superhelix density-independent activity against
the DNA at values of $-\sigma$ below 0.06 - 0.08 and/or acidic pH optimums. Thus, there
would be "background" nicking to an unacceptable extent and/or acid-catalyzed
damage to the DNA, and such attempts to probe more stringently the effects of
supercoiling on the activity of these enzymes would not be feasible.

A further result of the study with Alteromonas nuclease was the cleavage
detected in PM2 form I° DNA that has been rendered very highly positively supercoiled with ethidium bromide. Although no cleavage was detected in DNA with σ
values from -0.02 to 0.15, cleavage in the presence of increased enzyme concentrations was noted with σ above 0.15 and increased in rate with increasing σ.
These observations, together with control experiments in the same study, also
show that the Alteromonas nuclease is active at concentrations of ethidium
bromide sufficient for near-saturation binding to the DNA[32].

The cleavage of negatively supercoiled DNA is consistent with thermodynamic
considerations that favor the existence of unstacked or weakly hydrogen-bonded
bases in such DNA over their presence in nonsupercoiled (form II or form I°)
DNA[39,40]. Interestingly, the existence of locally denatured bases at both
negative and positive superhelix densities was predicted in a more recent theoretical study, in which the change in the molecular twist from its unstressed
(e.g., form II DNA) value is partitioned between local denaturation and smooth
twisting so as to minimize the conformational free energy.[41] This theory pre-
dicts that local denaturation will occur at negative superhelix densities near
-0.02 (the same as the threshold value for nicking of negatively supercoiled

DNA by the _Alteromonas_ nuclease) and again at positive superhelix densities near 0.11 (nicking of positively supercoiled DNA by the _Alteromonas_ nuclease becomes detectable somewhere between 0.15 and 0.19). The work with the _Alteromonas_ nuclease has implications for models[14,31,42,43] that describe the regions in supercoiled DNA that are susceptible to the action of such enzymes, some of which are discussed by Lau and Gray[32].

D. Duplex DNA Containing Single-Strand Breaks

Both the PM2 form I and form II DNAs, in a mixture produced by heating a sample of form I DNA until approximately 40% of the molecules were nicked, were readily converted to form III DNA by crude preparations of _Alteromonas_ nuclease. Phage $\lambda b_2 b_5 c$ DNA containing a small number of thermally induced breaks per molecule displayed molecular weight reduction when sedimented at neutral pH after exposure to such nuclease preparations, whereas intact samples of this DNA showed no effects after incubation with enzyme under the same conditions. Coliphage T5 DNA, which contains naturally occurring strand breaks at well-defined locations in the molecule[44-48], underwent molecular weight reduction when exposed to crude enzyme preparations, under conditions in which the exonuclease activity against linear duplex DNA would not have been detectable in the assay used[4]. Such "nick-specific" activity has been reported for the S_1 and snake venom nucleases[12,21,25-27].

In more recent studies with purified nuclease preparations, the form II DNA in a mixture of form I° and form II DNAs, produced by heating a sample of form I° DNA, was converted to form III DNA which was then degraded exonucleolytically[18]. This mixture contained an average of 0.8 nicks per molecule from the Poisson distribution and an average of 0.7 nuclease-sensitive sites per molecule. A sample of PM2 form II DNA, which arose from the spontaneous nicking of [^{3}H]thymidine-labeled form I DNA and contained 92% of molecules with one strand break and 8% with two or more breaks, was cleaved by the nuclease in a reaction whose velocity decreased to a very low value with approximately 12% of the form II DNA remaining[18]. It is not known whether this failure to cleave all the nicked circular DNA arises from the inhibitory effects of the duplex termini generated in the cleavage, which are also substrates, or the chemical nature of some of the breaks produced in the above manner. It is evident, in any event, that strand breaks are effectively used as substrate sites by the nuclease.

E. Nonsupercoiled Closed Circular DNA Containing Covalent Lesions

The extreme resistance of form I° DNA to cleavage by the Alteromonas nuclease allows stringent tests of the ability of the nuclease to cleave at sites
of covalent modifications in duplex DNA as long as such alterations do not
necessarily introduce strand breaks. Form I° DNAs may readily be produced from
the respective form I DNAs by the action of Type I topoisomerases[49].

Irradiation of PM2 form I° DNA with ultraviolet light gives rise to nuclease-
sensitive sites[18,50]. Legerski et al.[18], assuming the photoproducts to be cyclo-
butylpyrimidine dimers, found that at approximately 7 pyrimidine dimers per molecule, 7% of these lesions served as sites of attack by the Alteromonas nuclease.
Harless[50] compared the efficiency of the hydrolysis of ultraviolet-irradiated PM2
form I° DNA by the pyrimidine dimer-specific endonuclease induced by T4 phage (T4
endonuclease V)[51,52] with that catalyzed by the Alteromonas nuclease. Under conditions of incubation similar to those used by Legerski et al.,[18] it was found
that this nuclease cleaved DNA irradiated to a given extent at about 15% of the
frequency (in very long incubations) observed with T4 endonuclease V. It is
possible that the Alteromonas nuclease recognizes the distortion of duplex DNA
structure caused by the presence of a pyrimidine dimer, but not with high efficiency. Alternatively, another type of ultraviolet irradiation-induced lesion
of unknown structure, such as those shown to remain in duplex DNA when the pyrimidine photoproducts are removed by photoreactivating enzyme[53], could be the
sites of attack by the Alteromonas nuclease.

The carcinogen N-acetoxy-N-2-acetylaminofluorene introduces nuclease-
sensitive sites into PM2 form I° DNA, although the ratio of substrate sites
to the estimated number of modified deoxyguanosine nucleotide residues[54] was
small (0.011 and 0.022 at the two extents of reaction used)[18]. Nitrous acid,
which both deaminates cytosine and adenosine residues and introduces interstrand cross-links into duplex DNA[55], creates lesions in form I° DNA that are
efficiently cleaved, if it is assumed that the sites of cross-linking, and not
the sites of deaminated bases, are recognized by the nuclease[18].

Methylmethanesulfonate, which acts predominantly to methylate the N7 position of deoxyguanosine residues in duplex DNA[56], gives rise to nuclease-sensitive
sites in PM2 form I° DNA that are apparently apurinic sites resulting from the
spontaneous release of methylated guanine on incubation at 30°C in the nuclease
reaction buffer[13]. The rate of cleavage by purified nuclease of form I° DNA
thus modified was greatly increased after heating to remove quantitatively
alkylated purines[57]; this suggests that apurinic sites could comprise at least

some of the nuclease-sensitive sites. This was examined further by incubating methylmethanesulfonate-treated DNA in buffer alone and with enzyme, after which both sets of samples were subjected to treatment with alkali under conditions expected to result in quantitative hydrolysis of the phosphodiester bonds at apurinic sites[58]. The percentage of form I° DNA remaining as a function of time of incubation with either buffer or nuclease yielded very similar curves; hence, the loss of methylated purines was probably taking place on incubation in the Ca^{2+}- and Mg^{2+}-containing buffer and the substrates for the observed enzyme activity were probably the resulting apurinic sites. Methylated purines are stabilized at high pH with respect to hydrolysis of the glycoside linkage[59], so incubation in alkali is expected to result in no substantial damage but cleavage at apurinic sites.

The DNA sample in the above work modified to the least extent contained an average of 6.4 apurinic sites per molecule after heating to remove methylated purines[13]. In earlier work with PM2 DNA containing a somewhat lower number of apurinic sites per molecule introduced directly by heating at acid pH[58], conclusive evidence of cleavage by purified nuclease from the same preparation was not obtained. With a partially purified preparation of the F form of the Alteromonas nuclease, it has been shown that PM2 form I° DNA with an average of only 2.1 apurinic sites per molecule (produced by heating under acidic conditions) contains substrate sites for the enzyme. The rate and extent of cleavage clearly increased with increasing number of apurinic sites per molecule of form I° DNA. The incubation temperature for these studies was 20°C instead of 30°C, because the F form of the nuclease was observed to cleave nontreated form I° DNA to a significant extent at 30°C in extended incubations of enzyme concentrations near 120 units/ml (a unit is as defined elsewhere)[4,60]. The curves of percent form I° DNA versus time of incubation did not level off fully at the longest incubation time, so it was not feasible to assess accurately the ratio of nuclease-sensitive to apurinic sites. However, this ratio is at least 0.2 and appears to increase with increasing number of apurinic sites per molecule; this indicates that the cleavage of apurinic sites catalyzed by the F form of the nuclease is rather efficient.

N-Methyl-N-nitrosourea, another methylating carcinogen, produces sites in PM2 form I° DNA sensitive to the F form of the nuclease. Cleavage of the form I° DNA purified from nicked DNA in the carcinogen reaction mixture depended, both in rate and extent, on the time of reaction with a fixed concentration of carcinogen. When the modified DNA was heated to remove methylated purines, the rate and extent of the reaction were much greater owing to formation of apurinic sites as expected. However, when samples incubated with nuclease or with buffer alone were treated with alkali to hydrolyze the chain at apurinic sites that might have been generated by incubation with buffer, as in the case of methylmethanesulfonate-treated DNA, the samples treated with buffer alone did not show a detectable fraction of sites rendered alkali-labile, whereas the nuclease-treated sample showed a definite loss of closed circular DNA over the same period. This cleavage of form I° DNA was greater in initial rate and extent with samples that reacted with carcinogen for a longer time. These results demonstrate that there are lesions in DNA treated with carcinogen that serve as substrate sites for the nuclease and that are not apurinic sites generated in the course of incubation. The reason for the loss of methylated purines on incubation in buffer from methylmethanesulfonate-reacted DNA and not for that which was reacted with the nitroso compound is unknown, but it could be due to the 10°C difference in incubation temperature or to the differences in the distribution and types of methylated products yielded by the two compounds[56].

F. RNA

Crude preparations of the nuclease were shown to hydrolyze readily ribosomal RNA and t-RNA[4], but it was not demonstrated that these activities resided in the same protein species as the other activities. Preparations of the F form, purified with affinity chromatography and gel filtration chromatography as described later, readily hydrolyze 28 S rat liver ribosomal RNA as monitored by hyperchromicity. It is thus likely that the ribonuclease activities are expressed by the same enzyme as that possessing the deoxyribonuclease activities described above. More information on this potentially useful nuclease activity is not available.

III. PHYSICIAL PROPERTIES OF PARTIALLY PURIFIED PREPARATIONS

Little information is available on the properties of the protein itself, owing to the use of incompletely purified preparations in the reported studies. The purified S form preparation used in most of the work described above displayed a single band on electrophoresis using the method of Dietz and Lubrano[61] in heavily loaded nondenaturing polyacrylamide gels, but denatured samples gave rise to two bands in sodium dodecyl sulfate-polyacrylamide gel electrophoresis experiments carried out with Laemmli's procedure[62]. Purified F form preparations yield a very similar pattern in denaturing gels[63]. These results suggested a subunit structure for the enzyme, but this did not stand up to further scrutiny. The combined molecular weights of the two subunits would be much greater than that estimated for the nondenatured protein from gel filtration chromatography[63]. In addition, the relative amounts of material in the two bands revealed by staining with Coomassie blue[64] varied from preparation to preparation[63], and small quantities of nuclease have been obtained that display only a single band in the denaturing gels.

Assuming that the more intensely staining of the two bands in the denaturing gels represents the nuclease, the weight of both S and F forms from that technique using commercially available protein molecular weight standards is near 73 kilodaltons[63]. The molecular weight (F form) estimated from chromatography on Sephadex G-150 in the presence of commercial protein standards is 83 kilodaltons[63]. The Alteromonas nuclease (both S and F forms) is thus taken to be a single polypeptide chain with a molecular weight in the range above. Better molecular weight determinations from sedimentation equilibrium require preparations free of the apparent persistent contaminant noted above.

The two bands noted on electrophoresis in denaturing gels (S form) have been tested for carbohydrate content by staining such gels using the periodic acid-Schiff's base procedure described by Segrest and Jackson[65]. These gels showed undetectable to perhaps barely detectable amounts of carbohydrate staining in either band by visual inspection, whereas pancreatic DNAse I (Sigma) was stained to a readily detectable extent. The mixture of proteins making up bovine DNAse contains 4.6% carbohydrate by weight[66,67]. If the Alteromonas nuclease contains carbohydrate moieties, these could not account for more than about 2% of the mass of the molecule.

IV. DEPENDENCE OF CATALYTIC BEHAVIOR ON pH, TEMPERATURE, AND SOLVENT COMPOSITION

A. Stability

The Alteromonas nuclease is very stable on storage in solution both in crude and in purified form. Samples purified in 1976 were still in use in late 1980. A highly dilute (4.5 units/ml) purified preparation of S form nuclease stored at 4°C in CAM buffer[4] displayed no loss in its initial nicking rate against PM2 form I DNA over a 6-wk period[32]. A sample of partially purified F form nuclease, stored as above, showed constant (in units per milliliter) activity against single-stranded DNA when assayed[4,60] several times over a 6-mo period.

Freezing (-20°C) and thawing of both crude and purified (S form) samples in CAM buffer resulted in loss of activity against single-stranded DNA that was 1st-order, within the scatter of the data, with respect to the number of freezing-thawing cycles. It was estimated that 50% of the initial activity would be lost after 18 and 28 cycles of freezing and thawing for the crude and the purified preparation, respectively. These numbers should not be taken to indicate a markedly greater sensitivity of crude preparations to freezing and thawing, inasmuch as the standard deviations in the least-squares slopes of plots of ln(activity) versus number of cycles make such an interpretation dubious. It is clear that the nuclease can suffer several cycles of freezing and thawing without marked loss of activity, but its solutions should not be frozen and thawed routinely.

B. Effects of pH

A major advantage of the Alteromonas nuclease over other single-strand-specific nucleases, such as the Aspergillus[60,68] and mung bean[69] enzymes, is that its activity is optimal near neutral pH. Figure 2 shows the pH dependence of the activity against single-stranded DNA, the exonuclease activity against linear duplex DNA, and the endonuclease activity that cleaves supercoiled closed circular DNA in panels a, b, and c, respectively, for purified S form nuclease. The reaction velocities for the degradation of linear duplex DNA were determined from sedimentation coefficients in alkali and use of the Studier[10] equation as done for PM2 DNA by Gray et al.[4] These rates were based on at least four time points in each digestion. Reactions proceeded to greater than 50% digestion of the duplex in all cases. Velocities were constant over the entire course of the incubation, except for the reaction at a pH of 6, which decreased in rate in the later stages of the digestion--suggesting slow inactivation of the enzyme. The initial velocity was used in this case. Optimal pH for activity is near 8.8 for denatured DNA, but close to 8.0 for the nominally duplex substrates. A pH optimum near 8.0 has been found independently for the cleavage of PM2 form I DNA by Harless[50]. The pH optimum for the cleavage of denatured calf thymus DNA was also determined in the presence of 0.35 and 1.35 M NaCl and was unchanged from the value observed at 0.63 M NaCl (Figure 2a). Assays at higher pH values than those of Figure 2 were precluded by the precipitation of $Mg(OH)_2$.

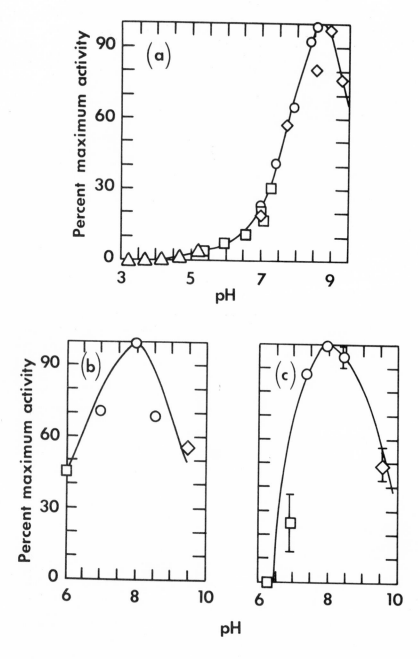

Fig. 2. Effect of pH on purified S from Alteromonas nuclease activity against (a) single-stranded calf thymus DNA, (b) linear duplex T7 DNA, (c) PM2 form I

DNA. Acidic or basic solutions of each buffer--containing the required concentrations of NaCl, MgSO$_4$, CaCl$_2$, and EDTA--were mixed to provide a pH near that
desired. DNA, in 100 ml NaCl-1 mM EDTA (pH, 7), and enzyme (0.05 volume) in CAM
buffer[4] were then added. The pH of the reaction mixture was read directly with
a small electrode, except for the experiments with PM2 form I DNA, in which components were mixed in the same ratios as for reactions for the other DNAs in
which the pH was directly measured. The reaction mixtures had the final composition given in the note to Table 1, except for the buffer, which was present at
10 mM (Tris or glycine concentration). DNA concentrations were: (a) near 150
µg/ml, (b) 36 µg/ml, and (c) 56 µg/ml. Incubation was at 30°C. Aliquots of the
mixtures were withdrawn at appropriate intervals, made 50 mM in EDTA (pH, 8), and
analyzed: (a) acid-soluble degradation products were monitored photometrically,[4]
(b) sedimentation coefficients were measured in alkaline solution and converted
to molecular weights,[4] and (c) cleavage of form I DNA was determined by cutting
and weighing the bands from photoelectric scanner tracings from alkaline band
sedimentation experiments[4,32]. Buffers used were: Δ, Tris-acetate; □, Tris-
maleate; O, Tris-HCl; and ◇, glycine-NaOH. Error bars in (c) indicate the range
of 2 or more determinations.

C. Effect of Temperature

The profile of enzyme activity as a function of temperature could satisfactorily be determined only with the single-stranded DNA substrate because
of the loss of stability of the duplex structures with increasing temperature,
which is expected to result in substrate sites for the nuclease that are not
present at lower temperatures. Substantial nicking of PM2 form I° DNA was noted
with S form nuclease at 37°C. The profile for the purified S form enzyme in
Figure 3 shows that the usual incubations at 30°C are done at only 13% of the
maximal nuclease activity as monitored for the denatured DNA substrate. Incubations to hydrolyze single-stranded substrates quantitatively can profitably
be done at temperatures up to the optimal temperature of 60°C. Although very
little activity at 0°C is observed (Figure 3), this activity is probably real
because the duplex exonuclease activity was shown to be expressed at this
temperature[4].

D. Effects of Electrolytes

The Alteromonas nuclease has a remarkable resistance to inactivation in
the presence of high concentrations of 1:1 electrolytes. Figure 4 shows the
dependence of the activity of S form enzyme on ionic strength in experiments
in which the concentration of sodium chloride was varied. Denatured calf
thymus DNA served as the substrate. The activity became maximal between 0 and
2 M added NaCl and the nuclease retained approximately 40% of its maximal
activity in the presence of 4.4 M NaCl.

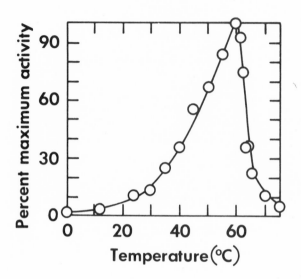

Fig. 3. Dependence of purified S form Alteromonas nuclease activity on temperature with denatured calf thymus DNA (approximately 150 μg/ml) as substrate.
Reaction mixture composition was as in the note to Table 1. All components but
enzyme were mixed and the pH was adjusted to 8.1 at the temperature of the experiment (with the pH meter standardized at that temperature). The pH was monitored throughout the reaction after the addition of 0.05 volume of enzyme in CAM
buffer. Where the addition of the enzyme solution resulted in a detectable
shift in pH, the small correction required was estimated from Figure 2a. The
release of acid-soluble degradation products was monitored photometrically.[4]

With CsCl, a more soluble electrolyte than NaCl, a profile displaying an
optimal activity (denatured DNA substrate) between 0.5-1.8 M CsCl was obtained
for the same enzyme preparation (data not shown). In the presence of 7 M CsCl,
the activity was 27% of the maximal value. The duplex exonuclease activity was
also tested in this regard, with phage T7 DNA as substrate, and was found to
decrease with increasing CsCl concentration with no indication of a maximum.
The activity decreased with increasing ionic strength more sharply than for the
denatured DNA substrate, although activity was present at molarities of CsCl as
high as 5.5. This was shown in extended (up to 20 h) incubations at 30°C at an
enzyme concentration that shortened the T7 DNA duplex to 60% of the full genome
length in 4 h of incubation in the presence of 0.4 M CsCl (0.6 M total ionic
strength). In the very extended digests, there was evidence of a low degree of
introduction of scissions away from the duplex termini, in that significant
trailing material appeared in the alkaline band sedimentation profiles.

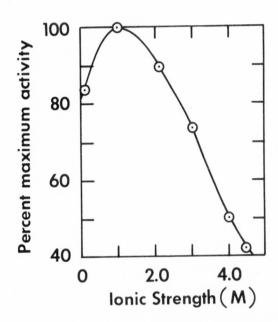

Fig. 4. Dependence of the activity of purified S form Alteromonas nuclease on
ionic strength with denatured calf thymus DNA (150 µg/ml) as substrate. Ionic
strength was varied by varying the NaCl concentration. The composition of the
mixture was otherwise as in the note to Table 1, and the incubations were performed at 30°C. The ionic strength of the mixture in the absence of added NaCl
was 0.12 M. A fraction of the volume of the reaction mixture was reserved for
solid NaCl and water. The amounts of water required to achieve this fraction in
each case were calculated by interpolation of tabulated (Handbook of Chemistry
and Physics, CRC Press, 57th ed.) data for the volume of water displaced by NaCl
as a function of NaCl concentration. The refractive index of an aliquot of each
solution was read before equilibration at 30°C and the addition of 0.025 volume
of enzyme solution. The NaCl concentration was calculated from refractive index
data in the same handbook tabulation and corrected for the addition of enzyme.
The molarity of NaCl calculated from the refractive index in each case was in
excellent agreement with the intended value.

E. Effects of Protein Denaturing Agents

The resistance of the Alteromonas nuclease to denaturation by agents that
normally cause the loss of biological activity of proteins is pronounced. Gray
et al.[4] showed that the enzyme is active against both linear duplex and supercoiled DNA in the presence of 5% (w/v) sodium dodecyl sulfate and can be preincubated with the detergent without loss of activity if Ca^{2+} and Mg^{2+} are present

190

at 12.5 mM before the addition of detergent. The effect of urea on the ability
of the purified S form enzyme to hydrolyze single-stranded DNA is shown in Figure 5. In these experiments, enzyme was added to a mixture of all other components immediately before incubation. Over 40% of the maximal activity is
still observed in the presence of 6.6 M urea. Experiments in which the nuclease
is preincubated with urea before addition of substrate have not been carried out.

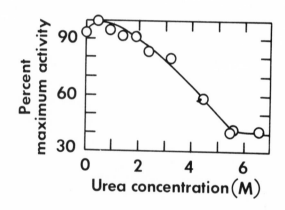

Fig. 5. Effects of urea on the activity of purified S form nuclease against
single-stranded DNA. The reaction mixtures were as in the note to Table 1,
except for the presence of urea at the indicated concentrations. Solid urea and
water constituted a constant fraction of the reaction mixture volume, and the
required amounts of water were determined by the procedure described in the caption of Figure 4. Urea concentrations calculated from refractive indexes were
in excellent agreement with the intended concentrations. All components were
mixed before addition of 0.025 volume of nuclease solution, and incubation at
30°C followed immediately.

F. Divalent Metal Cation Requirements

The enzyme exhibits no activity against single-stranded and nominally duplex substrates after the molar concentration of a chelating agent (EDTA) exceeds that of Ca^{2+}. Activity is not restored by adding an excess of Ca^{2+} to
solutions of enzyme thus inactivated[4]. This property is useful when sequential

digestions with the <u>Alteromonas</u> nuclease and, e.g., a restriction enzyme are to be carried out in which the <u>Alteromonas</u> enzyme must be inactivated[15].

Very low concentrations (0.2 mM) of Ca^{2+} appear to confer the full activity of the nuclease on single-stranded DNA, and there is no substantial dependence of reaction velocities on Ca^{2+} concentration up to 50 mM for this substrate (data not shown). In contrast, Harless[50] noted a very marked increase in the rate of cleavage of PM2 form I DNA with increasing Ca^{2+} concentration from 1 mM to approximately 10 mM; above the latter concentration, the reaction rate becomes somewhat independent of this variable, with a possible maximum between 15 and 20 mM. The reaction conditions used in this laboratory (note to Table 1) differed from those of Harless[50] only in that Harless used 0.5 M NaCl instead of 0.6 M NaCl. <u>S</u> form nuclease was used in both studies.

The activities are reduced, but not eliminated, when samples are dialyzed into buffers from which Mg^{2+} has been omitted. This effect is reversible, with activity restored on addition of Mg^{2+} salts. The reduction of the activity of purified <u>S</u> form nuclease seems far more pronounced in the reaction with single-stranded substrates than with a nominally duplex substrate (PM2 form I DNA). Denatured calf thymus DNA is hydrolyzed at zero concentration of Mg^{2+} at only 1.5% of the rate at 15 mM Mg^{2+} (data not shown), whereas Harless[50] observed that PM2 form I DNA was cleaved in the absence of Mg^{2+} at 13% of the rate measured in the presence of 15 mM Mg^{2+}. For both substrates, the reaction velocities increase markedly with Mg^{2+} concentration between 0 and 10 mM and become weakly dependent on the concentration of this cation above approximately 15 mM. The effects of Ca^{2+} and Mg^{2+} concentration have not been investigated for the exonuclease activity against linear duplex DNA.

A tentative interpretation of the dependence of nuclease activity on concentrations of Ca^{2+} and Mg^{2+} is that Ca^{2+} may be a prosthetic group required for nuclease activity and that the enzyme is most active against MgDNA substrates. The dependence of the rate of cleavage of form I DNA on Ca^{2+} concentration up to about 10 mM may indicate an interesting effect of this cation on the weakly hydrogen-bonded or unstacked bases that are presumed to exist, at least transiently, in supercoiled DNA to render it susceptible to attack by single-stranded nucleases and chemical reagents that normally react only with single-stranded DNA and do not react with form II or form I° DNA.

V. KINETIC PARAMETERS AND MECHANISTIC ASPECTS OF ACTIVITIES AGAINST SINGLE-STRANDED AND LINEAR DUPLEX DNA

Michaelis constants and maximal reaction velocities have been determined carefully only for the purified $\underline{S}$ form nuclease in the case of single-stranded and linear duplex DNA substrates. When ^{32}P-labeled single-stranded circular ØX174 viral DNA was used as substrate and the extent of reaction was determined measuring acid-soluble radioactivity, the value $K_m = 1.2 \pm 0.2 \times 10^{-5}$ mol of nucleotide per liter (average of values determined at three nuclease concentrations) was obtained at 30°C in reaction mixtures of the composition given in the note to Table 1. The maximal reaction velocity (V_m) was linear with enzyme concentration and had a value of 8.1×10^{-8} mol of nucleotide per liter per minute with 0.21 unit of enzyme per milliliter of reaction mixture.

Legerski $\underline{et}$ $\underline{al}$.[15] presented an integrated rate equation for the duplex exonuclease activity. The derivation[13] is based on a constant value of substrate concentration (molar concentration of duplex termini) and assumes a constant reaction velocity, which was seen to be the case with the crude[4] and the purified $\underline{S}$ form[13] nuclease. The equation is

$$M_t = M_0 - 2M_n V_m t/(K_m + S_0) \qquad (1)$$

where M_t is the molecular weight of a linear duplex DNA, initially of molecular weight M_0, after $\underline{t}$ minutes of digestion; M_n is the weight of a sodium nucleotide, taken as 330 daltons; and S_0 is the molar concentration of duplex termini. For reactions at 30°C in the 0.6 M NaCl reaction buffer (note to Table 1) and in the same buffer with the concentration of NaCl changed to 0.2 M, the values 2×10^{-6} and 3×10^{-6} mol of nucleotide per liter per minute were obtained, respectively, for an enzyme solution containing 28 units of activity per milliliter against single-stranded DNA.[15] The value of $K_m = 2 \times 10^{-8}$ mol of termini per liter may be used satisfactorily for both sets of conditions for exonuclease digestions.

However, with $\underline{F}$ form enzyme purified by affinity and gel filtration chromatography as described below, estimates of V_m (adjusted to a nuclease activity of 28 units/ml) and K_m become 1.7×10^{-5} mol of nucleotide per liter per minute and 5×10^{-9} mol of termini per liter, respectively, in the 0.2 M NaCl buffer near 30°C. At very low concentrations of duplex termini, so that S_0 may be neglected with respect to K_m, the initial rate of removal of nucleotides from duplex termini is thus approximately 23 times greater with the $\underline{F}$ form nuclease than with the $\underline{S}$ form at a fixed amount of activity against single-stranded DNA. Although detailed kinetic studies have not been made on all samples, it is clear that nuclease purified with affinity chromatography reproducibly displays much faster kinetics of degradation of duplex DNA, per unit of activity against single-stranded DNA, than nuclease purified with ion-exchange and gel filtration chromatography in the procedures described below. Another difference observed with the F form nuclease is that the velocity of the reaction to shorten linear duplex DNA may decrease markedly with time; this suggests inhibition of the nuclease by the released nucleotides. Whether only two forms of the enzyme exist or a wide range of kinetic parameters will be observed with various preparations is not known. However, the kinetic parameters determined for the S form by Legerski et al.[15] accurately predicted the extent of degradation of linear duplex DNA for a later S form preparation.

It is clear that the Alteromonas nuclease possesses an endonuclease activity (e.g., to initiate the degradation of single-stranded circular and nominally duplex circular DNA substrates). However, nearly random endonucleolytic attack, such as with bovine pancreatic DNAse I, will produce oligonucleotide fragments (Figure 6a). The absence of oligomeric species in the intermediate products of digestion of single-stranded DNA by the Alteromonas nuclease (Figure 6b) argues for a predominantly exonucleolytic mode of attack, in which an endonucleolytic break, if introduced, is followed by sequential removal of nucleotides from single-stranded ends.

The products of intermediate digests of linear duplex DNA (Figure 6c) are mononucleotides and remaining high molecular weight DNA (experiments have been done in which the amount of high molecular weight material relative to mononucleotides was much greater than in Figure 6c). These results suggest a mechanism in which either nucleotides are released one at a time from each strand or, if oligonucleotides are released in the course of the reaction, they are very rapidly degraded to mononucleotides.

VI. PURIFICATION OF ALTEROMONAS NUCLEASE

A. Enzyme Assay

Assays for nuclease activity in all stages of purification are conveniently done by monitoring photometrically the release of nucleotides from single-stranded DNA. Calf thymus DNA (Sigma, Type I) is dissolved at a nominal (by weighing) concentration of 2.5 mg/ml in BE buffer (100 mM NaCl, 20 mM Tris-HCl, and 1 mM EDTA at a pH of 8.1) by stirring overnight in the cold and is then alkali-denatured and neutralized as described elsewhere[71]. After dialysis into

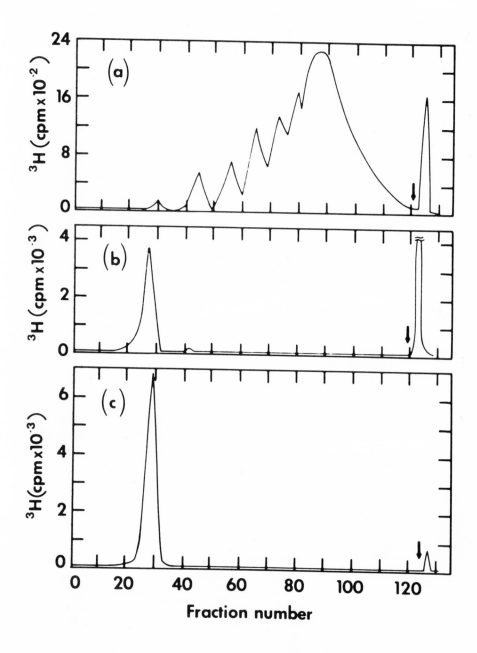

Fig. 6. Intermediate products of digestion of [3H]thymidine-labeled phage T7 DNA with (a) pancreatic DNAse I, and (b and c) puridied S form

Alteromonas nuclease. The T7 DNA was in the duplex form in (a) and (c) and was denatured in (b). Alteromonas nuclease digests were done as in the note to Table 1, except that the reactions were terminated before the development of limiting hyperchromicity. DNAse I digests were at 37°C in 15 mM NaCl, 10 mM Tris-HCl, 3 mM MgCl$_2$ and 1 mM EDTA (pH, 8.1) and were terminated by the addition of EDTA to 10 mM. Reaction mixtures were desalted on columns of Bio-Gel P-2, and all fractions containing substantial radioactivity were pooled, made 7 M in urea and 20 mM in Tris-HCl (pH, 7.8), and adsorbed to columns of DEAE-cellulose equilibrated in this buffer. Elution was with linear gradients running from the urea/Tris-HCl buffer above to the same buffer containing 0.3 M NaCl. An application of the buffer containing 2 M NaCl was made at the fraction indicated by the arrow in each panel, which eluted the high molecular weight material. The peak at left in (b) and (c) corresponds to 5'-mononucleo-tides, as confirmed by chromatography of [^{3}H]thymidine 5'-monophosphate. The near absence of mononucleotides in (a) is consistent with the mode of action of bovine pancreatic DNAse I[70].

BE buffer, appropriate amounts of 5 M NaCl, 0.5 M MgSO$_4$ or MgCl$_2$, and 0.5 M CaCl$_2$ (all these stock solutions contain 20 mM Tris-HCl and 1 mM EDTA at a pH of 8.1) are added to yield a stock solution that when diluted with 0.1 volume of CAM buffer (100 mM NaCl, 5 mM CaCl$_2$, 5 mM MgSO$_4$, 20 mM Tris-HCl, and 1 mM EDTA at a pH of 8.1)[4], gives the reaction mixture composition of the notes to Table 1. The concentration of single-stranded DNA in the stock solutions made up as above is approximately 650 µg/ml. Nine-tenths volume of the above mixture equilibrated at 30°C is mixed with 0.1 volume of enzyme in CAM buffer and 0.4 ml aliquots are withdrawn at appropriate intervals and mixed with 0.04 ml of 0.5 M sodium EDTA (pH, 8-8.5). Eight-tenths of a milliliter of 10% (w/v) HClO$_4$ is added, and the mixture, after standing on ice for not more than a few minutes, is passed through a nitrocellulose filter (13 mm diameter, 0.22 µm pore size, Millipore or Amicon Corp.) attached to a syringe[4,60]. This volume is convenient for filtration directly into a 1 cm path length spectrophotometer cuvette with a 4 mm wide cavity. The value of A^{260} is obtained. Blanks for experiments with crude nuclease preparations should be made up by addition of enzyme solution to the reaction mixture after the EDTA has been added. This will ensure that any acid-soluble ultraviolet-absorbing material in the nuclease preparations will not contribute to the net A^{260} when blank values are subtracted.

The unit definition of Vogt[60] is used, in which 1 unit will release 1 µg of nucleotides in 1 min. This unit is adapted here to the "standard" reaction conditions of 30°C in the above buffer. According to Vogt[60], 16.7 µg of nucleotides per milliliter of 6.7% (w/v) HClO$_4$ has an A^{260} value of 0.55. This means that the net A^{260} per minute multiplied by 30.3 and by the total reaction mixture volume (after addition of EDTA and $HClO_4$) yields the total number of units of activity expressed in the reaction mixture. The number of units per milliliter in the undiluted enzyme preparation is then calculated.

The reaction usually displays a constant velocity until the A^{260} of the released nucleotides exceeds approximately 1, after which a decrease in rate is often observed. Nonconcentrated culture supernatant usually contains 1-4 units of activity per milliliter, necessitating long (overnight) incubations to release sufficient acid-soluble material for reliable photometric assays when undiluted culture supernatant is used as 0.1 volume of the assay mixture.

B. Purification of "Fast" Form

Modified[4] AMS-4 medium[72] is made up with Difco Vitamin Assay Casamino Acids. Medium is inoculated with 0.02 volume of an overnight culture of Alteromonas espejiana BAL 31 and grown with vigorous aeration for at least 48 h, as enzyme is produced by stationary-phase cells. An aeration device containing a fritted (medium porosity) glass gas dispersion tube (Pyrex 39525-20M or equivalent) is recommended. The devices used in this laboratory are rubber stoppers that fit glass or plastic carboys (8-50 l) and are penetrated by three glass tubes, of which one is affixed to the gas dispersion tube, one serves as exhaust, and one is designed to deliver samples of the culture into test tubes when the exhaust tube is temporarily blocked.

Cells are removed from the culture fluid by centrifugation (10,000 g for 15 min) or, for cultures large enough that centrifugation is inconvenient, by filtration in a Millipore Pellicon cassette ultrafiltration apparatus (Millipore Corp.) with EGWP filters. Loaded with 10 filters, the apparatus can remove the cells from approximately 50 l of culture in a working day. It is important to make the clarified supernatant 1 mM in NaN_3 to inhibit secondary cell growth, which can be substantial even in the cold.

The same apparatus is used to concentrate the supernatants when volumes greater than 20 l are being processed. A large (1-2 l) pressure ultrafiltration device connected to a reservoir has been used for smaller volumes with a UM-10 (Amicon) or equivalent filter[4]. The Millipore PTGC membrane packet is used in the cassette device, with a new filter containing 5 ft^2 of usable area able to concentrate at least 75 l of cleared culture supernatant 100-fold in a day. The very expensive filter packets are reusable and should be cleaned and stored according to the manufacturer's advice, although it is our experience that such a packet will give rise to concentration of the supernatant at an unacceptably low rate and will retain appreciable amounts of enzyme activity after several hundred liters of supernatant have been passed through it. The manufacturer's instructions for flushing the filter after concentration of the material should be followed to recover as much nuclease activity as possible. Less than 1% of the nuclease activity is present in the filtrate discarded in the concentration process.

Further concentration of large preparations to 100-200 ml has been done by dialysis against dry sucrose, followed by dialysis against CAM buffer, or by use of vacuum-mediated concentration-dialysis devices like those available from Bio-Molecular Dynamics (Beaverton, OR). It has been found more recently, however, that the concentrate from the Millipore ultrafiltration apparatus can readily be further concentrated at least 10-fold by precipitation with acetone. One volume of ice-cold acetone is added to the concentrate (also on ice) and the mixture is allowed to stand near 0°C for 20 min before brief centrifugation (5,000-10,000 $\underline{g}$) in the cold. After removal of the supernatant, residual acetone is evaporated with a stream of air or nitrogen, and the precipitate is resuspended in CAM buffer. Recovery of the nuclease activity, as monitored in the assay using single-stranded DNA as substrate, is virtually complete. Much of the turbid material in these suspensions may be removed, without appreciable loss of enzyme activity, by centrifugation at 100,000 $\underline{g}$ for 1-2 h.

The key step in obtaining nuclease that displays "fast" kinetics of degradation of duplex DNA is affinity chromatography on columns of 5'-AMP covalently coupled through the C8 position to a hexane spacer arm, which is in turn coupled to agarose (P-L Laboratories, Cat. No. 5494).[63] Agarose with 5'-AMP covalently coupled through the N6-amino position is also a satisfactory resin. The column (0.8 X 10 cm) is equilibrated in CAM buffer, and the concentrated nuclease sample, either in CAM buffer or in the cell culture medium, is added. Up to 30,000 units in a volume of at least 120 ml may be placed onto a column of the above size with over 90% of the activity being bound and without the column being plugged by debris. The column is washed with several bed volumes of CAM buffer containing a total of 1 M NaCl and 1 mM NaN_3. Any turbid material in the preparation will have passed through the column by completion of the above steps. Elution of the nuclease itself is done with several bed volumes of the 1 M NaCl buffer above containing 15 mM 5'-AMP. The activity will elute beginning with the first fraction (fraction volume, near 1 ml) containing AMP and should be completely eluted by one bed volume thereafter.

Photometric assay of nuclease activity eluted with the AMP solution is hampered by the high background absorbance due to this nucleotide and also by its inhibitory effect on the nuclease activity. If it is desired to assay these fractions for nuclease activity or to terminate purification at this stage, chromatography on Sephadex G-25 columns (2.5 X 45 cm) eluted with CAM buffer will effectively remove the nucleotide. Nuclease preparations purified only to this stage may be used in such applications as exonucleolytic shortening of duplex DNA inasmuch as there appear to be no important interfering nuclease activities, such as might introduce strand breaks in linear duplex DNA molecules away from the termini.

A suitable step for further purification is gel-exclusion chromatography on Sephacryl S-200 (Pharmacia). If the enzyme to be purified is from the affinity chromatography step, samples should be dialyzed for 1 h into CAM buffer to reduce the NaCl concentration. Up to 10 ml of nuclease preparation is layered onto a 2.5 X 85 cm column, and elution of 2 ml fractions is carried out with CAM buffer. Some ultraviolet-absorbing material with very little nuclease activity elutes in the void volume, with the nuclease eluting reproducibly at a K_D value[73] of 0.26 or when a volume equivalent to about 44% of the sum of the included and excluded volumes has been eluted. This resin may be used at a flow rate high enough (see manufacturer's instructions) for the column to be run in several hours. The AMP from the affinity-column elution step is removed, as well as some contaminating protein.

Purification of the nuclease to yield only a single band on electrophoresis of denatured protein in denaturing polyacrylamide gels has been accomplished with linear polyacrylamide gradient slab gels (4-30%) purchased from Pharmacia. Up to 2 mg of total protein is layered on a single 8 X 8 X 0.3 cm gel, and electrophoresis is carried out as described elsewhere[61], except that the electrophoresis buffer contains 0.5 mM $CaCl_2$. The cytochrome c marker is caused to migrate to the bottom of the gel at a constant voltage of approximately 15 V/cm before the gel is removed, and slices approximately 2 mm wide are excised from each edge. Slices are stained for 5 min in Coomassie blue G-250 solution[64] and are destained with agitation in 5% (v/v) methanol-7% (v/v) acetic acid for 10 min or until the partially stained bands may be seen. The slices are aligned with the rest of the gel, so that the regions containing significant protein may be excised and homogenized with a glass tissue homogenizer in CAM buffer. Homogenized gel is removed by centrifugation and re-extracted with CAM buffer. The band containing the bulk of the nuclease activity will migrate roughly halfway through the gel under the above conditions for electrophoresis. Better recovery would probably be obtained by electrophoretic removal of the enzyme from minced gel slices, although this technique has not been tested in this application.

C. Purification of "Slow" Form

The concentrated culture supernatant is dialyzed into 0.25 X CAM buffer before adsorption to a 2.5 X 45 cm column of DEAE-cellulose (Whatman DE 52) equilibrated with the same buffer. Up to 500 ml of concentrated supernatant has been loaded onto such a column. The column is best loaded with a pump, because pigmented material adsorbs to the top of the column and retards flow. After washing with 100 ml of 0.25 X CAM buffer, the column is eluted with a 800 ml linear gradient of the above buffer to the same buffer containing 1 M NaCl. The nuclease activity reproducibly elutes near 0.4 M NaCl[4]. If further purification is desired, the remaining steps are as for the F form nuclease after concentration of the pooled active fractions from the ion-exchange column. Preparations subjected only to the ion-exchange chromatography step above should not contain interfering nuclease activities.

VII. DETECTION OF ALTERED HELICAL STRUCTURE IN NONSUPERCOILED CLOSED CIRCULAR DNA

Use of the F form nuclease is recommended in this application. For detection of cleavage with gel electrophoretic techniques[18], [^{3}H]thymidine-labeled PM2 form I DNA should be prepared[18] and converted to form I° DNA by the action of DNA topoisomerase I. This enzyme is readily prepared from mammalian cell nuclei[18,74,75], and the relaxed closed circular DNA is easily purified[18,37,76].

The details of modification of the form I° DNA with the agent to be tested for its ability to modify closed circular DNA covalently clearly depend on its solubility in aqueous solvents, reactivity, etc. Reactions may be carried out in mixed organic-aqueous solvents (20% (v/v) ethanol was used for reaction of duplex and form I° DNAs with N-acetoxy-N-2-acetylaminofluorene[18,77,78]) for compounds poorly soluble in water. NaCl at 50-100 mM in the reaction mixtures seems to stabilize the DNA against nicking. If there is an effective method to quench the reaction, this should be used; if not, mixtures should be placed on ice and subjected to ethanol precipitation or solvent extraction, followed in all cases by dialysis into BE buffer to remove residual reagent. If substantial nicking (greater than approximately 30%) is caused by the reaction, separation of the nicked from closed circular forms by centrifugation in propidium diiodide-CsCl buoyant density gradients[37] should be carried out.

Nuclease reactions should be carried out at 20°C in the buffer described in the note to Table 1 or in that buffer with the NaCl concentration changed to 1 M; 1 µg/ml ethidium bromide should also be present. This confers high stability of the DNA helix with respect to NaCl concentration, as well as optimal activity of the nuclease against single-stranded substrates (Figure 4). The intercalating dye ensures that negative superhelical turns will not be present in the substrate; ethidium at this level presents no problem with respect to inhibition of the nuclease[32]. The Alteromonas nuclease should be present at 50-150 units/ml to provide a stringent test for substrate sites in the modified DNA. Control reactions (nonmodified DNA) with F form nuclease at 120 units/ml in the 0.6 M NaCl buffer show greater than 80% PM2 form I° DNA remaining after 4 h at 20°C.

The gel electrophoresis assay to detect cleavage of radioactively labeled form I° DNA has been described in detail[18]. Form II DNA comigrates with form I° DNA in this technique, but will not interfere after a few minutes of incubation with nuclease under the above conditions, because of its rapid conversion to form III DNA and further degradation by attack at the termini. The presence of form II DNA in samples in the absence of nuclease digestion can be determined with the electrophoretic procedure with 0.06 µg/ml of ethidium bromide present in the gels[18]. Aliquots from reaction mixtures should be taken with care, because the determination is based only on the total amount of form I° DNA present, and not on the relative amounts of two forms.

VIII. CONTROLLED SIZE REDUCTION OF LINEAR DUPLEX DNAS THROUGH TERMINALLY DIRECTED HYDROLYSIS

For removal of several hundred nucleotides or more from the termini of a duplex DNA, the F form should be used. The S form may be better used if a much smaller number of nucleotides is to be removed, so that it is less likely that the nuclease will have to be diluted to a concentration at which there are fewer enzyme molecules than duplex termini in the very limited digests. There may be a minimal number of nucleotides that can be removed if the exonuclease reaction takes place in a "quasiprocessive" manner; e.g., several or more nucleotides are removed without the dissociation of the enzyme from the substrate. If the nucleotides removed processively during a single enzyme-substrate encounter are removed very rapidly compared with the time required for a dissociated enzyme molecule to reassociate and initiate a new round of degradation, a minimal number of nucleotides that can feasibly be removed might be observed. The overall guanine + cytosine content of the DNA and the actual sequences encountered by the enzyme will affect the extent of degradation and are expected to play a more important role in the removal of small numbers of nucleotides. It has been shown (Talmadge et al.[22] and K. Talmadge, private communication) that sequences containing "runs" of G-C base pairs can retard the action of the nuclease and that most of the shortened sequences obtained terminate in G-C base pairs. A dependence of nuclease reaction rate on guanine + cytosine content has been shown in much more extensive degradations by Legerski et al.[15].

Equation (1) with the appropriate kinetic parameters, taking V_m to be proportional to the number of units per milliliter, should be used to estimate reaction conditions for pilot digests. The 0.2 M NaCl buffer, for which kinetic data are available for both forms of the nuclease, is more convenient to use than the more concentrated salt solutions, if later digestion with other enzymes or ethanol precipitation is intended. Exonuclease activity is readily monitored in agarose or composite agarose-polyacrylamide gels[15,22]. The bands become broad with continued degradation owing to the asynchrony of the reaction, so that extended digestions (e.g., duplexes are shortened by 25% or more) will give rise to very broad bands in gels[15]. In such cases, analytical band[4] or preparative zone sedimentation techniques in alkaline solvents are more useful in estimating the extents to which duplexes have been shortened by the action of the Alteromonas nuclease. The extent of degradation in the more extended digestions can be monitored photometrically, as described for the assay of hydrolysis of single-stranded DNA. Measurement of acid-soluble radioactivity proved a convenient assay for radioactively labeled substrates and should be useful at lower extents of degradation than the photometric assay.

ACKNOWLEDGMENTS

Work from the authors' laboratories was supported by grants GM-21839 from the National Institute of General Medical Sciences and CA-11761 and CA-16527 from the National Cancer Institute. J. L. H. was supported during part of these studies by Postdoctoral Fellowship CA-00695 from the National Cancer Institute.

REFERENCES

1. Espejo, R. T. and Canelo, E. (1968) Virology, 34, 738-747.
2. Espejo, R. T., Canelo, E. and Sinsheimer, R. L. (1969) Proc. Natl. Acad. Sci. U.S.A., 63, 1164-1168.
3. Laval, F. (1974) Proc. Natl. Acad. Sci. U.S.A., 71, 4965-4969.

4. Gray, H. B., Jr., Ostrander, D. A., Hodnett, J. L., Legerski, R. J. and Robberson, D. L. (1975) Nucleic Acids Res., 2, 1459-1492.
5. Espejo, R. T. and Canelo, E. (1968) J. Bacteriol, 95, 1887-1891.
6. Chan, K. Y., Baumann, L., Garza, M. M. and Baumann, P. (1978) Int. J. Syst. Bacteriol., 28, 218-222.
7. Tomlinson, R. V. and Tener, G. (1963) Biochemistry, 2, 697-702.
8. Chen, P. S., Toribara, T. Y. and Warner, H. (1956) Anal. Chem., 28, 1756-1758.
9. Vinograd, J., Lebowitz, J., Radloff, R., Watson, R. and Laipis, P. (1965) Proc. Natl. Acad. Sci. U.S.A., 53, 1104-111.
10. Studier, F. W. (1965) J. Mol. Biol., 11, 373-390.
11. Sgaramella, V. (1972) Proc. Natl. Acad. Sci. U.S.A., 69, 3389-3393.
12. Beard, P., Morrow, J. F. and Berg, P. (1973) J. Virol., 12, 1303-1313.
13. Legerski, R. J. (1977) Ph.D. Dissertation, University of Houston.
14. Wang, J. C. (1974) J. Mol. Biol., 87, 797-816.
15. Legerski, R. J., Hodnett, J. L. and Gray, H. B., Jr. (1977) Nucleic Acids Res., 5, 1445-1464.
16. Mertz, J. E. and Davis, R. W. (1972) Proc. Natl. Acad. Sci. U.S.A., 69, 3370-3374.
17. Koller, T., Sogo, J. M. and Bujard, H. (1974) Biopolymers, 13, 995-1009.
18. Legerski, R. J., Gray, H. B., Jr. and Robberson, D. L. (1977) J. Biol. Chem., 252, 8740-8746.
19. McDonell, M. W., Simon, M. N. and Studier, F. W. (1977) J. Mol. Biol., 110, 119-146.
20. Roberts, T. M., Kacich, R. and Ptashne, M. (1979) Proc. Natl. Acad. Sci. U.S.A., 76, 760-764.
21. Pritchard, A. E., Kowalski, D. and Laskowski, M., Sr. (1977) J. Biol. Chem., 252, 8652-8659.
22. Talmadge, K., Stahl, S. and Gilbert, W. (1980) Proc. Natl. Acad. Sci. U.S.A. 77, 3369-3373.
23. Backman, K., Ptashne, M. and Gilbert, W. (1976) Proc. Natl. Acad. Sci. U.S.A., 73, 4174-4178.
24. Mechali, M., de Recondo, A.-M. and Girard, M. (1973) Biochem. Biophys. Res. Commun., 54, 1306-1320.
25. Germond, J.-E., Vogt, V. M. and Hirt, B. (1974) Eur. J. Biochem., 43, 591-600.
26. Wiegand, R. C., Godson, G. N. and Radding, C. M. (1975) J. Biol. Chem., 250, 8848-8855.
27. Pritchard, A. E. and Laskowski, M., Sr. (1978) J. Biol. Chem., 253, 6606-6613.
28. Pritchard, A. E. and Laskowski, M., Sr. (1978) J. Biol. Chem., 253, 7989-7992.
29. Bauer, W. R. and Vinograd, J. (1968) J. Mol. Biol., 33, 141-172.
30. Bauer, W. R. (1978) Ann. Rev. Biophys. Bioeng., 7, 287-313.
31. Woodworth-Gutai, M. and Lebowitz, J. (1976) J. Virol. 18, 195-204.
32. Lau, P. P. and Gray, H. B., Jr. (1979) Nucleic Acids Res., 6, 331-357.
33. Wang, J. C. (1974) J. Mol. Biol. 89, 783-801.
34. Pulleyblank, D. E. and Morgan, A. R. (1975) J. Mol. Biol., 91, 1-13.
35. Liu, L. F. and Wang, J. C. (1975) Biochim. Biophys. Acta, 395, 405-412.
36. Wang, J. C. (1969) J. Mol. Biol., 43, 25-39.
37. Hudson, B., Upholt, W. B., Devinny, J. and Vinograd, J. (1969) Proc. Natl. Acad. Sci. U.S.A., 62, 813-820.
38. Gray, H. B., Jr., Upholt, W. B. and Vinograd, J. (1971) J. Mol. Biol., 62, 1-19.
39. Bauer, W. R. and Vinograd, J. (1970) J. Mol. Biol., 47, 419-435.
40. Hsieh, T.-S. and Wang, J. C. (1975) Biochemistry, 14, 527-535.

41. Benham, C. (1979) Proc. Natl. Acad. Sci. U.S.A., 76, 3870-3874.
42. Beerman, T. and Lebowitz, J. (1973) J. Mol. Biol., 79, 451-470.
43. Lebowitz, J., Garon, C. G., Chen, M. and Salzman, N. P. (1976) J. Virol. 18, 205-210.
44. Abelson, J. and Thomas, C. A., Jr. (1966) J. Mol. Biol., 18, 262-291.
45. Bujard, H. (1969) Proc. Natl. Acad. Sci. U.S.A., 62, 1167-1174.
46. Jaquemin-Sablon, A. and Richardson, C. C. (1970) J. Mol. Biol., 47, 477-493.
47. Hayward, G. S. and Smith, M. G. (1972) J. Mol. Biol., 63, 397-407.
48. Bujard, H. and Hendrickson, H. E. (1973) Eur. J. Biochem., 33, 517-528.
49. Wang, J. C. and Liu, L. F. (1979) in Molecular Genetics, Part III, Taylor, J. H., ed., Academic Press, New York, pp. 65-88.
50. Harless, J. (1979) M. S. Thesis, University of Texas Health Science Center at Houston, Graduate School of Biomedical Sciences.
51. Friedberg, E. C. and King, J. J. (1971) J. Bacteriol., 106, 500-507.
52. Friedberg, E. C. (1975) in Molecular Mechanisms for Repair of DNA, Vol. 5B, Hanawalt, P. C. and Setlow, R. B., eds., Plenum Press, New York, pp. 125-134.
53. Feldberg, R. S. and Grossman, L. (1976) Biochemistry, 15, 2402-2408.
54. Kriek, E., Miller, J. A., Juhl, U. and Miller, E. C. (1967) Biochemistry, 6, 177-182.
55. Becker, E. F., Zimmerman, B. K. and Geiduschek, E. P. (1964) J. Mol. Biol., 8, 377-391.
56. Singer, B. (1975) Prog. Nucleic Acid Res. Mol. Biol., 15, 219-284.
57. Ljungquist, S., Andersson, A. and Lindahl, T. (1974) J. Biol. Chem., 249, 536-540.
58. Lindahl, T. and Andersson, A. (1972) Biochemistry, 11, 3618-3623.
59. Kohn, K. and Spears, C. (1967) Biochim. Biophys. Acta, 145, 734-741.
60. Vogt, V. M. (1973) Eur. J. Biochem., 33, 192-200.
61. Dietz, A. A. and Lubrano, T. (1967) Anal. Biochem., 20, 246-257.
62. Laemmli, U. K. (1970) Nature, 227, 680-685.
63. Nees, D. W. (1978) Senior Honors Thesis, University of Houston.
64. Weber, K. and Osborn, M. (1969) J. Biol. Chem., 244, 4406-4412.
65. Segrest, J. P. and Jackson, R. L. (1972) in Methods in Enzymology, Vol. 28B, Ginsburg, V., ed., Academic Press, New York, pp. 54-63.
66. Price, P. A., Liu, T.-Y, Stein, W. A. and Moore, S. (1969) J. Biol. Chem. 244, 917-923.
67. Liao, T.-H. (1974) J. Biol. Chem., 249, 2354-2356.
68. Ando, T. (1966) Biochim. Biophys. Acta, 114, 158-168.
69. Johnson, P. H. and Laskowski, M., Sr. (1970) J. Biol. Chem., 245, 891-898.
70. Sinsheimer, R. L. (1954) J. Biol. Chem., 208, 445-459.
71. Horiuchi, K. and Zinder, N. D. (1972) Proc. Natl. Acad. Sci. U.S.A., 69, 3220-3224.
72. Espejo, R. T., Canelo, E. S. and Sinsheimer, R. L. (1971) J. Mol. Biol., 56, 597-621.
73. Gelotte, B. (1960) J. Chromatog., 3, 330-342.
74. Hancock, R. (1974) J. Mol. Biol., 86, 649-663.
75. Germond, J. E., Hirt, B., Oudet, P., Gross-Bellard, M. and Chambon, P. (1975) Proc. Natl. Acad. Sci. U.S.A., 72, 1843-1847.
76. Radloff, R., Bauer, W. R. and Vinograd, J. (1967) Proc. Natl. Acad. Sci. U.S.A., 57, 1514-1521.
77. Miller, E. C., Juhl, U. and Miller, J. A. (1966) Science, 153, 1125-1127.
78. Fuchs, R. P. P. and Daune, M. (1971) FEBS Lett., 14, 206-208.

S_1 NUCLEASE OF ASPERGILLUS ORYZAE

George W. Rushizky

Laboratory of Nutrition and Endocrinology, NIAMDD, NIH,
Bethesda, Maryland 20205

S_1 nuclease was first described by T. Ando in 1966. After partial purification, the enzyme hydrolyzed single-stranded RNA and DNA to nucleoside 5'-monophosphates. Under the right conditions, double-stranded nucleic acids were completely resistant to the enzyme[1].

Other enzymes with similar properties[2] have been found in Neurospora crassa[3], mung beans[4], Penicillium citrinum[5], and yeast[6], but S_1 nuclease has been more widely used because it is easy to prepare in large amounts from commercially available amylase powders derived from Aspergillus oryzae.

I. PURIFICATION

A single chromatographic step on DEAE-cellulose was found to remove most nuclease activity not specific for single-stranded nucleic acids[7]. However, use of such enzyme preparations also led to disagreements -e.g., there was a questioned report that S_1 nuclease could hydrolyze duplex DNA as well as, although much slower than, single-stranded DNA[8,9].

The first extensive purification of the enzyme was that described by Vogt[10]. It involved five steps: heating to 70°C, salting out with $(NH_4)_2SO_4$, and column chromatography on DEAE-cellulose, Sulfo-Sephadex, and Sephadex G-100. The yield was 27%, and the enzyme was reported to be 90% pure. This degree of purification yielded an enzyme capable of hydrolyzing single-strand nucleic acid regions in DNA-DNA or DNA-RNA hybrids. However, the S_1 nuclease still contained RNases[10]. This was not surprising, because Aspergillus oryzae produces not only S_1 nuclease, but also RNase T_1 and RNase T_2. Such contaminants would not interfere with DNA-RNA hybridization, in which RNases are routinely added for the removal of nonhybridized RNA[12].

To remove RNases, chromatography of S_1 nuclease on DEAE-cellulose was carried out at a pH at which the three enzymes could be separated from each other; in addition RNase T_2 was removed by a CM-cellulose step in which the RNase was bound to the adsorbent and S_1 nuclease was not. This procedure yielded an S_1 nuclease preparation suitable for studies of viral RNA[13] and tRNAs[14,15].

S_1 nuclease was further purified[16] by taking advantage of its glyco-protein properties[17]. Thus, chromatography on concanavalin-A Sepharose and on phenyl Sepharose was added to published purification procedures[10,13] to yield product purified 1,600-fold with a yield of 32%. This S_1 nuclease was free of RNases and the DNase specific for native DNA[18] and free of contaminating proteins, as deduced from SDS-polyacrylamide gel electrophoresis[16]. It was also found to contain an inherent nucleotidase activity on 3'-ribonucleotides and to a lesser extent on 2'-ribonucleotides - a property shared with the single-strand-specific nucleases from mung beans[4] and Penicillium[5].

II. PHYSICAL PROPERTIES

The enzyme is assayed by measuring the release of acid-soluble products from heat-denatured DNA, single-stranded viral DNA, or RNA. The units of enzyme activity are usually those defined by Vogt[10]. The reported ratio of activities toward single-stranded DNA and RNA ranges from 2 (Oleson and Sasakuma[16], Rushizky et al.,[13]) to about 5-7 (Vogt[10]). This discrepancy may derive from the presence of duplex regions in the substrate. For example, single-stranded fd DNA was hydrolyzed twice as fast as heat-denatured calf thymus DNA[13], and heating of RNA for 2 minutes at 100°C doubled the RNase activity of S_1 nuclease compared to DNase activity[10].

A) Size. The molecular weight of the enzyme derived from SDS-polyacrylamide gel electrophoresis[19] was first reported to be 32,000 daltons[10]. In sucrose gradients at high or low salt concentrations at a pH of 4.6 and 7.0, the nuclease activity sedimented at about 3.3 S. For a globular protein, this value implied a molecular weight of 36,000; thus, the enzyme exists as a monomer[10]. Oleson and Sasakuma[16] obtained a molecular weight of 38,000 by SDS-gel electrophoresis - closer to the molecular weights of 39,000 and 44,000 of mung bean nuclease[4] and nuclease P_1[17], respectively. Glycoproteins containing more than 10% carbohydrate behave anomalously[1] in SDS-gel electrophoresis[20], and S_1 nuclease was found to contain 18% carbohydrate[16]; hence, additional methods of analysis should be used.

B) Isoelectric point. Isoelectric focusing[21] gave a pI of 4.3-4.4 for crude preparations of the enzyme[13], whereas highly purified S_1 nuclease[16] showed two minor forms with pI values of 3.35 and 3.53 and a major form (69%) with a pI of 3.67. Such multiple forms may be due, in part, to degradation of S_1 nuclease during the preparation of the commercial amylase used as starting material or to the heating to 70°C during the purification of the enzyme.

III. CATALYTIC PROPERTIES

A note of caution is in order. The pH, metals, salt concentration, and temperature may affect the enzyme and/or its substrates. Furthermore, commercial S_1 nuclease preparations may contain a double-strand-specific nuclease removed by further purification[29-31], and commercial substrates may contain various amounts of metals.

A) _pH Optimum_. The pH optimum[10] is 4.0-4.3, with half-maximal rates at 3.3 and 4.9. Higher pH values (4.6-5.0) have been used to avoid possible nicking of DNA substrates by acid depurination.

The pH optimum of mung bean nuclease for native DNA falls off more rapidly than that for single-stranded DNA[22]. By analogy, amplification of the single-strand activity of S_1 nuclease should also increase at a pH 0.3-0.5 units above the optimum, as has been noted in several instances, such as the specific hydrolysis of the anticodon loop of tRNAs[24].

B) _Metal Dependence_. As observed by Ando[1], S_1 nuclease requires Zn^{+2} for activity. Co^{+2} and Hg^{+2} can replace zinc, but are less effective[10,25]. Metal concentrations of 0.01-1.0 mM are optimal.

Dialysis at a pH of 4.3 against 1 mM EDTA completely inactivates the enzyme, and addition of zinc can reactivate[10] it to at least 70%.

C) _Salt_ _Concentration_. S_1 nuclease is optimally active at 0.1 M Nacl and relatively insensitive to ionic strength between 0.01-0.2 M NaCl, and degrades DNA at 55% of the maximal rate in 0.4 M NaCl[7]. As expected from the increase in the fraction of helical DNA with increasing ionic strength, the single-strand specificity is higher at higher salt concentrations than the double-strand specificity[10]. With SV40 DNA, one group of breaks was noted in 0.25 M NaCl, and two groups of breaks at 0.075 M NaCl. Between 0.01 and 0.25 M NaCl, there was a fivefold decrease in S_1 nuclease activity[8].

D) _Denaturing Agents_. The enzyme has remarkable stability. Thus, 60% of the RNA and all the DNA in HeLa cell lysates were degraded in 9 M urea-0.1% SDS at 45°C. The enzyme retained 7% of its maximal acitivity, and cell proteins suitable for gel electrophoresis were isolated[26]. In the absence of proteins, S_1 nuclease loses activity in the presence of SDS at more than 0.04%.

As a tool to study low-melting regions in DNA[27], the enzyme was also found to be active in 60% formamide, 50% dimethylsulfoxide, 30% dimethylformamide, and 2% formaldehyde[28].

E) _Temperature_. The enzyme has been used at 0-65°C. The single-strand activity of S_1 nuclease increased tenfold between 37°C and 65°C, but the effect of temperature on double-strand activity has not been studied[32].

F) _Stability_. In the presence of Zn^{+2}, the enzyme is stable[10] in 50% glycerol at -20°C.

IV. SPECIFICITY AND APPLICATIONS

In optimal conditions, the difference in the rate of hydrolysis between single- and double-stranded nucleic acids has been estimated to be 75,000-fold[25]. Details and examples of applications are presented below.

A) Cleavage of Single-Stranded Ends. S_1 nuclease readily removes tails from hybrids, such as the 12 nucleotides at the cohesive ends of lambda DNA. This is measured by inability to form H-bonded circles[25,33].

As an example, repetitive sequence elements from three sea urchin DNAs were prepared by reassociation to a Cot of 40, digestion with S_1 nuclease, and ligation of synthetic Eco RI sites to the ends. After cloning in plasmids, labeling, and strand separation, these fragments were hybridized with 800- to 900-nucleotide long unlabeled DNA. Estimates of frequencies were made at genomic or cloned DNA excess and used to relate the three DNAs to each other[46].

Similarly, removal of tails was used after the preparation of cDNA from the mRNA for preuteroglobin. Membrane-bound polysomes of induced endometrium were the source of the mRNA from which the cDNA was prepared with reverse transcriptase[47]. Such cDNA probes have been used to study gene frequencies, mRNA metabolism, and transcription of chromatin in vitro. After cloning, this yields a DNA of more limited sequence heterogeneity than DNAs prepared from total or enriched cellular DNA species[48,49].

B) Cleavage of Loops. Constructed heteroduplexes of known size and location of loops present two possible sites of hydrolysis: in the strand containing the loop and in the strand opposite the loop[25]. At low enzyme concentrations, alkaline sucrose gradient centrifugation revealed hydrolysis of the loop, but only 23% of the opposite strands were cut. An enzyme excess of 100-fold increased the latter value to 60%.

Cleavage at loops, as well as trimming of tails, permits better quantitation of nucleic acid hybridization than only chromatography on hydroxyapatite[7,50,51]. The latter suffers from the inability to distinguish between tails joined covalently to regions of reassociated complexes and duplex DNA itself, so that such single-stranded sequences are eluted in the double-stranded fraction. A disadvantage in the use of S_1 nuclease is that nonpaired regions are lost.

As an example, S_1 nuclease was used for the determination of the size distribution of resistant DNA duplexes in reassociation products[52]. The results of chromatography on hydroxyapatite indicated a short-period interspersion pattern in the starfish genome[53].

The extent of hybridization after S_1 nuclease treatment can also give

information on transcriptional controls. Thus, with vaccinia virus, hybridization of polyadenylated cytoplasmic RNA of infected cells was carried out
with denatured viral DNA probes. Early RNA hybridized to 26% and late RNA to
42% of vaccinia DNA[54].

With polyoma virus, up to 90 base pairs in sequences between the origin of
replication and the initiation codon for early protein synthesis were found to
be nonessential. This result was obtained by treating viable mutant DNAs,
generated by reinfection of susceptible cells, with restriction endonucleases
and then with S_1 nuclease[55].

S_1 nuclease was also used to find DNA regions with anomalous configura-
tions, such as single-stranded segments capable of binding regulatory proteins.
The lac operator was postulated to have single-stranded sites in the absence
of lac repressor[56], but this result was not confirmed by others with S_1
nuclease[57] or other methods[58].

C) <u>Cleavage at Nicks</u>. The enzyme will cut DNA at a single mismatched
base[34]. However, not all molecules containing nicks are cut. Thus, T5 DNA
with genetically-defined single-base nicks revealed cuts by S_1 nuclease
after alkaline sucrose gradient centrifugation, but even in the presence
of excess enzyme, molecules with nicks remained[25].

Nevertheless, T5 DNA could be denatured, partially renatured, and digested
with small amounts of S_1 nuclease to yield six fragments correlated in size
with the pattern expected from the sites of nicks in the DNA. After annealing
to ^{32}P end-labeled tRNA species, it was found that tRNA^{arg} hybridized exclusively to only one of the fragments[59].

In another application of nick hydrolysis, long-term labeled nuclear DNA
from various developmental stages of sea urchins was examined for single-
stranded regions susceptible to S_1 nuclease by sedimentation in neutral
and alkaline sucrose gradients. Size classes of DNA were found to change
with development to morula, blastula, and gastrula stages[60].

Treatment with S_1 nuclease also provided a sensitive method for measuring gamma irradiation-induced single-strand breaks and repair in eukaryotic
DNAs. Such measurement was obtained by assay of the fraction of duplex DNA
resistant to S_1 nuclease after irradiation and alkaline denaturation[61].

However, S_1 nuclease could not be used to excise a block of DNA between
mismatched single bases in two different heteroduplexes[62].

D) <u>Cleavage of Ultraviolet-Irradiated DNA</u>. Lambda DNA, irradiated to
have 50 pyrimidine dimers and with about half the strands intact, revealed

no intact strands after S_1 nuclease treatment and alkaline sucrose gradient centrifugation, which gave a number average molecular weight of 1.5×10^6. Few double-strand cuts were revealed by neutral sucrose gradient centrifugation[25]. Similar results were noted by others[35].

After trimethylpsoralen treatment and UV-irradiation, the extent of cross-linking in DNA was measured with S_1 nuclease after a denaturation-renaturation cycle[63].

E) Cleavage of Superhelical DNA. S_1 nuclease also hydrolyzed double-stranded DNA strained by superhelix formation. This action depends more on ionic strength and temperature than does the hydrolysis of single-stranded DNA and involves two steps, to the relaxed circular and then to the linear form. The intermediate can be isolated by neutral sucrose gradient centrifugation. This has been observed with ØX174 RF1 DNA[8,29,36] and polyma DNA[37]. In all cases, cleavage of superhelical DNA was limited to the one or two cuts needed to convert the strained to the relaxed forms. The sites cut were distributed throughout the molecule, at random with G4 and G14 phages[38], but not with SV40, where it was observed only in one or two small regions[8]. To identify the sites of hydrolysis, later restriction endonuclease treatment was use to identify new fragments[37] found in about 15% of all molecule with SV40 DNA[25], or by polynucleotide kinase labeling of S_1 nuclease digests. The regions where S_1 nuclease cleaves superhelical DNA are believed to be those where the DNA doubles back on itself to form a partially denatured site[39].

F) Hydrolysis of RNA Hybrids. For RNA, resistance to S_1 nuclease was found to decrease from double-stranded DNA to poly A and poly U, and then to cellular and viral RNAs. A general agreement between hyperchromicity and S_1 nuclease restance was observed, which was highly dependent on temperature and salt concentration[40,41]. Viral RNA resistance was also studied in more detail with S_1 nuclease prepared free of contaminating RNases. Over 70% of MS2 RNA, which has a helical content in vitro of about 65%, was converted to stable oligomers of 25,000 daltons or more. These compounds were stable to further hydrolysis by 20-fold higher S_1 nuclease concentrations. No such stable oligomers were isolated when the RNA was first denatured by formaldehyde unless the reaction was carried out under conditions that favor helical structure (0.25 M NaCl at 0°C)[13].

By analogy, S_1 nuclease was expected to generate stable oligomers from both sides of the protruding anticodon loop of tRNAs, i.e., tRNA halves. This was indeed observed[42], and at a pH of 5.7 yields of up to 92% halves, stable to 50-fold-higher enzyme concentrations, were obtained[24].

RNA: DNA hybrids, however, were found not to be completely equivalent to DNA: DNA hybrids. S_1 nuclease treatment of RNA: DNA hybrids labeled with ^{32}P and ^{33}P produced compounds containing 20% more RNA than DNA[44]. The excess RNA could be removed with RNase T_1 or with pancreatic RNase. This discrepancy, not found with use of the <u>Neurospora</u> single-strand-specific nuclease[45], was presumed to be due to overlapping double-stranded tails of looped-out RNA[44].

Small RNAs have been mapped with S_1 nuclease and other RNases in the presence of different anions and cations, at various temperatures and ionic strengths, and with end-labeled molecules[64,65], as a variation of the rapid gel sequencing technique of Donis-Keller et al.[66]. This procedure can reveal conformational properties of 4S and 5S RNAs or their precursors[67-71] and has not yet been fully explored. Results so obtained may be affected by preferential hydrolysis of a minor, rather than a major, conformation; and accessibility of hydrolysis sites could be limited, in part, to enzymes smaller than S_1 nuclease, such as RNase T_1.

SUMMARY

S_1 nuclease has found many uses in the analysis of the structure of nucleic acids, and more new applications, such as the mapping of splice points[72] of early mRNAs in SV40, will undoubtedly be found.

V. REFERENCES

1. Ando, T. (1966) Biochem. Biophys. Acta 114, 158-168.
2. Kowalski, D., and Laskowski, M., Sr. (1975) in Handbook of Biochemistry and Molecular Biology, third ed., Fasman G.D., ed., Nucleic Acids, Vol. 2, CRD Press, Cleveland, Ohio, pp. 491-531.
3. Linn, S., and Lehman, I.R. (1965) J. Biol. Chem 240, 1287-1293.
4. Kowalski, D., Kroeker, W.D., and Laskowski, M., Sr. (1976) Biochemistry 15, 4457-4463.
5. Fujimoto, M., Kuninaka, A., and Yoshino, H. (1974) Agr. Biol. Chem. 38, 777-783.
6. Lee, S.Y., Nakao, Y., and Bock, R.M. (1968) Biochem. Biophys. Acta 151, 126-136.
7. Sutton, W.D. (1971) Biochem. Biophys. Acta 240, 532-538.

8. Beard, P., Morrow, J.F., and Berg, P. (1973) J. Virol, 13, 1303-1313.

9. Godson, G.N. (1973) Biochem. Biophys. Acta 308, 59-67.

10. Vogt, V.M. (1973) Eur. J. Biochem. 33, 192-200.

11. Egami, F., and Nakamura, K. (1969) in Microbial Ribonucleases, Kleinzeller, A., ed., Springer Verlag, N.Y., pp. 18-33.

12. Gillespie, D., and Spiegelman, S. (1965) J. Mol. Biol. 12, 829-842.

13. Rushizky, G.W., Shaternikov, V.A., Mozejko, J.H., and Sober, H.A. (1975) Biochemistry 14, 4221-4226.

14. Flashner, M.S., and Vournakis, J.N. (1977) Nucleic Acids Res. 4, 2307-2311.

15. Rushizky, G.W., and Mozejko, J.H. (1977) Anal. Biochem. 77, 562-566.

16. Oleson, A.E., and Sasakuma, M. (1980) Arch. Biochem. Biophys., in press.

17. Kuninaka, A. (1976) in Microbial Production of Nucleic Acid-related Substances (Ogata, K., ed.) Wiley, N.Y., pp. 75-86.

18. Rushizky, G.W., and Whitlock, J.P. (1977) Biochemistry 16, 3256-3261.

19. Weber, K., and Osborn, M. (1969) J. Biol. Chem. 244, 4406-4412.

20. Segrest, J.P., and Jackson, R.L. (1972) in Methods in Enzymology (Ginsburg, V., ed.) Vol. 28, Academic Press, N.Y., pp. 54-63.

21. Wrigley, C.W. (1971) in Methods in Enzymology (Jakoby, W.B., ed.) Vol. 22, Academic Press, N.Y., pp. 559-564.

22. Kroeker, W.D., Kowalski, D., and Laskowski, M. (1976) Biochemistry 15, 4463-4468.

23. Kowalski, D., Kroeker, W.D., and Laskowski, M. (1976) Biochemistry 15, 4457-4463.

24. Tal, J. (1975) Nucleic Acids Res. 2, 1073-82.

25. Wiegand, R.C., Godson, G.N., and Radding, C.M. (1975) J. Biol. Chem. 250, 8848-55.

26. Zechel, K., and Weber, K. (1977) Eur. J. Biochem. 77, 133-139.

27. Kedzierski, W., and Laskowski, M. (1973) J. Biol. Chem. 248, 1277-80.

28. Hutton, J.P., and Wetmur, J.G. (1975) Biochem. Biophys. Res. Commun. 66, 942-948.

29. Godson, G.N., (1973) Biochem. Biophys. Acta 308, 59-67.

30. Johnson, J.D., and St. John, T. (1974) Fed. Proc. 33, 1521.

31. Hahn, W.E., and Van Ness, J. (1976) Nucleic Acids Res. 3, 1419-23.

32. Rushizky, G.W., unpublished observations.

33. Wu, R., and Ghangos, J. (1975) J. Biol. Chem. 250, 4601-4.

34. Shenk, T.E., Rhodes, C., Rigby, P.W.J., and Berg, P. (1975) Proc. Natl. Acad. Sci. 72, 989-993.

35. Shishido, K., and Ando, T. (1979) Biochem. Biophys. Res. Commun. 59, 1380-88.

36. Mechali, M., de Recondo, A-M., and Girard, M. (1973) Biochem. Biophys.

Res. Commun. 54, 1306-1320.

37. Germond, J.E., Vogt, V.M., and Hirt, B. (1974) Eur. J. Biochem. 43, 591-600.

38. Godson, G.N. (1974) Virology 58, 272-289.

39. Wang, J.C. (1974) J. Mol. Biol. 87, 797-816.

40. Shishido, K., and Ando, T. (1972) Biochem. Biophys. Acta 287, 477-484.

41. Shishido, K., and Ikeda, Y. (1970) J. Biochem. Tokyo, 759-767.

42. Harada, F., and Dahlberg, J.E. (1975) Nucleic Acids Res. 2, 865-871.

43. Shishido, K., and Ando, T. 1975) Agric. Biol. Chem. 39, 673-81.

44. Wittelsberger, S.C., and Hansen, J.N. (1977) Nucleic Acids Res. 4, 1829-35.

45. Rabin, E.Z., Mustard, M., and Fraser, J.J. (1968) Can. J. Biochem. 46, 1285-91.

46. Moore, G.P., Scheller, R.H., Davidson, E.H., and Britten, R.J. (1978) Cell 15, 649-660.

47. Arnemann, J., Heins, B., and Beato, M. (1979) Eur. J. Biochem. 99, 361-7.

48. Efstratiadis, A., Maniatis, T., Kafatos, F.C., Jeffrey, A., and Vournakis, J. (1975) Cell 4, 363-78.

49. Monahan, J.J., McReynolds, L.A., and O'Malley, B.W. (1976) J. Biol. Chem. 251, 7355-62.

50. Crosa, J.H., Brenner, D.J., and Falkow, S. (1973) J. Bacteriol. 115, 904-11.

51. Britten, R.J., and Kohne, D.E. (1968) Science 161, 528-37.

52. Smith, M.J., Britten, R.J., and Davidson, E.H., (1975) Proc. Natl. Acad. Sci. 72, 4805-9.

53. Smith, M.J., and Boal, R. (1978) Can. J. Biochem. 56, 1048-54.

54. Boone, R.F., and Moss, B. (1978) J. Virol. 26, 554-69.

55. Bendig, M.M., and Folk, W.R. (1979) J. Virol. 32, 530-5.

56. Chan, H.W., and Wells, R.D. (1974) Nature 252, 205-9.

57. Marians, K.J., and Wu, R. (1976) Nature 260, 360-3.

58. Wang, J.C., Barkley, M.D., and Bourgeois, S. (1974) Nature 251, 247-9.

59. Desai, S.M., Hunt, C., Locker, J., and Weiss, S.B. (1978) J. Biol. Chem. 253, 6544-50.

60. Case, S.T., Mongeon, R.L., and Baker, R.F. (1974) Biochem. Biophys. Acta 349, 1-12.

61. Sheridan, R.B., and Huang, P.C. (1977) Nucleic Acids. Res. 4, 299-318.

62. Dodgson, J.B., and Wells, R.D. (1977) Biochemistry 16, 2374-79.

63. Ben-Hur, E., Prager, A., and Biklis, E. (1979) Photochem. Photobiol. 29, 921-4.

64. Wurst, R., Vournakis, J., and Maxam, A. (1978) Biochemistry 17, 4493-99.

65. Wrede, P., Wurst, R., Vournakis, J., and Rich, A. (1979) J. Biol. Chem. 254, 9608-16.

66. Donis-Keller, H., Maxam, A., and Gilbert, W. (1977) Nucleic Acids Res. 4, 2527-31.

67. Bikoff, E.K., LaRue, B.F., and Gefter, M.L. (1975) J. Biol. Chem. 250, 6248-55.

68. Khan, M.S., and Maden, B.E. (1976) FEBS Lett. 15, 105-10.

69. Nichols, J.L., and Welder, L. (1979) Biochem. Biophys. Acta 561, 435-44.

70. Manale, A., Guthrie, C., and Colby, D. (1979) Biochemistry 18, 77-83.

71. Barber, C.J., and Nichols, J.L. (1979) Can. J. Biochem. 56, 357-64.

72. Berk, A.J., and Sharp, P.A. (1978) Proc. Natl. Acad. Sci. 75, 1274-78.

Analysis of Nucleic Acid Structure by RNase H

Robert J. Crouch

Laboratory of Molecular Genetics
National Institue of Child Health and Human Development
National Institutes of Health
Bethesda, Maryland 20205

I. INTRODUCTION

"Ribonuclease H" is the name given to enzymes that can degrade the RNA portion of DNA-RNA hybrids[1,2]; some papers have used the name "hybridase"[3]. RNase H was originally discovered in Hausen's laboratory[1,2] during examination of various fractions from DEAE-cellulose columns for factors that affect eukaryotic RNA polymerases, specifically RNA polymerase II of calf thymus. Transcription of single-stranded DNA by RNA polymerase leads to the formation of DNA-RNA hybrids[4]. In the presence of RNase H, the transcript is degraded; this gives RNase H the appearance (in an assay for acid-insoluble products) of being an inhibitor of RNA polymerase II.

Is RNase H peculiar to calf thymus ? The discovery of RNase H in calf thymus quickly led to its discovery in a wide variety of other organisms, including E. coli[3,5-9], yeast[10,11], KB cells[12], Krebs II ascites cells[13], and avian myeloblastosis virus[12,14-16]. In fact, two forms of RNase H have been reported in a number of cell types[11,13,17].

Although no definitive _in vivo_ function has been established for RNase H, three roles can be envisioned: removal of RNA primers in DNA replication[18,19], trimming of RNA primers of DNA replication to select for the most efficient priming (i.e., cleaving RNA at specific places to generate the most efficient primers[20] and eliminating incorrect primers[21,22]), and termination of transcription at specific sites (which can occur when DNA-RNA hybrids are formed during the course of transcription). Inasmuch as several cells have more than one form of RNase H, it seems possible that RNase H has multiple functions. It is hoped that the function of RNase H in E. coli can be determined with the help of a recently isolated mutant that exhibits lowered concentrations of RNase H[23]. No unusual property of the mutant has yet been observed; this may be attributed to residual RNase H (approximately 30% of normal) in the mutant.

In vitro functions or uses of RNase H have been developed almost from the time of its discovery. This paper is intended to point out these uses and some potential uses and stimulate the development of other uses of RNase H.

II. ENZYMATIC PROPERTIES OF RNase H

Even though RNase Hs from various organisms--or from a single organism--do not exhibit identical properties, many characteristics are shared. For example, most kinds of RNase H are endonucleolytic[7,12,24]; this property has been exploited in many uses of RNase H in nucleic acid analysis. Another characteristic is a requirement for a divalent metal ion. Usually, Mg^{2+} is the preferred ion[7,12], but Mn^{2+} sometimes gives a slightly higher rate of degradation[24,25]. Monovalent cations also exert an influence on the rate of reaction[7], but the monovalent-salt sensitivity varies more from one source to another[7,12]. RNase H from E. coli exhibits no strong preference for the sequence of RNA[9] that is cleaved, except in the presence of dextran[26]. Dextran inhibits the cleavage of poly A in poly rA·poly dT by E. coli RNase H even though the RNA of ØXDNA-RNA hybrids continues to be degraded in the presence of large excesses of dextran. Dextran has also been found to inhibit calf thymus RNase H[13]. The products of extensive digestion of poly rA·poly dT with most kinds of RNase H are a series of oligomers of adenylic acid terminated in 5'-phosphate moieties (Figure 1)[7,12,27].

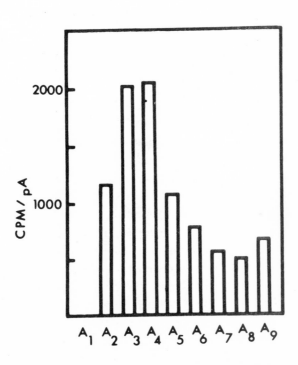

Figure 1. Product distribution of E. coli RNase H digestion of poly rA·poly dT. [^{32}P]Poly rA·poly dT was degraded with RNase H, and samples were subjected to electrophoresis on polyethyleneimine cellulose. Oligonucleotides were located by autoradiography. Radioactivity of each spot was determined, and cpm/pA residue was calculated to indicate relative molar yields of each product. A_1, 5' AMP,; A_2, pApA; and so forth.

Analysis of degradation products of ØXDNA-RNA by E. coli RNase H suggests that a similar distribution is generated. Owing to the nature of analysis of the products, a histogram of the results cannot be drawn. What is interesting about the oligomers produced in the reaction is that their distribution is apparent early in the reaction, as well as late (i.e., these products seem to be both initial and final products), and that changes in temperature or ionic strength (either monovalent or divalent metal ions) do not alter the distribution[27]. Such results suggest that the products result from an inherent property of the enzyme and do not reflect the stability of the DNA-RNA hybrid.

As pointed out above, the RNA in DNA-RNA hybrids is by definition the substrate for RNase H. Nevertheless, it has been important to define the limits of detection of a DNA-RNA hybrid by RNase H. For example, is a single ribonucleotide in a DNA molecule recognized by RNase H as a DNA-RNA hybrid? A subquestion is related to the junction between a DNA-RNA hybrid and a DNA-DNA duplex. Is the last ribonucleotide of such a molecule removed by RNase H?

One of the proposed functions of RNase H--removal of RNA primers in DNA replication--requires complete removal of the RNA primer, including the ribonucleotide to which the first DNA nucleotide is connected. The experiments to determine whether this last ribonucleotide is removed by RNase H rely on "nearest-neighbor" transfer of the 5' phosphate of the deoxyribonucleotide to the ribonucleotide as a result of alkaline digestion (see Figure 2). Cleavage by RNase H of the phosphodiester bond between the ribonucleotide and its adjacent deoxyribonucleotide would result in a molecule in which no label could be transferred to a ribonucleotide. With RNase H from KB cells, it has been possible to demonstrate the cleavage of the RNA-DNA junction[28]. It should be pointed out that substantial amounts of enzyme are required; if smaller amounts are used, substantial transfer of the phosphate from the deoxyribonucleotide to the ribonucleotide is observed in pNp. These diphosphorylated ribonucleotides are generated when RNase H cleaves at the phosphodiester bond between the ribonucleotides preceding the first deoxyribonucleotide. For E. coli RNase H, the results are contradictory: different reports suggest that E. coli RNase H can and cannot cleave the phosphodiester bond at the RNA-DNA junction[7,8].

Cleavage of a single ribonucleotide in what is otherwise a normal duplex DNA molecule by RNase H remains unclear. Vinograd and co-workers[29] demonstrated that chick embryo RNase H (isolated by Keller and Crouch) attacks the ribonucleotides in mitochondrial DNA. What is unclear is whether there is only one or a few ribonucleotides at these points.

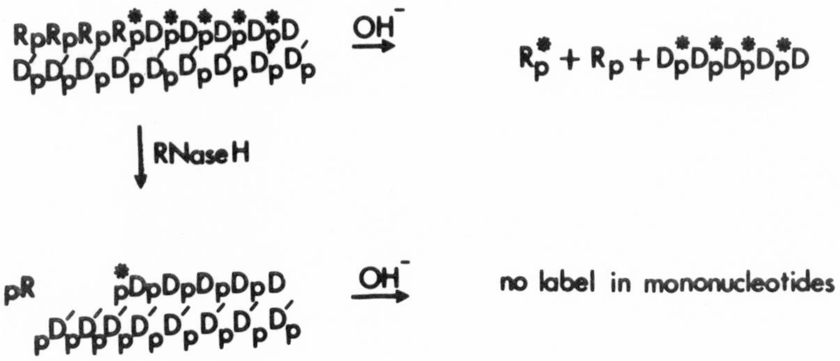

$$R_pR_pR_pR_p^*D_p^*D_p^*D_p^*D_p^*D_p^*D'_p$$
$$D'_pD'_pD'_pD'_pD'_pD'_pD'_pD'_pD'_pD'_p$$

$\xrightarrow{\text{OH}^-}$

$$R_p^* + R_p + D_p^*D_p^*D_p^*D_p^*D$$

$\downarrow$ RNase H

$$pR \quad {}^*\!pD_pD_pD_pD_pD$$
$$pD'_pD'_pD'_pD'_pD'_pD'_pD'_pD'_pD'_p$$

$\xrightarrow{\text{OH}^-}$ no label in mononucleotides

Figure 2. Nearest-neighbor transfer of phosphate from deoxyribonucleotide to ribonucleotides. D and D', deoxyribonucleotides (D' represents complementary nucleotides). R, ribonucleotides. Asterisks, labeled phosphate moieties.

Experiments detailing the requirement for the size of DNA required for recognition by RNase H are limited to those of Donis-Keller[30], which are described below. In those experiments, oligomers of four deoxyribonucleotides were sufficient. However, the stability of the DNA-RNA hybrid might be reflected in the assays, and the question of smaller runs of deoxyribonucleotides in DNA-RNA hybrids remains unanswered.

III. IN VITRO APPLICATIONS OF RNase H

A. Demonstration of DNA-RNA Hybrids

Once the specificity of RNase H was recognized, several investigators applied its recognition properties to demonstrate that some molecules contained DNA-RNA hybrids. In the example of RNA in plasmid DNA[12] or RNA in mitochondrial DNA[29], RNase H converts covalently closed supercoils to the open circular form and, as indicated in Figure 3, analysis of the products after denaturation reveals a single-stranded circle and a linear single-stranded fragment. Analysis of RNase H cleavage of such plasmids, in fact, was used to demonstrate the endonucleolytic character of cellular RNase H[12] and the exonucleolytic properties of retroviral reverse transcriptase RNase H[12].

222

Figure 3. Detection of RNA in covalently closed circular DNA. Treatment with RNase H of covalently closed circular DNA that contains RNA (wavy line) results in a relaxing of the supercoils. Denaturation of the product generates a single-stranded circle and a linear molecule.

One of the most sophisticated studies of the detection of specific DNA sequences or transcripts thereof was developed with RNase H. Detection of synthesis of specific mRNA by hybridization to cDNA followed by S1 nuclease digestion was a common method. The major problem encountered in such a procedure was high blank values--often equal to the amount of mRNA detected. When a step was added (i.e., treatment of the hybrids with RNase H), a more reliable technique became available[31]. In the procedure of Jacquet et al.[31], labeled cDNA is hybridized to mRNA, treatment of the hybrids with RNase H exposes single-stranded DNA, and degradation of the single-stranded DNA with a nuclease specific for single strands results in the liberation of small labeled deoxyribonucleotides, which indicate the presence of DNA-RNA hybrids.

B. Removal of Poly A

Some of the earliest experiments with RNase H dealt with polyadenylated mRNAs. Because of the heterogeneity in the length of poly A on mRNA molecules, the bands of mRNA are not as narrow as they would otherwise be. Removal of poly A on mRNAs can be accomplished with oligo dT or poly dT (see Figure 4) and RNase H; the result is a sharper band of mRNA in gel electrophoresis[32].

Elimination of poly A from mRNA by the same technique has also been used to study the role of poly A on mRNA. Deadenylated polio RNA is not nearly as infectious as polio RNA that has not been treated with RNase H in the presence of oligo dT.[33]

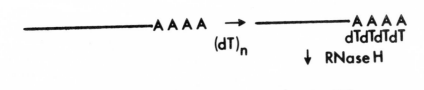

Figure 4. Removal of poly A from mRNA. Poly A removal from the 3' terminus of mRNA can be accomplished through annealing of oligo dT or poly dT to the poly A sequences and treatment with RNase H.

The procedure used in my laboratory for isolation of RNase H from E. coli uses sucrose-gradient centrifugation[6]. Sucrose contains dextran as a contaminant and can inhibit removal of poly A in the above procedure, but the concentration of RNase H is usually sufficient to preclude such a problem.

C. Specific Fragmentation of RNA with RNase H

A general procedure for fragmenting RNA to generate precise products for analysis has been desired for a number of years. Many viral RNAs are polycistronic, and subgenomic fragments would be extremely useful in determining the interaction or interdependence of different regions of these genomes. For these large RNA molecules, the generation of smaller fragments is also essential for sequence analysis. Until recently, sequencing of long RNA molecules has been accomplished by partial digestion with various ribonucleases that have limited base specificity (most commonly RNase T1). A second method for generating small fragments can complement and expand the techniques that already exist. RNase III[34-36] has been used with some success in producing specific fragments of polio RNA and ribosomal RNA. Donis-Keller has demonstrated the usefulness of RNase H in producing small, specific fragments of RNA molecules of varied sizes[30].

This procedure relies on annealing short oligodeoxyribonucleotides (see Figure 5) to the RNA to be cleaved and then digesting with RNase H (Donis-Keller has used both E. coli and calf thymus RNase H). As mentioned previously, tetrameric deoxyribonucleotides can be used. Specificity of clevage sites might be expected to increase with increasing length of the oligodeoxyribonucleotide. However, under the conditions used by Donis-Keller, hexameric deoxyribonucleotides permit cleavage at sites homologous to 4,5, or 6 of the nucleotides of the hexamer used. Some sites seem to be unavailable for cleavage, even though they have sequences homologous to the oligodeoxyribonucleotide, presumably because of secondary structure in the RNA. If a variety of oligomers are used, a wide range of cleavage sites can be generated; this makes the technique extremely valuble. Satellite tobacco necrosis viral RNA (approximately 1,240 nucleotide) has been fragmented with this procedure to permit substantial sequence information to be obtained. It is interesting that calf thymus RNase H and E. coli RNase H are not equivalent in their recognition of DNA-RNA hybrids in the technique of Donis-Keller. Some hybrids require more of one enzyme to cleave the same hybrid. In addition, the precise ribonucleotide cleaved in a given DNA-RNA hybrid depends on the source of the RNase H.

Figure 5. Site-specific fragmentation of RNA with RNase H in the presence of oligodeoxyribonucleotides. Segments A, B, and C of the RNA (wavy lines) can be separated after cleavage of the RNA in the presence of oligodeoxyribonucleotide (short straight lines) and RNase H.

D. Detection and Isolation of Heterogeneous Termini with RNase H

One of the more difficult problems in analyzing RNA structure results from heterogeneity in the termini of the RNAs under examination. Wellauer's laboratory has been studying α-amylase mRNA from mice and has discovered that two mRNAs with different lengths are present in such a system[37]. The size difference is insufficient to permit good separation of the intact mRNAs. However, if the RNAs are cleaved with RNase H in the presence of a restriction fragment of DNA that comes from a region close to the 5' terminus, smaller, more readily separable RNAs are produced. In Wellauer's laboratory, these mRNA fragments are subjected to electrophoresis on methylmercury-agarose gels, transferred to diazotized paper and detected by hybridization with "nick-translated" DNA[38].

In my laboratory, we are dealing with a similar problem, but at a slightly different level. The 5.8S rRNA from chickens has at least three different 5' termini[39]. Two of them differ from each other by a single nucleotide (i.e. one of these 5.8 rRNAs has an additional nucleotide at its 5' terminus)[40]. Labeling the 5' termini of the mixture of these two 5.8S rRNAs and then sequencing by partial nuclease digestion[41,42] produces a doublet pattern in autoradiograms: each doublet represents the common sequence between the two 5.8S rRNAs whose 5' termini are separated from the 3' terminus of the fragments by one nucleotide. In the case in which the 5' termini (or, for that matter, the 3' termini) are labeled, fragmentation of the RNA with RNase H and DNA from a restriction fragment near the appropriate end can produce short (15-25) oligonucleotides that can be readily separated for sequence analysis (see Figure 6).

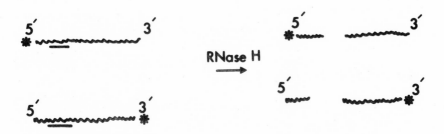

Figure 6. Cleavage of terminally-labeled RNA to produce smaller, more readily analyzable RNA. Terminally-labeled RNA (wavy lines) is specifically cleaved in the presence of DNA from restriction fragments (straight lines).

E. Future Uses of RNase H

Antibodies against RNase H could be extremely useful in detecting DNA-RNA hybrids _in vivo_ and in studying DNA-RNA hybrids with the electron microscope. It should be possible to detect which end of an mRNA is polyadenylated when specific oligodeoxyribonucleotides are annealing to an RNA and to increase sen-

226

sitivity in detecting DNA-RNA hybrids. A second use of antibodies to RNase H could be the purification of DNA that is complementary to a single species of RNA or to a set of RNAs. For example, when a library of cloned DNA fragments is available, hybridization of a specific RNA or a class of RNAs to DNA from the library under conditions in which R-loops are formed and then "trapping" of the DNA-RNA hybrid with RNase H and antibodies to RNase H could result in a substantial purification of DNA that is homologous to the RNA being hybridized. "Trapped" DNA-RNA hybrids could be separated from unhybridized DNA by using the protein A affinity for IgG. Transformation or transfection of the DNA should then yield a highly select set of DNA fragments. All of these experiments with RNase H antibodies depend on the ability to bind RNase H to hybrids without nucleolytic attack. Preliminary experiments have suggested that RNase H binds to DNA-RNA hybrids in the presence of EDTA without degradation of the RNA[43].

IV. RETROVIRAL REVERSE TRANSCRIPTASE RNase H

RNase H of retroviral reverse transcriptase has the unique ability to attack RNA in DNA-RNA hybrids in an exonucleolytic manner[12,16]. Most of the techniques for using RNase H described above rely on the endonucleolytic activity of RNase H. Thus, RNase H of reverse transcriptase has not been used to detect nucleic acid structure. One interesting property of reverse transcriptase RNase H is that the exonucleolytic activity can proceed from either the 5' terminus or the 3' terminus[16].

V. ENZYMES THAT DEGRADE BOTH DNA AND RNA OF DNA-RNA HYBRIDS

The ability to degrade the RNA of a DNA-RNA hybrid selectively was the basis for establishing that an enzyme is an RNase H[1,2]. Othr enzymes have been shown to degrade RNA in DNA-RNA hybrids, but they also degrade the DNA portion. Two enzymes from E. coli, DNA polymerase I[7,12] and exonuclease III,[12,44] can attack either the DNA or the RNA of a DNA-RNA hybrid or a specific DNA in a DNA duplex form. The 5' —> 3' exonucleolytic activity of DNA polymerase I is thought to be responsible for the degradation of RNA in DNA-RNA hybrids. Exonuclease III can start at the 3' terminus of either DNA-RNA hybrids or DNA-DNA duplexes and degrade in an exonucleolytic manner toward the 5' end. It has been suggested by Weiss[45] that exonuclease III has a recognition site for duplex structures and a second site for cleavage. Clearly, RNase H must have similar characteristics.

REFERENCES

1. Stein, H. and Hausen, P. (1969) Science 166, 393-395.
2. Hausen, P. and Stein, H. (1970) Eur. J. Biochem. 14, 278-283.
3. Henry, C. M., Ferdinand, F.-J. and Knippers, R. (1973) Biochem. Biophys. Res. Commun. 50, 603-611.
4. Chamberlin, M. and Berg, P. (1964) Cold Spring Harbor Symp. Quant. Biol. 28, 67-76.
5. Miller, H. L., Riggs, A. D. and Gill, G. N. (1973) J. Biol. Chem. 248, 2621-2624.
6. Crouch, R. J. (1974) J. Biol. Chem. 249, 1314-1316.
7. Berkower, I., Leis, J. and Hurwitz, J. (1973) J. Biol. Chem. 248, 5914-5921.
8. Darlix, J. L. (1975) Eur. J. Biochem. 51, 369-376.
9. Robertson, H. D. and Dunn, J. J. (1975) J. Biol. Chem. 250, 3050-3056.
10. Wyers, F., Sentenac, A. and Fromagoet, P. (1973) Eur. J. Biochem. 36, 270-281.
11. Wyers, F., Sentenac, A. and Fromagoet, P. (1976) Eur. J. Biochem. 69, 377-383.
12. Keller, W. and Crouch, R. (1972) Proc. Natl. Acad. Sci. USA 69, 3360-3364.
13. Cathala, G., Rech, J., Huet, J. and Jeanteur, P. (1979) J. Biol. Chem. 254, 7353-7359.
14. Molling, K., Bolognesi, D. P., Bauer, H., Büsen, W., Plassman, H. W. and Hausen, P. (1972) Nature (London) New Biol. 234, 240-243.
15. Baltimore, D. and Smoler, D. F. (1972) J. Biol. Chem. 247, 7282-7287.
16. Leis, J. P., Berkower, L. and Hurwitz, J. (1973) Proc. Natl. Acad. Sci. USA 70, 466-470.
17. Tashiro, F. and Mita, T. (1976) Eur. J. Biochem. 65, 123-130.
18. Keller, W. (1972) Proc. Natl. Acad. Sci. USA 69, 1560-1564.
19. Watson, J. D. (1972) Nature (London) New Biol. 239, 197-201.
20. Itoh, T. and Tomizawa, J. (1979) Cold Spring Harbor Symp. Quant. Biol. 43, 409-417.

21. Sumida-Yasumoto, C., Ikeda, J.-E., Benz, E., Marians, K. J., Vicuna, R., Sugrue, S., Zipursky, S. L. and Hurwitz, J. (1979) Cold Spring Harbor Symp. Quant. Biol. 43, 311-329.
22. Vicuna, R., Hurwitz, J., Wallace, S. and Girard, M. (1977) J. Biol. Chem. 252, 2524-2533.
23. Carl, P., Bloom, L. and Crouch, R. J. (1980) J. Bacteriol. 144, 28-35.
24. Haberkern, R. C. and Cantoni, G. L. (1973) Biochemistry 12, 2389-2395.
25. Büsen, W. and Hausen, P. (1975) Eur. J. Biochem. 52, 179-190.
26. Dirksen, M.-L. and Crouch, R. J., manuscript in preparation.
27. Crouch, R. J., manuscript in preparation.
28. Keller, W. and Crouch, R. J., unpublished observations.
29. Grossman, L. I., Watson, R. and Vinograd, J. (1973) Proc. Natl. Acad. Sci. USA 70, 3339-3343.
30. Donis-Keller, H. (1979) Nuc. Acids Res. 7, 179-192.
31. Jacquet, M., Groner, Y., Monroy, G. and Hurwitz, J. (1974) Proc. Natl. Acad. Sci. USA 71, 3045-3049.
32. Vournakis, J. N., Efstratiadis, A. and Kafatos, F. C. (1975) Proc. Natl. Acad. Sci. USA 72, 2959-2963.
33. Spector, D. H. and Baltimore, D. (1974) Proc. Natl. Acad. Sci. USA 71, 2983-2987.
34. Westphal, H. and Crouch, R. J. (1975) Proc. Natl. Acad. Sci. USA 72, 3077-3081.

35. Nomoto, A., Lee, Y. F., Babich, A., Jacobsen, A., Dunn, J. J. and Wimmer, E. (1979) J. Mol. Biol. 128, 165-177.
36. Stewart, M. L., Crouch, R. J. and Maizel, J. V., Jr. (1980) Virology 104, 375-397.
37. Hagenbuchle, D., Bovey, R. and Young, R. A. (1980) Cell 21, 179-187.
38. Wellauer, P., personal communication.
39. Earl, P., Seidman, S. and Crouch, R. J., unpublished observations.
40. Khan, N. S. N. and Maden, B. E. H. (1977) Nuc. Acids. Res. 4, 2495-2505.
41. Donis-Keller, H., Maxam, A. M. and Gilbert, W. (1977) Nuc. Acids Res. 4, 2527-2538.
42. Simoncsits, A., Brownlee, G. G., Brown, R. S., Rubin, J. R. and Guilley, H. (1977) Nature 269, 833-836.
43. Crouch, R. J., unpublished observations.
44. Weiss, B., Rogers, S. G. and Taylor, A. F. (1978) In P.C. Hanwalt, E. C. Friedberg and C. F. Fox (ed.) DNA Repair Mechanism, Academic Press, Inc. New York.
45. Weiss, B. (1976) J. Biol. Chem. 251, 1896-1901.

LABELING OF EUKARYOTIC MESSENGER RNA 5' TERMINUS WITH PHOSPHORUS -32: USE OF TOBACCO ACID PYROPHOSPHATASE FOR REMOVAL OF CAP STRUCTURES

Raymond E. Lockard,[*] Lauren Rieser,[+] and John N. Vournakis[+]

[*]Department of Biochemistry, The George Washington University Medical School, Washington, D.C. 20037

[+]Department of Biology, Syracuse University, Syracuse, NY 13210

I. INTRODUCTION

Recombinant DNA technology, has made it possible to isolate large quantities of DNA corresponding to any single gene. Such technology has generated a persistent need for new and efficient methods of screening recombinant clones and mapping gene structure. Although uniform phosphorus-32 labeling of complementary DNA and nick-translated cloned cDNA has often been the method of choice for screening recombinant clones and mapping restriction sites in many genes[1,2], in vitro labeling of RNA 5' ends has also proved useful for such purposes[3]. Labeling the 5'-termunus of a eukaryotic cellular or viral mRNA with phosphorus-32 in vitro may not at first appear straightforward, owing to the presence of the $m^7G^{5'}$ ppp^5-cap structure blocking the terminus; however, this structure can be efficiently removed with the enzyme tobacco acid pryrophosphatase (TAP) isolated from cultured tobacco cells[4-6], and the 5' terminus of the mRNA labeled to a high specific activity with phosphorus-32 through the use of T_4 polynucleotide kinase and $[\gamma-^{32}P]$ ATP. The mRNA so labeled can be used for screening clones and for mapping gene structure, and the labeling permits analysis of the nucleotide sequences in the mRNA. Sequence analysis allows one to map the beginning of an mRNA sequence within the DNA nucleotide sequence of the corresponding cloned gene and can help in the discrimination of intervening sequences from the coding regions.

Tobacco acid pyrophosphatase was first purified and characterized by Shinshi et al.[4]. It has a broad specificity for hydrolyzing pyrophosphate bonds in a variety of substrates, including ATP, NAD, poly(ADP-ribose), inorganic pyrophosphate, and $m^7G^{5'}ppp^{5'}$-caps[4-6]. The molecular weight of the native enzyme has been estimated to be 280,000 by gel filtration on Sephadex G-200, and SDS polyacrylamide gel electrophoresis has indicated that it is composed of subunits with molecular weights of 75,000[4]. TAP's substrate specificity and enzymologic properties, including its acidic pH optimum of 6 and its apparent lack of divalent cation requirement, distinguish it from other plant and animal pyrophosphatases[7,8]. TAP can be rapidly purified of contaminating phosphodiesterases and phosphatases and used for decapping minute quantities of eukaryotic mRNA, which can then be labeled in vitro to a high specific activity with phosphorus-32.

II. MATERIALS AND METHODS

A. Materials

Cultured tobacco cells (Nicotiana tabacum var. Wisconsin 38) used for preparation of TAP were purchased from the Alton Jones Cell Science Center, Lake Placid, NY 12946. TAP can be purchased from Bethesda Research Laboratories, Inc. (Cat. Number 8007). Calf intestinal alkaline phosphatase (CIAP), bacterial alkaline phosphatase (BAP), and T_4 polynucleotide kinase were obtained from Boehringer Mannheim and used without further purification. $[\gamma-^{32}P]$ ATP was either purchased from New England Nuclear (>2,000 Ci/mmol) or enzymatically synthesized by the method of Johnson and Walseth[9] to achieve specific activities of >7,000 Ci/mmol. Diethylpyrocarbonate (DEP) was purchased from Sigma Chemical Company. Polyethyleneimine precoated plastic sheets, Macherey-Nagel Cel 300 PEI were obtained from Brinkman Instruments, Inc. $[^{14}C]$ GTP (300 mCi/mmol) was purchased from New England Nuclear (Cat. Number NEC-432) Trituim-labeled vesicular stomatitis virus (VSV) mRNA was a kind gift from T. Nielson and C. Baglioni. DEAE cellulose (DE-52) and phosphocellulose (P-11) were obtained from Reeve Angel, Inc.

B. Methods

1. ATP Hydrolysis Assay for TAP

Each enzyme fraction eluting from the P-11 column was assayed for pyro-phosphatse activity in a 10-μl reaction mixture containg 1 μl of 10X assay buffer (0.5 M NaOAc, at a pH of 6.0; 0.1 M β-mercaptoethanol), 8 μl of [γ-^{32}P] ATP (3 pmol; 1,000 Ci/mmol) containing either in $\underline{A}$ 1 nmol of unlabeled ATP or in $\underline{B}$ 5 nmole of unlableled ATP, and 1 μl of enzyme fraction. The reaction mixture was incubated for 1 h at 37°C, after which 1 μl from each reaction tube was removed and spotted on a plastic-backed PEI cellulose thin-layer plate developed either in 0.8 M LiCl-0.8 M acetic acid or in 0.75 M KH$_2$PO$_4$ (pH, 3.5). One unit of TAP is defined as the activity re-quired to release 1 nmol of phosphorus -32 from [γ-^{32}P]ATP in 1 h at 37°C.

2. Chemical Removal of m^{7}G$^{5'}$ppp$^{5'}$-Cap Structures from Eukaryotic mRNA by β-Elimination

Preparation of reagents and the method for chemical removal of caps from mRNA molecules by β-elimination were identical with those published by Lockard et al.[10].

3. Enzymatic Removal of m^{7}G$^{5'}$ppp$^{5'}$-Cap Structures Using TAP for Microgram Quantities of mRNA

Ten units of TAP was routinely used to decap 0.25 A$_{260}$ units of mRNA in 10 μl of buffer containing 50 mM NaOAc (pH, 6.0) and 10 mM mercaptoethanol. The reaction mixture was incubated for 1 h at 37°C. As an option, DEP can be added at the beginning of the reaction to minimize breakdown of the mRNA. DEP (stored frozen in 0.5-ml aliquots at -80°C) was used to prepare a fresh 0.5% solution in water. An aliquot was immediately removed and added to the TAP reaction mixture to give a final DEP concentration of 0.01%. After the TAP incubation, enough 0.5 M Tris-HCl (pH, 8.5) was added to give a final concentration of 75 mM. BAP was then added to a final concentration of 0.1 unit/ml. Alternatively, CIAP can be used at a concentration of 0.1 unit/ml. Additional DEP can be added at this stage to give a final concentration of 0.015% in the reaction mixture. After incubation of 45 min at 37°C, the reaction mixture was immediately deproteinized with phenol, and the mRNA precipitated with ethanol, as previously described[10].

4. 5'-End Labeling of mRNA with Phosphorus-32

Decapped and dephosphorylated mRNA was labeled with phosphorus-32 through

the use of $[\gamma\text{-}^{32}P]ATP$ (2,000-7,000 Ci/mmol) and T_4 polynucleotide kinase. Up
to 0.40 A_{260} unit of mRNA can be labeled in a 10-μl reaction mixture containing 25 mM Tris-HCl (pH, 8.5), 10 mM $MgCl_2$, 10 mM DTT, 2-4 nmol of $[\gamma\text{-}^{32}P]ATP$,
and 2-4 units of T_4 polynucleotide kinase. After incubation for 30 min at
37°C, the reaction mixture is brought to a final concentration of 25 mM EDTA,
to which 30 μl of loading solution (0.1% bromophenol blue and 0.1% xylene
cyanole in 100% formamide) is added; the mixture is incubated at 37°C for
5 min before loading on a preparative polyacrylamide gel.

5. Enzymatic Removal of $m^7G^{5'}ppp^{5'}$-Cap Structures Using TAP for Nanogram Quantities of mRNA

Minute quantities of mRNA can effectively be decapped, dephosphorylated,
and 5'-terminally labeled with phosphorus-32 without introducing a phenol
extraction procedure to eliminate phosphatase activity, as is detailed in
Figure[8].

6. Preparative Polyacrylamide Gel Electrophoresis For RNA Purification

$5'^{32}P$-end labeled rabbit globin mRNA was resolved on a 7% polyacrylamide
slab gel (20 cm X 20 cm X 0.2 cm) run in TBE buffer (90 mM Tris-Borate at a
pH of 8.3; 4 mM EDTA) and 7 M urea at 30°C[10]. Labeled mouse α and β globin
mRNA were resolved on a 3.5% polyacrylamide slab gel run in TBE buffer and
7 M urea at 30°C[11].

7. Electrophoretic Elution of Labeled RNA From Preparative Polyacrylamide Gels

The radioactivity in the end-labeled RNA bands was located by autoradiography and excised from the gel. The labeled RNA was then recovered by electrophoretic elution into dialysis bags with a Model 1200 Disc Electrophoresis
Apparatus from Ames. Co., Elkhart, IN. The electrophoresis was performed in
pasteur pipettes with shortened tips that were plugged with 3% polyacrylamide
in TBE gel buffer minus urea. The plugged elution tubes were prerun overnight at 4 mA/tube in elution buffer (40 mM Tris-acetate buffer, at a pH of
7.8; 2 mM EDTA). Dialysis bags containing 0.5-1.0 ml of elution buffer were
then attached, and the excised gel bands were inserted into elution tubes and
electrophoresed for 4-8 hours at 10 mA/tube at 4°C. The labeled mRNA was removed from the dialysis sac by centrifugation into a 15-ml Corex tube to which
0.5 A_{260} unit of tRNA carrier was added. The RNA was made 0.5 M in LiCl, 3
volumes of 95% ethanol were added, and the RNA was precipitated at -80°C for
30 minutes. The precipitated RNA was collected by centrifugation at 10,000

RPM for 10 minutes at 4°C, and the pellet was dissolved in 100 µl of 0.5 M NaOAc (pH, 5.5). The dissolved RNA was transferred to a plastic Eppendorf tube to which 400 µl of ethanol was added, and the RNA was reprecipitated at -80°C for 20 minutes. The RNA was again pelleted at 10,000 RPM the supernatant was removed, the pellet was washed once with 95% ethanol, and the labeled RNA was dissolved in an appropriate volume for immediate use.

III. METHODS AVAILABLE FOR REMOVAL OF CAP STRUCTURES

One consistent structural feature found on most eukaryotic viral and cellular mRNAs is the methylated $m^7G^{5'}ppp^{5'}$-cap structure[12]. Its discovery necessitated the development of methods that could quantitatively remove this group, thereby leaving the mRNA intact and unmodified for mRNA translational studies. A chemical procedure initially used by many laboratories involved, first, sodium periodate oxidation of the 2'-, 3'-diol groups in the ribose ring of 7-methylguanosine as shown in Figure 1, and, second, treatment with aniline to remove the m^7G residue via β-elimination[13]. The aniline procedure, however, often results in degradation of the mRNA and appears also to introduce nonspecific modifications that impair the mRNA's translatability[14,15]. A modified procedure that uses cyclohexylamine, rather than aniline[16,17] for the β-elimination, as shown in Figure 1, is considerably more reproducible and leads to no detectable modifications of the mRNA[18]. Both chemical procedures are, however, time-consuming and require reasonably large amounts of mRNA. With the relatively recent observation that $m^7G^{5'}ppp^{5'}$-cap structures could be enzymatically removed using tobacco acid pyrophosphatase (TAP)[5,6], an alternative approach for preparing decapped mRNA both for functional studies and for obtaining $5'-^{32}P$-end labeled mRNA was quickly realized. TAP can efficiently remove $m^7G^{5'}ppp^{5'}$-caps from mRNA molecules by hydrolyzing either of the two pyrophosphate linkages on the cap structure, as indicated in Figure 1. The enzyme is commercially available from Bethesda Research Laboratories, Inc., or can be rapidly purified from cultured tobacco cells, as is described here.

IV. PROCEDURE FOR RAPID PURIFICATION AND ASSAY OF TAP

A. Purification Scheme

Figure 2 details a rapid and simple method for purifying TAP from tobacco cells grown in suspension culture[19,20]. This constitutes a substantial abbreviation of that previously published by Shinshi et al.[4] and will generate up to 8 X 10^3 units of TAP from 10 g of cells free of ribonuclease and phosphatase activities. Caution must be exercised never to freeze enzyme fractions, because freezing and thawing entail almost complete loss of enzymatic activity.

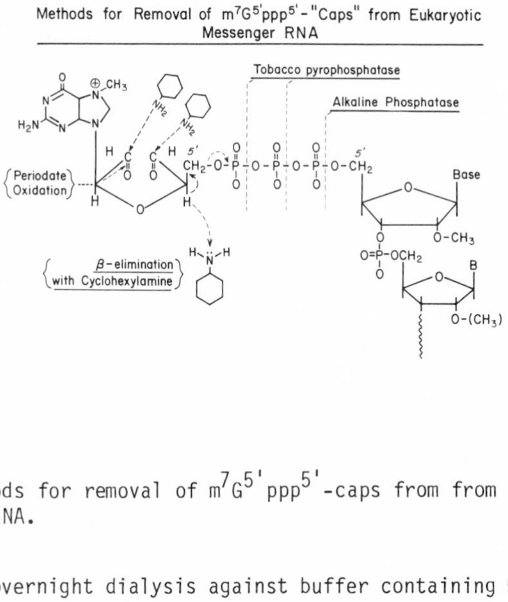

Figure 1. Methods for removal of m^{7}G$^{5'}$ppp$^{5'}$-caps from from eukaryotic messenger RNA.

However, after overnight dialysis against buffer containing 50% glycerol (Figure 2 step 17), one can concentrate the individual enzyme fractions eluting from the phosphocellulose column to 25% of their original volume. No detectable loss of enzymatic activity is observed after storage in 50% glycerol at -20°C for over a year.

B. <u>ATP Hydrolysis as an Assay for TAP Enzymatic Activity</u>

Enzyme fractions can be conveniently assayed for pyrophosphatase activity by using [γ-^{32}P]ATP for a substrate. ATPase activity can be followed for each individual fraction by removing 1 µl from the assay incubation mix-

236

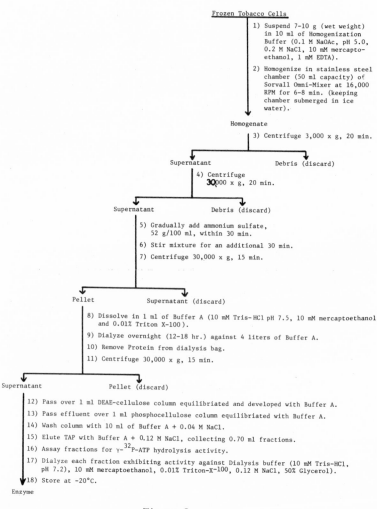

Figure 2

Frozen Tobacco Cells

1) Suspend 7-10 g (wet weight) in 10 ml of Homogenization Buffer (0.1 M NaOAc, pH 5.0, 0.2 M NaCl, 10 mM mercapto-ethanol, 1 mM EDTA).

2) Homogenize in stainless steel chamber (50 ml capacity) of Sorvall Omni-Mixer at 16,000 RPM for 6-8 min. (keeping chamber submerged in ice water).

Homogenate

3) Centrifuge 3,000 x g, 20 min.

Supernatant   Debris (discard)

4) Centrifuge 30,000 x g, 20 min.

Supernatant   Debris (discard)

5) Gradually add ammonium sulfate, 52 g/100 ml, within 30 min.

6) Stir mixture for an additional 30 min.

7) Centrifuge 30,000 x g, 15 min.

Pellet   Supernatant (discard)

8) Dissolve in 1 ml of Buffer A (10 mM Tris-HCl pH 7.5, 10 mM mercaptoethanol and 0.01% Triton X-100).

9) Dialyze overnight (12-18 hr.) against 4 liters of Buffer A.

10) Remove Protein from dialysis bag.

11) Centrifuge 30,000 x g, 15 min.

Supernatant   Pellet (discard)

12) Pass over 1 ml DEAE-cellulose column equilibriated and developed with Buffer A.

13) Pass effluent over 1 ml phosphocellulose column equilibriated with Buffer A.

14) Wash column with 10 ml of Buffer A + 0.04 M NaCl.

15) Elute TAP with Buffer A + 0.12 M NaCl, collecting 0.70 ml fractions.

16) Assay fractions for γ-^{32}P-ATP hydrolysis activity.

17) Dialyze each fraction exhibiting activity against Dialysis buffer (10 mM Tris-HCl, pH 7.2), 10 mM mercaptoethanol, 0.01% Triton-X-100, 0.12 M NaCl, 50% Glycerol).

18) Store at -20°C.

Enzyme

ture and spotting it on a plastic-backed PEI cellulose thin-layer plate, which
is then developed in 0.8 M LiCl-0.8 M acetic acid. The disappearance of the
ATP spot into P_i is followed by autoradiography, as shown in Figure 3, and the
amount of hydrolysis for each fraction is measured by cutting out the product
and substrate spots and counting them separately by liquid scintillation. In
Figure 3, it can be seen that fractions eluting from the phosphosphocellulose
column 2-4 contained most of the enzymatic activity and were individually dialyzed against dialysis buffer. These fractions were again assayed, and their
titer was determined by diluting the $[\gamma\text{-}^{32}P]ATP$ present in the assayed solution
with increasing amounts of unlabeled ATP. The specific activity of the enzyme
is expressed in ATPase units: 1 unit hydrolyzes 1 nmol of ATP in 1 hr at 37°C.

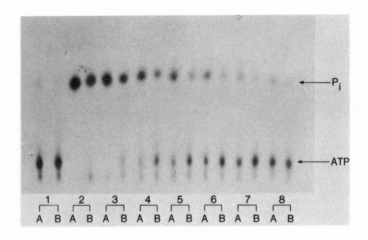

Figure 3. Autoradiogram of PEI-cellulose thin-layer plate after ATPase assay
for TAP activity. Numbers 1-8 represent 0.7-ml fractions eluting
from the phosphocellulose column. One-microliter aliquots were removed from each assay reaction mixture and spotted on PEI-cellulose
plate, which was then developed in 0.8 M LiCl-0.8 M acetic acid. A,
assay reaction mixture containing $[\gamma\text{-}^{32}P]ATP$ (3 pmol 1,000 Ci/mmol)
with 1 nmole of unlabeled ATP. B, assay reaction mixture containing
same quantity of $[\gamma\text{-}^{32}P]ATP$ with 5 nmol of unlabeled ATP.

C. $m^7G^{5'}ppp^{5'}$-Cap Hydrolysis as an Assay for TAP Enzymatic Activity

Enzyme preparations showing comparable ATP hydrolysis activities can differ markedly in their ability to hydrolyze $m^7G^{5'}ppp^{5'}$-cap structures. Although we are not certain why various preparations differ in their substrate specificity, such differences stress the need for an alternative assay to express cap hydrolysis adequately. The data shown in Figure 4 represent such an assay, with mRNA having a tritum label in the 7-methyl position of the m^7G cap nucleotide. The figure also compares the rate of mRNA cap hydrolysis with ATP hydrolysis by TAP.

VSV messenger RNA was specifically labeled with tritium in the cap and purified by the procedure of Toneguzzo and Ghosh[21]. The mRNA was hydrolyzed with TAP, and aliquots were chromatographed on PEI cellulose developed in 1.0 M LiCl to identify the main product as $[^3H]m^7GMP$ (data not shown). TAP hydrolysis was measured by trichloroacetic acid precipitation on nitrocellulose filters. Figure 4 summarizes these studies. Decapping of even small amounts of VSV mRNA (1.1 pmol) proceeds to completion in 60 min. The rate increases dramatically with a 50-fold (50-pmol) addition of rabbit globin mRNA (mixed α and β). This indicates that the substrate concentration was increased by adding capped globin mRNA and implies that the observed rate of hydrolysis represents the rate of decapping of globin mRNA, where the $[^3H]VSV$ capped mRNA acts as a convenient radioactive tracer (marker). This is the result expected, on the basis of classical enzyme kinetics, and highlights the demonstrated need for long incubation times and excess enzyme in the decapping of nanogram quantities of mRNA, as discussed elsewhere.

The rate of hydrolysis of ATP in a reaction containing 50 pmole of $[\gamma-^{32}P]ATP$ is identical with the rate observed in the decapping of globin mRNA in a reaction that also contains about 50 pmol of substrate, as indicated in Figure 4. This suggests that the unit of enzyme activity defined on the basis of hydrolysis of ATP is a direct and valid measure of the mRNA cap-hydrolysis activity when rabbit golbin mRNA is the substrate. One should be cautious in generalizing this observation to include all mRNAs, in as much as it is known that TAP cannot decap all mRNAs with equal efficiency (data not shown). Such factors as the conformation of mRNA in the 5' region may affect the accessibility of the cap to the enzyme. Experiments designed to assess the effect of various amounts of ribosomal and transfer RNA on the cap hydrolytic activity of TAP show that no inhibition of decapping

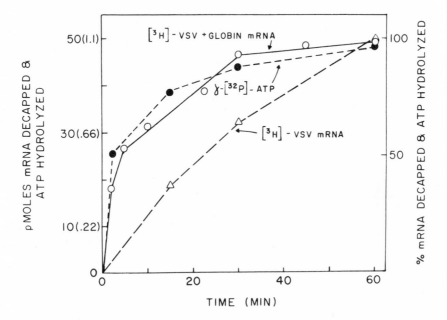

Figure 4.   Comparison of rates of hydrolysis of VSV mRNA, VSV + rabbit globin
mRNA, and ATP by TAP. All reactions were at standard conditions
(50 mM NaOAc at pH of 6.0, 10 mM mercaptoethanol, and incubation
at 37°C for times indicated). The reaction of TAP with [^{3}H]VSV
mRNA (--Δ--) contained 1.1 pmol of mRNA and 5 u of enzyme in 5 μl
total volume; with [^{3}H]VSV + globin mRNA (--o--) contained 2.2
pmol of VSV mRNA, 50 pmol of rabbit globin mRNA, and 10 u of enzyme in 10-μl total volume; with [γ-^{32}P]ATP (--o--) contained 50
pmol of ATP and 10 u of enzyme in 10 μl total volume. Reactions
containing [^{3}H]VSV mRNA were analysed for cap removal by trichloroacetic acid precipitation on filters and counting in a liquid
scintillant. The [γ-^{32}P]ATP reaction was assayed by PEI-cellulose chromatography and scintillation counting. Numbers in
parenthesis along left ordinate refer to [^{3}H]VSV curve (--Δ--).

is observed with up to 100-fold excess rRNA or tRNA (data not shown).

D. GTP Hydrolysis by TAP: Evidence of a Preferred Cleavage Site

A series of experiments were undertaken to determine the exact site at which TAP cleaves the 5' cap during the decapping reaction. A cap analogue, $[^{14}C]$GTP (labeled in the base), was used as substrate in a typical TAP reaction, and PEI cellulose thin-layer chromatography was used to identify and measure the various reaction products. Figure 5 presents the data obtained in one such experiment. The primary conclusion from these data is that TAP cleaves between the β phosphates. Initially, GDP accumulates in substantial excess over GMP is the final product. The insert in Figure 5 illustrates the primary and secondary sites attacked by TAP in this cap analogue. This experiment suggests that two positions are not equally accessible to the enzyme, probably owing to geometric constraints within its active site.

V. 5'END-LABELING OF EUKARYOTIC mRNA WITH PHOSPOROUS-32

Figure 6 outlines the general procedure for cap removal, dephosphorylation, and 5'-terminus labeling of eukaryotic mRNA. Depending on the amount of mRNA available, the procedure can be tailored for end-labeling either microgram or nanogram amounts of RNA. If reasonable quantities of mRNA are available (≥ 0.25 A_{260} units), the phenol extraction procedure may be used to remove alkaline phosphatase activity after decapping and dephosphorylation. As detailed earlier, this procedure has the advantage of allowing relatively large quantities of decapped mRNA to be end-labeled in a relatively small volume (0.5 A_{260} unit 10 μl of reaction mixture), thereby conserving on the amount of $[\gamma-^{32}P]$ATP required for efficient labeling. The presence of any small RNA fragments in the mRNA preparation will effectively compete for $[\gamma-^{32}P]$ATP in the reaction mixture. Hence, it is always advisable to have a 3-10 fold excess of $[\gamma-^{32}P]$ATP present in the kinase reaction mixture over the number of picomoles of mRNA molecules calculated be present.

Figure 7 shows a comparison of 5'-^{32}P-end -labeled rabbit globin mRNA decapped either chemically with cyclohexylamine for the β-elimination or enzymatically with TAP. Both methods result in negligible breakdown of the globin mRNA. Labeling is usually 20-40% more efficient with the chemicaly decapped globin mRNA, owing either to partial denaturation of the mRNA after chemical treatment (thus making the 5'-terminus more accessible during in vitro labeling) or to a more reproducible quantitative removal of cap structures by this method.

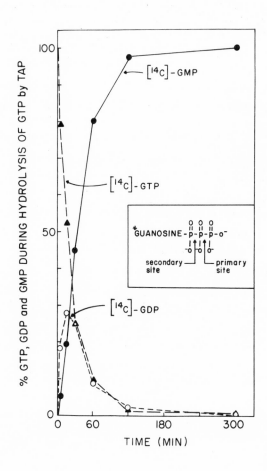

Figure 5.   Hydrolysis of GTP of tobacco acid pyrophosphatase. 180 pmol [14C]
GTP (75 mCi/nmol) was hydrolyzed in reaction containing 50mM NaOAc
pH, 5.5), 10 mM mercaptoethanol, 1.0 mM MgCl₂, and 0.5 U of TAP in
20-µl total volume at 37°C. Aliquots (2 µl) were removed at indicated times and were spotted on PEI-cellulose thin-layer plates,
which were developed in 0.75 M KH₂PO₄ (pH, 3.5). Positions of
GTP, GDP, and GMP were located by autoradiography, and spots were
counted in liquid scintillant. CPM obtained in this way were used
to determine percentage of each species.

SCHEME FOR 5'-END GROUP LABELING AND
SEQUENCING OF EUKARYOTIC mRNA

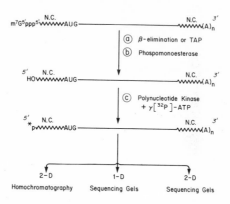

Figure 6.   Scheme for 5' terminus-labeling and sequencing of eukaryotic mRNA.

If the quantity of mRNA is limiting, the phenol extraction procedure
used to eliminate the phosphatase activity before end-labeling with polynucleotide kinase can be dispensed with. The chelator nitrilotriacetic
acid (NTA) has a higher affinity for Zn^{2+} than to Mg^{2+} ions[22] and hence
can selectively complex the Zn^{2+} atoms required for alkaline phosphatase
activity[23] without attenuating the enzymatic activity of polynucleotide
kinase in the later reaction. Figure 8 outlines a method for 5'-terminal
labeling of nanogram amounts of mRNA. After incubation with alkaline
phosphatase, the reaction mixture is brought to a final concentration of
5 mm with NTA (step 9) and incubated for 20 min at room temperature to
allow time for complexation of the Zn^{2+} atoms. Boiling for 1-2 min will
leave the mRNA intact while abolishing the phosphatase activity (step 11).
Addition of excess Mg^{2+} (step 12) will neutralize any residual NTA present
before labeling with polynucleotide kinase. This three-step decapping and
labeling procedure is fast and efficient and allows one to layer the reaction mixture directly onto a preparative polyacrylamide gel (step 17-19)

after the T_4 kinase reaction. Figure 9 shows an autoradiogram of a polyacrylamide gel with 0.50 µg of mouse globin mRNA labeled by three-step method.

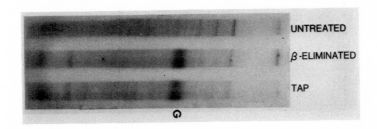

Figure 7.   Autoradiogram of 5'-[^{32}P]-end-labeled rabbit globin mRNA decapped
either chemically or enzymatically with TAP and electrophoresed on
a 7% polyacrylamide slab gel in 7 M urea. Untreated globin mRNA
(2 pmol) bkgd; chemically decapped globin mRNA (2 pmol), 10.5%
labeling.

The procedure detailed in Figure 8 can accommodate up to 0.30 unit of
mRNA without modification of the specific reaction conditions. The labeling
efficiency is usually 20-40% greater by this method than by the phenol extraction procedure, perhaps becasue of the boiling and quick-cooling of the RNA
(step 11), which may allow the 5' termini to become more accessible for
labeling during the T_4 kinase reaction. The percentage of the mRNA molecules
in a reacton mixture that actually become labeled can be anywhere from 10-60%,
depending on the nature of the mRNA species being labeled, the concentration
of [γ-^{32}P]ATP in the reaction mixture, and whether the phenol extraction or
the three-step method is used for eliminating phosphatase activity. Prepa-

244

5'-[^{32}P]-END LABELING NANOGRAM AMOUNTS OF MESSENGER RNA

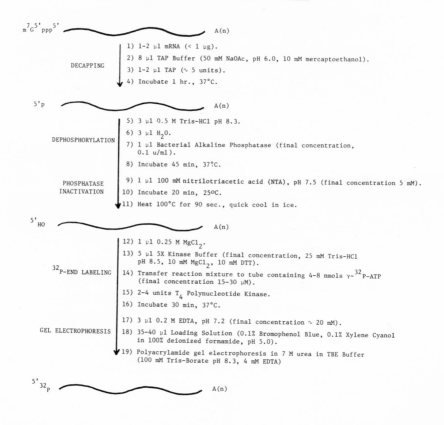

Figure 8. 5'-terminal-labeling of nanogram amounts of messenger RNA with p-32.

ration of [γ-^{32}P]ATP by the method of Johnson and Walseth[9] can result in ATP specific activities of >7,000 Ci/mmol, which will markedly improve the specific activity of mRNA obtained from any of the above labeling protocols. mRNA can be labeled to a specific activity of 1-10 X 10^6 DPM/pmol with the three step method and the very high-specific-activity [γ-^{32}P]ATP. Any 5'-^{32}P-end-labeled mRNA species resolved on a preparative polyacrylamide gel can be conveniently recovered by first locating the RNA band by autoradiography and excising the radioactive RNA from the gel, and then electroeluting the radioactive RNA into a dialysis sac with a tube gel electrophoretic apparatus. Electroeluted end-labeled RNA can then be quickly precipitated with ethanol and used directly.

Figure 9. Autoradiogram of 0.50 µg of [^{32}P]-end labeled mouse globin mRNA, enzymatically decapped and labeled by the procedure detailed in Figure 8 and electrophoresed on 3.5% polyacrylamide slab gel run in 7 M urea.

VI. CAP REMOVAL AND 5'-END-LABELING WITH PHOSPHORUS-32 IN THE PRESENCE OF THE RNAse INHIBITOR DIETHYLPYROCARBONATE (DEP)

Ribonuclease activity is a frequent problem in 5'-end-labeling of any RNA molecule with phosphorus-32. Ribonuclease can be a serious contaminant in both the alkaline phosphatase and polynucleotide kinase preparations, as well as in the RNA preparation itself. DEP, a known inhibitor of many ribonucleases[24], can be included in both the TAP and phosphatase reaction mixtures to inhibit RNAse activity. Figure 10 demonstrates how effective the inclusion of DEP can be in inhibiting RNAse activity in late adenovires 2 mRNA preparations form HeLa cells 32 h after infection. Figure 10A shows an autoradiogram of a fingerprint analysis of an RNAse T_1 digest of 5'-^{32}P-labeled late Ad 2 mRNA, without the inclusion of DEP in the TAP and alkaline phosphatase enzymatic reactions. The 5'-^{32}P undecanucleotide ^{32}P-m^{6}Am CUCUCUUCCG$_p$, representing the 5'-terminal T_1 fragment of late Ad 2 mRNA[25], indicated by the arrow, is representative of <0.1% of the total DPMs present in the multitude of T_1 fragments in the fingerprint. With the inclusion of DEP in both enzymatic reactions, however, at a final concentration of 0.015%, shown in Figure 10B, essentially all the degradation is abolished and the 5'-^{32}P undecanucleotide now accounts for 8.0% of the total radioactivity in the fingerprint, and it is in agreement with the estimated amount of Ad 2 mRNA in HeLa cell polysomes during lytic infection[26]. DEP at concentrations used in this work (0.015%) produces no detectable base modifications of the mRNA. The overall labeling efficiency, however, is reduced by 25-50% with its inclusion. Because obtaining undergraded 5'-^{32}P-end-labeled mRNA is a priority, this reduction in labeling efficiency is insignificant.

VII. CELL-FREE TRANSLATION OF MESSENGER RNA AS AN ASSAY FOR CAP REMOVAL BY TAP

Experiments to test the effectiveness of TAP in hydrolyzing the m^{7}G cap nucleotide from VSV and rabbit globin mRNA were shown in Figure 4. It was concluded that usually close to 100% of the mRNA molecules are decapped under the conditions studied. The effectiveness of TAP in removing the 5' cap can also be assessed by using a cell-free translation assay. Such a study is summarized in the data shown in Figure 11. In this case, the ribonuclease inhibitor DEP was included in the decapping reaction mixture, to reduce transient nuclease activity. DEP is known to inhibit TAP by about 40-50% under the conditions

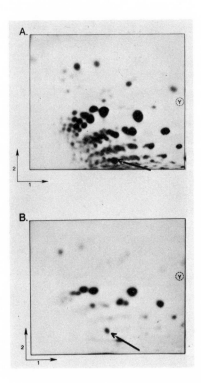

Figure 10. Autoradiogram of fingerprint analysis of RNAse T_1 digest of $5'-^{32}P$-end labeled late Ad 2 mRNA. First dimension: cellulose/acetate electrophoresis in pyridinium/acetate buffer (pH, 3.5) in 7 M urea. Second dimension: chromatography on polyethleneimine cellulose, thin-layer plates, with 3.0 M pyridinium formate buffer (pH, 3.4). Circled "Y" indicates position of methyl orange dye. Arrow indicates $5'-^{32}P$ undecanucleotide representing $5'$-terminal T_1 fragment of late Ad 2 mRNA. A, no DEP. B, 0.015% DEP.

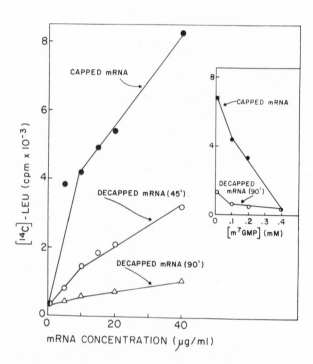

Figure 11. Translation of capped and decapped rabbit globin mRNA in rabbit reticulocyte lysate cell-free system. Incorporation of [14]C Leu into protein is plotted vs. mRNA concentration. Fifty-microliter (μl) reactions include 10 μl of micrococcal nuclease treated lysate, 10 μl of standard reaction mix (27), and optimal concentrations of $MgCl_2$ and KCl (each optimized for this lysate and for capped and decapped mRNAs, separately). Reactions were incubated at 34°C for 40 minutes CPM were determined by trichloroacetic acid precipitation on filters and counting in a toluene-based scintillation fluid. Insert shows inhibition of translation caused by increasing concentrations of m[7]GMP, in reactions containing mRNA at 10 μg/ml and incubated at 34°C for 40 minutes. mRNA decapping was done in reactions containing 50 mM NaOAc (pH, 6.0), 10 mM mercaptoethanol, 500 pmol of rabbit globin mRNA, 0.01% diethylpyrocarbonate, and 50 units of TAP in 25-μl total volume; incubation at 37°C for times indicated. Decapped mRNA was repurified by phenol, ether extraction, and oligo (dT)-cellulose chromatography before use in translation system.

used here (data not shown). Thus, the rate of decapping is lower in these
experiments than in the ones shown earlier, and complete decapping does not
occur in 45-60 minutes, as was shown in Figure 4. Figure 11 demonstrates
that the activity of globin mRNA as a template for translation drops dramatically on decapping and is a function of the reaction time with TAP. Incubation
of mRNA with DEP in the absence of TAP had a negligible effect translatability.
Also, it is seen that translation is inhibited by addition of increasing concentrations of the cap analogue m^7GMP (Figure 11, insert) and that the degree of inhibition is greatly reduced when decapped mRNA is used as a template. There is residual translational activity of globin mRNA after 90 minutes of reaction with TAP. This is due either to the presence of small amounts
of capped mRNA molecules or to the intrinsic low rate of translation of decapped
mRNA.

VIII. SUMMARY

In recent years, there has been a growing appreciation of the potential
applications of $5'$-^{32}P-end-labeled mRNA, not only for screening recombinant clones and mapping gene structure, but also for revealing possible nucleotide sequence and structural signals within mRNA molecules themselves, which
may be important for eukaryotic mRNA processing and turnover and for controlling differential rates of translational initiation. Three major problems,
however, have retarded progress in this area, lack of methods for efficient
and reproducible removal of $m^7G^5ppp^{5'}$-cap structures, which maintain the integrity of an RNA molecule; inability to generate a sufficient amount of labeled
mRNA, owing to the limited availability of most pure mRNA species; and the
frequent problem of RNA degradation during _in vitro_ end-labeling owing to RNAse
contamination. The procedures presented here permit one to decap and label
minute quantities of mRNA, effectively. Tobacco acid pyrophosphatase is relatively efficient in removing cap structures from even nanogram quantities of
available mRNA, and enough radioactivity can be easily generated from minute
amounts of intact mRNA with very high-specific-activity $[\gamma^{32}P]ATP$ and the inhibition of ribonuclease contamination with diethylpyrocarbonate. These procedures can be modified and applied to almost any other type of RNA molecule
as well. In Section III of this volume, we explore in detail how effectively
$5'$-end-labeled mRNA can be used not only for nucleotide sequence analysis, but
also for mapping mRNA secondary structure.

ACKNOWLEDGMENTS

This work was supported in part through the Medical Foundation of Boston from the Charles A. King Trust to R.E.L.; and by grants from the N.I.H., N.S. F., Alton-Jones Foundation, and Syracuse University Senate Research Committee to J.N.V. We thank U.L. RajBhandary (M.I.T.), in whose laboratory much of this work was carried out; George Pavlakis for TAP purification; Peggy Pinphanichakarn for performing the cell-free translation assays; and Rebecca Inman for preparation of the manuscript. This work is dedicated to the memory of Mrs. Jean Jaffee.

REFERENCES

1. Grunstein, M. and Hogness, D.S., Proc. Natl. Acad. Sci. USA, 72, 3961-3965 (1975).
2. Maniatis, T., Jeffrey, A., and Kleid, D.G., Proc. Natl Acad. Sci. USA, 72, 1184-1189 (1975).
3. Heckman, J.E. and RajBhandary, U.L., Cell, 17, 583-595 (1979).
4. Shinshi, J., Miwa, M., Kato, K., Noguchi, M., Matsushima, T., and Sugimura, T., Biochemistry, 15, 2185-2190 (1976).
5. Shimotohno, K. and Miura, K., FEBS Letters, 65, 254-257 (1976).
6. Efstratiadis, A., Vournakis, J.N., Donis-Keller, H., Chaconas, G., Dougall, D.K., and Kafatos, F.C., Nucleic Acids Res., 4, 4165-4174, (1977).
7. Lerch, B. and Wolf, G., Biochem. Biophys. Acta, 258, 206-218 (1972).
8. Dolapchiev, L.B., Sulkowski, E., and Laskowski, M., Biochem. Biophys. Res. Comm., 61, 273-281 (1974).
9. Johnson, R.A. and Walseth, T.F., Advances in Cyclic Nucleotide Research, Vol. 10, 135-167 (1979).
10. Lockard, R.E., Alzner-Deweerd, B., Heckman, J.E., MacGee, J., Tabor, M.W., and RajBhandary, U.L., Nucleic Acids Research, 5, 37-52 (1978).
11. Pavlakis, G.N., Lockard, R.E., Vamvakopoulos, N., Rieser, L., RajBhandary, U.L. and Vournakis, J.N., Cell, 19, 91-102 (1980).
12. Shatkin, A.J., Cell 9, 645-653 (1976).
13. Fraknkel-Conrat, H. and Steinschneider, A., Methods in Enzymology, eds. Grossman, L. and Moldave, K. (Academic Press, NY) Vol. 12, Part B, 243-246 (1967).
14. Rose, J.K. and Lodish, H.F., Nature, 262, 32-37 (1976).
15. Furuichi, Y., LaFiandra, A., and Shatkin, A.J., Nature, 266, 235-239 (1977).
16. Keith, G., and Gilham, P.T., Biochemistry, 13, 3601-3607 (1974).
17. Lockard, R.E., and RajBhandary, U.L., Cell, 9, 747-760 (1976).
18. Lockard, R.E., and Lane, C.D., Nucleic Acids, Res., 5, 3237-3247 (1978).
19. Gamborg, O.L., Plant Physiol., 45, 373-375 (1970).
20. Murachige, T. and Skoog, F., Plant Physiol., 15, 473-497 (1962).
21. Toneguzzo, F. and Ghosh, H.P., J. Virol., 17 483-491 (1976).
22. Blaedel, W.J. and Meloche, V.W., in Elementary Quantitative Analysis (Harper and Row, NY), 576-582 (1963).
23. Plocke, D.J. and Valhee, B.L., Biochemistry, 1, 1039-1049 (1962).
24. Solymosy, F., Fedorcsak, I., Gulyas, A., Farkas, G.L. and Ehrenberg, L., Eur. J. Biochem., 5, 520-527 (1968).

25. Lockard, E.R., Berget, S.M., RajBhandary, U.L. and Sharp, P.A., J. Biol. Chem., 254, 587-590 (1979).
26. Flint, S.J. and Sharp, P.A., J. Mol. Biol, 106, 749-771 (1976).
27. Koneck, D., Kramer, G., Pinphanichakarn, P., and Hardesty, B., Arch. Biochem. and Biophys., 169, 192-199 (1973).

5' End Labeling of RNA with Capping and Methylating Enzymes

Bernard Moss

Laboratory of Biology of Viruses
National Institute of Allergy and Infectious Diseases
National Institutes of Health
Bethesda, Maryland 20205

254

I. INTRODUCTION

The specificity of the capping and methylating enzymes from vaccinia virus make them suitable for radioactively labeling 5' di- or triphosphate ends of RNAs or capped ends after chemical removal of the terminal 7-methylguanosine (m^7G) residue[1]. In particular, for identification, genome mapping and sequencing of the original 5' ends of transcripts, capping has distinct advantages over alternative labeling methods.

The cap structure of viral[2-5] and eukaryotic[6-10] mRNAs consists of an m^7G residue connected from the 5' position through a triphosphate bridge to the 5' position of the next nucleotide which is frequently methylated at the 2' position of the ribose (N^m). Enzymes that catalyze cap formation are present in purified virus particles from cytoplasmic polyhedrosis virus[11], vaccinia virus[12], reovirus[13], and vesicular stomatitis virus[14]. However, only those from vaccinia virus have been obtained in an active soluble form capable of end labeling full length RNA molecules[15]. Moreover, for cytoplasmic polyhedrosis virus[16] and vesicular stomatitis virus[17], there is evidence that capping is closely coupled to transcription. While alternative eukaryotic sources would seem to offer advantages of convenience and economy for large scale enzyme purification, the capping enzymes of HeLa cells[18-20], rat liver[21], and wheat germ[22] have not yet been purified as extensively as that of vaccinia virus. In addition, the apparent specificity of the purified eukaryotic enzymes for di- rather than triphosphate ended RNA[20-22] may make separate RNA triphosphatase treatment necessary prior to capping. For these reasons, the capping and methylating enzymes of vaccinia virus are the only ones that have been used for specific end labeling thus far.

A large number of enzymes involved in transcription and possibly other functions are packaged within the core of vaccinia virus and are released in an active form by treatment with sodium deoxycholate, dithiothreitol, and NaCl[15,23]. Among the reactions catalyzed by the soluble extract, the following four are pertinent to this review.

$$pppN(pN)n \longrightarrow ppN(pN)n + Pi \qquad (i)$$

$$GTP + ppN(pN)n \rightleftharpoons G(5')pppN(pN)n + PPi \qquad (ii)$$

$$AdoMet + G(5')pppN(pN)n \longrightarrow m^7G(5')pppN(pN)n + AdoHcy \qquad (iii)$$

$$AdoMet + m^7G(5')pppN(pN)n \longrightarrow m^7G(5')pppN^m(pN)n + AdoHcy \qquad (iv)$$

Vaccinia virus RNA guanylyltransferase and RNA (guanine-7-)methyltransferase activities, which catalyze reactions (ii) and (iii) respectively, have been shown to exist as an enzyme complex[24-27]. This complex, which has been extensively purified, has a molecular weight of approximately 127,000 and polypeptide subunits of 95,000 and 32,000. The RNA triphosphatase[28], which catalyzes step (i), has recently been shown to be part of the same multifunctional complex containing RNA guanylyltransferase and RNA (guanine-7-)methyltransferase activities[29]. For simplicity, I refer to this complex as capping enzyme. Because of the extremely high RNA triphosphatase activity, RNAs containing tri- or diphosphates are capped at similar rates[29]. Significantly, however, RNAs containing less than two phosphates are not acceptors[25], a feature that makes it possible to specifically end label RNA even if nicked. At least qualitatively, the capping enzyme has not been shown to exhibit sequence specificity and homopolyribonucleotides and RNAs from a variety of sources containing terminal purines or pyrimidines have been capped. However, there is insufficient data to assume that all RNAs in a mixture will be capped at the same rate. In this regard, attempts to cap QB RNA were unsuccessful (unpublished), presumably due to extensive base pairing at the 5' end of this molecule. Preliminary studies on the effect of chain length indicated that ribonucleoside diphosphates were capped at a very low rate but that dinucleotides such as ppGpC or pppGpC were readily capped, although less efficiently than polymers[26,27].

Of the four common ribonucleoside triphophates, only GTP can serve as the donor in step (ii)[26]. The reaction consists of the transfer of a GMP residue with the liberation of pyrophosphate. Recent studies indicate that the enzyme can also catalyze GTP-pyrophosphate exchange via a GMP-enzyme intermediate[30]. dGTP is also a donor; however, neither GDP nor m^7GTP are used by the capping enzyme[26]. The absence of activity with m^7GTP indicates that methylation must follow guanylylation of RNA as indicated by the above order of reactions. A divalent cation is required and optimal activity is obtained with 0.5 to 2.5 mM $MgCl_2$, 60 mM NaCl, and pH of 7.5 to 7.8[25,27]. The Km for GTP is about 20 μM. As indicated in step (ii), the RNA guanylyltransferase reaction is readily reversible[25-27]. This reverse reaction is prevented when step iii occurs and the added guanosine is methylated. S-adenosylmethionine (AdoMet) is not required for capping although high concentrations increase the percentage of RNA capped presumably because of the above considerations[25].

As indicated in step (iii), the RNA (guanine-7-)methyltransferase activity of the capping enzyme uses AdoMet as the methyl donor and RNA ending in G(5') pppN - as an acceptor[25,26]. Dinucleoside triphosphate caps that are unattached to RNA, i.e. G(5')pppA, and even GTP are methylated but at much lower rates than capped RNA[26]. Neither GTP nor Mg^{2+} is required for methylation and the enzyme exhibits a broad pH optimum around neutrality[25]. As expected, the reaction is severely inhibited by S-adenosylhomocysteine (AdoHcy).

RNA (nucleoside-2'-)methyltransferase, which catalyzes step (iv), has been purified essentially free of RNA guanylyltransferase and RNA (guanine-7)-methyltransferase activities. The enzyme has a single subunit with a molecular weight of 38,000[31]. The specificity of the RNA (nucleoside-2'-)methyltransferase for RNA ending in m^7G(5')pppN- provides evidence for the reaction order indicated above[32]. Dinucleoside triphosphate caps of the type m^7G(5')pppN that are unattached to RNA are poor acceptors and do not compete with capped RNA[32]. Divalent cations are not required and the pH optimum is 7.5. The Km for AdoMet is approximately 2 μM and AdoHcy is a potent inhibitor.

II. PREPARATION OF VACCINIA VIRUS CAPPING ENZYME

A. General Considerations

At present, neither capping enzyme nor vaccinia virus are commercially available. Since the enzyme purification procedure is relatively simple, the major problem for biochemists or molecular biologists is to obtain sufficient amounts of virus. For this reason, a virus purification procedure adapted from that of Joklik[33,34] and used in my laboratory is outlined below.

B. Purification of Vaccinia Virus

Although vaccinia virus is safe to work with, vaccination of laboratory workers, avoidance of direct physical contact and aerosols, as well as good microbiological technique, is recommended. Large preparations of vaccinia virus (strain WR) are conveniently grown in HeLa cell suspension cultures. The virus inoculum is prepared by incubating a crude cell lysate containing vaccinia virus, [usually 2 X 10^9 to 2 X 10^{10} plaque forming units (PFU)/ml in Eagle's medium containing 2% horse serum] with 0.12 mg/per ml of crystallized trypsin for 30 min at 37°C with frequent vortexing to eliminate clumps. Growing HeLa cells are then centrifuged and resuspended at a concentration of 2×10^7 per ml in Eagle's spinner medium containing 5% horse serum. The cells are incubated with 3 plaque forming units (PFU) of virus per cell at 37°C for 30 min with constant stirring. At the end of this time, the cells are diluted 40-fold with fresh medium containing 5% horse serum and incubated at 37°C for 24 to 48 hr. The cells are harvested by low speed centrifugation and resuspended at 5×10^7 per ml in cold 0.01 M Tris-HCl (pH 9). The remainder of the procedure is carried out at 0 to 4°C. A Dounce homogenizer is used for cell disruption and nuclei are removed by centrifugation at 200 X g for 5 min. The supernatant is saved and the nuclear pellet is resuspended in a minimal volume of Tris buffer and centrifuged once again. Supernatants are combined in a tube immersed in crushed ice and dispersed by sonication for eight 15 sec. intervals at maximum power using a Branson Sonifier. Approximately 25 ml of this cytoplasmic fraction is layered over an equal volume of 36% (w/v) sucrose in 0.01 M Tris-HCl (pH 9) and centrifuged in an SW 25.2 rotor at 20,000 rev/min (49,200 X g) for 45 min. The pellet is resuspended by sonication in 5 ml of 1 mM Tris-HCl (pH 9), layered over a linear 25 to 40% (w/v) sucrose gradient in the latter buffer, and centrifuged in an SW 25.2 rotor at 13,500 rev/min (22,400 X g) for 45 min. The thick white virus band is collected by puncturing the side of the tube with a needle and syringe, then diluted with two volumes of 1 mM Tris-HCl (pH 9) and centrifuged at 20,000 rev/min for 30 min. The pellet is resuspended by sonication in 3 ml of 1 mM Tris-HCl (pH 9) and subjected to a second round of sucrose gradient centrifugation. Since a significant amount of virus aggregates during each gradient centrifugation step, the yield can be increased by resuspending virus pellets from several tubes and reapplying them to gradients. An average yield is approximately 40 mg (1 light scattering unit at 260 nm corresponds to about 64 µg of protein) of two-times banded virus from 6 l of cells. Virus is routinely stored at -70°C for periods of more than six months.

C. Capping Assays

Since the capping enzyme exists as a complex of RNA triphosphatase, RNA guanylyltransferase, and RNA (guanine-7-)methyltransferase activities, column fractions can be monitored in several different ways depending on the substrates available.

1. RNA Guanylyltransferase Assay. The RNA guanylyltransferase assay depends on the presence of a di- or triphosphate-ended RNA acceptor. Large amounts

of di- and triphosphate-ended poly(A) can be made by chemical addition of phosphates to nuclease P1 nicked poly(A)[19,24]. However, smaller amounts of triphosphate ended poly(A) may be more conveniently prepared using Escherichia coli RNA polymerase[20,35]. Although well-defined natural RNAs ending in di- or triphosphates such as satellite tobacco necrosis virus RNA, influenza virus RNA, and uncapped vaccinia virus or reovirus RNAs have been used as substrates[1,36], they are difficult to obtain in large quantities. Although not tested by the author, crude soluble RNA preparations from yeast have been reported to contain a fraction, 10 to 20 nucleotides in length, that is an acceptor for eukaryotic capping enzyme[21,37].

Activity is routinely assayed in 50 μl reactions containing 25 mM Tris-HCl (pH 7.8), 2 mM $MgCl_2$, 1 mM dithiothreitol, 25 μM [8-^{3}H]GTP (10 Ci/mmol) and 20 pmol of di- or triphosphate-ended poly(A). After 30 min at 37°C, the product is either trichloroacetic acid precipitated by standard methods or more conveniently 40 μl samples are spotted onto 2.5 cm diameter DE 81 filters held above a styrofoam board with pins. The still moist DE 81 filters are washed three times for 15 min intervals with 5% NaH_2PO_4, once for 5 min with water, twice for 5 min with ethanol:diethyl ether (1:1) and once with diethyl ether[38]. The washing procedure is facilitated by dropping the filters, with pins attached, into a wire cage suspended in a beaker containing a rotating magnetic bar at the bottom. One unit of enzyme catalyzes the capping of 1 pmol of RNA at 37°C in 30 min.

2. RNA (Guanine-7-)Methyltransferase Assay. The RNA (guanine-7-)methyltransferase assay is convenient because GTP is a substrate[26,31]. The Km for GTP is quite high, however, compared to the capped ends of RNAs. The reaction mixture contains 50 mM Tris-HCl (pH 7.6), 1 mM dithiothreitol, 10 mM GTP and 1 μM Ado[methyl-^{3}H]Met (11.6 Ci/mmol) in a final volume of 0.1 ml. After incubation with enzyme at 37°C for 30 min, 80 μl is applied to a DE-81 filter. Unincorporated Ado[methyl-^{3}H]Met is removed by washing four times with 25 mM ammonium formate, once with H_2O, twice with ethanol:ether (1:1), and once with ether.

3. RNA Triphosphatase Assay.[28,29] RNA triphosphatase activity is measured in 25 μl reaction mixtures containing 50 mM Tris-HCl (pH 8.4), 2.5 mM $MgCl_2$, 2 mM dithiothreitol, and 1 pmol of [γ-^{32}P]labeled poly(A) (2,500 cpm/pmol) prepared with E. coli RNA polymerase. After incubation at 37°C for 5 min, the samples are spotted on PEI thin layer sheets and chromatographed in 0.75 M $KH_2(PO_4)$ at pH 3.4. Pi is located by autoradiography and the spot is cut out and counted.

4. GTP-Pyrophosphate Exchange Assay[30]. The GTP-pyrophosphate exchange assay, like the one for RNA (guanine-7-)methyltransferase, does not require the

preparation of special substrates. Reaction mixtures (50 μl) contained 60 mM Tris-HCl (pH 8.4), 10 mM dithiothreitol, 5 mM MgCl$_2$, 0.2 mM GTP and 1 mM [^{32}P] PPi (20-50 cpm/pmol). After 30 min at 37°C, reactions are terminated by adding 0.5 ml of cold 10% trichloroacetic acid, 0.1 ml of sodium pyrophosphate and 0.2 mg of bovine serum albumin. Acid-insoluble material is removed by centrifugation in a microcentrifuge and the supernatants added to 0.1 ml of a 30% suspension of Norit A charcoal. Norit-adsorbed material is collected on glass fiber filters, washed with 5% trichloroacetic acid, and the Cerenkov counts determined.

D. Purification of Capping Enzyme

1. **Preparation of Virus Cores.** 40 to 100 mg of purified vaccinia virus is incubated in 10 ml of 0.05 M Tris-HCl (pH 8.4), 0.5% Nonidet P-40 detergent, 0.05 M dithiothreitol for 30 min at 37°C with frequent shaking. The material is then layered over 1.5 ml of 36% (w/v) sucrose in 0.01 M Tris-HCl (pH 8.4), 1 mM dithiothreitol in three SW 50.1 tubes and centrifuged at 25,000 rev/min (58,400 X g) for 30 min at 4°C. The supernatant and sucrose layers are carefully removed and discarded and the pelleted virus cores are resuspended in 5 ml of 0.01 M Tris-HCl (pH 8.4).

2. **Disruption of Virus Cores.** Five ml of 0.5 M Tris-HCl (pH 8.4), 0.1 M dithiothreitol, 0.4% sodium deoxycholate and 0.5 M NaCl is added to 5 ml of resuspended cores and the mixture is incubated on ice for 30 min. The viscosity of the released DNA can be reduced by brief sonication while the tube is immersed in an ice water mixture. (This step is avoided when RNA polymerase is also to be purified). Insoluble structural proteins are removed by centrifugation at 136,000 X g for 60 min. If the viscosity of the DNA is not reduced by sonication, the supernatant is passed through a 23 gauge needle.

3. **DEAE-Cellulose Chromatography.** The supernatant, containing about 30% of the total core protein, is then diluted and adjusted to contain 0.2 M NaCl, 10% glycerol, 0.1% Triton X-100, and 1 mM EDTA and passed at 20 ml/hr through a DEAE-cellulose column (3 X 1.5 cm) equilibrated with buffer A [0.2 M NaCl, 0.2 M Tris-HCl (pH 8.4), 1 mM dithiothreitol, 0.1 mM EDTA, 0.1% Triton X-100, and 10% glycerol] to remove DNA. After washing the column with additional Buffer A, the protein-containing flow-through material is diluted to reduce the NaCl and Tris-HCl concentrations to 0.05 M and adjusted to maintain the previous concentrations of dithiothreitol, glycerol, Triton X-100, and EDTA (Buffer B). This is then applied to a DEAE-cellulose or DEAE-Biogel column (5 X 1 cm) which is equilibrated with Buffer B. Capping and methylating enzymes do not bind to the column and are obtained in the flow-through. The vaccinia virus RNA polymerase binds

to the column and can be eluted if desired[39,40]. To aid in the pooling of flow-through fractions from both DEAE columns, 2 µl samples are spotted on cellulose acetate strips which are then stained for protein by immersion in 0.1% Coomassie Blue "R" in 50% methanol, 10% acetic acid for 3 min at room temperature followed by two brief washes in 50% methanol, 10% acetic acid.

Such preparations of crude enzyme have been successfully used for labeling the ends of RNAs in several studies. However, it should be realized that at this stage nucleic acid-dependent nucleoside triphosphatases[41], poly(A) polymerase[42], single-strand-specific DNase[43], protein kinase[44], DNA unwinding protein[45], low levels of RNase[46] and a 5'-phosphate polyribonucleotide kinase[47] are all present. In the presence of ATP, the latter enzyme might add a second phosphate to the end of RNA containing one phosphate and thereby generate a new end that could be capped. Although the latter reaction should not occur in the absence of ATP, further purification of the capping enzyme is recommended.

4. <u>DNA Cellulose Chromatography</u>. Chomatography on DNA cellulose provides a useful procedure for further purification[29]. The sample from the second DEAE-cellulose or DEAE-Biogel column is applied at 25 ml/hr to a (1 X 15 cm) column of denatured calf thymus DNA cellulose equilibrated with buffer B. (Routinely, we connect the second DEAE column directly to the DNA cellulose column since the same buffer is used for both). After washing with two column volumes, a 640 ml linear gradient from 0.05 M to 0.4 NaCl in buffer B is applied at 25 ml/hr. The capping enzyme complex elutes as a sharp peak at approximately 0.1 M NaCl followed closely by DNA-dependent ATPase[29,41]. Enzyme purified to this stage contains trace amounts of RNA (nucleoside-2'-)methyltransferase but appears to be free of other interfering activities. However, a variety of other columns including phos-phocellulose can be used if purification to near-homogeneity is desired[24,29,35]. Yields of 15 to 20,000 units (assay 1) are obtained from 100 m g of virus.

III. PREPARATION AND LABELING OF RNA

A. Preparation of RNA

RNA with a di- or triphosphate end can be capped directly without further preparation. If the RNA is naturally capped, the m^{7}G residue can be removed by periodate oxidation and β-elimination[48,49]. This is carried out in 0.1 to 0.2 ml of freshly prepared 0.9 mM sodium periodate (greater than 100-fold molar excess), 0.15 M sodium acetate (pH 5.3) in the dark at 0°C. After 30 min, the RNA is precipitated two times with ethanol. If RNA carrier is needed, alkaline phos-

phatase treated tRNA may be added prior to precipitation. Approximately 0.1 to
0.2 ml of a solution containing 0.3 M distilled aniline and 0.01 M acetic acid
adjusted to pH 5 with concentrated HCl is used to dissolve the RNA. After 3 hr
in the dark at room temperature, the RNA is either precipitated repeatedly with
ethanol or filtered through a G-50 Sephadex column. Except for removal of the
terminal m^7G (and the 3' terminal nucleoside), there is no apparent damage to
the RNA which is translatable after recapping[50]. Presently, there is no alternative enzymatic procedure for removing the cap and leaving two or three terminal phosphates on the RNA. Both tobacco acid pyrophosphatase[51,52] and potato
acid pyrophosphatase[53,54] apparently leave a single phosphate. A novel enzyme
that adds one or two phosphates to the 5' end of a polyribonucleotide with one
phosphate has been isolated from vaccinia virus[47]. Whether this enzyme can
effectively phosphorylate full-length molecules and whether it can be prepared
in useful amounts is not known yet. A cap-specific pyrophosphatase activity
isolated from HeLa cells[55,56] and chick embryos[57] has been reported to cleave
$m^7G(5')pppN$ to yield pm^7G and ppN; however, only isolated caps or oligonucleotides up to eight to ten units in length are substrates.

B. Labeling of RNA

The conditions for capping are similar to those described in the assay for
RNA guanylyltransferase. Optimal enzyme concentrations are best determined
empirically since the units obtained with different assays and different RNAs
have not been standardized. RNA in pmole amounts can be quantitatively capped
using excess enzyme and concentrations of GTP above the Km value[27,29]. However,
5 to 20 µM concentrations of $[\alpha-^{32}P]GTP$ (400 Ci/mmole) are frequently used to
conserve isotope. Under these conditions, capping does not usually exceed 30%
of theoretical. With the above specific activity isotope, there will be approximately 900,000 dpm per pmol of capped RNA (e.g. 0.3 µg of an RNA of 1,000 nucleotides). Higher specific activity GTP is available but has not been used in my
laboratory. Addition of AdoMet (100 µM) will increase the extent of capping up
to 2-fold presumably by preventing pyrophosphate inhibition. However, we usually
use AdoHcy in place of AdoMet if the 5' end of the RNA is to be sequenced since
incomplete methylation or alkaline degradation of m^7G could lead to artifactual
heterogeneity.

Labeling can also be performed with $[^3H]GTP$. However, since ^3H is commonly
in the 8-position of guanine, AdoMet must not be used since methylation in the
7 position will lead to loss of the isotope.

Another alternative is to label with Ado[methyl-^{3}H]Met (1 to 5 µM) in the presence of unlabeled GTP. Under these conditions, high concentrations of GTP (1 mM) can be used and capping is nearly quantitative[1].

Reactions are usually stopped after 30 min at 37°C by addition of EDTA to 4 mM and sodium dodecyl sulfate to 0.2%. After phenol-chloroform extraction, the RNA is recovered by filtration through a column of G-75 Sephadex equilibrated with 50 mM triethylamine bicarbonate and either dried by lyophilization or ethanol precipitated. Proof that the label is in the cap structure is obtained by digestion of a sample with nuclease P1 and alkaline phosphatase. Digestion is carried out in a volume of 50 µl containing 25 µg of nuclease P1 in 0.01 M ammonium acetate (pH 6) for one hr at 37°C. The pH is raised by addition of 5 µl of 0.5 M Tris-HCl (pH 8.5) and the material is incubated with 10 µg of bacterial alkaline phosphatase for one hr at 37°C. The released cap structures, e.g., G(5')pppN or m^{7}G(5')pppN are then analyzed by high voltage paper electrophoresis at pH 3.5[58] with appropriate markers (P. L. Biochemicals).

IV. APPLICATIONS

A. Genome Mapping

One of the most powerful and useful applications of capping enzyme is for mapping transcription initiation sites. This can be accomplished by hybridizing 5' end labeled RNA to DNA restriction fragments that have been immobilized on nitrocellulose membrane[59] or diazotized paper[60]. As described above, either naturally di- or triphosphate-ended RNA or decapped mRNAs can be labeled. Capping can be carried out on total RNA, polyadenylated RNA or RNA preselected by hybridization. Controlled alkaline hydrolysis[61] of the cap-labeled RNA can be used to increase the efficiency of hybridization to DNA fragments encoding the 5' ends of the RNAs. With vaccinia virus DNA, hybridization was carried out in sealed plastic bags containing 5 X SSC (SSC is 0.15 M sodium chloride, 0.015 M sodium citrate) immersed in a rocking 60°C water bath for 24 to 36 hr. Alternatively, hybridization is carried out with formamide at lower temperatures. The precise conditions of hybridization varies with different RNA-DNA hybrids. Unhybridized RNA is removed by repeated washes with 2 X SSC at 60°C. Treatment with 10 µg of RNase A per ml in 2 X SSC for 15 min at 25°C removes overhanging RNA tails. After RNase treatment, the filters are washed three times with 25 ml of 2 X SSC containing 0.5% SDS and three times with 2 X SSC. The dried filters are placed in contact with X-ray film and fluorography is carried out with the use of an appropriate intensifying screen. This approach served to identify the transcription initiation sites of Xenopus laevis preribosomal RNA[62], fibroin mRNA of Bombyx mori[63], early vaccinia virus transcripts[64,65] and early herpesvirus transcripts[66].

An alternative approach based on the nuclease S1 procedure of Berk and Sharp[67] has also been used[65]. Cap labeled vaccinia virus RNA was hybridized to cloned DNA fragments under conditions favoring DNA-RNA hybridization. Following nuclease S1 treatment, the 0.3 ml samples were digested with 25 ng of RNase A at 25°C for 10 to 15 min to reduce the background of unhybridized RNA[65]. The reaction was terminated by addition of sodium dodecyl sulfate to 0.25% and two successive phenol:chloroform (1:1) extractions. After ethanol precipitation, RNA-DNA hybrids were resolved by agarose gel electrophoresis and fluorographed. The location of the 5' end of the transcript can be determined from the size of the nuclease-resistant fragment.

B. Sequence Determination

RNA that has been 5' end labeled with capping enzyme may be sequenced using base-specific ribonucleases in the same general manner as RNA labeled with polynucleotide kinase. This procedure has been used to determine the first few nucleotides of pre-ribosomal rRNA of Xenopus laevis[62] and of RNAs of satellite necrosis virus and influenza virus[36]. Using the rapid gel method, the first 164 residues beyond the cap of brome mosaic virus RNA were sequenced[68]. In addition, the first few residues were determined independently by partial digestion with nuclease P1 followed by mobility shift analysis using two dimensional cellulose acetate electrophoresis and homochromatography.

The 5'-terminal nucleotides of RNAs with caps that are not 2'-O-methylated, e.g., plant and plant viral RNA, Sindbis virus RNA, and Newcastle disease virus RNA[10], can be labeled with purified vaccinia virus RNA (nucleoside-2'-)methyltransferase or with the DEAE-cellulose flow through containing this enzyme[1]. In this manner, the 5'-ends of tobacco mosaic virus RNA[1], brome mosaic virus RNA[32,68] and potato virus X RNA[69] were identified.

C. Protein Synthesis and Transcription Studies

One of the first uses of the vaccinia virus capping enzyme was for protein synthesis studies. By chemically removing the m[7]G residue from vaccinia virus mRNAs and then replacing it enzymatically, the role of the cap structure in ribosome binding and translation was confirmed[50]. In particular, the readdition of m[7]G but not unmethylated G resulted in the recovery of translatability. Similarly, enzymatic capping of prokaryotic mRNAs resulted in their efficient translation in a eukaryotic cell-free system[70]. However, enzymatic addition of caps to satellite necrosis virus RNA, which is naturally uncapped, does not affect the rate or extent of formation of protein synthesis initiation complexes unless fragmented RNA is used[71,72].

The capping and methylating enzymes of vaccinia virus have also been useful in establishing the primer specificity of the influenza virion transcriptase. Evidence has been obtained that both 7-methylguanosine and 2'-O-methylribonucleoside are required[73,74].

REFERENCES

1. Moss, B. (1977) Biochem. Biophys. Res. Commun., 74, 374-383.
2. Wei, C.-M. and Moss, B. (1975) Proc. Nat. Acad. Sci., 72, 318-322.
3. Furuichi, Y., Morgan, M., Muthukrishnan, S. and Shatkin, A.J. (1975) Proc. Nat. Acac. Sci., 72, 362-366.
4. Furuichi, Y. and Miura, K.I. (1975) Nature (London) 253, 374-375.
5. Abraham, G., Rhodes, D.P. and Banerjee, A.K. (1975) Cell, 5, 51-58.
6. Wei, C.-M., Gershowitz, A. and Moss, B. (1975) Cell, 4, 379-386.
7. Perry, R.P., Kelley, D.E., Frederici, K. and Rottman, F. (1975) Cell 4, 387-394.
8. Adams, J.M. and Cory, S. (1975) Nature, 255, 28-33.
9. Furuichi, Y., Morgan, M., Shatkin, A.J., Jelinek, W., Salditt-Georgieff, M. and Darnell, J.E. (1975) Proc. Nat. Acad. Sci., 72, 1904-1908.
10. Banerjee, A.K. (1980) Microbiol. Rev. 44, 175-199.
11. Furuichi, Y. (1974) Nucleic Acids Res., 1, 802-809.
12. Wei, C.-M. and Moss, B. (1974) Proc. Nat. Acad. Sci., 71, 3014-3018.
13. Shatkin, A.J. (1974) Proc. Nat. Acad. Sci., 71, 3204-3207.
14. Rhodes, D.P., Moyer, S.A. and Banerjee, A.K. (1974) Cell 3, 327-333.
15. Ensinger, M.J., Martin, S.A., Paoletti, E. and Moss, B. (1975) Proc. Nat. Acad. Sci., 72, 2525-2529.
16. Furuichi, Y. (1978) Proc. Nat. Acad. Sci. 75, 1086-1090.
17. Abraham, G. and Banerjee, A.K. (1976) Virology, 71, 230-241.
18. Wei, C.-M. and Moss, B. (1977) Proc. Nat. Acad. Sci. 74, 3758-3761.
19. Venkatesan, S., Gershowitz, A., and Moss, B. (1980) J. Biol. Chem., 225, 2829-2834.
20. Venkatesan, S. and Moss, B. (1980) J. Biol. Chem., 255, 2835-2842.
21. Mizumoto, K. and Lipmann, F. (1979) Proc. Nat. Acad. Sci. 76, 4961-4965.
22. Keith, J., Venkatesan, S., Gershowitz, A. and Moss, B., Biochemistry, manuscript submitted.
23. Moss, B. (1978) In Molecular Biology of Animal Viruses, Vol. 2, D.P. Nayak, ed., Marcel Dekker, Inc., New York, pp. 849-890.
24. Martin, S.A., Paoletti, E. and Moss, B. (1975) J. Biol. Chem., 250, 9322-9329.
25. Martin, S.A. and Moss, B. (1975) J. Biol. Chem., 250, 9330-9335.
26. Martin, S.A. and Moss, B. (1976) J. Biol. Chem., 251, 7313-7321.
27. Monroy, G., Spencer, E. and Hurwitz, J. (1978) J. Biol. Chem., 253, 4490-4498.
28. Tutas, D.J. and Paoletti, E. (1977) J. Biol. Chem., 252, 3092-3098.
29. Venkatesan, S., Gershowitz, A. and Moss, B. (1980) J. Biol. Chem., 255, 903-908.

30. Shuman, S., Surks, M., Furneaux, H. and Hurwitz, J. (1980) J. Biol. Chem., 255, 11588-11598.
31. Barbosa, E. and Moss, B. (1978) J. Biol. Chem., 253, 7692-7697.
32. Barbosa, E. and Moss, B. (1978) J. Biol. Chem., 253, 7698-7702.
33. Joklik, W.K. (1962) Biochem. Biophys. Acta, 61, 290-301.
34. Joklik, W.K. (1962) Virology, 18, 9-18.
35. Monroy, G., Spencer, E. and Hurwitz, J. (1978) J. Biol. Chem., 253, 4481-4489.
36. Moss, B., Keith, J.M., Gershowitz, A., Ritchey, M.B. and Palese, P. J. Virol., 25, 312-318.
37. Laycock, D.G. (1976) Fed. Proc., 36, 770.
38. Blatti, S.P., Ingles, C.J., Lindell, T.J., Morris, P.W., Weaver, R.F., Weinberg, F. and Rutter, W.J. (1970) Cold Spring Harbor Symp. Quant. Biol., 35, 649-657.
39. Baroudy, B.M. and Moss, B. (1980) J. Biol. Chem., 255, 4372-4380.
40. Spencer, E., Shuman, S. and Hurwitz, J. (1980) J. Biol. Chem., 255 5388-5395.
41. Paoletti, E., Rosemond-Hornbeak, H. and Moss, B (1974) J. Biol. Chem., 249, 3273-3280.
42. Moss, B., Rosenblum, E.N. and Paoletti, E. (1973) Nature New Biol., 245, 59-63.
43. Rosemond-Hornbeak, H., Paoletti, E. and Moss, B. (1974) J. Biol. Chem., 249, 3287-3291.
44. Kleiman, J.H. and Moss, B. (1975) J. Biol. Chem., 250, 2420-2429.
45. Bauer, W.R., Ressner, E.C., Kates, J. and Patzke, J.V. (1977) Proc. Nat. Acad. Sci., 74, 1841-1845.
46. Paoletti, E. and Lipinskas, B.R. (1978) J. Virol., 26, 822-824.
47. Spencer, E., Loring, D., Hurwitz, J. and Monroy, G. (1978) Proc. Nat. Acad. Sci., 75, 4793-4797.
48. Fraenkel-Conrat, H. and Steinschneider, A (1967) In Methods in Enzymology, Vol. 12B. Grossman, L. and Moldave, K. ed., Academic Press, New York, pp. 243-246.
49. Moss, B. and Koczot, F. (1976) J. Virol. 17, 385-392.
50. Muthukrishnan, S., Moss, B., Cooper, J.A. and Maxwell, E.S. (1978) J. Biol. Chem., 253, 1710-1715.
51. Shinshi, H., Miua, M., Sugimura, T., Shimotohno, K. and Miura, K.-I. (1976) FEBS Lett. 65, 254-257.
52. Shimotohno, K., Kodama, Y., Hashimoto, J. and Miura, K.-I. (1977) Proc. Nat. Acad. Sci., 74, 2734-2738.
53. Kole, R., Sierakowska, H. and Shugar, D. (1976) Biochim. Biophys. Acta, 438, 540-550.
54. Zan-Kowalczewka, M., Bretner, M., Sierakowska, H., Szczesna, E., Filipowicz, W. and Shatkin, A.J. (1977) Nucleic Acids Res., 4, 3065-3081.
55. Nuss, D.L., Furuichi, Y., Koch, G. and Shatkin, A.J. (1975) Cell, 6, 21-27.
56. Nuss, D.L. and Furuichi, Y. (1977) J. Biol. Chem. 252, 2815-2821.
57. Lavers, G.C. (1977) Molec. Biol. Reports, 3, 413-420.
58. Brownlee, G.G. (1972) In Laboratory Techniques in Biochemistry and Molecular Biology, Vol. 3, Part 1, Work, T.S. and Work E., ed., American Elsevier Publishing Co., Inc., New York. pp. 1-265.
59. Southern, E.M. (1975) J. Mol. Biol., 98, 503-517.
60. Wahl, G.M., Stern, M. and Stark, G.R. (1979) Proc. Nat. Acad. Sci., 76, 3683-3687.
61. Evans, R.M., Fraser, N., Ziff, E., Weber, J., Wilson, M. and Darnell, J.E. (1977) Cell, 12, 733-739.
62. Reeder, R.H., Sollner-Webb, B. and Wahn, H.C. (1977) Proc. Nat. Acad. Sci., 74, 5402-5406.

63. Tsujimoto, Y. and Suzuki, Y. (1979) Cell, 16, 425-436.
64. Wittek, R, Cooper, J.A., Barbosa, E. and Moss, B. (1980) Cell, 21, 487-493.
65. Venkatesan, S. and Moss, B. (1981) J. Virol., in press.
66. Mackem, S. and Roizman, B., Proc. Nat. Acad. Sci., in press.
67. Berk, A.J. and Sharp, P.A. (1977) Cell, 12, 721-732.
68. Ahlquist, P., Dasgupta, R., Shih, D.S., Zimmern, D. and Kaesberg, P. (1979) Nature, 281, 277-282.
69. Sonenberg, N., Shatkin, A.J., Ricciardia, R.P., Rubin, M. and Goodman, R.M. (1978) Nucleic Acids Res., 5, 2501-2512.
70. Paterson, B.M. and Rosenberg, M. (1979) Nature, 279, 692-696.
71. Smith, R.E. and Clark, J.M. (1978) Biochemistry, 18, 1366-1371.
72. Brooker, J. and Marcus, A. (1977) FEBS Letters, 83, 118-124.
73. Plotch, S.J., Bouloy, M. and Krug, R.M. (1979) Proc. Natl. Acad. Sci., 76, 1618-1622.
74. Bouloy, M., Plotch, S.J. and Krug, R.M. (1977) Proc. Natl. Acad. Sci., 77, 3952-3956.

SEQUENCE AND STRUCTURE ANALYSIS OF END-LABELED RNA WITH NUCLEASES

John N. Vournakis, James Celantano, Margot Finn, Raymond E. Lockard[*],
Tanaji Mitra, George Pavlakis[+], Anthony Troutt
Margaret van den Berg, and Regina M. Wurst[‡]
Department of Biology, Syracuse University,
Syracuse, New York 13210

Present address: [*]Department of Biochemistry
The George Washington University Medical School
2300 Eye St., Washington, D.C. 20037

[+]National Institutes of Health
Recombinant DNA Unit
Bethesda, Maryland 20014

[‡]Department of Biology, M.I.T.,
Cambridge, Massachusetts 02139

I. INTRODUCTION

Recent advances in the development of rapid gel sequencing methods for DNA[1,2] have provided a new strategy and impetus for the attempt to establish parallel methods for RNA sequence and structure analysis. RNA sequencing had required milligram quantities of purified species and monumental effort before the advent of the classical radioactive sequencing techniques pioneered by Sanger and colleagues in the mid-1960s[3],[4]. This approach requires the purification of uniformly ^{32}P-labeled RNA molecules after in vivo labeling, degradation by base-specific ribonucleases, two-dimensional separation of the resulting oligonucleotides (fingerprinting), and sequence analysis involving identification of overlapping nucleotides, to establish a unique sequence[4]. The first complete RNA sequence obtained in this way was of E. coli 5S rRNA by Brownlee and co-workers[5]. The culmination of the application of this method was the complete sequence of the 3,569 nucleotides of MS2 RNA[6]. Although small amounts of purified RNA can be sequenced by the oligonucleotide fingerprinting method it remains technically cumbersome and time-consuming.

Two new approaches for sequencing RNA that have appeared recently are based on the degradative[1] and the copying[2] methods of DNA sequencing. The degradative approach is simply the RNA analogue of the Maxam-Gilbert DNA sequencing method[1],[7]. Donis-Keller et al.[8] and Simoncits et al.[9] used base-specific enzymes to generate partial digests of $5'$-^{32}P-end-labeled 5.8S rRNA and tRNA molecules. Peattie[10] used chemical reactions for the partial hydrolysis of $3'$-^{32}P-end-labeled 5S and 5.8S rRNA. In the copying approach, the RNA to be sequenced is the template to which a specific DNA primer (often a restriction-enzyme fragment) is hybridized, and the primer is extended with a DNA synthesizing enzyme under special conditions to allow sequencing. Brownlee and Cartwright[11] used d(pT$_{10}$-G-C), reverse transcriptase, and DNA polymerase I (Klenow fragment) to sequence the 3' region of chicken ovalbumin mRNA with a method analogous to the "plus-minus" method of Sanger and Coulson[12] for sequencing DNA. Sures et al.[13] used specific restriction fragments and avian myeloblastosis virus (AMV) reverse transcriptase to sequence the 5' noncoding regions of the sea urchin histone mRNAs with a method based on the "2'3'-dideoxy" DNA sequencing technique of Sanger et al[2]. Ghosh et al.[14] made cDNA reverse transcripts with uniformly labeled specific restriction fragments of various regions of SV40 mRNAs and sequenced these regions with Maxam-Gilbert DNA sequencing protocols[7].

The degradative approach to RNA sequencing is limited by the physical dimensions of the polyacrylamide gels used to separate the oligonucleotides obtained after the enzymatic or chemical partial hydrolyses. About 300 nucleotides can be read if "standard"[7] sequencing gels and both 5'- and 3'-end labeling are used. The use of thin long gels can extend this limitation to molecules of about 1,000 nucleotides[15-18]. The recent demonstration by Donis-Keller[19] that RNAse H can be used to generate RNA fragments of specific length gives promise for the applicability of the degradative method to RNAs of any length. The copying approach does not, in principle, have a size limitation, provided that enough primer molecules are available for sequencing of overlapping regions of 300-400 nucleotides along the RNA. It also has the advantage that RNAs can be sequenced without extensive purification.

The degradative and copying approaches cannot be used to locate modified nucleotides, except for the 2'-0-methyl derivatives[20], and must therefore be coupled with other techniques for identifying modified bases[21]. Randerath and co-workers[22,23] have combined chemical degradation, end-labeling, and thin-layer chromatography to develop a RNA sequencing method in which both modified and major nucleotides are identified. This approach is particularly useful for sequencing transfer RNAs. It requires highly purified material in greater quantities than are needed for the other methods, and it is subject to the same limitations discussed earlier for the degradative approach.

RNA structure studies that use enzymes as structure probes have appeared in the biochemical literature for more than 15 years[24-31]. Much of the early work was focused on locating particular non-base-paired nucleotides by the use of base-specific ribonucleases like T1 and U2, given that these enzymes will hydrolyze unpaired nucleotides at a greater rate than nucleotide bases in hairpin helices. The secondary structure ("flower") model of MS2 RNA is based on such studies[6]. A major advance in the use of enzymes to study nucleic acid structure came with the discovery of structure-specific nucleases, such as S1 nuclease, the single-strand-specific enzyme isolated from Aspergillus oryzae[32,33]. The use of S1 in studies of RNA structure has been extensive[34-38]. Other single-strand-specific[39-41] and, recently, double-strand-specific[42,43] enzymatic activities have been reported. In general, the structure-specific enzymes have little or no base specificity.

The availability of new sequencing technology and of potentially useful enzymatic probes stimulated the development of a new method for the analysis of RNA structure[20,44]. This method combines the end-labeling and gel sequencing techniques[1,8,9] with the use of structure-specific nucleases to identify the positions of both nonpaired[20] and paired[43] nucleotides along the primary structure of RNA molecules. The method has been applied to several systems and, has generated considerable new information on the conformational properties of tRNAs[45], ribosomal RNAs[46,47], and rabbit globin mRNA[17,18,48]. A major limitation of the use of this technique is that, although base-paired nucleotides can be identified directly, the exact set of nucleotides in particular base-paired regions cannot be unequivocally assigned. Two other methods exist for analyzing RNA structure that use enzymes and can in some instances definitively identify sequences that are base-paired in an RNA molecule. The first, developed by Ross and Brimacombe[49], uses S1 digestion to hydrolyze uniformly labeled RNA, two-dimensional gel electrophoresis to separate and purify the two strands of individual helical regions that survive S1 digestion, and classical fingerprinting to sequence the strands of each region. They have applied this technique to study the structure of E. coli 16S rRNA. The other method, developed by Rabin and Crothers[50], involves cross-linking base-paired regions with a derivative of psoralen, partial hydrolysis with T1 ribonuclease and gel electrophoresis to separate the cross-linked helical segments, and sequencing of the individual segments after photoreversal of the psoralen cross-links and separation of strands. These authors have obtained direct proof for the existence of the 5'-3' stem region proposed to exist in all prokaryotic 5S rRNAs[50].

This paper presents a general method for using enzymes in the analysis of RNA sequence and structure in solution. Its focus is primarily on the techniques developed in this laboratory and used since 1977[17,18,20,40,43-48]. The method has the potential to determine the location of paired and unpaired bases in RNA sequences and to detect conformational transitions induced by changes in the physical state of the RNA in solution. Figure 1 is a flow diagram that

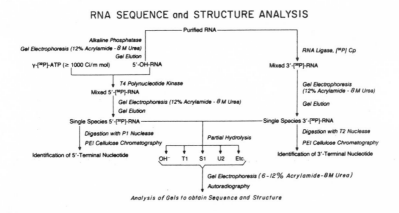

Figure 1. Flow diagram for RNA sequence and structure analysis.

summarizes the general approach. We include new data describing the isolation and properties of a double-strand-specific ribonuclease from cobra venom. Although the method for mapping structure described below is rapid and can yield useful data, it should be used together with other approaches to construct the best possible structural models of RNA.

II. MATERIALS

Chemicals were purchased from the following sources: brewers yeast tRNA^{phe} from Boehringer-Mannheim; yeast unfractionated tRNA from Bethesda Research Laboratories Inc.; acrylamide and bisacrylamide (electrophoresis grade) from Bio Rad laboratories; acrylamide (97% pure) from Polysciences, Inc.; amberlite MB-1 mixed-bed resin, diethylpyrocarbonate (DEP), dimethyldichlorosilane, and HEPES buffer from Sigma Chemical Co.; DEAE-cellulose and CM-cellulose from Reeve Angel, Inc., Sephadex from Pharmacia Chemicals; urea (ultrapure) from R-Plus Laboratories; and [5'-^{32}P]ATP (>2,000 Ci/mmol) and [5'-^{32}P]pCp (>2,000 Ci/mmol) from New England Nuclear, Inc.

Enzymes were obtained as follows: S1 nuclease and cobra venom ribonuclease were purified and characterized in this laboratory as described below. Calf intestine alkaline phosphatase (CIAP) was purchased from Boehringer-Mannheim, then rendered nuclease-free as described elsewhere[51] and assayed with the p-nitrophenyl phosphate method[52]. T1 (Sankyo), U2(R-Plus Laboratories), P1 (P-L Biochemicals), T2 (Sigma Chemical Co.), pancreatic RNAse A (Worthington Biochemical Corp.), and CL3 (Bethesda Research Laboratories Inc.) nucleases were used without further purification. Physarum nuclease, isolated from Physarum polycephalum by the method of Pilly et al.[53], was a gift of H. Donis-Keller. B. cereus endonuclease was purified as described elsewhere[54] and was a gift of J. Heckman. T4 RNA ligase was a gift of N. Pace. Mung bean nuclease was obtained as a gift from M. Laskowski Sr. and by purchase from P. L. Biochemicals and was used without further purification.

Equipment and supplies were purchased from various sources: PEI-cellulose thin-layer sheets (Macherey-Nagel cel300PE1) from Brinkman Instruments; gel spacers, templates, and electrophoresis supplies from Dan-Kar Plastic Products; and x-ray film, exposure holders, and developing chemicals from Eastman Kodak Co.

III. ENZYMES: PURIFICATION AND ASSAYS OF ACTIVITY

S1 nuclease was purified from α-amylase powder (Sigma Chemical Co.) by a modification of the method of Rushizky et al.[33]. A second DEAE-cellulose step was used after CM-cellulose chromatography. The DEAE-cellulose was developed both times with a 0.1-0.3 M gradient of NaCl in 50 mM NaAc (pH, 5.0) and 1 mM $ZnSO_4$. The activity of the enzyme was determined on denatured calf thymus DNA by the assay of Vogt[32]. The structure specificity was assayed with [^{3}H] poly(U) and [^{3}H]poly(A) (10-30 Ci/µmol P, Miles Laboratories) as substrates[36,55] by TCA filter precipitation in reactions containing 4-5 x 10^3 dpm [^{3}H]polymer unfractionated tRNA, 40 mM NaAc (pH, 4.5), 0.2 M NaCl, 10 mM $ZnSO_4$, and 6 x 10^{-3} U/µg RNA S1 incubated at 37°C[55]. Enzyme preparations were found to be free of contaminating RNAse T1 and T2 activities[46]. S1 is stored in 50 mM NaAc (pH, 5.0), 0.2 M NaCl, and 1 mM $ZnSO_4$ at 0.6 U/µl. There is no loss of activity in 2-3 yr either at -20°C in storage buffer plus 50% glycerol or at 4°C over a drop of chloroform.

Cobra venom ribonuclease, a structure-specific enzyme that recognizes base-paired regions and tertiary interaction in yeast tRNA^{Phe}[43], was purified by the procedure of Vassilenko and Rait[42] with modifications. Two hundred milligrams of lyophilized venom (Sigma Chemical Co) of Naja oxiana was dissolved in 1 ml of 50 mM Tris-HCl (pH, 7.0), and centrifuged at 5,000 g for 10 min at 2°C in a Beckman J21B centrifuge. The supernatant was applied to a Sephadex G-75 column (60 x 3 cm) and was eluted with 50 mM Tris-succinate (pH, 5.6). Five-milliliter fractions were collected, and the absorbance of each fraction at 280 nm was read. They were assayed for RNAse activity in 5-µl reactions containing 10 mM Tris-HCl (pH, 7.5), 5 mM $MgCl_2$, 100 mM NaCl, 5 µg of unfractionated tRNA, and 1 µl of column fraction. After incubation at 37°C for 30 min, reactions were chilled on ice and loaded onto a 40 x 33 x 0.15 cm 20% polyacrylamide-8.3 M urea slab gel after the addition of an equal volume of dyes and urea solution - 8 M urea, 0.1% xylene cyanol (XC), and 0.1% bromophenol blue (BPB) - to each reaction. The gel was stained with methylene blue after electrophoresis at 400 V for 2h. Fractions that caused the hydrolysis of tRNA were further studied to locate the double-strand-specific ribonuclease by partial hydrolysis of [5'-^{32}P]tRNA^{Phe}, as described below. The enzyme is stored in 20 mM Tris-HCl (pH, 7.0), and 50% glycerol at -20°C.

The extent of structure specificity of mung bean nuclease, a single-strand-specific endonuclease[39], with RNA was assayed with the [^{3}H]poly(U) and [^{3}H]poly (A) reactions described above for S1 nuclease. An enzyme-to-substrate ratio of 6 x 10^{-4} U/µg of RNA and a buffer containing 0.10 M NaAc (pH, 5.0), 10 µM $ZnAc_2$, 0.001% Triton X-100, and 1 mM L-cysteine were used, with a unit defined as by Kowalski et al.[56]. The enzyme was found to be capable of hydrolyzing only non-base-paired poly(U), and not poly(A), which is base-paired under these conditions (see Figure 5)[36]. The enzyme was stored at -20°C in 10 mM NaAc (pH, 5.0), 0.1 mM $ZnAc_2$, 1 mM L-cysteine, and 50% glycerol at a concentration of 0.3 U/µl.

All other enzymes used in this work were purified and assayed by either the company from which they were purchased or the colleague from whom they were obtained as a gift, and they were used with no further purification. These enzymes retained their specificities over 2 yr when stored at -20°C as follows: physarum nuclease (10 mM NaAc, pH 4.5; and 40% glycerol); T1 (0.1 U/µl; 10 mM Tris-HCl, pH 7.4; and 1 mM EDTA); and pancreatic RNAse A (0.2 µg/µl; 10 mM Tris-HCl, pH 7.4; and 1 mM EDTA). U2 (1 U/µl) was stored in 20 mM NH_4Ac (pH, 4.5) at -20°C and is less stable than the other enzymes. The B. cereus enzyme (4 U/ml) was stored at -80°C in 10-µl aliquots in 20 mM NaAc (pH, 4.8). Aliquots that were thawed and refrozen lost their pyrimidine specificity. CIAP was stored at 4°C in 15 mM Tris-HCl (pH, 7.4), 75 mM KCl, and 25% glycerol. T4 polynucleotide kinase was stored at 5^{-20}U/µl in 50 mM Tris-HCl (pH, 7.6), 5 mM DTT, and 50% glycerol at -20°C.

IV. PURIFICATION OF RNA

The RNA used in the sequencing and structure studies illustrated here was either purchased or which we purified. Bakers yeast tRNA[Phe] (Boehringer-Mannheim) was used with no further purification. Ribosomal RNAs (5.8S and 5S rRNAs) from a variety of sources (D. melanogaster, Tetrahymena thermophila, rabbit, B. stearothermophilus, Bombyx mori, and E. coli) were purified by either of two standard methods: The phenol-chloroform technique or the guanidinium chloride technique, which are well described elsewhere[46]. Total RNA was extracted from whole animal[46], cultured organisms[40,47], or red-blood-cell preparations[43]. In some cases, polysomal RNA was purified after salt precipitation of polysomes[40,46] with either the phenol or the guanidinium technique. Rabbit globin mRNA was obtained as described by Pavlakis et al.[17] and by Lockard et al.[57].

The individual RNA species were purified to homogeneity by preparative gel electrophoresis. The 5S and 5.8S rRNAs were purified free of other RNAs by electrophoresis on 8% polyacrylamide-8.3 M urea slab gels (6 mm thick). RNA bands were visualized by UV shadowing[55] and were eluted by crushing and shaking of the gel slice in a buffer containing 0.3 M NH_4Ac (pH, 4.5) and 1 mM EDTA for 6 h at 37°C. RNA solutions were separated from gel pieces by passage through a plastic syringe plugged with sterile siliconized glass wool and a 26 G needle. The RNA was then precipitated with 3 volumes of ethanol at -20°C for more than 8 h. These samples were used directly for end-labeling before sequencing.

V. PREPARATIVE DEPHOSPHORYLATION

All RNA samples were dephosphorylated before end-labeling.

A. Dephosphorylation of tRNA:

Four nanomoles of yeast tRNA^{Phe} were incubated in 45 µl of 25 mM Tris-HCl (pH, 8.3) with 4.5 x 10^{-2} U of CIAP at 55°C for 30 min. The reaction mixture was phenol-extracted and ethanol-precipitated. An equal volume of loading buffer (0.05% BPB, 0.05% XC, 10 M urea, and 40 mM EDTA) was added, and the mixture was heated at 50°C for 3 min. The solution was loaded immediately on a 12% polyacrylamide-7 M urea slab gel (20 cm x 20 cm x 1.5 mm) polymerized from 12% (w/v) acrylamide, 0.4% (w/v) bis, 7 M urea, 50 mM Tris-borate (pH, 8.3) and 1 mM EDTA. The reservoir buffer was 50 mM Tris-borate (pH, 8.3) and 1 mM EDTA. Electrophoresis was conducted at 40-60 watts (constant power) until the XC had migrated 12 cm. The RNA bands were visualized with UV[55]. Full-length tRNA was eluted from crushed gel slices into 1 ml of elution buffer (0.3 M NH$_4$Ac and 0.1 mM EDTA) by shaking at 37°C for 6 h. The eluate was passed through a plastic syringe with a siliconized glass-wool plug and a 26 G needle. The RNA was precipitated from ethanol twice and dried. Dephosphorylated RNA was dissolved in water and measured by its absorbance at 260 nm, with an extinction coefficient of 86.6 cm^{-1}-mol^{-1}(tRNA). The RNA was stored at -20°C before end-labeling.

B. Dephosphorylation of Picomole Quantities of rRNA:

rRNA (50-100 pmol) was dephosphorylated in 25 mM Tris-HCl (pH, 8.3) with 10^{-5} U/pmol CIAP at 50°C for 30 min in a total volume of 100 µl. The RNA was extracted once with an equal volume of phenol-chloroform (1:1 v/v). The organic phase was back extracted, and the aqueous phases were combined and ether extracted four times. Remaining ether was removed by evaporation, and the RNA was precipitated with 3 volumes of ethanol and stored at -20°C before end-labeling.

C. Dephosphorylation of Globin mRNA:

Globin mRNA was dephosphorylated after the enzymatic removal of the 5' cap with tobacco acid pyrophosphatase (TAP), as described by Lockard et al.[57] elsewhere in this volume.

VI. PREPARATIVE END-LABELING OF RNA

Both 5'- and 3'-end-labeling procedures were developed as modifications of published procedures, specifically to optimize the labeling of particular RNA species.

A. 5'-Labeling of tRNA:

One nanomole of phosphatase-treated tRNA was 5'-end-labeled in 25 µl of 50 mM Tris-HCl (pH, 9.0), 10 mM $MgCl_2$, 10 mM β-mercaptoethanol, 10% glycerol, and 10-µg/ml BSA (Pentex) containing 2.5 nmol of [$-^{32}P$] ATP (100 Ci/mmol) and 12 U of kinase. Glycerol and BSA were included at the suggestion of Silberklang et al.[58]. After 1 h of incubation at 37°C, [5'-^{32}P]tRNA was purified electrophoretically by modification of the procedure described above. Superior purification of full-length tRNA suitable for sequence and structure analysis was achieved if end-labeled RNA was electrophoresed successively at 50-60 watts (constant power) on two 40-cm-long slab gels until XC had migrated at least 30 cm on each gel. The first gel was polymerized from 19.3% (w/v) acrylamide, 0.7% (w/v) bis, 8.3 M urea, 50 mM Tris-borate (pH, 8.3) and 1 mM EDTA. The reservoir buffer was 50 mM Tris-borate (pH, 8.3) and 1 mM EDTA. The second gel was polymerized from 9.5% (w/v) acrylamide, 0.5% bis, 8.3 M urea, 100 mM Tris-borate (pH, 8.3), and 2 mM EDTA. The reservoir buffer was 100 mM Tris-borate (pH, 8.3), and 2 mM EDTA. After both electrophoreses, labeled tRNA was located by autoradiography and recovered. Elution with 0.2 M NH_4Ac and 1 mM EDTA for 2-6 h has proven successful in preventing degradation of the [^{32}P] RNA. The presence of Mg^{2+} or other divalent cations in the elution buffer can fragment the radiolabeled RNA. Because end-labeled RNA remained intact longer if not frozen, it was stored in water at 4°C.

B. 5'-Labeling of rRNA:

Fifty to one hundred picomoles of dephosphorylated RNA was end-labeled with 2- to 3-times molar excess of [5'-^{32}P]-ATP in a 10- to 20-µl reaction containing 50 mM Tris-HCl (pH, 8.9), 10 mM $MgCl_2$, 10 mM dithiothreitol, 5% glycerol, and 4.5 U of polynucleotide kinase, incubated at 37°C for 30-60 min. An equal volume of a solution containing 9 M urea, 10 mM EDTA, 0.1% XC, and 0.1% BPB was added to the reaction mixture, and it was loaded onto an 8% polyacrylamide-8.3 M urea gel (20 x 40 x 0.15 cm) and electrophoresed until the XC migrated more than 30 cm. Radioactive bands were located by autoradiography and were excised, and the [^{32}P] RNA was eluted as described above. The RNA was ethanol-precipitated, dissolved in water, and stored at 4°C.

C. 5'-Labeling of Globin mRNA:

Globin mRNA was 5'-end-labeled and purified by methods described elsewhere[17].

D. 3'-Labeling of tRNA and rRNA:

All RNAs were dephosphorylated as described above to provide 3'-OH ends to which [5'-^{32}P]Cp can be ligated with T4 RNA ligase[59]. Fifty to one hundred picomoles of RNA was incubated with an equimolar amount of [5'-^{32}P]Cp in a 10-µl reaction of 50 mM HEPES (pH, 8.3), 10 mM MgCl$_2$, 5 mM dithiothreitol, a 4-times molar excess of ATP over the RNA, and 2-3 U of T4 ligase at 4°C for 24-96 h. At 24 h and at intervals thereafter, portions of the reaction mixture (0.1 µl) were withdrawn and chromatographed on PEl-cellulose in 0.75 M KH$_2$PO$_4$ buffer (pH adjusted to 3.5 with phosphoric acid). Thin-layer plates were autoradiographed to locate the positions of 3'-labeled product (at the origin), [5'-^{32}P] Cp, and ^{32}P$_i$. These spots were excised and counted, and the percentage incorporation of pCp into RNA was determined. Typically, more than 50% of the [^{32}P]Cp was ligated to the RNA. 3'-Labeled RNAs were electrophoresed, eluted, and stored as described for 5'-end-labeled samples. In all cases, both 5'- and 3'-labeled RNAs were mixed with unfractionated carrier tRNA after gel elution to a final concentration of 20 µg/ml of RNA, to facilitate ethanol precipitation.

VII. END-ANALYSIS OF RNA

End-analyses of labeled RNAs were routinely performed, to assay the purity of the end-labeled RNA preparations.

A. 5'-End Analysis:

5'-end-labeled samples of RNA were digested exhaustively with P1 nuclease in 10-µl reactions containing 2,500-cpm [5'-^{32}P]RNA, 0.2 M NaAc (pH, 4.5), and 1 µg of P1 incubated at 70°C for 2 h. Digests were spotted onto PEl-cellulose and chromatographed in 1.0 M LiCl to identify the labeled 5'-monophosphates. Chromatograms were autoradiographed and counts were measured by liquid scintillation to determine the purity of labeled samples.

B. 3'-End Analysis:

3'-end-labeled samples were digested exhaustively with T2 RNAse in 10-µl reactions containing 2,500-cpm labeled RNA, 0.02 M NH$_4$ AC (pH, 4.5), and 10 U of T2 incubated at 37°C for 2 h. PEl-cellulose chromatography was used to quantitate the purity of samples, as above.

VIII. RNA SEQUENCE AND STRUCTURE ANALYSIS PROTOCOLS

A. Partial Digestions with Base-Specific Enzymes.

The method by which partial digests were achieved in urea has been described elsewhere[8,20]. Ribonuclease T1 at 2×10^{-3} U/µg of RNA and ribonuclease U2
at 10^{-2} U/µg of RNA produced suitable partial digests for all substrates. Pancreatic ribonuclease generated partial digests at 9×10^{-5} U/µg. The activity
of the Phy I preparation has not been determined. Appropriate enzyme to substrate ratios were obtained by serial dilutions of the enzyme. The pyrimidine-
specific enzyme from B. cereus was used for 8 min at 60°C in the same reaction
buffer used for the other sequence reactions, except that urea and dyes were
added only after digestion. When assayed on $[^3H]$poly(C), the nuclease had an
activity comparable with the 4 U/ml of RNAse A. By this criterion, the B. cereus
enzyme at 4×10^{-3} U/µg of RNA generated partial digests suitable for sequence
analysis. The C-specific enzyme CL3 purified from chicken liver by Levy and
co-workers[60,61] was used at an enzyme to substrate ratio of 1.5 U/µg of RNA
(unit as defined by Bethesda Research Laboratories, Inc.) in a final buffer
containing 10 mM $NaPO_4$ (pH, 6.5). Samples for successive loadings were digested
simultaneously and stored at -20°C before electrophoresis.

B. Limited Alkaline Hydrolysis.

$[^{32}P]$RNA was partially hydrolyzed in 10 µl of 50 mM $NaHCO_3/Na_2CO_3$ (pH, 9.2),
and 1 mM EDTA containing 2.5 µg of unfractionated tRNA carrier by heating at
90°C for 7 min. An equal volume of 0.1% BPB, 0.1% XC, and 9 M urea was then
added, and samples were loaded on the gel immediately or stored at -20°C.

C. Suggested Sequencing Protocol.

The standard sequencing reaction mix is made up by combining the following
components: 50 µl of 10 M urea, 2 µl of 1 M sodium citrate (pH, 5.0), 1 µl of
0.1 M Na_2EDTA, 4 µl of 0.1% XC + 0.1% BPB, and adjustable amounts of unfractionated tRNA carrier (10 µg/µl), $[^{32}P]$RNA, and water to a final volume of 75
µl. The amounts of the last three components are determined by adjusting so as
to achieve a concentration of 0.25 µg/µl of RNA and 1,000-2,000 cpm (Cerenkov)
per microliter in the final mixture. The buffer used to dilute RNAses A, T1,
and U2 for use in the sequencing reactions consists of 100 µl of 10 M urea, 36
µl of water, 4 µl of 1 M sodium citrate (pH, 5.0), 2 µl of 0.1 M Na_2EDTA, and 8
µl of dyes (0.1% XC + 0.1% BPB), for a final volume of 150 µl. We have achieved
improved reproducibility and accuracy with U2 RNAse if pH 3.5 sodium citrate is

used in both of the above solutions[69]. The enzymes are diluted as follows immediately before use: 1 μl of a stock of T1 - 0.1 U/μl in 10 mM Tris-HCl (pH, 7.4), and 1 mM EDTA - is added to 50 μl of the enzyme dilution buffer; 1 μl of a stock of U2 - 1 U/μl in 20 mM NH$_4$Ac (pH, 4.5) - is added to 50 μl of the enzyme dilution buffer; 2 μl of the stock of RNAse A - 0.2 μg/μl in 10 mM Tris-HCl (pH, 7.4), and 1 mM EDTA - is added to 20 μl of the enzyme dilution buffer; then 2 μl of this solution is added to 20 μl of the enzyme dilution buffer. The Physarum, B. cereus, and CL3 enzymes are used undiluted. Final enzyme reacions include: T1, U2, A (4 μl of reaction mix plus 1 μl of diluted enzyme); Physarum (4.25 μl of reaction mix plus 0.75 μl of diluted enzyme); B. cereus (4.5 μl of reaction mix that does not include urea but is otherwise identical with that used above for T1, etc., plus 0.5 μl of undiluted enzyme); and CL3 (4.5 μl of reaction mix that includes 1-1.5 μg of RNA, 5,000-10,000 cpm of [^{32}P]RNA in 10 mM NaPO$_4$ at a pH of 6.5 plus 0.5 μl of enzyme, 1.75 U). All reactions are incubated at 60°C, except CL3, which is incubated at 37°C, for 5-15 min. An equal volume of dyes and urea is added to the B. cereus and CL3 reactions on completion. It is emphasized that 60°C should be used to minimize base-pairing and allow maximal access of enzymes to their specific cleavage sites. Alkaline hydrolysis reactions include 1.2 μg of RNA and 5,000-10,000 cpm (Cerenkov) per 5 μl and are performed as described above. Samples are either stored at -20°C or loaded directly onto gels. The above protocols, especially the enzyme-to-substrate ratios and reaction times, are given here as suggested starting points. The exact conditions for achieving optimal partial digestons may vary considerably for different enzyme preparations and RNA molecules of different size.

D. Partial Digestions with Structure-Specific Enzymes.

Digestions with S1 were performed in 5- to 10-μl reactions containing 40 mM NaAc (pH, 4.5), 10 mM ZnSO$_4$, 0.2 M NaCl, [^{32}P]RNA (5,000-10,000 cpm), unfractionated tRNA carrier, and an enzyme to substrate ratio of 0.006-0.06 U/μg of RNA. Samples were preincubated for at least 10 min at 37°C and were digested with S1 for the indicated times. Reactions were stopped with an equal volume of "stop" solution (9 M urea, 10 mM EDTA, 0.5 μg/μl of tRNA carrier, 0.1% XC, and 0.1% BPB), and were stored at -80°C.

Mung bean nuclease digestions were performed in 5- to 10-μl volumes in a standard buffer containing 0.1 M NaAc (pH, 5.0), 10 μM ZnAc$_2$, 0.001% Triton X-100, 1 mM L-cysteine, [^{32}P]RNA (5,000-10,000 cpm), unfractionated tRNA carrier, and enzyme-to-substrate ratios of $1.5^{-6} \times 10^{-3}$ U/μg of RNA were used. Incubations were at several temperatures for the indicated times. Reactions were stopped with an equal volume of the same solution used for S1 reaction and could also be frozen at -80°C before electrophoresis.

T1 RNAse structure-specific reactions were performed in 5 µl of 10 mM sodium citrate or NaAc (pH, 5), 0.2 M NaCl, and 10 mM MgCl$_2$ for 5 min at 37°C with a final RNA concentration of 0.65 µg/µl and 10,000 cpm of [^{32}P]RNA. Enzyme to substrate ratios were 2-30 x 10^{-5} U/µg of RNA. Reactions were stopped by freezing at -80°C after addition of dyes and urea solution (9 M urea, 0.05% XC, and 0.05% BPB).

Cobra venom ribonuclease reaction were performed in 5-µl volumes and contained 10 mM Tris-HCl (pH, 7.5), 2 mM MgCl$_2$, 200 mM NaCl, 1.8 µg of total RNA containing 5,000-10,000 cpm of [^{32}P]RNA, and 1 µl of enzyme (1:10 to 1:100 dilution of stock enzyme solution). After incubation at 0-70°C for different times, reactions were stopped by adding an equal volume of a solution containing 50 mM sodium citrate (pH, 3.5), 8 M urea, 20 mM EDTA, and dyes. EDTA completely inhibits the enzyme due to its Mg^{2+}-dependence (see Fig. 6).

It is essential that kinetic characteristics - i.e., time and enzyme to substrate ratios - be varied, in order to identify correctly the non-base-paired (with S1, T1 and mung bean nucleases) and the base-paired (with cobra venom ribonuclease) nucleotides. The presence of degradation in the minus enzyme control reaction will also severely limit the reliability of the interpretation of data.

IX. GEL ELECTROPHORESIS

A. Purification of Acrylamide for Sequencing Gels.

To reduce the cost of gel sequencing, we routinely purify acrylamide as follows: 800 g of bulk acrylamide (Polysciences, Inc.) is dissolved in distilled, deionized water to a final volume of 2 L to make a 40% w/v solution. This is stirred with 80 g of amberlite MB-1 mixed-bed resin for 2-3 h. The solution is then filtered sequentially through Whatman 3 and Whatman 1 filters and stored at 4°C in dark bottles. The 40% solution is used directly in the preparation of sequencing gels. This simple procedure results in a 30-fold savings in the cost of acrylamide.

B. One-dimensional Polyacrylamide Gels.

Polyacrylamide gels were poured and run as described elsewhere[1,7,8,20]. Samples were loaded onto slab gels (40 x 33 x 0.15 cm) of either 20% acrylamide-8.3 M urea or 10% acrylamide-8.3 M urea. The 10% gels were polymerized from

9.5% (w/v) acrylamide, 0.5% (w/v) bisacrylamide, 8.3 M urea, 100 mM Tris-borate (pH, 8.3), and 2 mM EDTA; and the reservoir buffer was 100 mM Tris-borate (pH, 8.3), and 2 EDTA. For increased resolution, 0.4 and 0.30 thick 20%, 15%, 10%, and 6% gels were used, as described by Sanger and Coulson[15]. In some cases, gels 90 cm long were used (see Fig. 10) to increase the length of nucleotide sequences that could be read in a single run. Typically, electrophoresis was terminated when BPB had migrated 21 cm in the 20% gels and XC had migrated 30 cm in the 10% gels. This allowed the resolution of oligonucleotides up to 40 bases long, and of oligonucleotides 40 bases long and longer, respectively. For optimal resolution, and to minimize "smiling" and other forms of band distortion, gels are routinely prerun with dyes and urea solution in each slot until the BPB has migrated more than 50% of the total gel length. This dispels a salt front that forms at the initiation of electrophoresis. Gels are run at constant power adjusted to keep the gel temperature at 55-65°C. This requires 40-70 W for both 40-cm- and 90-cm-long thin (0.3-0.4-mm-thick) gels.

C. Two-dimensional Polyacrylamide Gels.

Two-dimensional gels can be used for the accurate resolution of the pyrimidines (see Fig. 3). Partial alkaline digests are carried out as described for one-dimensional sequencing gels on at least 150,000 cpm of [5'-^{32}P]RNA. 20 μl of loading solution, containing 7 M urea, 25% sucrose, 0.1% BPB, and 0.3% XC is added. Electrophoresis in the first dimension is in a 10% polyacrylamide slab gel run at a pH of 3.5, and in the second dimension in a 20% polyacrylamide slab gel run at a pH of 8.3. The gel solution for the first dimension contains 10% acrylamide, 0.5% bisacrylamide, 7 M urea, and 25 mM citric acid. The solution is deaerated for 30 min at reduced pressure and polymerized as described by DeWachter and Fiers[62]. Electrophoresis is carried out at 200 V until the BPB dye migrates 23 cm (for analysis of oligonucleotides up to 40 base pairs long) or 33 cm (for analysis of oligonucleotides 40 base pairs long and longer). After electrophoresis, a strip 1 cm wide and at least 2 cm long is cut from an appropriate region of the gel, and placed lengthwise 1 cm from the bottom between two glass plates.

The gel solution - 20% acrylamide, 1% bisacylamide, 7 M urea, and 90 mM Tris-borate (pH, 8.3) - for the second dimension is poured in two stages. In the first stage, enough gel solution is poured to cover the upper edge of the first-dimension gel strip. To ensure rapid polymerization, this gel solution contains 0.25% ammonium persulfate. After polymerization of the gel block, the remainder of the gel solution is poured and allowed to polymerize. The direction of electrophoresis is from bottom to top with 90 mM Tris-borate (pH, 8.3) and 4 mM EDTA as the running buffer. Electrophoresis is carried out within 1 h after polymerization at 300-400 V at room temperature for various periods, depending on the region of the first-dimension gel being analyzed. For analysis of oligonucleotides 4-40 base pairs long, electrophoresis is carried out until the BPB migrates about 23 cm; for oligonucleotides 25-60 base pairs long, until the XC was run off the gel.

X. ANALYSIS OF DATA

After gel electrophoresis, x-ray films were exposed by gels at -20°C for times between 8-96 h, depending on the amount of phosphorus-32 RNA label used in the experiment. In some cases, Dupont Lightening-Plus Intensifying Screens were used to abbreviate exposure times, although this procedure generally resulted in blurring of bands. Ambiguities were resolved by additional sequence or structure studies. Recurring problems - such as double bands, band contraction, and persistent band faintness - were encountered and dealt with as described below.

Computer-generated predictions of secondary structure were obtained with a modified version of the program developed by Pipas and McMahon[63], based on the empirical rules of Tinoco <u>et al.</u>[64] and run on an IBM 370 computer. The structure-specific enzymatic susceptibility data obtained with S1, mung bean, T1, and cobra venom RNase enzymes were used as input information in the structure-modeling program[43,46-48].

XI. SOME RNA SEQUENCING RESULTS

The general approach use for sequencing RNA is illustrated in the flow diagram in Figure 1. Examples of results obtained with this method are shown in most of the other figures. Figure 2 displays a portion of the <u>D. melanogaster</u> 5.8Sa rRNA-previously called m 5.8S rRNA[46] - sequence, as determined with a set of enzymes (T1, U2, Phy M, and <u>B. cereus</u> enzymes). Donis-Keller[66] defined PhyM to be the U plus A-specific endoribonuclease activity observed when the <u>Physarum polycephalum</u> enzyme (often called Phy I) is used in reactions containing urea at a pH of 5.0 and 60°C. The data in Fig. 2 show that PhyM does not cleave at C, and only lightly at G residues. Sequencing results are included, below, for yeast tRNA[Phe] (Figs. 4,5,7, and 8), rabbit 5S rRNA (Fig. 6), and rabbit globin messenger RNA (Figs. 3 and 9). Figure 9 also illustrates that more than 160 nucleotides of sequence can be "read out" after a single electro-

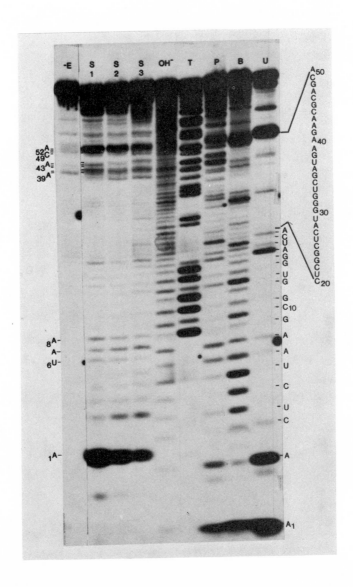

Fig. 2. Electrophoresis of partial enzymatic digestions for sequence and structure analysis of 5-^{32}P-end-labeled <u>D. melanogaster</u> 5.8Sa rRNA on 20% polyacrylamide-8.3 M urea gel. Lanes from left to right: -E, minus enzyme control; S(1), S(2), and S(3), S1 nuclease, 0.1 U/μg of RNA, 0.01 U/μg of RNA, 5 min, and 0.01 U/μg of RNA, 1 min, respectively; OH, alkaline digestion, 3 min at 90°C; T, T1 RNase, 5 X 10 U/μg^{-4} of RNA, 3 min; P, PhyM RNase, 10^{-2} U/μg of RNA, 10 min. Approximately 20,000 cpm (Cerenkov) of labeled RNA was loaded per slot. Positions of cleavage by S1 are shown along left side, and nucleotide sequence is shown along right side.

284

phoresis run on a 6%, thin (0.3 mm), 90-cm-long gel[17].

One of the most difficult problems in sequencing end-labeled RNA is to distinguish U and C residues. Several alternative methods have been devised to deal with this problem, including the new sequencing technique of Tanaka et. al.[66], which is designed after the method of Stanley and Vassilenko[16] and is

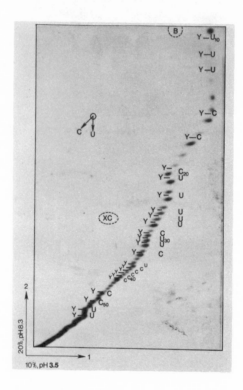

Fig. 3. Autoradiogram of a partial alkaline digest on 5'-^{32}P-labeled rabbit β-globin mRNA as analyzed by two-dimensional polyacrylamide-gel electrophoresis. Circled B and XC indicate location of bromophenol blue and xylene cyanol tracking dyes, respectively. Numbers on right indicate size of radioactive oligonucleotide. Pyrimidines (Y) in β mRNA sequence determined here are underlined as follows:

^{32}P-Am C(m)A C U U G C U U U U G A C A C A A C U G U G U U U A
10 ... 20

C U U G C A A U C C C C C A A A A C A G A C A G A
30 ... 40 ... 50

A U G G U G
met   val

very similar to that of Gupta and Randerath[22]. We present two alternatives for solving this problem. Figure 3 shows the result of an experiment in which 5'-^{32}P-labeled rabbit β-globin mRNA was partially digested with alkali and the digest was subjected to two-dimensional gel electrophoresis. Electrophoresis in the first dimension was on a 40-cm-long gel, and only a 20-cm-long strip from it was used for electrophoresis in the second dimension. The region from which the gel strip was cut and the time of electrophoresis in the first dimension determine the size range of the partial fragments being analyzed. It is useful to apply the same partial digest in two parallel slots for the first dimension and to include an overlap of about 6 cm or more among the two vertical gel strips that are later used for electrophoresis in the second dimension. The principle behind the separation of the homologous oligonucleotides observed in Figure 3 is similar to that in two-dimensional homochromatography[58]. The mobility shifts between two homologous oligonucleotides that differ by a C or A residue can be easily distinguished from those which differ by a U or G residue. In Figure 3, mobility shifts due to U or G are almost vertical, whereas those due to C or A are at a sharp angle. A second, more direct way to distinguish among Us and Cs is indicated in Figure 7. Here, the C-specific chicken liver enzyme CL3, recently characterized by Levy and co-workers[60,61], is used to locate directly the C residues in 5'-^{32}P-labeled 5S rRNA from rabbit. The enzyme is 60 times more sensitive to Cs than to Us in general, although it seems to have a greater specificity for some Us (note that CL3 digests U12 and U53 strongly). This enzyme has made it possible to use standard gel-ladder sequencing technology with much greater confidence than was possible prior to its purification.

Several enzymes other than those described here have been found useful in RNA sequencing studies on gels. Krupp and Gross[69] have found conditions under which S. aureus nuclease cleaves all pyrimidine bonds more uniformly than pancreatic RNase A. They have also demonstrated that N. crassa endonuclease can digest all phosphodiester bonds, except C-N bonds, in 7 M urea and generates 3'-OH oligonucleotides, making this enzyme potentially useful in parallel with the structure-specific S1, mung bean, and cobra venom enzymes. In our experience, of the sequencing enzymes, the T1 and U2 (at a pH of 3.5) RNases are the most reliable and reproducible. The pyrimidine-specific enzymes are more difficult to work with, and several enzymes (PhyM, B. cereus, RNase A, and RNase Cl3) must be used to produce an accurate RNA sequence.

XII. STRUCTURE ANALYSIS WITH SINGLE- AND DOUBLE-STRAND-SPECIFIC ENZYMES

The development and use of an accurate method for identifying structural properties of RNA molecules with enzymes as probes requires attention to a considerable number of characteristics, as follows: RNA purity is crucial, and essentially nondegraded end-labeled molecules are required to avoid confusion in assigning bands. Reaction time and enzyme to substrate ratio must be varied to distinguish bands that represent structure specific cleavage from artifacts that result from overdigestion of the end-labeled molecule. Variation in physicochemical properties--such as pH, ionic strength, divalent-cation concentration, and temperature,--can result in the identification of conformational stability properties of the molecule. A varied set of enzymes should be used--

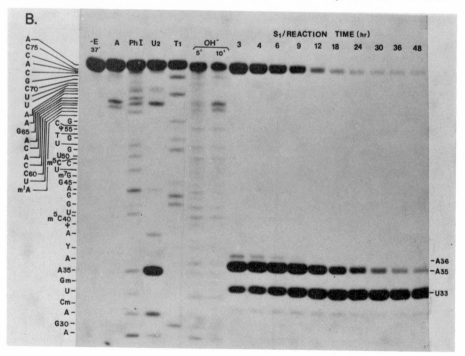

Fig. 4. Kinetics of digestion of $5'-^{32}P$-tRNA^{Phe} by S1 nuclease, during the initial 2 h of incubation. The sequencing reactions in the seven left-most gel lanes were done according to Wurst, et. al.[20] Digests with S1 nuclease (20,000 cpm/lane) were withdrawn from a reaction mixture (40 mM NaAc pH 4.5, 0.2 M NaCl, 10 mM ZnSO$_4$) incubated at 37°C containing 6×10^{-3} U/µg RNA. Electrophoresis was in a 10% polyacrylamide-8.3 M urea gels, run until the XC tracking dye migrated 30 cm. S1 cleavage positions are indicated along the right side.

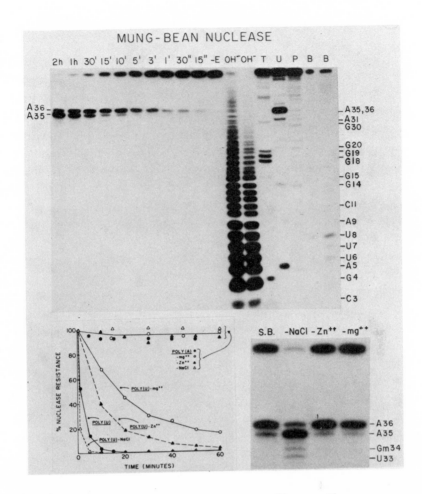

Fig. 5. Top Panel: Kinetics of digestion of $5'$-(^{32}P)-tRNA^{Phe} by mung-bean nuclease, during initial 2 h of incubation. Sequencing track in the right-most seven lanes were done as described.[20] Digests with mung-bean nuclease were withdrawn from a reaction mixture containing standard buffer (see methods) incubated at 23°C containing 1.5×10^{-3} U/µg RNA. Electrophoresis was in a 20% polyacrylamide-8.3 M urea gel, until the BPB dye migrated 21 cm. Mung-bean nuclease cleavage positions are indicated on the left side.

Lower Left Panel: Digestion of $[^{3}H]$poly(U) and $[^{3}H]$ poly(A) with mung-bean nuclease. Digestions were at 23°C in standard buffer, as described in methods and in buffers in which Zn^{2+}, Mg^{2+}, and NaCl were removed, respectively. Points were obtained by TCA precipitation in triplicate assays.

Lower Right Panel: Digestion of $5'$-(^{32}P)-tRNA^{Phe} by mung-bean nuclease in standard buffer (SB), and in the absence of Zn^{2+}, Mg^{2+} and NaCl, respectively. This figure is a blow-up of the region of an autoradiogram showing the anti-codon loop nucleotides. Reactions were incubated at 23°C for 15 min and contained 1.5×10^{-3} U/µg RNA. Electrophoresis was in a 20% polyacrylamide-8.3 M urea gel run until the BPB dye migrated 21 cm.

e.g., S1 and mung bean nucleases, T1 RNase, and cobra venom RNase--in parallel lanes to maximize the structural information obtained. Studies on fragments can highlight the existence of secondary structure that may be inaccessible in the full-length species, owing to tertiary structure. A variety of gel-electro-phoresis conditions (e.g., % acrylamide from 6 to 20% and different running times) can be used to obtain the sequence of long RNA molecules. Several exam-ples of results obtained by varying these parameters are included in Figures 2 and 4-9. Much of the information in these figures was obtained with yeast tRNA^{Phe}, used as a well-characterized model system.

Figures 4 and 5 illustrate the kind of kinetic data that can be obtained with S1 and mung bean nucleases in experiments with [5'-^{32}P] tRNA^{Phe}. These enzymes generate 3'-OH oligonucleotides, so care must be taken to ensure the correct assignment of band positions (this is clearly illustrated in Fig. 2, where it is seen that A1 migrates about 1.5 bands more slowly in the S1 lanes than in the U2 lane). In both the S1 and the mung bean nuclease studies, the most sensitive sites are the neighboring adenines (A36 and A35) in the anticodon loop. A band is seen to accumulate in the S1 experiment (Fig. 4) at position U33. The two As are roughly equally accessible to enzymatic digestion in short times. However, A36 is clearly more exposed to mung bean nuclease, because it is cleaved first, and no U33 band is detected. This highlights that these two single-strand-specific enzymes have different properties. Although they both attack nonpaired nucleotides, they do not cleave identically, pro-bably because of the details of their interactions with the three-dimensional structure of the tRNA molecule. Figure 4 includes data on the effect of chan-ging the ionic environment in mung bean nuclease reactions (bottom panels). It is seen that the enzyme maintains its specificity for non-base-paired RNA, inas-much as it cannot digest poly(A) under any of the conditions tested. However the rate of hydrolysis of poly(U) is affected. Removal of Zn^{2+} or Mg^{2+} (with EDTA) decreases the rate, whereas removal of NaCl increases it. The bottom-right panel shows the striking result that bands at Gm34 and U33 accumulate in the absence of NaCl. Similar studies by Wurst[55] show that removal of Zn^{2+} decreases, removal of NaCl increases, and addition of Mg^{2+} does not affect the rate of S1 hydrolysis. Because removal of NaCl leads to weakening of base-stacking, the observed increase in rate by the two enzymes is not surprising. The Zn^{2+} effect is also explicable, in that both S1 and mung bean nucleases are zinc metallo-enzymes.

Results of studies with the double-strand-specific cobra venom RNase (CVR) are shown in Figure 6. These data highlight the usefulness of this enzyme as a

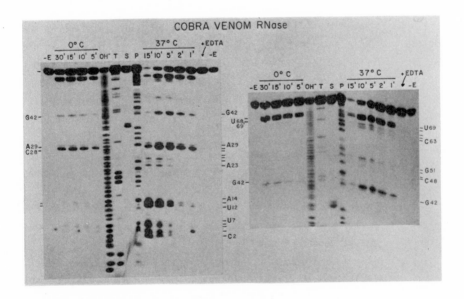

Fig. 6. Cobra venom RNase digestions of 5'-^{32}P-end-labeled yeast tRNA^{Phe}. Left and right panels include identical reactions. Reactions were incubated at either 0°C or 37°C, for the indicated times. Lanes include from left to right: -E, minus enzyme control; 0°C cobra venom RNase reactions, 1:200 enzyme dilution in standard reaction (see text); OH$^-$, alkaline hydrolysis, 5 min at 90°C, 10,000 cpm; T, T1 RNase, 1.6 X 10^{-5} U/µg of RNA, 15 min, 7,000 cpm; S, S1 nuclease 6 X 10^{-3} U/µg of RNA, 37°C, 20 min, 7,500 cpm; P, PhyM RNase, 5 min, 7,000 cpm; 37°C cobra venom RNase reactions, as above for 0°C reactions; + EDTA, standard reaction mixture made 20 mM in EDTA before addition of cobra venom enzyme. Electrophoresis in left and right panels was in 20% and 10% polyacrylmide-8.3 M urea gels until XC migrated 11 and 30 cm, respectively. Positions at which enzyme digests yeast tRNA^{Phe} are indicated along sides.

complement to the "single-strand" probes. CVR can digest tRNA[Phe] at several temperatures (Fig. 6), is completely inhibited by EDTA (Fig. 6) because it requires divalent-cations[43], and generates 3'-OH oligonucleotides. It is easy to purify free of other RNase activities. It will not digest double-stranded DNA, but it will hydrolyze triple-stranded RNA[43]. The positions of cleavage of tRNA[Phe] by this enzyme (Fig. 6) are all within the stem regions of the cloverleaf secondary-structure model of this molecule. Not all base-paired nucleotides are digested, and differences in the intensity of cleavage exist, probably owing to structural details imposed by the tertiary structure of the tRNA molecule. It is seen that the enzyme has fewer sites of cleavage at 0°C than at 37°C. In particular, the acceptor stem (C2-U7 and A67-U69), D stem (U12-A14 and A23-C25), anticodon stem (C27-A29), and T[4] C stem (C48-G51 and C63-G65) become accessible to the enzyme at the higher temperature. One particularly interesting cleavage is at C48, a residue involved in a tertiary interaction. Further details are published elsewhere[43]. Figure 7 is a composite of reactions on the same gel with rabbit 5'-^{32}P-labeled 5S rRNA as substrate. These data show that S1 and CVR can be used in parallel to identify paired and nonpaired regions in 5S rRNA. Regions of clustered bands in CVR lanes are generally blank in S1 lanes, and vice versa (see NP and BP symbols along the left side of the figure). This is the expected result, and the data agree with a general model for eukaryotic 5S rRNA secondary structure[47].

The data in Figure 8 are results obtained with partial-length molecules to locate base-paired regions in yeast tRNA[Phe]. The fragments were obtained by chemical cleavage of tRNA[Phe] at M[7]G45[55]. Comparison of the cleavage patterns in Figures 4 and 8 shows that the previously resistant bases in the D and T[4]C loops (Fig. 4) due to stable tertiary structure become accessible to the probe in the fragments (Fig. 8). Thus, it is possible to infer both secondary and tertiary structural properties of an RNA by combining analyses of data on end-labeled full-length and fragment molecules.

Finally, Figure 9 shows the power of the gel technique for sequence and structure analysis of long molecules. A long (90 cm), thin (0.3mm) gel was used to display S1 and T1 structure probing enzymatic reactions, plus PhyM, T1, and alkaline hydrolysis sequencing reactions of 5'- ^{32}P-labeled rabbit α-globin mRNA[17]. The use of this technique and the long gels made it possible to distinguish interesting differences in the structure at the AUG initiator region and at intervening sequence splice points between rabbit α- and β-globin mRNAs[17,18,48].

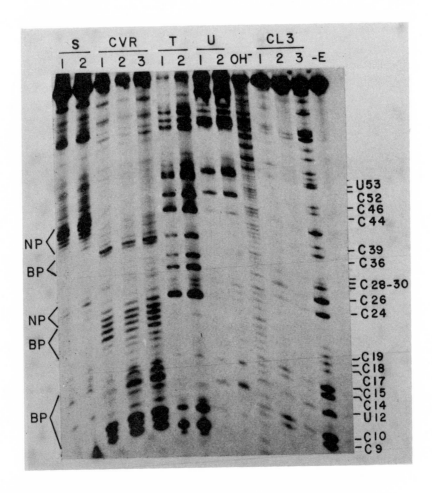

Fig. 7. Electrophoresis of partial enzymatic digestions for sequence and structure analysis of 5'-^{32}P-end-labeled rabbit 5S rRNA. Lanes from left to right are: S, S1 nuclease, 1(10 min) and 2 (20 min) digestions, at 0.12 U/µg of RNA, 37°C, 10,500 cpm; CVR, cobra venom RNase, 1 (10 min, 1,200 enzyme dilution--see text); 2 (10 min, 1,000 enzyme dilution), 3 (20 min, 1,000 enzyme dilution), 37°C, 10,000 cpm; T, T1 RNase, 1 (10 min) and 2 (20 min), 1.6 X 10^{-5} U/µg of RNA, 16,000 cpm; U, U2 RNase, 1 (5 min) and 2 (10 min), 3.2 X 10^{-2} U/µg of RNA, 16,000 cpm; OH$^-$, alkaline digest, 4.5 min at 90°C, 16,000 cpm; CL3, chicken liver RNase, 1 (0.15 U/µg of RNA, 15 min), 2 (0.30 U/µg) of RNA, 30 min), 3 (1.50 U/µg of RNA, 15 min), 37°C, 12,000 cpm; and -E, minus enzyme control. Electrophoresis was on a 10% polyacrylamide-8.3 M urea gel run until XC migrated 15 cm. The NP and BP symbols along left side refer to nonpaired and base-paired regions. Positions of cleavage by CL3 RNase are indicated along right side.

292

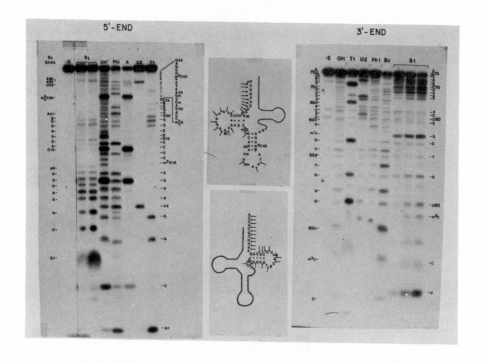

Fig. 8. Structure analysis with S1 nuclease of fragments of yeast tRNA[Phe]. The 5'-end (1-45) and 3'-end (47-76) fragments were each 5'-end-labeled and digested (8,000 cpm/lane). S1 nuclease digests contained electrophoretically purified 5S rRNA as carrier and enzyme concentrations of 3 X 10⁻³ U/μg of RNA (left panel) and 6 X 10⁻³ U/μg of RNA (right panel). Sequencing reactions were as described (20). Electrophoresis was on a 20% polyacrylamide-8.3 M urea gel run until XC migrated 6 cm (left) and 3 cm (right).

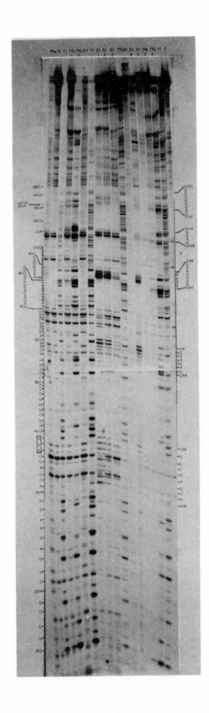

Fig. 9. Sequence and structure analysis of 5'-^{32}P-end-labeled rabbit α-globin messenger RNA on 90-cm long, 0.3 thick 10% polyacrylamide-8.3 M urea gel. Electrophoresis was at 55-60 watts for 6 h. Both S1 nuclease and T1 RNase reactions were used for structure analysis, and a standard set of sequencing reactions with T1, U2, PhyM, and alkali are included. Sequence of mRNA is indicated along left side, and sites of nonpaired bases as derived from T1 and S1 structure-specific reactions are indicated along right side[17].

XIII. SUMMARY: LIMITATIONS, PROBLEMS, AND PERSPECTIVES

The RNA sequencing and structure analytic technique described here has some potential sources of error. We are aware of the following limitations of the method. It cannot be used to identify modified nucleotides. The general approach cannot now be used for structure studies of long RNA molecules (over ∿ 1,200 nucleotides). Although base-paired nucleotides can be located within an RNA sequence, this method does not definitively identify which nucleotides are paired to one another. There are other methods[15,22,23,68] for identifying modified bases in an RNA, and they should be used in concert with the method presented here for sequencing. The RNA size limitation is potentially solvable for sequencing studies by application of the RNase H fragmentation technique of Donis-Keller[19]. However, the size limitation is much more serious for structure studies, since it is critical to study the full-length molecule in its "native state." Although fragment studies can be useful (see Fig. 8 and discussion above), the native structure of full-length molecules cannot be deduced from fragments alone. We are investigating new technologies that would allow the direct structure analysis of very long RNA molecules.

Several problems have been encountered and solved during the last 3 years, including the following: contraction of bands on gels due to altered migration of oligonucleotides containing stable hairpin helices that are not melted by the denaturing conditions use; identification of true sequence heterogeneity when it exists; determination of the length of oligonucleotides generated by enzymes (S1, mung bean, and CV RNase) that leave 3'-OH after cleavage; and identification and resolution of doublet bands, particularly those obtained on digestion of 3'-^{32}P-labeled molecules. The band contraction problem, which has plagued both RNA and DNA sequence studies and has been discussed in several articles[7,12], is identified by the existence of abnormally short distances between neighboring oligonucleotides in a particular region of the gel. This is usually easily spotted in the alkaline digest lanes. The problem is solved by increasing the temperature of the gel, by adding additional denaturing chemicals (such as formamide) to the gel, or by repeating the experiment with RNA labeled at the opposite end.

If contraction is observed with 5'-labeled RNA, it will be absent in that region of the molecule in studies with 3'-labeled species.

Sequence heterogeneity can be detected and verified by our sequencing method[46,47]. It is useful to apply another technique, such as that of Tanaka et. al.[67], to substantiate the existence and exact location of heterogeneity in sequence. The counting problem mentioned above for enzyme digests that generate 3'-OH oligonucleotides can be solved by careful and painstaking gel "reading." Recently, Krupp and Gross[68] have demonstrated that N. crassa nuclease can be used to generate a complete set of 3'-OH oligonucleotides, thus making use of this enzyme mandatory in parallel with the structure-specific reactions. We typically depend on data from RNases A, PhyM, B. cereus, and CL3 to clarify ambiguities and turn to 2D gel electrophoresis, if needed, as a final resort. In some cases, a pattern of doublets has been observed in experiments with 3'-labeled molecules[46]. This occurs, for example, when T1 RNase is used to digest an RNA with a G at its normal 3'-end before labeling with $[5'-^{32}P]Cp$. Such doublet patterns do not interfere with sequencing, provided that one is fully aware of their origin[46].

The structure mapping technique described here is new, and work is continuing to develop new probes. Considerable effort is needed to develop an understanding of the mechanism of several of the enzymes, e.g., the mung bean, S1, and cobra venom nucleases. The utility of the method in studies of protein-RNA interactions has not been fully explored. The method for structure analysis is direct, but the structural models based on the data are inferred. The method is semiquantitative. It can be made fully quantitative with considerable effort, and this would provide a useful measure of the 'relative accessibility' of regions of an RNA molecule to a particular structure probe. The method is not applicable to studies of the dynamics of RNA conformational changes, but can be used to determine the average structure at equilibrium in a given physical state. It give the kind of information that can clarify and extend interpretation of data obtained with sophisticated physical methods.

ACKNOWLEDGMENTS

The work reported here was supported by grants from The National Institutes of Health (GM22280), The National Science Foundation (PCM80004620), and the Syracuse University Senate Research Fund. We give special thanks to A. Maxam for stimulating the development of the line of work discussed. We also thank W. Curtis, L. Rieser, M. Lane, M. Michelackakis, N. Vamvakopoulos, R. Slepecky,

D. Sullivan, and C. Vary for technical and intellectual assistance. We are particularly grateful to G. Ventura and K. Vournakis for help in preparing this manuscript.

REFERENCES

1. Maxam, A. and Gilbert, W. (1977), Proc. Natl. Acad. Sci. USA 74, 560-564.
2. Sanger, F., Nicklen, O., and Coulson, A. R. (1977), Proc. Natl. Acad. Sci. USA 74, 5463-5467.
3. Sanger, F., Brownlee, G. G. and Barrell, B. G. (1965), J. Mol. Biol. 13, 373-398.
4. Brownlee, G. G. (1972), In "Laboratory Techniques in Biochemistry and Molecular Biology", (T. S. and E. Work, eds), Vol. 3, Part I, Determination of Sequences in RNA, pp. 1-265, North-Holland, Amsterdam.
5. Brownlee, G. G., Sanger, F. and Barrell, B. G. (1968), J. Mol. Biol. 34, 379-412.
6. Fiers, W., Contreras, F., Duerinck, F., Haegeman, G., Iserentant, D., Merregaert, J., MinJou, W., Molemans, F., Raeymaekers, A., Van den Berghe, A., Volckaert, G. and Ysebaert, M. (1976), Nature 260, 500-507.
7. Maxam, A. M. and Gilbert, W. (1980), in "Methods in Enzymology," (L. Grossman and K. Moldave, eds), Vol. 65, Part I, Nucleic Acids, pp. 499-560, Academic Press, New York.
8. Donis-Keller, H., Maxam, A. and Gilbert, W. (1977), Nucleic Acids Res. 4, 2527-2538.
9. Simoncits, A., Brownlee, G. G., Brown, R. S., Rubin, J. R. and Guilley, H. (1977), Nature 269, 833-836.
10. Peattie, D., (1979), Proc. Natl. Acad. Sci. USA 76, 1760-1764.
11. Brownlee, G. G. and Cartwright, E. M. (1977), J. Mol. Biol. 114, 93-117.
12. Sanger, F. and Coulson, A. R. (1975), J. Mol. Biol. 94, 441-448.
13. Sures, I., Levy, S. and Kedes, L. H. (1980), Proc. Natl. Acad. Sci. USA 77, 1265-1269.
14. Ghosh, P. K., Reddy, V. B., Piatak, M., Lebowitz, P. and Neissman, S. M. (1980), in Methods in Enzymology, (L. Grossman and K. Moldave, eds), Vol. 65, Part I, Nucleic Acids, pp. 580-595, Academic Press, New York.
15. Sanger, F. and Coulson, A. R. (1978), FEBS Lett. 87, 107-110.
16. Stanley, J. and Vassilenko, S. (1978), Nature 274, 87-89.
17. Pavlakis, G., Lockard, R., Vamvakopoulos, N., Rieser, L., RajBhandary, U. L. and Vournakis, J. N. (1980), Cell 19, 91-102.
18. Pavlakis, G. and Vournakis, J., unpublished results.
19. Donis-Keller, H. (1979), Nucleic Acids Res. 7, 179-192.
20. Wurst, R. M., Vournakis, J. N. and Maxam, A. (1978), Biochemistry 17, 4493-4499.
21. RajBhandary, U. L., Heckman, J. E., Yin, S. and Alzner-DeWeerd, B. (1979), in "Transfer RNA: Structure, Properties and Recognition," (P. Schimmel D. Soll and J. Abelson eds), pp. 3-17, Cold Spring Harbor Monograph, Cold Spring Harbor, New York.
22. Gupta, R. C. and Randerath, K. (1979), Nucleic Acids Res. 6, 3443-3458.
23. Randerath, K., Gupta, R. C. and Randerath, E. (1980), in "Methods in Enzymology," (L. Grossman and K. Moldave, eds) Vol 65, Part I, Nucleic Acids, pp. 638-680, Academic Press, New York.
24. Wagner, E. K. and Ingram, V. M. (1966), Biochemistry 5, 3019-3027.
25. Brownlee, G. G. and Sanger, F. (1969), Eur. J. Biochem. 11, 395-399.
26. Jordan, B. R. (1971), J. Mol. Biol. 55, 423-439.

27. Vigne, R. and Jordan, B. R. (1971), Biochimie 53, 981-986.
28. Vigne, D., Jordan, B. R. and Monier, R. (1973), J. Mol. Biol. 76, 303-311.
29. Feunteun, J. and Monier, R. (1971), Biochimie 53, 657-660.
30. Erdman, V. A. (1976), in Progress in Nucleic Acids Research and Molecular Biology" (W. E. Cohn, ed), Vol. 18, pp. 45-90, Academic Press, New York.
31. Ehresmann, C., Stiegler, P., Maekie, G. A., Zimmermann, R. A., Ebel, J. P. and Fellner, P. (1975), Nucleic Acids Res. 2, 265-278.
32. Vogt, V. M. (1973), Eur. J. Biochem. 33, 192-200.
33. Rushizky, G. W., Shaternikov, U. A., Mozejko, J. H. and Sober, H. A. (1975), Biochemistry 14, 4221-4226.
34. Harada, F. and Dahlberg, J. E. (1975), Nucleic Acid Res. 2, 865-871.
35. Khan, M. S. and Maden, B. E. H. (1976), FEBS Lett, 72, 105-110.
36. Vournakis, J. N., Flashner, M., Katopes, M., Kitos, G., Vamvakopoulos, N., Sell, M. and Wurst, R. M. (1976), in "Progress in Nucleic Acids Research and Molecular Biology" (W. E. Cohn and E. Volkin, eds), Vol. 19, mRNA: The Relation of Structure to Function, pp. 233-252, Academic Press, New York.
37. Flashner, M. S. and Vournakis, J. N. (1977), Nucleic Acids Res. 4, 2307-2319.
38. Barber, C. and Nichols, J. L. (1978), Can. J. Biochem. 56, 357-364.
39. Laskowski, M. (1980), in "Methods in Enzymology", (L. Grossman and K. Moldave, eds), Vol. 65, Part I, Nucleic Acids, pp. 263-276, Academic Press, New York.
40. Finn, M., Wurst, R. M. and Vournakis, J. N., unpublished data.
41. Fraser, M. J. (1980), in "Methods in Enzymology," (L. Grossman and K. Moldave, eds), Vol. 65, Part I, Nucleic Acids, pp. 255-263, Academic Press, New York.
42. Vassilenko, S. K. and Rait, V. K. (1975), Biokhimiya 40, 578-583.
43. Mitra, T., Pavalakis, G. and Vournakis, J. (1980), Nucleic Acids Res., submitted.
44. Wurst, R. M. and Vournakis, J. N. (1977), J. Cell Biol. 75, 36a.
45. Wrede, P., Wurst, R., Rich, A. and Vournakis, J. N. (1979), J. Biol. Chem. 254, 9608-9616.
46. Pavlakis, G., Jordan, B. R., Wurst, R. and Vournakis, J. N. (1979), Nucleic Acids Res. 7, 2213-2238.
47. Pavlakis, G., Katopes, M. and Vournakis, J. N. (1980), J. Biol. Chem., submitted.
48. Vary, C., Pavlakis, G., Morris, A. and Vournakis, J. N. (1980), Nature, submitted.
49. Ross, A. and Brimacombe, R. (1979), Nature 281, 271-277.
50. Rabin, D. and Crothers, D. M. (1979), Nucleic Acids Res. 7, 689-703.
51. Efstratiadis, A., Vournakis, J. N., Donis-Keller, H., Chaconas, G., Dougall, D. and Kafatos, F. (1977), Nucleic Acids Res. 4, 4165-4174.
52. Garen, H. and Levinthal, C., (1960), Biochem. Biophys. Acta 38, 470-483.
53. Pilly, D., Niemeyer, A., Schmidt, M. and Bargetzi, J. P. (1978), J. Mol. Biol. 253, 437-445.
54. Lockard, R. E., Alzner-DeWeerd, B., Heckman, J. E., MacGee, J., Trabor, M. W. and RajBhandary, U. L. (1978), Nucleic Acids Res. 5, 37-56.
55. Wurst, R. M. (1979), PhD. Thesis, Syracuse Univ., Syracuse, NY.
56. Kowalski, D., Kroeker, W. D. and Laskowski, M. (1976), Biochemistry 15, 4457-4463.
57. Lockard, R., Rieser, L. and Vournakis, J. N., see this volume, Chapter 10.
58. Silberklang, M., Gillum, A. M. and RajBhandary, U. L. (1977), Nucleic Acids Res. 4, 4091-4108.
59. England, T. E. and Uhlenbeck, O. C. (1978) Nature 275, 560-561.
60. Levy, C. C. and Karpetsky, T. P. (1980), J. Biol. Chem. 255, 2153-2159.

298

61. Boguski, M. S., Nieter, P. A. and Levy, C. C. (1980), J. Biol. Chem. <u>255</u> 2160-2163.
62. DeWachter, R. and Fiers, W. (1972), Anal. Biochem. <u>49</u>, 184-197.
63. Pipas, J. M. and McMahon, J. E. (1975), Proc. Natl. Acad. Sci. USA <u>72</u>, 2017-2021.
64. Tinoco, I., Borer, D. N., Dengler, B., Levine, M., Uhlenbeck, O. C., Crothers, D. M. and Gralla, J. (1973), Nature N. B. <u>246</u>, 40-41.
65. Donis-Keller, H. (1980) Nucleic Acids Res. <u>8</u>, 3133-3142.
66. Tanaka, Y., Dyer, T. A. and Brownlee, G. G. (1980) Nucleic Acids. Res. <u>8</u>, 1259-1272.
67. Krupp, G. and Gross, H. J. (1979) Nucleic Acids Res. <u>6</u>, 3481-3490.

PHAGE T4 POLYNUCLEOTIDE KINASE

WILLIAM R. FOLK
Department of Biological Chemistry
University of Michigan Medical School
Ann Arbor, Michigan 48109

I. STRUCTURE OF PHAGE T4 POLYNUCLEOTIDE KINASE

Phage T_4 polynucleotide kinase is widely used as a reagent to phosphorylate 5' hydroxyl termini of nucleic acids and nucleotides. Several newly developed sequencing procedures call for polynucleotides labeled at their termini, and posphorylation with T_4 polynucleotide kinase is one way to achieve such label-ing.[1-4] T_4 polynucleotide kinase is also extensively used in the chemistry of synthetic oligonucleotides,[5] in the measurement of endonuclease cleavage of DNAs,[6-8] and in the 5' end-group analysis of oligonucleotides.[9]

A review of prokaryotic and eukaryotic polynucleotide kinases was published in 1978.[10] This review focuses on the more recent studies of the in vitro acti-vities and in vivo functions of T_4 polynucleotide kinase.

Phage T4 polynucleotide kinase appears to be an oligomer of four subunits. Active enzyme elutes from gel filtration columns as a protein of 140,000 ± 10% daltons.[11] The subunit molecular weight, as determined by SDS gel electrophore-sis and by sedimentation equilibrium, is 33,000 ± 5%.[12] Several oligomeric species (monomers, dimers, tetramers, and perhaps higher-order oligomers) have been observed in sedimentation studies in which the variables included ionic strength, pH, and the presence or absence of spermine, ATP, and 3'-dTMP. The oligomeric forms are favored in solutions of high ionic strenth (0.1 M KCl) and in solutions that contain spermine or substrates.[12] Phosphorylation by T4 poly-nucleotide kinase is enhanced by the addition of KCl or spermine to the reaction components.[13]

N-terminal amino acid analysis of homogeneous T4 polynucleotide kinase yielded phenylalanine as the sole amino acid. That result and the size homogen-eity of the monomers strongly suggest that the tetramer is composed of identical polypeptide chains.[11,12] Each monomer has two cysteine residues, one of which reacts much more readily with 5,5'-dithio-bis(2-nitrobenzoic acid) than the other. The lack of effect of substrates on the reactivity of this cysteine suggests that it is not in an active site. However, phosphorylation activity depends on the presence of sulfhydryl reducing agents; in their absence, the protein precipitates in solutions of low ionic strength. Thus, reduced cysteines are vitally important for maintaining an active conformation.[12]

Circular dichroic measurements indicate that the protein has a relatively large amount of α-helix content.[12] Ultraviolet-absorption spectra indicate the presence of one or several exposed tryptophan residues: there is a pronounced shoulder in the spectrum at 290 nm. On excitation at 280 nm, T4 polynucleotide kinase fluoresces, with an emission maximum of 340 nm.[12] Substrates quench this fluorescence, and the quenching has been used to measure the association constants for ATP, 3'-dTMP, inorganic phosphate and (Pi) (Table 1). The observed values agree well with the apparent association constants obtained from kinetic data.

TABLE 1

Substrate	Association Constant[a] K_a (M^{-1})	Michaelis Constant (M)
ATP	7.9×10^{5b} 7.0×10^{5c}	4×10^{-5} (with 3'dTMP)[d] [21,22] 1.5×10^{-4} (with calf thymus[21,22] DNA[d]
3' dTMP	4.8×10^{5b} 6.8×10^{5c} [21,22]	5×10^{-5d} [21,21]
Pi	7.2×10^{2b} [12]	
Calf thymus DNA (treated with micrococcal nuclease)		1.8×10^{-5} (hydroxyl termini)[d] [21]

	V_{max} relative to dT(pT)$_4$	Apparent Michaelis Constant (M)
ATP		14×10^{-5} (with calf thymus DNA)[21] 26×10^{-5} (with phage T$_7$ DNA)[21] 1.3×10^{-5} (with dT(pT)$_9$)[21]
3' dTMP	10.5	2.2×10^{-5} [21]
dT(pT)$_4$	1.0	0.55×10^{-5} [21]
dT(pT)$_9$	2.8	0.18×10^{-5} [21]
dT(pT)$_{14}$	2.5	0.35×10^{-5} [21]
Calf thymus DNA (micrococcal nuclease)	16.7	0.76×10^{-5} [21]
Mouse tRNA	0.3	0.006×10^{-5e} [23]
E. coli tRNA	0.5	0.135×10^{-5e} [23]

a) Determined by fluorescence quenching.[12]
b) 50 mM Tris-HCl (pH, 8.0), 5 mM MgCl$_2$, and 10 mM β-mercaptoethanol.
c) 20 mM potassium phosphate (pH, 7.6), 5 mM MgCl$_2$, and 10 mM β-mercaptoethanol.
d) 60 mM Tris-HCl (pH, 8.0), 9 mM MgCl$_2$, and 15 mM β-mercaptoethanol.
e) 60 mM Tris HCl (pH, 8.0), 9 mM MgCl$_2$, 15 mM β-mercaptoethanol, and 66 μM ATP.

Biochemical and genetic evidence indicates that, in addition to phosphory-
lating 5'-hydroxyl termini, T4 polynucleotide kinase acts on 3'-phosphoryl termini as a phosphatase. The 3' phosphatase and the 5'-polynucleotide kinase activities copurify through numerous steps, and substrates of the 5'-polynucleotide
kinase (ATP, $(Up)_4U$) protect both activities from thermal inactivation. A product of both reactions (pU_5) protects neither activity, whereas a substrate of
the 3' phosphatase but a product of the 5'-polynucleotide kinase $(pU)_5p$ protects
only the 3'-phosphatase activity.[14] Evidence for the association of both activities in one protein has come from genetic studies showing that numerous pseT
mutants of phage T4 do not express either activity.[15] One mutant, however,
expresses the 5'-polynucleotide kinase without the concomitant 3'-phosphatase
activity (pseT1;[15-17]); so it is likely that the two activities are distinct
functions of the same protein.

II. ENZYMATIC ACTIVITIES

A. Phosphorylation of 5'-hydroxyl oligonucleotides:

The forward reaction catalyzed by T4 polynucleotide kinase can be written
schematically as:

5'-OH (DNA, RNA, nucleotide) + NTP——>5'-P (DNA, RNA, nucleotide) + NDP,
where N is adenosine, cytosine, guanosine, or perhaps other nucleosides.[9,18,19]
This reaction has been studied with numerous polynucleotide substrates.
Components of the reaction buffer appear to significantly affect the activity of
the enzyme with such substrates. In the phosphorylation of DNA digested by
micrococcal nuclease (DNA fragments containing 5'-hydroxyl termini), the enzyme
has a broad pH dependence, with maximal activity between pH 6.5 and pH 8.5.[9,14,18]
It is activated by numerous salts (NaCl, KCl, CsCl, LiCl, and NH_4Cl), with
maximal phosphorylation activity at approximately 0.125 M. Polyamines (e.g.,
spermine and spermidine) increase the phosphorylation activity approximately
threefold, but their effect is not additive with salt.[13]
The initial rate of phosphorylation of single-stranded oligonucleotides in
low ionic strength buffer is maximal at pH 9.5, but the final amount of phosphorylated product is higher at pH 8.0 with 0.1 M KCl and 1.5 mM spermine.[13,20]
Both KCl and spermine promote the formation of oligomers of the enzyme monomer
and are thereby likely to increase the amount of active enzyme present in solution. Studies of the kinetics of phosphorylation of DNA suggest that these
activators alter the reaction mechanism from an ordered sequential type (in

which binding of DNA occurs before binding of ATP) to a rapid-equilibrium, random type, as well as modifying the apparent V_{max} and K_m.[13,19,21]

In contrast with the phosphorylation of single-stranded oligonucleotides or micrococcal-nuclease-digested DNA, well-defined duplex oligonucleotides containing blunt or internal 5'-hydroxyl termini are phosphorylated at lower rates and to lesser extents in the presence of 0.125 M KCl.[22] Spermine has little effect on either the rate or the extent of phosphorylation of these oligonucleotides. They are poor substrates for T4 polynucleotide kinase because of the duplex character of the 5'-hydroxyl termini; apparently, any increased enhancement in enzyme activity due to salt or spermine is negated by the stabilizing effect these agents exert on the structure of the DNA termini.

The Michaelis constant for ATP varies somewhat with the DNA substrate that is phosphorylated. With calf thymus and phage T7 DNAs as substrates at 2×10^{-4} M, a K_m^{app} of approximately 2×10^{-4} M was observed; with a single-stranded oligonucleotide, the K_m^{app} was 1.3×10^{-5} M; with 3' dTMP as a substrate the K_m^{app} was 4×10^{-5} M. Addition of spermine slightly decreased the K_m^{app} for ATP, but addition of salt had no effect.[13,21] With these substrates, rapid and essentially complete phosphorylation can be achieved, as long as the ATP concentration is 0.1-1 µM (by necessity, in excess of the concentration of 5'-hydroxyl termini). In contrast, with duplex DNA's that have blunt or internal 5'-hydroxyl termini, the rate and extent of phosphorylation depend on much higher concentrations of ATP. Only at 20 µM ATP (1,000-fold excess over the number of 5'-hydroxyl termini) is complete phosphorylation achieved. This dependence on high concentrations of ATP may reflect steric hindrance of an ATP binding site by the DNA's that lack single-stranded 5'-hydroxyl termini, and possibly the use of a less favorable reaction mechanism in which ATP binds to the enzyme before DNA.[13,19,21,22]

Through kinetic studies, the true Michaelis constant for calf thymus DNA, $K_{DNA\ c.t}$, was found to be 1.8×10^{-5} M (5'-hydroxyl termini), and the true V_{max} was 30 nmol/min.[21] From these values, a turnover number of 25,000/min can be calculated. The apparent K_m and V_{max} values for a variety of 3'-phosphoryl nucleotides and single-stranded oligodeoxyribonucleotides have also been measured. It was observed that, in general, oligonucleotides varying in chain length from 5 to 15 have lower K_m and V_{max} values than mononucleotides (Table 1; 21). It is likely that chain length and the greater charge density of the oligonucleotides enhance binding.

It is particularly interesting that phosphorylation of the 5'-termini of tRNAs by T4 polynucleotide kinase proceeds with an apparent Michaelis constant one-fifth to one-hundredth that of the reaction with calf thymus DNA.[23] Most tRNAs have 5'-termini that are hydrogen-bonded to opposing bases on the CCA acceptor stem. Why DNA's with similar duplex termini are such poor substrates, whereas tRNAs appear to be much better ones, is unclear. The three-dimensional shape of the tRNA molecule may make it a particularly good substrate.

In a series of decanucleotides, each containing only one of the four bases, the apparent K_m varied fifteen-fold, with $dT(pT)_9$ having the lowest (1.8 µM) and $dG(pG)_9$ the highest value (29.6 µM).[21] The apparent V_{max} changed by a factor of ten in the same direction. These data suggest that T4 polynucleotide kinase exhibits some specificity toward the 5'-terminal nucleotide. This has been confirmed with one synthetic duplex oligonucleotide that contains two internal 5'-hydroxyl termini, one ending in G and the other in T. In the presence of 0.125 M KCl, the chain ending in T was phosphorylated approximately five-fold more rapidly than the chain ending in G. However, at lower salt concentration or in the presence of spermine, this difference was not observed, nor was any specificity observed with similar duplex oligonucleotides that had chains ending in G/A or A/T.[22]

B. Dephosphorylation of 5'-phosphoryl oligonucleotides:

In the presence of a nucleotide acceptor, T4 polynucleotide kinase dephosphorylates a 5'-phosphoryl oligonucleotide.[20]

5'-P (DNA, RNA, nucleotide) + NDP——>5'-OH (DNA, RNA, nucleotide) + NTP. Such dephosphorylation is the reversal of the phosphorylation reaction, and not simply a phosphatase activity. ADP is a good nucleotide acceptor; ATP will also function, albeit more poorly, forming adenosine tetraphosphate. At low pH, slow release of Pi has been observed, presumably due to a weak 5'-phosphatase activity of polynucleotide kinase or to a contaminating phosphatase.[20]

The dephosphorylation reaction has been characterized with single-stranded oligonucleotides and with duplex DNA's containing protruding 5'-phosphoryl termini.[6,20] The pH optimum is near 6.2; at pH 7.4, the apparent K_{ADP} is 204 µM and the V_{max} is 4.3 pmol/min-µg (the V_{max} for phosphorylation at this pH is 31.3 pmol/min ug). As expected, reduced thiols are essential for enzymatic activity.

C. Exchange Reaction

Concomitant dephosphorylation and phosphorylation of 5'-phosphoryl oligonucleotides in the presence of ADP and $[\gamma\text{-}^{32}P]$ ATP permits the direct labeling of 5'-phosphoryl oligonucleotides:[6,7,20,24]

5'-OH (DNA, RNA, nucleotide) + NTP $\rightleftharpoons$ 5'-P (DNA, RNA, nucleotide) + NDP.

This reaction is particularly useful for labeling DNA fragments generated by restriction endonucleases, most of which contain 5'-phosphoryl termini.[6,7] The exchange reaction has been studied with a single-stranded oligonucleotide and a variety of duplex DNAs as substrates. It occurs optimally at an ATP concentration of 12 μM and at an ADP concentration of 100 μM for oligonucleotides with external 5'-phosphoryl termini and 200 μM for those with blunt or internal 5'-phosphoryl termini. Spermine has little effect on the reaction, and increasing the ionic strength decreases the extent of exchange. The pH optimum for exchange is 6.2 to 6.6, similar to that for dephosphorylation.[7,20]

Both the initial rates and the extents of exchange mediated by T4 polynucleotide kinase differ for the types of 5'-phosphoryl termini with which the enzyme reacts. Phosphoryl groups of single-stranded DNA's or of duplex DNA's with external 5'-phosphoryl termini are exchanged more rapidly and to greater extents than phosphoryl groups of duplex DNA's with blunt or internal 5'-phosphoryl termini.[6,7] Whereas the former DNA's can be labeled quantitatively by using large concentrations of either ATP (for single-stranded substrates) or enzyme (for duplex DNA's with external termini), quantitative exchange of phosphorus -32 into blunt or internal 5'-phosphoryl termini has not been reported. Usually, only 10 to 15% of the theoretical maximum is achieved. The exchange of phosphate into nicks in DNA occurs at only 3 to 5% of the rate observed for exchange into the same sites that have been converted into Eco RI termini (external 5'-phosphoryl termini).[6] One tRNA (with an internal 5'-phosphoryl terminus) has been reported not to undergo the exchange reaction whereas if it is dephosphorylated by alkaline phosphatase, it will readily undergo phosphorylation by T4 polynucleotide kinase.[25] This latitude in preference by T4 polynucleotide kinase for the variety of termini at which it will catalyze the exchange reaction reflects both its ability to react with an oligonucleotide that has a 5'-phosphate, as well as the apparent equilibrium between phosphorylated oligonucleotide, ATP, and ADP.

Measurement of the separate rates of phosphorylation and dephosphorylation of a variety of single-stranded or duplex DNA's that have either external or internal 5'-phosphoryl termini indicates that the extent of phosphorylation achieved with a particular enzyme-to-DNA ratio need not reflect a true thermodynamic equilibrium in which the enzyme is functioning solely as a catalyst. For single-stranded DNA's or duplex DNA's with long single-stranded termini, the K_{eq} of phosphorylation equals the reciprocal of the K_{eq} for dephosphorylation, as expected for a true equilibrium. [At a pH of 8.0, the K_{eq} of phosphorylation is 50 for the single-stranded oligonucleotide dT(pT)$_8$[20]]. However, for duplex

DNA's with short external 5'-phosphoryl termini or with blunt or internal 5'-phosphoryl termini, a true equilibrium is not reached .[6,22,24] Instead, the extent of phosphorylation depends markedly on the concentration of enzyme. The number of enzyme molecules used in the exchange reaction is often equivalent to the number of termini being labeled. These observations have led to the suggestion that the quantity of enzyme-oligonucleotide complexes formed under such circumstances determines the maximal amount of phosphorylation.[6,22]

In spite of the difficulty in achieving complete phosphorylation of DNAs through exchange, if the ratio of enzyme to DNA is kept constant the extent of labeling accurately reflects the number of termini present; this has been used to measure the rate of appearance of new termini as a function of restriction endonuclease activity.[6,7,26]

In the exchange reaction, there appears to be a slight but measurable preference of T_4 polynucleotide kinase for termini ending in T over termini ending in A or C.[22] It is perhaps also significant that termini of DNAs containing 5-bromodeoxyuridine in place of thymidine are labeled approximately twice as well as are DNAs containing thymidine .[26] Such reactivity may be due to an increased single-stranded character of the terminal nucleotides.

D. Nucleolytic Activities:

T_4 polynucleotide kinase is reported to have the capacity to "decap" the 7-methylguanosine diphosphate moiety from the ends of eukaryotic mRNAs:

$$M^7G(5')ppp(5')Np^MNp(RNA) \longrightarrow M^7G(5')ppp + (5')pNP^MNp(RNA).$$

This methylated nucleotide is attached to the body of the mRNA through a 5'-5' triphosphate linkage: $m^7G(5')ppp(5')Np^MNp(RNA)$.[27] Cleavage of the β-pyrophosphate bond of the cap by T_4 polynucleotide kinase generates a 7-methylguanosine nucleoside diphosphate and an mRNA with a 5'-phosphoryl terminus. The cap structure might be similar to a reaction intermediate in the phosphorylation reaction in which 7-methylguanosine triphosphate is used as a phosphate donor and the 5'-hydroxyl terminus of RNA is the acceptor.[28]

One report has described a DNase activity associated with T_4 polynucleotide kinase,[30] but that has not been confirmed.

E. 3'-Phosphatase Activity:

Little has been reported about the 3'-phosphatase activity associated with T_4 polynucleotide kinase. It is presumably the same activity originally reported by Becker and Hurwitz.[29] It requires Mg^{2+} and acts on 3'-phosphoryl deoxyribonucleotides (but not 3'-phosphoryl ribonucleotides), as well as 3'-phosphoryl ribo- or deoxyribo-oligonucleotides.[14,17] Maximal activity occurs near a pH of 5.9. As previously mentioned, it appears to be a distinct and chemically separable function of the T_4 polynucleotide kinase protein.[14,15,16]

III. IN VIVO FUNCTIONS

Studies of the in vivo functions of T_4 polynucleotide kinase protein are complicated by its having multiple enzymatic activities. Virtually nothing is known about the role of the 5'-polynucleotide kinase activity in vivo. It is expressed by phage T_4 early after infection, but mutants lacking this activity appear normal on infection of laboratory strains of E. coli B or K_{12} in every respect tested, including genetic recombination and repair of DNA damage caused by phosphorus-32 decay and gamma and ultraviolet irradiation.[31] A similar 5'-polynucleotide kinase activity in uninfected E. coli has not been found.[9,15] Phage T_4 mutants that lack the 3'-phosphatase activity (pseT) have been isolated.[17] After the discovery that the 5'-polynucleotide kinase activity was physically associated with the 3'-phosphatase activity, these mutants were found to be partially or completely deficient in 5'-polynucleotide kinase activity as well.[15] The pseT mutants are distinguished by their not being able to grow on a hospital isolate of E. coli (CTr5x) or E. coli K_{12} lit mutants.[32] The former hosts express large amounts of 3'-phosphatase activity, so it is not for lack of this general enzymatic activity that the phages are unable to grow.

The pseT mutants are defective in true-late gene expression on E. coli K_{12} at a step before translation.[15,17] Mutant pseT 1-which has less than 1% of the normal 3'-phosphatase and 75% of the 5'-polynucleotide kinase activity,[42] synthesizes DNA that is abnormally small. Its DNA is not packaged properly, perhaps because of the lack of late gene products. Mutant pset T47 lacks 5' polynucleotide kinase activity but retains approximately 10% of the 3'phosphatase activity. It is phenotypically indistinguishable from mutant pset 1.[32] This suggests that the molecular defect, whatever its nature, may be caused by a reduction in either the 3'-phosphatase or the 5'-kinase activity.

Extragenic suppressors of the pseT mutations have been isolated, both on the phage genome and on the E. coli genome.[15,17,34] The phage suppressor is recessive, so expression of its wild-type gene activity may cause a requirement for a coupled 3'-phosphatase 5'-kinase activity; inactivation of the suppressor wild-type gene function permits the pseT mutants to grow. E. coli CTr5x can acquire the ability to grow pseT mutants (including deletion mutants) through the acquisition of strong amber suppressors.[15,17] This suggests that translational suppression of a secondary host mutation might compensate for lack of the pseT function. The functions, or lack thereof, induced by the suppressor mutations are unknown in both cases.

Most of the evidence indicates that the 5'-polynucleotide kinase-3'-phosphatase protein functions to process the termini of nucleic acids. However, whether the preferred substrate is DNA or RNA (or both) is not known. The pseT gene is near other phage T_4 genetic loci encoding proteins that interact with or modify nucleic acids[17,33]; futhermore, mutations in the T_4 encoded RNA ligase/tail fiber (gene 63) cause a phenotype that is very similar to, if not identical with, that of the pseT mutants,[33] so both proteins may act at similar steps in phage growth. A logical function for the 5'-polynucleotide kinase-3'-phosphate would be to modify oligonucleotides with 5'hydroxyl and 3' phosphoryl groups so that they become substrates for DNA and RNA ligases.[17,32,34,35] Why the pseT gene function is obligatory in some hosts, but not in others, is not obvious; presumably hosts differ in their content of nucleases capable of generating nonligatable termini. Recently, a phage T_4-infected permeabilized cell system has been used to demonstrate the incorporation of phosphorus-32 from [γ-^{32}P]ATP into the termini of polynucleotides, and the subsequent conversion of a small fraction of this label into phosphatase-resistant bonds.[32] Presumably, this pathway uses the T_4 polynucleotide kinase activity.

IV. PROCEDURES FOR THE USE OF PHAGE T_4 POLYNUCLEOTIDE KINASE.

A. Purification:

High yields of T_4 polynucleotide kinase are obtained from E. coli su$^-$ cells infected with phage T_4 nonsense mutants that are defective in DNA replication (amXF1 or am N122) and that exhibit delayed lysis (am N55 SP62). Alternatively, phage T_4 mutants that lack DNA polymerase (am 4313) or DNA ligase (am H39X) may be used to help to avoid potential contamination of T_4 polynucleotide kinase by these enzymes.[2,6,9,11,16,35,36,37] A mutant of phage T_4 which expresses high yields of altered polynucleotide kinase lacking the 3'-phosphatase activity (am N122 pseT) simplifies the purification of 3'-phosphatase-deficient 5'-polynucleotide kinase.[15,16]

Conventional purifications with streptomycin sulfate precipitation, ammonium sulfate fractionation, and column chromatography on DEAE-cellulose (or DEAE-Sephadex), phosphocellulose, and hydroxylapatite have been used to obtain homogeneous enzyme in 20-25% yield.[6,9,11,14,17] Procedures that permit the simultaaneous purification of T_4 polynucleotide ligase and T_4 DNA polymerase[37] or T_4 RNA ligase[9] have been described. Several procedures employing affinity columns (DNA cellulose[2]; Blue Dextran Sepharose[37]) have also been developed; these considerably reduce the number of steps required. It is important, in using the Blue Dextran-Sepharose affinity column, to start with an $RNase^+$ cell, so that the autolysis will work satisfactorily.[38]

T_4 polynucleotide kinase is stable in a purified form for several years when stored at -20°C in a buffer at neutral pH (KPO_4, Tris-HCl) with reduced thiol (DTT, mercaptoethanol) and 20% glycerol. ATP helps to stabilize the enzyme.[11,14]

B. Reaction Conditions:

    a) Phosphorylation of single-stranded oligonucleotides with 5'-hydroxyl termini.[13]

        60 mM Tris (pH, 8.0)

        9 mM $MgCl_2$

        15 mM β-mercaptoethanol or 10 mM DTT

        0.1 M KCl

        1.5 mM spermine

        ATP equal to or in excess of oligonucleotide termini

In the absence of KCl and spermine, a pH of 8.5-9.5 gives higher initial rates than a pH of 7.5-8.0.[11]

    b) Phosphorylation of duplex DNA's with 5'-phosphoryl termini.[22]

        60 mM Tris (pH, 8.0)

        9 mM $MgCl_2$

        15 mM β-mercaptoethanol or 10 mM DDT

        ATP in excess of oligonucleotide termini.

Spermine and KCl have little effect on the phosphorylation of duplex DNA's with external 5'-hydroxyl termini, and they inhibit the phosphorylation of blunt or internal 5'-hydroxyl termini. Excess ATP and enzyme increase the extent of phosphorylation, particularly with oligonucleotides that have blunt or internal termini.

To increase the single-strandedness of the termini, dimethylsulfoxide (15-30%) can be added to the reaction mixture,[40] or the DNA's can be subjected to a 3'>5' exonuclease [exonuclease III[41]; DNA polymerase I Klenow fragment[42]]. Alternatively, the DNA strands can be denatured before phosphorylation.[1]

c) Exchange labeling of DNAs with 5'-phosphoryl termini.[6,20,29]

   50 mM imidazole-HCl (pH, 6.6)

   19 mM $MgCl_2$

   4.5 mM DTT

   12 μM ATP

   100-200 μM ADP

   Autoclaved gelatin at 0.11 mg/ml

Single-stranded and duplex DNA's with external 5'-hydroxyl termini are labeled at 100 μM ADP; duplex DNA's with other termini are labeled at 200 μM ADP. Excess enzyme is required for maximal incorporation.

d) Decapping of mRNA's.[28]

   65 mM Tris-HCl (pH, 7.3)

   15 mM β-mercaptoethanol

   9 mM Mg acetate

e) Hydrolysis of 3'-phosphoryl groups[14]

   100 mM imidazole-HCl (pH, 6.0)

   10 mM $MgCl_2$

   10 mM β-mercaptoethanol

   200 μg/ml bovine serum albumin

V. REFERENCES

1. Maxam, A. and Gilbert, W. (1980), in (L. Grossman and K. Moldave, eds.) Methods in Enzymology, p. 499-560, Academic Press, New York.
2. Murray, K. (1973). Biochem. J. 569-583.
3. Simoncsits, A., Brownlee, G. G., Brown, R. S., Rubin, J. R. and Guilley, H. (1977). Nature 269, 833-836.
4. Seif, I., Khoury, G. and Dhar, R. (1980). Nucleic Acids Research 8, 2225-2240.
5. Khorana, H. G., Agarwal, I. L., Besmer, P., Buchi, H., Caruthers, M., Cashion, P. J., Fridkin, J., Jay, E., Kleppe, K., Kleppe, R., Kumar, A., Rajbhandary, U. L., Ramamoorthy, G., Sekiya, T., van de Sande, J. H. (1976). J. Biol. Chem. 251, 565-570.
6. Berkner, K. L. and Folk, W. R. (1977). J. Biol. Chem. 252, 3176-3184.
7. Berkner, K. L. and Folk, W. R. (1979). J. Biol. Chem. 254, 2561-2564.
8. Kroeker, W. D. and Laskowski, M., Sr. (1977). Analytical Biochemistry 74, 63-72.
9. Richardson, C. C. (1971), in (G. L. Cantoni and D. R. Davies, eds.) Procedures in Nucleic Acid Research, pp. 815-828. Harper and Row, New York.
10. Kleppe, K. and Lillehaug, J. R. (1978), in (A. Meister, ed.) Advances in Enzymology, 48, 245-275 John Wiley and Sons, New York.
11. Panet, A., van de Sande, H. H., Loewen, P. C., Khorana, H. G., Raae, A.J., Lillehaug, J. R., Kleppe, R. K., Kleppe, K. (1973). Biochem. 12, 5045-5049.
12. Lillehaug, J. R. (1977). Eur. J. Biochem. 73, 499-506.

13. Lillehaug, J. R. and Kleppe, K. (1975). Biochemistry 14.
14. Cameron, V. and Uhlenbeck, O. C. (1977). Biochemistry 16.
15. Sirotkin, K., Cooley, W., Runnels, J. and Synder, L. R. (1978). J. Mol. Biol. 123, 221-233.
16. Cameron, V., Soltis, D. and Uhlenbeck, O. C. (1978). Nucleic Acids Research 5, 825-834.
17. Depew, R. and Cozzarelli, R. (1974) J. Virology 13, 888-897.
18. Novogrodsky, A., Tal, M., Traub, A. and Hurwitz, J. (1968). J. Biol. Chem. 241, 2933-2943.
19. Sano, H. (1976). Biochim. Biophys. Acta 422, 109-119.
20. van de Sande, J. H., Kleppe, K. and Khorana, H. G. (1973). Biochem. 12, 5050-5055.
21. Lillehaug, J. R. and Kleppe, K. (1975). Biochemistry 14, 1221-1225.
22. Lillehaug, J. R., Kleppe, R. K., Kleppe, K. (1976). Biochemistry 15, 1858-1865.
23. Lillehaug, J. R. and Kleppe, K. (1977). Nucleic Acids Research 4, 373-380.
24. Chaconas, G., van de Sande, J. H., Church, R. B. (1975). Biochemical and Biophysical Research Communications 66, 962-969.
25. Allen, J. D. and Parson, S. M. (1977). Biochemical and Biophysical Research Communications 78, 28-35.
26. Berkner, K. L. and Folk, W. R. (1979). J. Biol. Chem. 254, 2551-2560.
27. Furuichi, Y., Morgan, M., Muthukrishnan, S., Shatkin, A. J. (1975). Proc. Nat. Acad. Sci. USA 72, 362-366.
28. Abraham, K. A. and Lillehaug, J. R. (1976). FEBS Letters 71, 49-52.
29. Becker, A. and Hurwitz, J. (1976). J. Biol. Chem. 242, 936-950.
30. Loewen, P. C. (1976). Nucleic Acids Research 3, 3133-3141.
31. Chan, V. L. and Ebisuzaki, K. (1970). Mol. Gen. Genetics 109, 162-168.
32. David, M., Vekstein, R. and Kaufmann, G. (1979). Proc. Nat. Acad. Sci. USA 16, 5430-5434.
33. Snyder, L. Personal Communication.
34. Cooley, W., Sirotkin, K., Green, R., Snyder, L. (1979). J. Bacteriology 140, 83-91.
35. Berkner, K. L. and Folk, W. R. Unpublished Observations.
36. Hughes, S. H. and Brown, P. R. (1973). Biochem. J. 131, 583.
37. Nichols, B. P., Lindell, T. D., Stellwagen, E. and Donelson, J. E. (1978). Biochem. Biophys. Acta. 526, 410-417.
38. Donelson, J., Personal Communication.
39. Uhlenbeck, O. Personal Communications.
40. Bosely, P., Moss, T., Machler, M., Portmann, R. and Birnstiel, M. (1979). Cell 17, 19-31.
41. Thomas, T. and Folk, W. R. Unpublished Observations.
42. Kronenberg, H. M., McDevitt, B. E., Majzous, J. A., Nathans, J., Sharp, P. A., Potts, J. T., Jr., Rich, A. (1977). Proc. Nat. Acad Sci. USA 76, 4981-4985.

T4 RNA LIGASE AS A NUCLEIC ACID SYNTHESIS
AND MODIFICATION REAGENT

RICHARD I. GUMPORT and OLKE C. UHLENBECK

The Department of Biochemistry
School of Chemical Sciences and School of Basic Medical Sciences
University of Illinois
Urbana, Illinois 61801
USA

I. INTRODUCTION

RNA ligase was reported in 1972 by Leis _et al._[1] as an activity induced by T-even bacteriophage upon infection of _E. coli_. The activity appears early after infection and is made in substantial amounts throughout the life cycle of T4[2]. RNA ligase has been identified as the product of T4 gene 63[3] but its function in infection remains unknown[4,5]. Several early reports of an analogous RNA ligase activity in eukaryotic cells[2,6,7] were probably incorrect[8]. Although the activity that joins tRNA half-molecules after the removal of the intervening sequence in yeast and in several other eukaryotes[9] is an RNA ligase, it shows considerable substrate specificity and probably proceeds by a mechanism different from that of T4 enzyme described here. Presumably, the "splicing" enzyme(s) involved in removing intervening sequences in mRNAs will show similar high specificity. Thus, the only general RNA joining activity known to date is produced by T-even bacteriophage and the T4 enzyme alone has been studied in detail.

T4 RNA ligase was first described as an enzyme catalyzing the ATP-dependent circularization of oligoribonucleotides[1,2]. Subsequently it was shown that an intermolecular joining reaction could occur as well[10-12]. RNA ligase catalyzes the formation of a 3'——>5' phosphodiester bond between the terminal 3'-hydroxyl of an acceptor oligonucleotide and the 5'-phosphate of a donor oligonucleotide:

$$\text{ATP} + \ldots\text{XpY-OH} + \text{pNp}\ldots\ldots \longrightarrow \ldots\text{XpYpNp}\ldots\ldots + \text{AMP} + \text{PP}_i.$$
$$\text{(acceptor)} \quad \text{(donor)} \qquad\qquad \text{(product)}$$

If the acceptor and donor are two different oligonucleotides, a longer intermolecular product is formed, whereas if they are part of the same oligonucleotide, a circular product is made. The free energy for formation of the

internucleotide phosphodiester bond is provided by the hydrolysis of ATP to
AMP and PP_i[1,2]. This activity is analogous to the ATP-dependent DNA ligases
from bacteriophage T4 and eukaryotic cells[13]. Both RNA ligase[7] and these DNA
ligases can catalyze an ATP-PP_i exchange reaction. Besides its preference for
RNA substrates, one of the most striking features of RNA ligase, which contrasts with the DNA ligases, is that it does not require a complementary
"template" strand to align donor phosphate with acceptor hydroxyl before reaction. In fact, a complementary nucleic acid strand often inhibits the RNA
ligase reaction[7]. Thus, the apposition of the single-strand donor and acceptor
occurs in RNA binding sites on the surface of the enzyme. This unusual property of RNA ligase and its surprising lack of specificity for different nucleotide sequences make it an extremely versatile tool for manipulating nucleic acid molecules.

This paper is intended as a guide to the use of T4 RNA ligase and not as
a comprehensive review. Thus, practical considerations are stressed. We
discuss the purification and assay of the enzyme, the types of substrates that
are active, and the reaction conditions required for optimal use. A summary of
the applications of RNA ligase to problems in biochemistry and molecular
biology is also included. A broader discussion of RNA ligase, including a more
detailed review of the reaction mechanism and some information about its role
in T4 infection, will appear elsewhere[14].

II. PURIFICATION

RNA ligase has been purified from E. coli cells infected with both wild-
type[7,11] and mutant strains of bacteriophage T4. Amber mutants defective in
DNA synthesis (DNA-negative or DO mutants) are a preferred source of the
enzyme because they allow the production of the phage-encoded early proteins,
but do not lyse the cells[2,15-18]. The mutant bacteriophage are grown on
E. coli CR63, an amber suppressor-containing strain. DO mutants that have been
used to infect nonpermissive strains of E. coli B as sources for enzyme purification are T4 amE4314 (gene 43⁻)[17], T4 amN82 (gene 44⁻)[15,19], and T4 amE10
(gene 45⁻)[16,18]. The cells are harvested 2-5 h after infection to maximize
accumulation of early proteins. RNA ligase activity in infected cells stored
at -70°C is stable for years. Strains containing regA, a specific regulatory
mutation, in conjunction with amber DO mutations have been reported to produce
7 times as much as wild-type bacteriophage and 2-3 times as much as the DO single mutants[16].

Because the enzyme may represent as much as 1% of the total protein in sonicated and centrifuged extracts of cells, substantial amounts of enzyme are readily obtainable[17]. Reported yields of physically homogeneous enzyme range from 7 to 20 mg/100 g of infected cells with yields of activity ranging from 10% to 22%. These preparations sometimes require further purification to remove trace contaminants of nucleases that might be expected to interfere with with the intended applications.

A. Assays

The enzyme can be assayed at all stages of purification by determining the circularization of $[5'-^{32}P]$-labeled oligoribonucleotides[1]. This is done indirectly by measuring the conversion of the label to phosphomonoesterase resistance as it is incorporated into a phosphodiester bond. The assay does not, however, distinguish between label in phosphodiester bonds and in pyrophosphate bonds, which may be formed by adenylylation of the 5'-phosphates[20]. A unit has been defined as the incorporation of 1 nmol of $[5'-^{32}P]pA_n$ into a phosphomonoesterase-resistant form in 30 min at 37°C[2]. For measurements of activity in the early stages of purification, it is important to correct for the considerable destruction of substrate by ribonucleases and phosphatases. The addition of known amounts of pure enzyme to the extract is a convenient method to calibrate these corrections.

Accurate comparison of activity measurements obtained by different laboratories with this circularization assay is difficult for several reasons. Because the activity is directly proportional to oligomer concentrations from at least 0.02 to 10 μM[7], the substrate is not saturating and its concentration must be specified. In addition, riboadenylylates of different chain lengths have often been used and this can significantly affect the apparent activity[21]. A final problem is that activity measurements are sometimes proportional to the amount of enzyme only over a narrow concentration range[15]. Thus, specific activities of highly purified enzyme have been reported at 2,400[15], 2,200[17], 6,500[16], 1,000[22], and 1,900[18] units/mg on the basis of the circularization assay. For these reasons and because physically homogeneous enzyme can be easily prepared, we recommend that enzyme concentration be expressed in molar rather than in activity terms.

After some initial purification, the adenylylation of the enzyme can be measured either directly with ATP labeled in its AMP portion or indirectly by an ATP-PP_i exchange reaction[7]. Because the formation of labeled adenylylated enzyme is a stoichiometric assay, more enzyme is consumed than in catalytic assays. The formation of acid-insoluble enzyme-[^{3}H]AMP is rapid (5 min at 0°)[23] and can be assayed by collection of the acid-insoluble precipitate on filters[7]. Thus, the proportion of active protein in a homogeneous preparation can be estimated. The proportion of adenylylated enzyme can also be determined by SDS-gel electrophoresis since it is easily separated from nonadenylylated enzyme[16]. The exchange reaction uses [^{32}P]PP$_i$ as the labeled substrate and is a reliable way of following the activity after contaminating ATPase and ATP-exchanging enzymes have been removed. A unit of exchange activity has usually been defined as the exchange of 1 nmol of [^{32}P]PP$_i$ into ATP in 30 min at 37°C[6,7,24,25]. Because different concentrations and ratios of ATP to PP$_i$ have been used by various investigators, there are also difficulties in comparing these units. However, we find approximately 30 exchange units per circularization unit[17,25].

B. Protein Isolation

Infected cells are opened by sonication[2,7,11,15,17] or with a French press[16]. Nucleic acids are partially removed with 0.7-1% (w/v) streptomycin sulfate and the enzyme is concentrated by ammonium sulfate precipitation. The streptomycin sulfate precipitate contains T4 polynucleotide kinase[17] and can be used as a source of this activity[26]. T4 DNA ligase is in the streptomycin sulfate supernatant and is also precipitated by the 55% saturated ammonium sulfate used to collect the RNA ligase[6,18].

A variety of chromatographic steps involving ion exchange, gel filtration, and adsorption have been used to purify the enzyme further. It binds to DEAE- and QAE- exchange resins[23] but not to phosphocellulose or Biorex-70 columns[17]. When the ATP-PP$_i$ exchange assay is used on the DEAE-cellulose column eluate, a peak of activity catalyzed by T4 DNA ligase elutes earlier than the one due to RNA ligase[6,18]. After these initial steps, different procedures have generally used gel-filtration[2,16,17] or hydroxylapatite chromatography[15]. Finally, further DEAE-column chromatography[7,15,17] or DNA-agarose chromatography[16] has been used to obtain physically homogeneous enzymes. Glycerol density-gradient velocity sedimentation[7] and isoelectric focusing have also been used to purify the enzyme[23]. Values of 5.8[27] and 6.1[23] have been observed for the isoelectric point of RNA ligase.

Immobilized dye columns can be used after a DEAE-cellulose step as a reliable method of removing both RNase and DNase contaminants[17,18]. The enzyme

binds to Affi-gel Blue and Matrex Gel Red A columns and can be eluted with NaCl
and ATP from the former and with NaCl alone from the latter. The Matrex Gel
Red A column separates free enzyme from adenylylated enzyme[18]. Contaminating
RNases have also been removed from the enzyme by chromatography on 2', 5'-ADP
Sepharose[22] and Blue-Dextran Sepharose[28]. A method currently in used in our
laboratory[18] follows an initial DEAE-cellulose step with Affi-gel Blue, Matrex
Gel Red A, and hydroxylapatite column chromatography steps to give RNA ligase
as greater than 90% homogeneous nonadenylylated protein. The enzyme is nearly
free of DNase and RNase activities and approximately 12 mg is obtained per
100 g of cells. For applications involving DNA, we recommend a second Affi-gel
Blue column to remove traces of an exonuclease activity[17].

The enzyme can be concentrated by ammonium sulfate precipitation, vacuum
dialysis[7] in collodion bags[17] or ultrafiltration with Amicon UM 10[15], PM 10[17],
or PM 30[18] membranes or solid polyethylene glycol[28]. When the enzyme is stored
concentrated (>1 mg/ml) in 50% glycerol at -20°C, it is stable for at least a
year. We found less than 10% loss of activity when the concentrated enzyme
(3.8 mg/ml) was left at room temperature for 5 d in 25 mM 4-(2-hydroxyethyl)-1-
piperazineethanesulfonic acid (HEPES) (pH, 7.5), 12.5 mM NaCl, 0.5 mM dithiothreitol (DDT), and 50% glycerol.

When a preparation of RNA ligase[17] which was greater than 90% homogenous
was dialyzed to equilibrium against 20 mM HEPES (pH, 7.5), 20 mM KCl, and
0.1 mM DTT, the ultraviolet spectrum showed λ_{max} = 278.5 nm, λ_{min} = 251 nm, and
A_{280}/A_{260} = 1.98. When acid-precipitated protein values were determined by the
method of Lowry with egg white lysozyme as the standard and a molecular weight
(Mr) of 43,000 for RNA ligase[17], the molar extinction coefficient for the
enzyme was 5.72×10^4 $M^{-1}cm^{-1}$ at 280 nm and $E_{1cm}^{1\%}$ was 13.3 at the same wave-
length.

III. MECHANISM

The mechanism of action of RNA ligase can be described in terms of a set
of covalent bonds which are broken and formed during the course of the reaction.
The availability of enzyme free of interfering activities and of a detailed
model for the analogous DNA ligase mechanism[13] has made this understanding
possible. The formal three-step mechanism, outlined below, provides a convenient and rational framework for the discussions of the applications of the
enzyme.

A. Covalent Intermediates

Two kinds of covalent complexes have been isolated from RNA ligase reaction mixtures. The first is an adenylylated enzyme that forms through the pyrophosphorolytic cleavage of ATP[7]. The AMP is probably bound the ϵ-amino group of a lysyl residue by a phosphoamide linkage[29]. The adenylylated enzyme can be discharged either by reversing the reaction with PP_i to reform ATP or by supplying a suitable oligonucleotide substrate to allow the overall reaction to continue in the forward direction[7]. The second covalent complex arises from the transfer of the adenylyl group from the enzyme to the 5'-phosphate of an oligonucleotide to form a structure of the general form Ado-5'PP5'-oligonucleotide[10,30]. This compound contains the AMP joined to the oligonucleotide by a phosphoanhydride bond. The pyrophosphate linkage can be split by reverse reaction with the enzyme to reform adenylylated RNA ligase and 5'-P oligonucleotide or by supplying an oligonucleotide with a 3'-hydroxyl group so that the reaction can continue to form a phosphodiester bond and eliminate AMP[20].

The existence of these covalent complexes has led to the formulation of the following three step mechanism for the RNA ligase reaction:

(1) $E + ATP \rightleftharpoons EpA + PP_i$,

(2) $EpA + pN_n \rightleftharpoons E[A-5'pp5'-N_n]$, and

(3) $E[A-5'pp5'-N_n] + M_n \rightleftharpoons M_npN_n + AMP + E$,

where E represents enzyme; EpA, adenylylated enzyme; pN_n, a 5'-phosphate terminated donor; M_n, a 3'-hydroxyl terminated acceptor; and M_npN_n, the product.

The two covalent complexes have both been shown to react with RNA ligase in the manner indicated; but, because thorough kinetic studies have not been carried out, their role as obligatory intermediates in the joining reaction is not established. Nevertheless, it is very likely that the above mechanism is correct. Each of the intermediates can accumulate in reaction mixtures indicating that the reaction does not necessarily proceed uninterruptedly on the enzyme surface.

B. ATP-Independent Reactions

The forward direction of the third step of the reaction mechanism provides an explanation for a class of RNA ligase reactions which does not require ATP. When any of a wide variety of β-substituted ADP derivatives (Ado-5'PP-X) is incubated with RNA ligase and an RNA acceptor, the P-X moiety is joined to the acceptor by a phosphodiester bond and AMP is eliminated (Table I). For example, when A_3C is reacted with NAD^+ (Ado-5'PP5'-Nir) in the absence of ATP, nicotinamide mononucleotide is joined to the A_3C to form A_3CpNir and AMP is released[31]. The β-substituted ADP derivatives are analogues of the adenylylated donor oligoribonucleotide and the reaction proceeds as described by the third step in the mechanistic scheme when the β-phosphate of ADP is joined to a primary hydroxyl of the X group by an ester bond.

C. Reverse Reactions

All three steps of the RNA ligase mechanism are readily reversible. The exchange of ATP and PP_i demonstrates the reversibility of the first step. The last (third) step has recently[36] been shown to be reversible; it was noted that incubation of 5'-AMP with an oligonucleotide having a 3'-terminal phosphate (such as A_3Cp) results in the removal of the 3'-terminal 3', 5'-bisphosphate (pCp) by the formation of the adenylylated intermediate (A-5'pp5'-Cp). This reformation of the phosphoanhydride bond from the phosphodiester bond does not require the presence of pyrophosphate. Although the preferred site of reaction is at a 3'-phosphorylated terminus, reversal at internal phosphodiester bonds can occur as well. The second step of the ligase mechanism is also reversible. Activated donors, such as A-5'pp5'-Cp, can react with RNA ligase to form adenylylated enzyme and release the pCp. Combinations of forward and reverse reaction steps can often lead to unusual products. For example, incubation of AMP, [5'-^{32}P]pCp, and A_3Cp with RNA ligase will lead to the efficient formation of A_3[3'—>5'-^{32}P]pCp[36]. Reverse reactions are generally suppressed by high ATP concentrations. As discussed later, the reverse reaction can be used synthetically[37] but it also sometimes introduces difficulties since in the usual intermolecular synthesis reactions, the commonly used 3'-phosphate blocking group for the donor facilitates the reversal of the third step[36].

D. Unexpected Reactions

While examining the substrate specificity of the ATP-independent reaction of β-substituted ADPs with oligoribonucleotides, we found an unanticipated product. For example, when we reacted the acceptor A_3C with ADP-p-nitrophenol and RNA ligase, $A_3C>p$ (A_3C terminated with a 2', 3'-cyclic phosphate) formed instead of the expected A_3Cp-p-nitrophenol[35]. AMP and p-nitrophenol were the other products. Several other activated donors yielded the same product (Table I). The reaction proceeds through the expected addition of the P-X

TABLE I

RNA LIGASE DONORS IN THE ATP-INDEPENDENT REACTION[a]

Active Substrates

ADP-adenosine
  -cytidine
  -guanosine
  -uridine
  -deoxyadenosine
  -deoxythymidine
  -ribose
  -nicotinamide riboside
  -nicotinimide α-riboside
  -8-bromoadenosine
  -2'-fluoroadenosine
  -2'-azidoadenosine
  -8,2'-o-cycloadenosine
  -8,2'-s-cycloadenosine
  -adenosine-2'(3')-0-aminoacyl

ADP-pantetheine
  -riboflavin
  -cyanoethanol
  -arabanosyladenine

  -o-nitrobenzyl alcohol
  -hexylamine
  -hexylamine-biotin
  -hexylamine-tetramethylrhodamine

  -p-nitrophenol[b]
  -4 methylumbelleferone[b]
  -fluorine[b]
  -p-methoxyphenol[b]
  -glucose[b]

Inactive Substrates

ADP

P2'-ADP-nicotinamide riboside
P3'-ADP-pantetheine

Inosine-5'PP5'-nicotinamide riboside
ε-Adenosine-5'PP5'-nicotinamide riboside
ε-Cytidine-5'PP5'-nicotinamide riboside
8-bromoadenosine-5'PP5'-8 bromoadenosine
2'-Fluoroadenosine-5'PP5'-2'-fluoroadenosine
2'-Azidoadenosine-5'PP5'-2'-azidoadenosine

[a]Data from England et al.[31], Hecht et al.[32], Ohtsuka et al.[33,34], Gumport et al.[35], and unpublished observations.

[b]The compound is a substrate but reaction with a ribonucleotide acceptor causes the product to decompose to a terminal 2', 3'-cyclic phosphate.

portion of the A-5'ppX to the acceptor but, because the X moiety is a good leaving group, it is rapidly and nonenzymatically elminated by a nucleophilic attack of the adjacent 2'-hydroxyl on the phosphate. When one is designing β-substituted ADPs as substrates for RNA ligase in attempts to modify the 3' ends of RNA, this reaction must be considered. Although we find that p-nitrophenylADP forms the cyclic phosphate-terminated acceptor, Ohtsuka et at.[33] demonstrated normal addition of the o-nitrobenzylphosphate group of o-nitrobenzylADP indicating that a methylene group interposed between the β-phosphate of ADP and the electron-withdrawing X group is sufficient to prevent its elimination. Even the strongly deactivated p-methoxyphenyl derivative of ADP yields some of the acceptor with cyclic phosphate termini as well as the expected product with the p-methoxyphenylphosphate added to the acceptor[35].

In intermolecular reactions of oligodeoxyribonucleotides where a 3'-phosphate was used as a blocking group on the donor, we found an unexpected modification of the 3'-phosphate[38,39]. In the presence of high ATP concentrations and the 3'-phosphate terminated oligodeoxyribonucleotide dT_4dCp, RNA ligase catalyzes the formation of dT_4dC-3'pp5'-A, where the AMP is linked to the 3'-phosphate by an anhydride bond. This reaction presumably occurs when the donor binds in the acceptor site in such a way that the 3'-phosphate is available to accept the AMP from the adenylylated enzyme. This hypothesis is supported by the observation that increasing concentrations of acceptor inhibit the formation of the 3'-modified product[38]. Because this modification requires high ATP concentrations and DNA joining reactions are run at low ATP concentrations, it does not seriously interfere with synthesis. It can, however, be responsible for minor products in the reaction mixtures[38].

IV. SUBSTRATE SPECIFICITY

A wide variety of molecules has been tested as substrates for T4 RNA ligase. This has allowed an estimate of the minimal substrate sizes, the composition requirements of the donor and acceptor, and the specificity of the ATP site. The available data are fragmentary and lack quantitation for two reasons. First, as detailed in the following section, the kinetics of RNA ligase reactions are sufficiently unusual that values of K_m and V_{max} for different donor and acceptor pairs are difficult to obtain. Second, most of the information on the relative efficiency of different donors and acceptors in the ligase reaction has been obtained by investigators interested in synthesizing particular oligomers rather than in studying the substrate specificity

of the enzyme. Thus, for example, little information is available where a
variety of acceptors is tested with a single donor under standard conditions
of donor, acceptor, enzyme, and buffer concentrations. Conditions reported
have generally been adjusted to maximize yield for a given pair of oligomer reactants. Nevertheless, since the relative efficiency of different coupling
reactions can strongly influence overall synthetic strategy, we will try to sum-
marize the available information and draw some tentative conclusions.

A. Circularization

Although most information on substrate specificity is for the intermolecular reaction, some data are available on the intramolecular circularization
reaction. Kaufmann et al.[21] have compared the kinetics of circularization of
pA_n for n = 6-100. They find that pA_8 is the shortest possible substrate for
circularization and that it reacts poorly. Maximal reaction rate is observed
for pA_{10} to pA_{16} with an apparent K_m of 1 μM and V_{max} of about 0.1 μmol/mg/h.
With longer chains rates are lower. These data are consistent with the
reaction rates being dominated by the increasing difficulty of juxtaposing
the termini of the longer adenylylate chains and by the strain induced by ring
closure with shorter chains. A similar dependence of the circularization rate
on chain length would be expected for other oligomers as long as no secondary
structure formed. For example, the shortest oligodeoxyribonucleotide that can
be circularized by RNA ligase is pdT_6[19].

In the circularization reaction, the donor and acceptor are parts of the
same molecule so the reaction is more rapid and requires lower enzyme concentrations than an intermolecular reaction. If the two types of reactions compete, the circularization reaction predominates[12]. For example, this is seen
in attempts to add pCp to the 3' terminus of E. coli $tRNA_f^{Met}$[40]. This tRNA has
a reactive 5'-phosphorylated terminus close to the 3' terminus. Thus, a
majority of the product formed is cyclic $tRNA_f^{Met}$ and little $tRNA_f^{Met}$-pCp is
produced even at high ratios of pCp to tRNA. These considerations also
emphasize the importance of blocking the 3' terminus of donors in the intermolecular reaction since the rapidity of the circularization reaction relative
to the intermolecular reaction would otherwise lead to cyclic donors instead
of the desired intermolecular product.

B. ATP

RNA ligase has a strong preference for ATP as a source of free-energy for phosphodiester bond formation. Although dATP can be substituted, a slower rate is observed in either the joining reaction[2] or the ATP-PP$_i$ exchange reaction[7]. No other ribonucleoside triphosphate can support the reaction[7]. The apparent K_m for ATP was found to be 0.2 µM as measured in a pA$_n$ circularization reaction[2] or 12 µM as measured in the ATP-PP$_i$ exchange reaction[7].

In the reversal of the RNA ligase reaction, AMP binds to the ATP site of the enzyme[36]. A similar strong specificity for adenosine is seen since other nucleoside monophosphates do not promote reversal. The apparent K_m for AMP in the reverse transfer reaction is 5 µM, indicating that AMP binds to ligase with an affinity similar to that of ATP and that the major contacts between the enzyme and the ATP may not involve the β- and γ-phosphates.

C. Donors

The minimal donor in the ATP-dependent RNA ligase reaction is a nucleoside 3', 5'-bisphosphate (pNp)[24,31,41]. Although nucleoside 5'-monophosphates have the phosphate necessary to form the internucleoside phosphodiester bond, they do not serve as donors for RNA ligase. Thus, a phosphate 3' to the nucleoside is essential for recognition by the donor binding site. Its precise location on the ribose is clearly important since nucleoside 2', 5'-bisphosphates are not substrates[41] and are only poor competitive inhibitors[22] of the reaction. However, inasmuch as pApAp is as good a donor as pAp[41], only a single charge on the 3'-phosphate of the pNp is necessary for binding to the enzyme. A higher charge density near the 3'-phosphate does not prevent the reaction, because pGpp is also a substrate, albeit a less efficient one[42]. It is not likely that the donor site on the enzyme extends beyond the initial pNp region. Comparison of the reaction rates of several pA$_n$p and pC$_n$Gp with a common acceptor[41] shows that longer oligonucleotide donors react with rates similar to those characteristic of their 5'-terminal pNp.

The minimal donor size described above for the complete reaction is a reflection of the binding of the donor to the enzyme for the adenylylation step, because the donor can be even smaller in the final transfer to the acceptor. This is indicated by the spectacular variety of β-substituted ADPs which function as donors in the ATP-independent reaction (Table I). Thus, once the donor is adenylylated, the majority of the donor portion of the molecule is no longer required for binding to the enzyme or for further reaction. Therefore, not only is the 3'-phosphate of the donor no longer required (A-5'pp5'-A is a donor), but the base is not required (A-5'pp-ribose is a donor), and even the ribose is not required (A-5'pp-cyanoethanol is a donor). The tight binding of the AMP portion of the β-substituted ADP to the ATP site on RNA ligase is presumably sufficient to allow these donors to react.

The strong specificity for adenosine in the ATP site noted for the ATP-dependent reaction is also observed with the β-substituted ADPs in the ATP-independent reaction. For symmetric molecules of the sort A-5'pp5'-N, the pN portion is always transferred to the acceptor and the pA is released. When the pA portion is altered in either the base or the ribose portions, no reaction occurs (Table I).

The nucleotide composition of the donor has only a modest effect on the rate of the ATP-dependent reaction. Comparison of the four ribonucleoside 3', 5'-bisphosphates with a common acceptor under identical conditions[24,41,44] reveals that the pyrimidine pNps are 2-10 times better donors than the purine pNps with pGp being the worst of the four. Modification of the 2'-hydroxyl of the ribose has little effect on efficiency of a pNp donor. Both pdNps and 2'-o-methyl pNps are excellent RNA ligase donors[41,43]. Since the donor site recognizes only the 5'-terminal pNp, these conclusions also apply to longer-oligomer donors. Although some neighbor effects appear to occur, the efficiency of an oligomer donor is determined primarily by the composition of the 5'-terminal nucleotide, as discussed above.

The availability of a convenient method to prepare pNps from their corresponding nucleosides by the use of pyrophosphoryl chloride[43] has allowed the synthesis of a wide variety of modified nucleoside 3', 5'-bisphosphates and their testing as substrates for RNA ligase. It is striking that every modified pNp tested thus far has been a successful substrate. Many of these are listed in Table II. Since the pyrimidine pNps are better substrates than the purine pNps, and pNps with three aromatic rings (pεAp, plinAp, pεGp, puGp) are among the worst donors, there may be an inverse correlation between base size and activity. It is interesting that the nonplanar base dihydrouracil as a pNp is also an effective substrate for RNA ligase indicating little specificity in the base portion of the donor binding site. The fact that nearly any modified nucleotide is a substrate for RNA ligase allows impressive freedom in the construction of modified nucleic acids.

The many β-substituted ADPs in Table I that act as donors have not been compared carefully for their relative efficiency as substrates. The possibility

TABLE II

NUCLEOSIDE 3',5'-BISPHOSPHATE DONORS IN THE ATP-DEPENDENT REACTION[a]

Uridine	Guanosine
Deoxyuridine	Deoxyguanosine
Deoxythymidine	2'-o-Methylguanosine
2'-o-Methyluridine	Inosine
Dihydrouridine	2-Aminopurine
Pseudouridine	Purine
4-Thiouridine	7-Methylguanosine
3-Methyluridine	1-Methylguanosine
Deazauridine	ε-Guanosine
5-Bromouridine	μ-Guanosine
5-Bromodeoxyuridine	
5-Fluorouridine	
5-Iodouridine	
Cytidine	Adenosine
Deoxycytidine	Deoxyadenosine
2'-o-Methylcytidine	ε-Adenosine
ε-Cytidine	lin-Benzoadenosine
5-Iodocytidine	N^6-Hexylaminoadenosine

---

[a]Data from England and Uhlenbeck[41], Barrio et al.[43], and unpublished data of N. Pace and of W. Wittenberg.

of affixing a variety of nonnucleotide groups to the 3' terminus of RNA with β-substituted ADPs allows additional synthetic versatility in the manipulation of nucleic acids.

In an early study of the substrate specificity of the circularization reaction with homopolymers of intermediate chain lengths (n = 17-37), Cranston et al.[7] showed that, whereas all homopolymers were active substrates, poly (U) was not as good as poly (A) or poly (C). In addition, the presence of a polymer complementary to the substrates either had little effect or inhibited the formation of phosphatase-resistant 5' termini. It was therefore concluded that the introduction of secondary structures probably inhibited the ligase reaction.

This point is seen more clearly by examining tRNAs as donors in the RNA ligase reaction.[40] The 5'-phosphate of yeast tRNA^{Phe} is almost totally inactive as a donor under a variety of conditions. Two experiments indicate that the poor reactivity is due to the base-pairing of the 5'-terminal nucleotide as part of the acceptor stem. First, a total ribonuclease T_1 digest of tRNA^{Phe} releases the 5'-nucleotide as pGp and makes it an active donor. Second, if tRNA^{Phe} is cleaved in the anticodon and the halves are separated, the isolated 5'-half molecule, with an unpaired 5'-terminal nucleotide, is also an active donor. Furthermore, in striking contrast to tRNA^{Phe}, the 5'-terminal phosphate of E. coli tRNA$_f^{Met}$ is quite reactive with RNA ligase, forming either a cyclic tRNA or an intermolecular adduct. This is presumably because the 5'-terminal nucleotide in this tRNA is not base-paired in the acceptor stem.

The presence of secondary or tertiary structure is thought to be the reason why several 5'-phosphorylated natural RNAs tested were not good donors with RNA ligase. This is the case for 5S RNA, where the 5'-terminal nucleotides are in a helix. When the molecule is cleaved at position 39, the 5'-terminal 39-nucleotide fragment is an excellent donor.[45,46] Several other RNAs including BMV-4 RNA, STNV RNA, and MDV-1 RNA showed no detectable donor activity in either an intramolecular or an intermolecular reaction, even though their T_1 digestion products were active.[47] Low yields (5-10% for the addition of A_3C acceptor to the 5' terminus of E. coli 16S and 23S rRNA have been observed and attempts to increase them have not been successful.[48] The poor reactivity of most large RNA molecules as donors has prevented experiments involving their intermolecular joining.

In contrast to structured RNAs, DNA restriction fragments are good donors. Higgins et al.[49] used a variety of oligoribonucleotide acceptors and showed that the 5'termini of restriction fragments generated by Eco RI or Hae III from the plasmid Col El could be extended in excellent yield. Somewhat surprisingly, little difference in yield was seen between Eco RI fragments, which have extended 5'-terminal phosphates, and Hae III fragments which have "blunt", base-paired 5'-terminal phosphates. Full-length linear ØX174 viral DNA was also a donor[49]. We have observed similar results with oligodeoxyribonucleotide acceptors bearing single ribonucleotides at their 3' termini with Taq I, Sal I, and Rsa I restriction-fragment donors[27]. Duplex formation between donors and acceptors in the intermolecular reaction of oligodeoxyribonucleotides inhibits the reaction[38].

Another class of polymer that is a donor in the RNA ligase reaction is poly (ADP-ribose)[50]. The A terminus of this molecule is a β-substituted ADP and thus is active in the ATP-independent reaction with an A_3C acceptor. By analogy, ADP ribosylated proteins would also be expected to be donors although none has been reported.

D. Acceptors

Several groups have tested a homologous series of acceptors with a given donor and obtained the uniform result that although dinucleoside monophosphates are generally inactive, trinucleoside diphosphates are effective RNA ligase acceptors[24,25,41]. Longer oligomers only rarely show much higher yields or rates; hence, the acceptor site on the enzyme probably contains binding sites for two phosphates and three nucleosides. As might be expected, some acceptors which are shorter than the trimer binding site are reactive (such as GpG and UpI)[51], but higher enzyme concentrations are required and low yields are obtained.

In contrast to the equal reactivity of RNA and DNA donors, oligodeoxyribonucleotides are much less effective acceptors than oligoribonucleotides. In one case where a direct comparison was made, dA_4 reacted at a rate about one two-hundredth that of rA_4 with a donor even though the reaction conditions were optimized for deoxynucleotide additions[38]. Other deoxynucleotide acceptors react at similarly slow rates resulting in maximal turnovers of 2 per hour for deoxynucleotide additions[38,52]. It is not clear why deoxynucleotide acceptors react so poorly. It is unlikely that the 2'-hydroxyl at the terminal nucleotide is needed in the joining reaction, since a ribonucleotide with a 2'-o-methyl in that position is an adequate acceptor. Although the addition

of a 3'-terminal ribonucleotide to an oligodeoxyribonucleotide greatly stimulates the acceptor activity of the molecule, it does not make it as good as a
pure oligoribonucleotide[27]. Similarly, the addition of a single deoxynucleotide to a ribonucleotide acceptor greatly decreases its reactivity as an acceptor, but it remains a better acceptor than a pure deoxynucleotide[27]. The other
nucleotide binding loci of the acceptor site presumably also prefer ribonucleotides.

It is clear that the nucleotide composition of the acceptor greatly
affects its efficiency in an intermolecular joining reaction. Comparison of
the homopolymeric acceptors with a common donor[24,41,49] gave the uniform result
that oligo (A)s were the best acceptors, oligo (C)s and oligo (I)s were somewhat worse, and oligo (U)s were very poor. The differences are significant:
more than 30 times more enzyme is required for addition of a donor to U_3 than
for comparable addition to A_3. Other trinucleotide acceptors gave intermediate
yields; hence, all three positions in the acceptor binding site probably con-
tibute to the efficiency of reaction. For example, AAG is a better acceptor
than AUG or UAG. Although the relative efficiencies of different trimers have
not been fully explored, the presence of a uridine in any one of the three
positions of the acceptor probably decreases its efficiency. Thus, strategies
for the synthesis of long oligomers should avoid this situation, if possible.
Similar but smaller composition effects on the relative reactivity of oligodeoxynucleotide acceptors have also been observed[35,38,52]. More complicated
sequence effects have not yet been evaluated.

The poor reactivity of DNA acceptors with RNA ligase has prevented substantial reaction at the 3' termini of restriction fragments. In contrast,
RNA molecules are generally excellent acceptors for RNA ligase. The reaction
between a pCp donor and the 3' terminus of yeast tRNA[Phe] acceptor has been
studied in considerable detail[40]. The reaction rates are comparable with those
of oligomer joining reactions and, if sufficient enzyme is added, quantitative
addition can be achieved. A wide variety of other RNAs had been tested for
reaction with RNA ligase and they are nearly always found to be active[53].
Reaction yields vary from 5% to 100%, but substrate concentrations and other
reaction conditions generally differ enough to prevent useful comparisons.
On the basis of the oligonucleotide data, one would expect that the sequence
at the 3' terminus of the RNA could affect yields substantially. The effects
of local RNA secondary structure appear to be less important for RNA ligase
acceptors than for donors. Even double-helical reovirus RNA can act as an
acceptor. It is also important to note that the "cap" structure at the

5' terminus of eukaryotic mRNAs is neither an acceptor nor a donor for RNA ligase.

In summary, the three substrate binding sites on the enzyme have different specificities. The donor site binds one nucleoside and the flanking 5'- and 3'-phosphates, the acceptor site probably binds three nucleosides and two internal phosphates, and the ATP site binds one nucleoside and at least one or more phosphates. The ATP site has the strongest specificity for the identity of the nucleoside. Both the donor and acceptor sites can accommodate a wide variety of sequences, but the donor has preference for pyrimidines and the acceptor for purines.

V. PRACTICAL CONSIDERATIONS

The attempts of investigators to improve the yields of the RNA ligase reactions used in their particular applications have involved a large number of variations in reaction conditions and components. Although no universally applicable set of conditions has been or is likely to be found, a few generalities can be discerned.

Low yields are sometimes caused by the accumulation of either or both of the covalent reaction intermediates discussed avove. The efficiency of the acceptor, i.e., how well the third or phosphodiester-bond forming step of the mechanism proceeds generally determines whether the intermediates build up. For example, in intramolecular circularization reactions of oligoriboadenylylates no intermediates are observed because pA_ns combine good acceptor composition and the ready juxtaposition of ends to allow a concerted reaction to occur. At the opposite extreme, in intermolecular reactions with deoxyoligoribonucleotide acceptors both intermediates often accumulate[20,25,38].

The buildup of enzyme-AMP decreases the concentration of free enzyme available to carry out the third step of the mechanism, thereby slowing the overall reaction[20,52]. When the third step is slowed, the adenylylated donor product of the second step can dissociate from the enzyme and accumulate. Various strategies have been devised to increase the efficiency of poor acceptors and thereby accelerate the third step of the mechanism. These have involved the manipulation of the reaction conditions and have often proved successful.

A further difficulty arises from the fact that both the initial velocities and the final yields in intermolecular reactions of oligoribonucleotides[55], of ribonucleoside 3, 5'-bisphosphates with oligoribonucleotides or RNA[40,41], or oligodeoxyribonucleotides[38] are proportional to enzyme concentration. This may be due to inactivation of the enzyme during the reaction[55]; however, all attempts to circumvent the difficulty by addition of stabilizing compounds were unsuccessful. The problem is usually overcome by increasing the enzyme concentration until maximal yields are obtained under conditions previously optimized with respect to other variables.

A. ATP and Oligonucleotide Concentrations

With RNA substrates, ATP concentrations slightly (by a factor of 1-5) in excess of those stoichiometrically required for joining are usually optimal[40,55]. For example, with 0.1 mM oligoribonucleotide acceptor and donor, no increase in the rate of reaction was observed on increasing the ATP from 0.2 mM to 1.5 mM[55]. ATP concentrations as low as 3 µM can be used when the acceptor is present at very low concentrations during 3' end-labeling reactions[56].

Conversely, increasing the ATP concentration with oligodeoxynucleotide acceptors decreases the initial velocity of the reaction while increasing the relative yield[25,38,52]. High ATP concentrations promote increased product formation by maintaining high concentrations of the adenylylated donor, which is rapidly formed in high yield[25]. However, any free enzyme arising from dissociation of the adenylylated donor will be immediately adenylylated by the excess ATP thereby lowering the free enzyme concentration and slowing the final step of the mechanism[25]. This problem is best overcome by maintaining a low but constant ATP concentration through the use of an ATP regeneration system[38,52]. When 2'-deoxynucleoside 3', 5'-bisphosphates are being added to deoxyoligomers, an ATP to donor ratio of 0.05 is optimal[52] whereas, if equimolar single-strand deoxyoligomers are reacting, any ATP to donor ratio between 0.1 and 0.9 suffices[38]. Using a low and constant ATP concentration optimizes the ratio of enzyme to adenylylated enzyme and provides the necessary concentrations of activated donor to achieve maximal yield. The regeneration system enzymes are commercially available and are free of DNase activities[52].

All the intermolecular reaction rates are increased by increasing the oligonucleotide substrate concentrations, as expected. Because the K_m values for acceptors and donors are probably in the millimolar range, the enzyme is not saturated with substrate during most applications. For equimolar joining of RNA oligomers, concentrations of 100 - 500 µM work well[55]. RNA acceptors at concentrations less than 10 nM have been successfully labeled with [5'-^{32}P]pCp[56]. For adding pdNps to DNA oligomers, concentrations as high as 1 mM, 8 mM, and 0.4 mM for acceptor, donor, and ATP, respectively, have been used[52]. Increasing rates and yields in equimolar additions of DNA acceptors and donors occur at concentrations from 50 μM to 1 mM. These reactions can also be driven with excess acceptor, e.g., 2.5 mM acceptor and 250 μM donor[38].

DNA restriction fragment donors at 0.2-1.0 μM have been successfully extended with oligonucleotide acceptors at 10-500 fold concentration excess[27,49].

In the ATP-independent RNA ligase reaction, riboacceptor concentrations of 10-1,000 μM and A-5'ppXs at 2-10 times the acceptor concentrations have been successfully used[31-34,40,57]. The emphasis has been either on obtaining high yields or on comparing yields with various activated donors; as a result, careful studies of the effects of substrate concentrations have not been done. In general, as in the case of the ATP-dependent reaction, increasing the substrate concentrations inproves yields.

B. Reaction Conditions

RNA ligase reactions have been run in Tris, glycylglycine, imidazole, and HEPES buffers. Glycylglycine and HEPES allowed the highest yields in intermolecular RNA oligomer reactions[55], and HEPES was better in reactions with DNA substrates[52]. For all applications, HEPES is the buffer most widely used[31,38,55,56,58].

All intermolecular RNA ligase reactions have a broad pH optimum from 7.2 to 8.4. With the oligoribonucleotide joining reaction of A_3C and pU_5p, the ratio of product to adenylylated donor formed was pH-dependent, with the highest ratio at a value of 8.3. The pH optimum for product formation was 7.9 and that for intermediate synthesis was 7.5. The value of 8.3 was chosen to optimize the product formation and suppress intermediate synthesis[55]. With poorer acceptors, the incubation temperature is lowered to facilitate reaction and this causes the optimal pH for maximal yield to be 7.5 at 4°C[56]. When DNA oligomers serve as acceptors, a pH of 8.3 is optimal in the absence of an ATP regenerating system and 7.9 is better in its presence[52]. The effects of pH can be significant and this variable should be examined carefully in a given application if high yields are required.

In reactions involving RNA acceptors, Mg(II) is the metal ion of choice and a concentration of 20 mM was found to optimize the ratio of product to adenylylated donor[55]. With DNA acceptors, Mg(II) will function[19,20], but

Mn(II) is required for better rates and maximal yields[25,38,52].

Lowering the temperature of RNA ligase incubation mixtures has allowed otherwise inactive acceptors to become reactive. In addition, with acceptors that react at 37°C, lowering the reaction temperature often increases the ratio of product to adenylylated donor. Lowering the temperature sometimes requires longer incubation times. Temperatures of 4-17°C have been found optimal for various applications. For end-labeling RNA, the temperature must be below 15°C[40,56]. With DNA acceptors, the first two steps of the mechanism occur perfectly well at 37°C but the third step can proceed only if the temperature is below 22°C[25,38]. Low temperatures also facilitate the ATP-independent reaction with some A-5'ppXs[57] and the extension of DNA restriction fragment donors[27]. The mechanism by which decreased temperature facilitates the reaction is unclear. It could stabilize the enzyme, an enzyme-substrate complex, or a conformation of the enzyme that can bind the oligonucleotides in such a way that a more concerted reaction takes place.

DTT and dimethylsulfoxide (DMSO) can substantially influence reaction yields. RNA ligase is inactivated by a sulfhydryl inactivating reagent[2] and high concentrations of DTT can increase yields[19,38,55]. Concentrations of 1-20 mM have been used in various applications and a concentration as high as 33 mM does not inhibit the intermolecular reaction of oligoribonucleotides[55]. The extent of some RNA ligase reactions can be increased severalfold by the addition of 10-20% (v/v) DMSO,[40,41,53] and the enzyme remains active in 40% DMSO[40]. In the end-labeling reaction of RNA with RNA ligase and pCp, the omission of 10% DMSO results in a decrease in yields by factor of 2-3[40,53]. In contrast, DMSO has no effect on the addition of pdNps to DNA oligomers[52], but at 10% it stimulates the joining of single-strand oligodeoxynucleotides[39]. The stimulation by DMSO is unlikely to be due to effects on the secondary structure of the substrates, inasmuch as the increases in yields can be demonstrated with substrates incapable of forming secondary structures, such as oligouridylylates[56].

In some reactions involving DNA acceptors, the polyamines spermine, spermidine, and putrescine and the DNA binding protein RNase A also increase rates and yields[38,52]. Although RNase A can disrupt secondary structures and thus promote reactions with base-paired donors and acceptors[38], it can also stimulate reactions where no secondary structures are possible, e.g., between dA$_5$ and pdNps[52]. The mechanism by which these cationic nucleic acid binding compounds stimulate the RNA ligase reaction with DNA substrates is unknown.

Ammonium ion is an inhibitor of the enzyme and if the protein is concentrated by ammonium sulfate precipitation, it should be thoroughly dialyzed before use. The ammonium and triethylammonium salts of the oligonucleotide substrates have been used successfully but, if very high concentrations are involved, it may be desirable to convert them to the sodium forms. Glycerol at 30-40% (v/v) does not inhibit the enzyme.

A final comment relates to the concentrations of RNA ligase required in various applications. Whenever intermolecular reactions are undertaken high concentrations of enzyme are needed. The levels used range from 0.1 μM (4.3 μg/ml) with the best RNA acceptors to 50 μM (2.2 mg/ml) in the least efficient DNA oligomer joining reactions. The worst RNA acceptors generally react satisfactorily with enzyme at less than 5 μM and the ATP-independent reactions proceed well at 3-10 μM. Such high enzyme concentrations, in some cases greater than those of their substrates[56], may be needed to facilitate the weak binding of poor acceptors. In spite of the high concentrations required, the ready availability of highly purified enzyme makes its use practical. For example, 1 mg of A_3C could be quantitatively converted to A_3CAp with only 36 μg of enzyme[55] and 1.6 mg of enzyme was required to synthesize approximately 0.5 μmol of phosphodiester bonds between various DNA oligomer acceptors and pdNps[35]. Immobilization of the enzyme may further increase its utility[59].

VI APPLICATIONS

RNA ligase is being used in two general ways. First, it is an important tool in the synthesis of oligonucleotides. While the enzyme is better suited for oligoribonucleotide synthesis, oligodeoxyribonucleotides can be made as well. Recent advances in molecular biology have created an "insatiable demand for oligonucleotides"[60]. Although organic synthesis techniques have made spectacular improvements over the last decade[61] and are the best methods for the preparation of sequences of intermediate lengths in large quantities, reaction yields decrease considerably as chain length increases. The availability of ligase permits construction of much longer fragments of single-stranded nucleic acids. A case can also be made for using RNA ligase for completely enzymatic syntheses of oligonucleotides. Not only do organic synthesis methods require special equipment and supplies, but enzyme reactions are also more familiar to the biochemist or molecular biologist who will often use the oligonucleotide. Furthermore, owing to the absence of chemical blocking groups, the mild reaction and workup conditions, and the high specificity of enzyme reactions, the products of enzymatic synthesis often tend to be more homogeneous than those of organic synthesis.

The second major area of application of RNA ligase is a nucleic acid modification tool. The enzyme can be used to add nucleotides to terminal or internal positions in RNA molecules. A variety of other compounds can also be added to RNA termini. This not only provides a convenient method of labeling RNA in vitro, but also aids in the study of the structure and function of RNA. This application needs considerable development, but we feel that the broad substrate specificity of RNA ligase should allow extremely detailed synthetic control over RNA molecules.

A. Oligoribonucleotide Synthesis

Since oligoribonucleotides of almost any size range can be efficiently joined by RNA ligase, this enzyme plays an essential role in the synthesis of RNA fragments. Because RNA ligase acceptors must be at least trinucleoside diphosphates, other methods must be used to prepare them. Small amounts of oligoribonucleotides can be made by purely enzymatic techniques. Controlled polymerization of nucleoside 5'-diphosphates on the 3'-terminus of a primer fragment is the most common method[62]. Several short oligomers can be made by reversal of nuclease catalyzed reaction[63]. A total nuclease digest of RNA or mixed copolymers can also be a useful source of several oligomer sequences[64,65]. Most short oligomer sequences can also be made by purely chemical methods. Although selective blocking of the 2'-hydroxyl and subsequent deblocking make the chemical synthesis of oligoribonucleotides considerably more difficult than oligodeoxyribonucleotides, substantial improvements in the methods have occurred[61]. Synthesis of triribonucleotides and tetraribonucleotides is routine and many longer sequences have been made.

Oligoribonucleotide donors can be as short as a nucleoside 3', 5'-bisphosphates, and a convenient method for the preparation of these compounds is available[43]. However, if longer donors are used, a removable blocking group on the 3' terminus of the donor is useful in preventing donor self-addition or cyclization. RNA ligase then joins the 5'-phosphorylated, 3'-blocked donor to an acceptor with 3'- and 5'-terminal hydroxyls to form a unique product with a 5'-hydroxyl and a 3'-blocked termini (Fig. I). Removal of the blocking group makes the product a new acceptor. Phosphorylation of the 5'-hydroxyl of the product with ATP and T4 polynucleotide kinase makes the product a new donor. Thus, synthesis with RNA ligase can proceed in either direction. Sequential

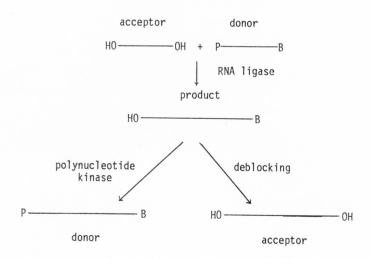

Fig. 1. Strategy of oligonucleotide synthesis. B, removable 3'-terminal blocking group.

ligations then lead to the synthesis of the final RNA fragment. It is general-ly advantageous to use a highly branched pathway by linking smaller fragments to make larger ones and then joining the larger ones to make the final sequence. Sequential addition of short fragments to one end of the chain would require very large quantities of the starting oligomers and would necessitate more difficult separations of products and unreacted substrates.

The choice of a 3'-terminal blocking group for the donor is important. The most obvious choice is a phosphate. Not only can terminal phosphates be easily removed with phosphomonoesterases, but they are also present on oligonu-cleotides prepared by ribonucleases or by the addition of a pNp with RNA ligase. However, 3'-phosphates have two shortcomings as donor blocking groups. First, the phosphorylation of the 5' terminus of donors is generally carried out by polynucleotide kinase and this enzyme has an endogenous 3'-phosphatase activi-ty that can remove the blocking group[26]; fortunately, the availability of a polynucleotide kinase produced by the Pset 1 mutant of T4 provides an enzyme which lacks the troublesome phosphatase activity[66]. The second shortcoming of 3'-phosphates as donor blocking groups is more serious. In several situations it has been demonstrated that 3'-phosphorylated termini are preferred locations for the reversal of the RNA ligase reaction[36], (see section IIIC). Through the reverse reaction, not only is the donor or product reduced in size by one nucleotide, but, once the 3'-blocking group is removed, forward reaction steps can lead to additional products which often accummulate in large amounts. Although reversal of ligase can, in principle, occur at any internucleotide bond, it is striking that under normal forward reaction conditions substantial reversal occurs only at 3'-phosphorylated termini. The rate of reversal is not high enough to prohibit the use of all 3'-phosphorylated donors, but it is not yet possible to predict when reversal will be a serious problem. One strategy to overcome the reverse reaction is to remove AMP enzymatically from the reaction mixture during the reaction. Another, more conservative approach is to discontinue the use of phosphates as 3'-blocking groups. Several other blocking groups have been shown to be effective in RNA ligase reactions, including the acid-labile O-(α-methoxyethyl)[30] and ethoxymethylidene[67] groups and the photo-labile o-nitrobenzyl group[33]. If the donor itself is a poor acceptor, a 3'-blocking group can be omitted entirely and an excess of acceptor can be included in the reaction to ensure the desired product[68].

RNA ligase has been used in the synthesis of several large RNA fragments. Ohtsuka, Ikehara, and colleagues have combined organic and enzymatic methods to complete the synthesis of the 76-nucleotide $tRNA_f^{Met}$ of $\underline{E. \ coli}$[58,67,69,91]. Reactions were generally carried out with approximately 0.1 mM oligomer concentrations and RNA ligase at 100 units/ml and the yields varied from 5% to 60%. Although aminoacylation activity of the $tRNA_f^{Met}$ has been demonstrated, only relatively small amounts of tRNA have thus far been prepared, and as a result, biochemical characterization is quite limited. The Nucleic Acids Collaboration Group of the Academia Sinica has prepared the 41-nucleotide 3' half of yeast $tRNA^{Ala}$ primarily by enzymatic methods[70]. Much higher ligase yields (40-90%) were observed in this case. Neilson and colleagues have used a combination of organic and enzymatic techniques to prepare several decanucleotide fragments of a mRNA intercistronic region and to demonstrate their activity in binding to ribosomes[37,71]. Finally, using entirely enzymatic methods, Studencki[65] has prepared several 13-nucleotide tRNA anticodon loops ($C_nGA_mG_3$) and Krug $\underline{et \ al.}$[72] have prepared a 21-nucleotide fragment of R17 RNA. Both kinds of fragments have been shown to have biologic activity. One of the anticodon loops can bind to 30S ribosomes in the presence of poly (U) and the R17 fragment binds R17 coat protein with the same affinity as a similar naturally isolated RNA fragment. Several other groups have prepared short oligomers with RNA ligase for various applications[73,74].

Whereas the above syntheses were all directed toward the preparation of biologically active RNA molecules for structure-function studies, another

potentially important use of RNA ligase is to provide a rapid method of producing oligonucleotide hybridization probes for molecular biologic applications.
Given the availability of suitable short fragments, RNA ligase can be used to
rapidly join them to produce small amounts of radiolabeled oligomers in the
size range 8-15. Sequences deduced from amino acid sequences or conceived
through other considerations can be prepared and used to hybridize to mixtures
of DNA or mRNA molecules and identify the molecule of interest. Because the
free energy of formation of the G-U pair is nearly zero for RNA-RNA and RNA-DNA
interactions, whereas that of the dG-dT pair is fairly positive for DNA-DNA
interaction, oligoribonucleotide probes have some advantage over oligodeoxyribonucleotide probes. In the only example to date, Seliger et al.[64] have
prepared a 15-residue fragment present in a viroid RNA sequence.

B. Oligodeoxyribonucleotide Synthesis

The combination of excellent organic synthetic technology and the poor
efficiency of deoxynucleotide acceptors renders RNA ligase generally less
useful for this application than for oligoribonucleotide synthesis. Nevertheless, the large number of applications for oligodeoxynucleotides in molecular
biology[75] and the relatively small amount of oligomer generally required make
RNA ligase a feasible adjunct in their synthesis. Furthermore, the wide
variety of nucleoside 3', 5'-bisphosphates that can be used as donors should
allow convenient preparation of deoxyoligomers containing modified nucleotides. One application illustrative of the particular synthetic capabilities
of RNA ligase is the synthesis of fragments of restriction enzyme sites that
contain a modified nucleotide in the recognition region[35].

Conditions for deoxynucleotide additions have been carefully optimized[18,25,38,52]. Although relatively large amounts of enzyme and long incubation times are required, reaction yields of 20-90% were obtained, depending
on the particular donor and acceptor pair[35]. Synthetic strategies involving
multiple ligations of DNA oligomers are similar to those for RNA oligomers.
The use of phosphates as 3'-blocking groups again introduces the possibility
of the reverse reaction. Secondary structures of fragments can be disrupted
with ribonuclease A to improve yields[35,38].

C. Reactions with RNA Molecules

RNA ligase promises to be exceptionally useful for understanding the relation of the structure of an RNA molecule to its biologic function. RNA mole-

cules with systematic changes in their sequence can be prepared and assayed for biologic function. Hypotheses concerning the importance of particular nucleotides in the folding of RNA or the "recognition" of RNA by a protein can be tested by introducing appropriate modifications. The most efficient way to prepare a variety of RNA molecules altered in a particular position in the sequence is to avoid synthesizing the whole molecule, but rather to use as much as possible of the natural RNA molecule in the synthetic scheme. Thus, efforts are directed toward the regions of the molecule that one wishes to change. RNA molecules and fragments of RNA molecules are excellent substrates for RNA ligase. The 3' termini are essentially always reactive and the 5' termini are reactive if secondary structure in the region has been eliminated.

Perhaps the most common application of RNA ligase to date has been in in vitro labeling of RNA. The 3' terminus of an RNA molecule is used as an acceptor and a radiolabeled nucleoside 3', 5'-bisphosphate as a donor. A review of this method has appeared recently[56]. Generally, pCp is used as the donor because it is the most effective substrate. Although radioactivity has usually been introduced into pCp with a [5'-^{32}P]phosphate, carbon-14 or tritium could be introduced in the cytosine, or iodine-125 could be introduced as 5-iodo-pCp[40]. To obtain RNA with a [^{32}P]phosphate as the terminal phosphate, [3'-β^{32}P]pGpp has also been used as a donor[42]. RNA molecules labeled at or near the 3' terminus have a wide variety of uses analogous to those of in vitro 5'-^{32}P-labeled RNA obtained with polynucleotide kinase. The recently developed chemical[76] and enzymatic[77] RNA sequencing methods require end-labeled RNA. Thus, RNA labeled with RNA ligase allows determination of the sequence of the RNA in the neighborhood of the 3' terminus. As examples, the precise distribution of the poly A chain lengths at the 3' termini of mRNAs[78] and the exact composition of the 3' termini of in vitro transcription products[79,80] can be easily determined. An alternative sequencing scheme involving partial hydrolysis of RNA before labeling with RNA ligase could prove particularly useful for sequencing RNAs that contain modified nucleotides[81]. Information about the secondary and tertiary structure in the neighborhood of the 3' termini of RNA can be obtained by partial digestion of 3'-^{32}P-labeled RNA with single-strand specific ribonucleases under conditions in which the structure is intact[82]. Even more detailed structural information can be obtained using the chemical sequencing reactions on 3'-^{32}P-labeled RNA in a similar fashion[83].

A powerful use for T4 RNA ligase is in the study of RNA-processing enzymes. N. Pace and colleagues have used RNA ligase to extend the termini of B. subtilis 5S RNA to make synthetic substrates for RNase M5, the enzyme that cleaves the 5S RNA precursor endonucleolytically in two positions to form mature 5S RNA. 5S RNA with a short oligonucleotide attached to its 5' terminus was an excellent substrates for RNase M5[45]. This provides a convenient precipitation assay for the processing enzyme. More recent experiments using RNA ligase to modify both the 3' and the 5' termini of 5S RNA have provided an even more detailed understanding of the substrate specificity of RNase M5[46]. We expect that synthetic substrates for processing enzymes prepared with RNA ligase will be useful in the identification and purification of other processing enzymes. The general difficulty of obtaining sufficient amounts of RNA precursors is the primary reason that more processing enzymes have not been isolated. Because it is characteristic of most processing enzymes not to recognize the structural elements outside the mature RNA, any small oligonucleotide attached to the terminus of the RNA is likely to allow it to serve as a substrate. For example, E. coli tRNA$_f^{Met}$ with an oligonucleotide attached to the 5' terminus serves as a substrate for E. coli RNase P, the enzyme responsible for tRNA maturation[84]. Furthermore, because it is difficult to accurately determine the concentration of naturally isolated tRNA precursors synthetically produced precursors should be valuable in obtaining precise kinetic measurements for processing enzymes.

T4 RNA ligase has been used to alter the 3' terminus of tRNA to examine functional events at that position. The terminal adenosine of tRNA can be removed by periodate oxidation, β-elimination, and alkaline phosphatase treatment and the resulting tRNA-CC can be used as an acceptor. In one study[85], the terminal adenosine of yeast tRNA^{Phe} was replaced with lin-benzoadenosine, a nucleoside that could not be inserted into that positon with tRNA nucleotidyl transferase. Hecht and co-workers[32] have prepared A-5'pp5'-A with a blocked amino acid on the 3'-hydroxyl of one of the adenosines. The blocked aminoacyl-AMP moiety was added to tRNA-CC using RNA ligase in the ATP-independent reaction. The blocking group could then be removed to result in tRNAs "chemically aminoacylated" with an incorrect amino acid or an amino acid analogue. This general procedure for replacing the terminal nucleotide of tRNA has been extended by Johnson and Bock[86] to alter the fourth nucleotide in from the 3' terminus of yeast tRNA^{Phe}. Four rounds of periodate oxidation followed by addition of pGp with RNA ligase and repair with tRNA nucleotidyl transferase resulted in a tRNA^{Phe} with a G in position 73, instead of an A. Because many RNA molecules are good acceptors with RNA ligase, nucleotide substitution experiments similar to those described above should be useful for learning more about the function of the 3' terminus of RNA molecules.

RNA ligase can also be used to make modifications at internal positions in an RNA chain. This was first shown by Kaufmann and Littauer[10], who were able to cleave yeast tRNA[Phe] at nucleotide Y-37 in the anticodon loop, remove the nucleotide, and in low yield, reseal the two half-molecules to prepare a tRNA[Phe] that lacked Y-37. More recently, a procedure has been developed to replace four nucleotides in the anticodon loop of yeast tRNA[Phe] with any oligonucleotide sequence in high yield[87]. Residues 34-37 are removed from yeast tRNA[Phe] by a chemical chain scission and a partial ribonuclease digestion resulting in two half-molecules that are base-paired to one another. RNA ligase is used first to add an oligomer to the 3' half-molecule and then to join the half-molecules. More than 30 different oligomers have been successfully inserted in this manner for functional studies. It is clear that these methods could be used to replace nucleotides in internal positions of other RNA molecules provided that several criteria can be met. First, a procedure must be available to make one or more specific cleavages in the RNA molecule to give defined fragments. A method employing RNase H and complementary deoxyoligo-nucleotides seems particularly promising in this regard[88]. Second, the normal termini of the RNA must be blocked or inaccessible to RNA ligase so that reactions occur only at the internal position. Third, there should be little secondary structure in the position of modification so that the enzyme can work efficiently. If these criteria are satisfied, any nucleotide sequence includ-ing those which contain modified residues can be inserted into the RNA. This should allow detailed studies of structure-function relationships in RNA.

D. Other Applications

RNA ligase can be used to prepare circular single-stranded oligonucleo-tides for functional studies. Kozak[89] prepared a synthetic circular mRNA for testing in a eukaryotic protein synthesis system. deHaseth and Uhlenbeck[90] prepared cyclic oligoadenylates to demonstrate a circular RNA binding domain for host-factor protein.

A potentially important use of the ATP-independent reaction of RNA ligase is to attach covalently nonnucleotide compounds to the 3' terminus of RNA. Both the biotin and fluorescent derivatives of ADP have been prepared (Table 1) and attached to the 3' terminus of yeast tRNA[Phe][35]. These will be used to detect DNA-RNA hybrids and thus offer an alternative to the use of radio-activity.

VII. CONCLUSIONS

T4 RNA ligase is an enzyme at a relatively early stage of development. It was discovered in 1972 and several years were required to purify the protein, determine its substrate specificity, and study its enzymological properties. Applications of RNA ligase are therefore all relatively recent and they are comparatively straightforward. However, as the ease of use and the broad substrate specificity of this enzyme become more familiar to nucleic acid chemists one can expect a variety of subtle applications in nucleic acid synthesis and modification.

ACKNOWLEDGMENTS

The research described from the authors' laboratories was supported by NIH grants GM 19442 and 25621 to R. I. G. and GM 19059 to O. C. U. We thank the many colleagues who have communicated their unpublished results to us.

REFERENCES

1. Leis, J., Silber, R., Malathi, V.G. and Hurwitz, J., Advances in the Biosciences, Vol. 8, G. Raspe, (1972), Ed., Pergamon Press, Elmsford, New York, p. 117.
2. Silber, R., Malathi, V.G. and Hurwitz, J. (1972) Proc. Natl. Acad. Sci. U.S.A., $\underline{69}$, 3009.
3. Snopek, T.J. Wood, W.B., Conley, M.P., Chen, P. and Cozzarelli, N.R.(1977) Proc. Natl. Acad. Sci. U.S.A., $\underline{74}$, 3355.
4. David, M., Vekstein, R. and Kaufman, G. (1979) Proc. Natl. Acad. Sci. U.S.A., $\underline{76}$, 5430.
5. Runnels, J., Soltis, D., Hey, T. and Snyder, L. in preparation.
6. Linne, T., Oberg, B. and Philipson, L. (1974) Eur. J. Biochem., $\underline{42}$, 157.
7. Cranston, J.W., Silber, R., Malathi, V.G. and Hurwitz, J. (1974) J. Biol. Chem., $\underline{249}$, 7447.
8. Bedows, E., Wachsman, J.T. and Gumport, R.I. 1975 Biochem. Biophys. Res. Commun., $\underline{67}$, 1100.
9. Ogden, R.C., Knapp, G., Peebles, C., Kang, H.S., Beckmann, J.S., Johnson, P.F., Fuhrmann, S.A. and Abelson, J.N. (1980), Transfer RNA: Biological Aspects, Soll, D., Abelson, J.N. and Schimmel, P.R. Eds., Cold Spring Harbor Laboratory, Cold Spring Harbor, New York, p. 173.
10. Kaufmann, G. and Littauer, U.Z. (1974) Proc. Natl. Acad. Sci. U.S.A., $\underline{71}$, 3741.
11. Walker, G.C., Uhlenbeck, O.C., Bedows, E. and Gumport, R.I. (1975) Proc. Natl. Acad. Sci. USA, $\underline{72}$, 122.
12. Kaufmann, G. and Kallenbach, N.R. (1975) Nature (Lond.), $\underline{254}$, 452.
13. Kornberg, A. (1980) "DNA Replication," W.E. Freeman, San Francisco, California, p. 261.
14. Uhlenbeck, O.C. and Gumport, R.I. (1981) The Enzymes Vol. 15, in preparation.
15. Last, J.A. and Anderson W.F. (1976) Arch. Biochem. Biophys., $\underline{174}$, 167.

16. Higgins, N.P., Geballe, A.P., Snopek, T.J., Sugino, A. and Cozzarelli, N.R. (1977) Nucleic Acids Res., 4, 3175.
17. Moseman McCoy, M.I., Lubben, T.H. and Gumport, R.I. (1979) Biochim. Biophys. Acta, 562, 149.
18. Gumport, R.I., Manthey, A.E., Baez, J.A., Moseman McCoy, M.I. and Hinton, D.M. (1981) PRC-FRG Joint Symposium on Nucleic Acids and Proteins, Shanghai, in press.
19. Snopek, T.J., Sugino, A.. Agarwal, K. and Cozzarelli, N.R. (1976) Biochem. Biophys. Res. Commun., 68, 417.
20. Sugino, A., Snopek, T.J. and Cozzarelli, N.R. (1977) J. Biol. Chem., 252, 1732.
21. Kaufmann, G., Klein, T. and Littauer, U.Z. (1974) FEBS Lett., 46, 271.
22. Sugiura, M., Suzuki, M., Ohtsuka, E., Nishikawa, S., Uemura, H and Ikehara, M. (1979) FEBS Lett., 97, 73.
23. Vasilenko, S.K., Veniyaminova, A.G., Yamkovoy, V.I. and Maiyorov, V.I. (1979) Bioorg. Khim., 5, 621.
24. Kikuchi, Y., Hishinuma, F. and Sakaguchi, K. (1978) Proc. Natl. Acad. Sci. USA, 75, 1270.
25. Hinton, D.M., Baez, J.A. and Gumport, R.I. (1978) Biochemistry, 17, 5091.
26. Cameron, V. and Uhlenbeck, O.C. (1977) Biochemistry, 16, 5120.
27. Baez, J.A. and Gumport, R.I., unpublished observations.
28. Hu, M., Wang, A., Hua, H., Chen, Y. and Xue, C. (1980) Sci. Reports Beijing Univ., 4, in press.
29. Juodka, B.A., Markuckas, A.Y., Snechkute, M.A., Zilinskiene, V.J. and Drigin, Y.F. (1980) Bioorg. Khim., in press.
30. Sninsky, J.J, Last, J.A., and Gilham, P.T. (1976) Nucleic Acids Res., 3, 3157.
31. England, T.E., Gumport, R.I. and Uhlenbeck, O.C. (1977) Proc. Natl. Acad. Sci. USA, 74, 4839.
32. Hecht, S.M., Alford, B.L., Kuroda, Y. and Kitano, S. (1978) J. Biol. Chem., 253, 4517.
33. Ohtsuka, E., Uemura, H., Doi, T., Miyake, T., Nishikawa, S. and Ikehara, M. (1979) Nucleic Acids Res., 6, 443.
34. Ohtsuka, E., Miyake, T., Nagao, K., Uemura, H., Nishikawa, S., Sugiura, M. and Ikehara, M. (1980) Nucleic Acids Res., 8, 601.
35. Gumport, R.I., Hinton, D.M., Pyle, V.S. and Richardson, R.W. (1980) Nucleic Acids Res., Symposium Series No. 7, 167.
36. Krug, M. and Uhlenbeck, O.C., in preparation.
37. Neilson, T., Kofoid, E.C. and Ganoza, M.C. (1980) Nucleic Acids Res., Symposium Series No. 7, 313.
38. Moseman McCoy, M.I. and Gumport, R.I. (1980) Biochemistry, 19, 635.
39. Brennan, C.A. and Gumport, R.I. unpublished observations.
40. Bruce, A.G. and Uhlenbeck, O.C. (1978) Nucleic Acids Res., 5, 3665.
41. England, T.E. and Uhlenbeck, O.C. (1978) Biochemistry, 17, 2069.
42. Simoncsits, A. (1980) Nucleic Acids Res., 8, 4111.
43. Barrio, J.R., Barrio, M.G., Leonard, N.J., England, T.E. and Uhlenbeck, O.C. (1978) Biochemistry, 17, 2077.
44. Kikuchi, Y., Hirai, K. and Sakaguchi, K. (1979) Nucleic Acids Res., Symposium Series No. 6, s189.
45. Meyhack, B., Pace, B., Uhlenbeck, O.C. and Pace, N. (1978) Proc. Natl. Acad. Sci. USA, 75, 3045.
46. Stahl, D.A., Meyhack, B. and Pace, N.R. (1980) Proc. Natl. Acad. Sci. USA, 77, 5644.
47. England, T.E. and Uhlenbeck, O.C., unpublished observations.
48. Cameron, V. and Uhlenbeck, O.C., unpublished observations.

344

49. Higgins, N.P., Geballe, A.P., and Cozzarelli, N.R. (1979) Nucleic Acids Res., 6, 1013.
50. England, T.E., Uhlenbeck, O.C., Miwa, M. and Sugimura, T., unpublished observations.
51. Collaboration Group of Nucleic Acid Synthesis, PRC-FRG Joint Symposium on Nucleic Acids and Proteins, Shanghai, (1981), in press.
52. Hinton, D.M. and Gumport, R.I., (1979) Nucleic Acids Res., 7, 453.
53. England, T.E. and Uhlenbeck, O.C. (1978) Nature (Lond.), 275, 560.
54. Ohtsuka, E., Nishikawa, S., Sugiura, M. and Ikehara, M. (1976) Nucleic Acids Res., 3, 1613.
55. Uhlenbeck, O.C. and Cameron, V. (1977) Nucleic Acids Res., 4, 85.
56. England, T.E., Bruce, A.G. and Uhlenbeck, O.C. (1980) Methods in Enzymology, 65, 65.
57. Richardson, R.W. and Gumport, R.I., unpublished observations.
58. Ohtsuka, E., Nishikawa, S., Fukumoto, R., Uemura, H., Tanaka, T., Nakagawa, E., Miyake, T. and Ikehara, M. (1980) Eur. J. Biochem., 105, 481.
59. Cashion, P., Javed, A., Sathe, G. and Ali, G. (1980) Nucleic Acids Res., Symposium Series No. 7, 173.
60. Abelson, J.N. (1980) Science, 209, 1319.
61. Koester, H. Ed., (1980) Nucleic Acids Res., Symposium Series No. 7, "Nucleic Acids Synthesis: Applications to Molecular Biology and Genetic Engineering", Information Retrieval, Ltd., London.
62. Thatch, R.E. "Procedures in Nucleic Acid Research", Vol I., Cantoni, G.L. and Davis, P.R., Eds., (1966) Harper and Row, New York, p. 520.
63. Khabarova, M.I., Smolyaninova, O.A., Bagdonas, A.S., Kovalenko, M.I. and Zhenodarova, S.M. (1978) Bioorg. Khim., 4, 740.
64. Seliger, H., Haas, B., Holupirek, M., Knable, T., Todling, G. and Phillipp, M. (1980) Nucleic Acids Res., Symposium Series No. 7, 191.
65. Studencki, A. and Uhlenbeck, O.C., in preparation.
66. Cameron, V., Soltis, D. and Uhlenbeck, O.C. (1978) Nucleic Acids Res., 5, 825.
67. Ohtsuka, E., Nishikawa, S., Markham. A.F., Tanaka, S., Miyake, T., Wakabayashi, T., Ikehara, M. and Sugiura, M. (1978) Biochemistry, 17, 4894.
68. Collaboration Group of Nucleic Acids Synthesis. (1978) Scientia Sinica, 21, 687.
69. Ohtsuka, E., Markham, A.F., Tanaka, S., Miyake, T., Nakagawa, T., Wakabayshi, T., Taniyama, Y., Fujiyama, K., Nishikawa, S., Fukumoto, R., Uemura, H., Doi, T., Tokunaga, T. and Ikehara, M. (1980) Nucleic Acids Res., Symposium Series No. 7, 335.
70. Wang, T.P. (1980) Nucleic Acids Res., Symposium Series No. 7, 325.
71. Neilson, T., Gregoire, R.J., Fraser, A.R., Kofoid, E.C. and Ganoza, M.C. (1979) Eur. J. Biochem., 99, 429.
72. Krug, M., deHaseth, P. and Uhlenbeck, O.C., in preparation.
73. Kikuchi, Y. and Sakaguchi, K. (1978) Nucleic Acids Res., 5, 591.
74. Zhenodarova, S.M., Klyagina, V.P., Smolyaninova, O.A., Soboleva, I.A. and Khabarova, M.I. (1978) Nucleic Acids Res., Special Symposium Series No. 4, s137.
75. Itakura, K. and Riggs, A.D. (1980) Science, 209, 1401.
76. Peattie, D.A. (1979) Proc. Natl. Acad. Sci. USA 76, 1760.
77. Lockard, R.E., Alzner-Deward, B., Heckman, J., MacGee, J., Tabor, M.W. and RajBhandary, U.L. (1978) Nucleic Acids Res., 5, 37.
78. Ahlquist, P. and Kaesberg, P. (1979) Nucleic Acids Res., 7, 1195.
79. Dunn, J.J. and Studier, F.W. (1980) Nucleic Acids Res., 8, 2219.
80. Calva, E. and Burgess, R.P. (1980) J. Biol. Chem., 255, 1017.
81. Gupta, R.C. and Randerath, D. (1979) Nucleic Acids Res., 6, 3443.

82. Vournakis, J., Pavlakis, G., Worst, R., Mitra, T. and Finn, M., this Vol.
83. Peattie, D.A. and Gilbert, W. (1980) Proc. Natl. Acad. Sci. USA, 77, 4679.
84. Cameron, V., unpublished results.
85. Hinterberger, M., Ph.D. Thesis (1980) Biochemistry Department, University of Illinois, Urbana.
86. Johnson, B., Ph.D. Thesis (1979) Biochemistry Department, University of Wisconsin, Madison.
87. Bruce, A.G. and Uhlenbeck, O.C., in preparation.
88. Donis-Keller, H. (1979) Nucleic Acids Res., 7, 179.
89. Kozak, M. (1979) Nature (Lond.), 280, 82.
90. deHaseth, P. and Uhlenbeck, O.C. (1980) Biochemistry, 19, 6138.
91. Ohtsuka, E., Doi, T., Uemura, H., Taniyama, Y. and Ikehara, M. (1980) Nucleic Acids Res., 8, 3909.

DISSECTION OF CHROMATIN STRUCTURE WITH NUCLEASES

Robert T. Simpson

Developmental Biochemistry Section
National Institute of Arthritis, Metabolism and Digestive Diseases
National Institutes of Health
Bethesda, Maryland 20205

I. INTRODUCTION

The last decade has seen a great increase in our understanding of the structure of chromatin - the complex of DNA, histones, and other proteins that is the packaged form of the genetic message in eukaryotic cells. Every step along the path to this increased understanding has been marked by the use of nucleases as probes for the structure of chromatin.

Before 1970, most envisioned chromatin as a compact form of DNA, likely to be superhelical and shrouded in histones; the histones thus shielded the nucleic acid from solution components and served as nonspecific repressors to preclude transcription of unneeded gene sequences. In 1971, Clark and Felsenfeld, using micrococcal nuclease (staphylococcal nuclease or nucleate 3'-oligonucleotido-hydrolase, E.C. 3.1.4.7), provided one of the first major pieces of evidence to cast doubt on this picture. They showed that isolated chromatin was digested rather readily by micrococcal nuclease, to a consistent end point where about 50% of the DNA was acid-soluble and 50% remained as fragments with a weight-average size corresponding to a length of about 100 base pairs (bp) of DNA.[1] Their interpretation of these data, that half the DNA interacts with protein and half is essentially free of protein, is now understood to be quantitatively incorrect, but it remains qualitatively valid.

Some 2 yr later, a major advance (unheralded at the time) was made by Hewish and Burgoyne.[2] Rodent liver nuclei contain an endogenous endonuclease that is activated by calcium and magnesium. This enzyme degrades chromatin DNA in situ to a series of fragments that are multiples of a common size, about 150,000-200,000 daltons. The interpretation of this finding by Burgoyne et al.[3] - "the bands appeared to be the result of a non-random spacing of proteins on the DNA such that a non-random series of sites susceptible to DNAase action was created" - was a far-sighted projection of our current view of the first level of organi-zation of chromatin. Noll[4] extended these observations and showed that other enzymes, specifically micrococcal nuclease, could cut chromatin DNA into frag-ments that were multiples of about 200 bp, thereby establishing the generality of the organizational pattern revealed in the earlier studies of Hewish and Burgoyne.

Two types of data not derived from nuclease digestion aided in the synthesis of the concept of a repetitive, subunit organization for chromatin. Olins and Olins[5] and Woodcock[6] described the morphology of chromatin gently lysed on an electron microscope grid as one in which beads, about 10 nm in diameter, were connected in tandem array by threads about the diameter of protein-free DNA. Some 17 yr earlier, Cruft, Mauritzen and Stedman had described the composition and properties of β-histone; under nonaggregating conditions, this "histone"

had a molecular weight of 57,000.[7] Its amino acid compostition was nearly
exactly that expected for an equal mixture of H3 and H4,[7] which were not recognized as two of the four inner histones at the time. Kornberg and Thomas[8]
used histones prepared by the salt-extraction method of Van der Westhuyzen
and Von Holt[9] and demonstrated the presence of a solution tetramer of H3
and H4, as well as association of H2A and H2B, probably into a dimer. The
H3 + H4 association was independently described by Roark et al.[10] Strong
interactions between these pairs of histones had been demonstrated in other
physical studies of Isenberg's group.[11]

Synthesis of these observations by Kornberg[12] led to the formal postu-
late that chromatin structure, at the first level of organization, involves
specific complexes of 180-200 bp of DNA with octamers of the four smaller histones, two each of H2A, H2B, H3, and H4. Rapidly, Sahasrabuddhe and Van Holde[13]
Noll,[14] and Weintraub and Van Lente[15] described the nucleoprotein equivalents
of the DNA fragments, demonstrating that discrete, compact DNA-histone complexes were released from nuclei by micrococcal nuclease.

One of the first indications of the structural organization of the chromatin subunit, or nucleosome, came from the observation of Noll[15] that DNAse
I (deoxyribonucleate 5'-oligonucleotidohydrolase, E.C.3.1.4.5) cut chromatin
DNA in a fashion that led to DNA fragments that were multiples of about 10
bases long. Noting that this is the helical repeat of DNA, Noll suggested
that the DNA might be wrapped on the outside of histones in the nucleosome;
the digestion pattern would then reflect the periodic availability of a given
strand of DNA when the cylindric molecule is laid on a surface.[15] Wrapping
of DNA around the histones was also suggested by Griffith's observation[16]
that the number of nucleosomes in the SV-40 minichromosome was nearly equal to
the number of superhelical turns in the DNA after removal of proteins. Using
a nuclease present in "untwisting extract" (now known to be DNA topoisomerase
I) to treat a reassociated complex of SV-40 DNA and histones, Germond et al.[17]
clearly demonstrated the equivalence between the number of nucleosomes on a
closed circular DNA and the number of superhelical turns in that DNA after
protein removal. Wrapping of the DNA around the histone octamer was also
suggested by several physical studies.

More detailed studies of the kinetics of digestion of chromatin by different nucleases and examination of the sizes of the products produced by micrococcal nuclease, DNAse I, and the endogenous liver endonuclease led to a model
for the chromatin subunit that included a tandem array of core particles, containing about 140 bp of DNA, with interspersed linker DNA regions,[18,19] in

350

contrast with the originally proposed association of the inner histones with the full length of the DNA repeat in chromatin.[12]

Precise localization of the sites within the nucleosome core particle (145 bp of DNA plus the histone octamer) at which nuclease cutting occurred was made possible by development of end-label mapping for the intact particle in our laboratory[20] and by Noll[21] and Lutter.[22] Conceptually analogous to the Maxam and Gilbert[23] DNA sequencing method, cutting-site mapping with DNAse I showed three important features of nucleosome DNA: cutting sites were spaced at 10-base intervals from the 5' end of the DNA and varied quite widely in their susceptibility to the nuclease, the pattern of cutting susceptibilities was essentially symmetric around the center of the nucleosome core-particle DNA at 70 bp from either end, and cutting sites with similar suscetibilities were spaced 80 bp apart along the length of the DNA. These solution chemical data and early x-ray diffraction data on crystals of core particles provided a large portion of the experimental information that generated the currently accepted model of the nucleosome core particle: an inner octamer of the histones sur-rounded by 1.75 turns of DNA in a shallow helical path, leading to a squat cylindric structure about 5 nm high and 11 nm in diameter.[24]

Nucleases have recently been used to probe features of chromatin segments that differ in function from the bulk of the nucleoprotein. Hybridization studies using cDNAs to mRNAs from a given cell type showed that actively transcribed genes were present in chromatin in nucleosomes.[25] However, studies of Weintraub and Groudine[26] and Garel and Axel[27] dramatically demonstrated a structural difference between such active chromatin and the structure of the major portion of the nuclear material. DNAse I, which prefer-entially cuts within nucleosomes, digests active gene sequences much more ra-pidly than inactive ones; this suggests that the internal structure of actively transcribed nucleosomes differs from that of nucleosomes whose DNA is inactive as a template for RNA polymerase. More recently, micrococcal nuclease has been used to excise active gene sequences preferentially from chromatin;[28,29] earlier methods used DNAse II (deoxyribonucleate 3'-oligonucleotidohydrolase, E.C.3.1.4.6.) for similar purposes.[30]

Although perhaps biased, this brief survey clearly indicates the important role that nuclease dissection of chromatin structure has played in the deve-lopment of the current paradigm of this complex entity. It seems obvious that the use of these enzymes (and others) will continue to be a major experimental approach to the unraveling of the structure-function relationships of chromo-somes. I attempt to provide here an applications "handbook" for the use of several nucleases to examine chromatin at different levels of organization. For each case, I describe how the enzyme is thought to act in the context of current ideas of chromatin structure, outline experimental methods, illustrate the types of results that have been obtained, and provide cautions in experimental applications and data interpretation. This presentation is not intended to be either a comprehensive review of chromatin structure (such reviews have been offered recently[31,32]) or a laboratory methods book; rather, I hope that it will serve as a guide in the use of the several enzymes in dissection of chromatin. References are cited selectively, and the bibliography is neither a full representation of the chromatin-nuclease literature nor an acknowledgment of all the research groups that have contributed in this subject.

II. PREPARATION OF CHROMATIN SUBUNITS WITH MICROCOCCAL NUCLEASE

A. Background

Chromatin DNA is organized as a series of tandemly repeated subunit structures, or nucleosomes,[33] along the length of the nucleic acid. Each structure consists of a core particle - a 145 bp length of DNA associated with an octamer of the four smaller histones - and a variable length of linker or bridging DNA between core particles.[18,19] Histone H1 is bound outside the core particle; a particle containing an additional 20 bp of DNA and one molecule of H1 or H5 is the chromatosome.[38] Micrococcal nuclease preferentially cuts chromatin DNA in the linker region. Its secondary preference is for either further endonucleolytic cuts in this region or exonucleolytic degradation of the linker DNA, which leads to formation of chromatosomes. Further digestion under H1 leads to release of this histone and to trimming, which forms core particles. Micrococcal nuclease also cuts within the core particle; this leads to a well-defined series of small DNA fragments,[39] which can apparently remain associated with histones, at least to some stage of the degradation. All these processes occur together; therefore, preparation of any individual level of chromosomal subunit must be optimized individually, and yields of material are generally much lower than they would be if the digestions were rigorously sequential and synchronous for all the nucleosomes in a given sample.

B. Experimental Methods

Nuclei are prepared in isotonic sucrose solutions; nonionic detergents are used to remove outer membranes and adherent cytoplasmic tags and divalent cations to stabilize nuclear structure.[40] For tissues with active endogenous endonucleases, polyamines can substitute for divalent cations, to prevent pre-mature digestion.[2] Digestions are usually performed at a pH of 8 - a compromise between chromatin stability and the pH optimum for micrococcal nuclease. Isotonic sucrose is generally present in the digestion buffer, which also contains Tris/Cl at 10 mM and $CaCl_2$ at 1-10 mM. Digestions are at 37°C and are terminated by cooling samples to 0°C and adding EDTA to twice the Ca^{2+} concentration.

Efficiency in the use of time and materials demands that experiments be run on a pilot scale with analysis by gel electrophoresis of DNA samples at different degrees of digestion. Both enzyme concentrations and time of digestion should be varied; because of the complexity of the digestion process, equal products of enzyme concentration and time may not yield identical results. Generally, the combination of more enzyme and a shorter time optimizes the production of the larger chromosomal particles, nucleosomes and chromatosomes; the converse is often true for core particle production. After removal of proteins by proteinase treatment and/or phenol-SDS extraction, DNA size is analyzed by electrophoresis on 1% agarose, 2.5% polyacrylamide, or composite gels for determination of nucleosome repeat length or on 5% polyacrylamide gels for determination of optimal conditions for production of monomeric subunits. A Hae III or Hinc II digest of ØX174 RF DNA provides convenient size standards up to about 1,000 bp.

For preparation of larger fragments and determination of the nucleosome repeat length, digestion at 0°C has been suggested;[37] the presumption is that the endonucleolytic activity of the nuclease depends less on temperature than does the exonucleolytic activity. Digestion in a medium of moderate ionic strength (0.1 M) has been suggested to reduce the rate of internal cutting of core particles.[41] For core particle preparation, on the basis of the background information above, brief digestion after removal of H1 seems, in theory, to be most expedient. A problem in this approach is that H1 extraction with NaCl requires concentrations of over 0.6 M; at this ionic strength, nucleosome sliding, leading to close-packed monomers, has been observed.[42,43] Use of the low-pH method for H1 removal[44] should surmount this problem and allow higher yield core particle preparations than are currently possible; to my knowledge, this approach for core particle preparation with quantitative measurements of yields has not been reported. Another method to improve yields of core particles involves digestion, fractionation, removal of H1 by salt extraction, and trimming of the mixture of core particles with variable-length tails of DNA by the use of either exonuclease III[45] or redigestion with micrococcal nuclease (P. Kunzler and L. W. Bergman, personal communication).

Lysis of nuclei to release nucleoprotein fragments is not achieved with equal facility for all nuclear types. After digestion with nuclease, monomeric and multimeric nucleosomes are released from Triton-washed HeLa cell nuclei by simple addition of EDTA. In contrast, chicken erythrocyte nuclei are not lysed by adding EDTA after digestion; they are pelleted and lyse by resuspension in 1 mM EDTA or, better, by overnight dialysis against the same. Sea urchin sperm nuclei are resistant to drastic treatment with chelating and reducing agents, mechanical disruption, and low ionic strength. We have been able to release chromatin fragments from these nuclei only by adjusting the sample to 0.9 M NaCl to dissociate the H1 equivalent from the digested chromatin. The most effective method for nuclear lysis must be determined separately for each nuclear type studied.

Preparative fractionation of the digests can be done by gel filtration on agarose matrices[19] or by sucrose gradient sedimentation. Isokinetic sucrose gradients (particle density, 1.51 g/cm^3 at 4°C) with a meniscus concentration of 5% (w/w) are convenient.[46] The ionic strength in the gradient affects the resolution obtained.[41] At low ionic strength (ca. 5 mM or less), nucleosomes, chromatosomes, core particles, and core particles with some internal nicking are are all soluble and will sediment in a broad monomeric peak. At 0.1 M NaCl, chromatosomes and particles with internal cutting are much less soluble; this leads to enhanced resolution of the gradients and a purer core particle preparation. At 0.6 M NaCl, H1 is removed, and all particles are again soluble; resolution of gradients at this ionic strength is decidedly inferior to that attained under either of the two previous conditions. Core particle preparations made at low ionic strength can be cleaned up by precipitation of 120 bp particles and chromatosomes with 0.1 M NaCl or KCl after isolation of the monomeric peak from a low ionic strength gradient.[47] When samples whose ionic strength differs from that in the gradient are used, we routinely layer the sample over a shelf of the same ionic strength, to allow partial separation of nucleosomes from molecules dissociated in the sample application buffer before the samples enter the altered environment in the gradient.

C. Results

Micrococcal nuclease digestion of chromatin has been used to characterize the first level of organization of DNA in numerous tissues and species. The biochemical repeat length for the nucleosome varies widely - about 165 bp for yeast;[48,49] 190-200 bp for a large number of tissues, summarized by Kornberg;[31] and 240-260 bp for sea urchin sperm.[50,51] Yet all of these tissues have a common core particle containing 145 bp of DNA associated with the four inner histones. The basic packaging of DNA in eukaryotes apparently is conserved at least as rigorously as the sequences of H3 and H4. Obviously, variability in nucleosome repeat length is related to differences in length of linker DNA between core particles. Differences in H1 are known for various tissues, also; the relationship of these two compositional variations is not understood.

Heterogeneity in the monomeric nucleosome population is even greater than the simple three species listed above. With high-resolution one- and two-dimensional gel electrophoretic analysis, many subspecies have been found that have various DNA lengths and various amounts of H1 and nonhistone proteins (see, for example and leading references, 52). Heterogeneity in chromatin at early stages of digestion would be expected to be the rule. For example, in chromatin with a repeat length of 195 bp, a monomer excised by cuts at the ends of its core-particle segment on either end would have 145 bp of DNA and no H1; one cut at the end of its nucleosome repeat length on either end would have 245 bp of DNA and one, two, or three molecules of H1, depending on the partitioning of the histone once its primary binding site had been partially cut; the spectrum of molecules should include everything between these two extremes.

The primary benefit of micrococcal nuclease digestion of chromatin has been the production of highly homogeneous nucleoprotein particles for physical study. In contrast with the heterogeneity of nuclei or chromatin, core particles prepared from a tissue poor in nonhistone proteins are a physically elegant combination of 145 bp of DNA and the octamer of H2A, H2B, H3 and H4. Intensive study of these particles by a variety of physical and chemical methods has led to a rather detailed understanding of the interactions of histones and DNA at this first level of packaging of the nucleic acid, as reviewed by McGhee and Felsenfeld.[32]

D. Precautions

In determination of repeat lengths, the size of the monomeric nucleosome DNA, even in short digestions, is somewhat less than the repeat length. The actual repeat length can be determined by extrapolation of data for various size fragments to zero time of digestion[18,19] or (probably more accurately) by measurement of differences in size between adjacent nucleosome multimeric DNA lengths.[37]

A potential problem not widely recognized is that micrococcal nuclease has marked compositional preferences for hydrolytic sites in DNA. As an example, in a study of semisynthetic chromatins formed from poly(dA-dT) and from poly(dG-dC), 100 times as much micrococcal nuclease was required for equivalent digestion of the G-C polymer as for the A-T polymeric complex with histones.[53] With random, native DNA from a cell, the effects of sequence variation will be randomized; this leads to a loss of resolution, but no serious misinterpretation of digestion patterns. In contrast, with sequence-defined DNA (satellite, cloned fragments, or small virus), if histone binding sites are phased on the DNA, sequence effects on rates of nuclease hydrolysis may be marked. Thus, one can envision a G-C rich region just outside the core particle segment, which would lead, because of its relative nuclease resistance, to an apparently stable core particle substantially longer than the 145 bp canonical core particle.

Micrococcal nuclease is catalytically inactive in the absence of calcium; apparently, however, it can still bind to its substrate DNA. Thus, it apparently cosediments partially with nucleosomes in sucrose gradients. This leads to rapid redigestion of isolated nucleosomes and is a particular problem when other enzymatic procedures are to be applied to isolated core particles, e.g., end labeling or digestion with other nucleases. Inclusion of EGTA in buffers during such procedures usually prevents the activity of any contaminating micrococcal nuclease from being a problem.

III. MAPPING NUCLEASE CUTTING SITES IN CHROMATIN PARTICLES

A. Background

DNA in the core particle is wrapped in a shallow helix, 1.75 turns of 80 bp circumference each, on the outside of a core of histones.[24] Digestion of nuclei, chromatin, or core particles with DNAse I,[15] DNAse II,[54] or micrococcal nuclease[39] leads to well-defined DNA fragments. Most clearly for DNAse I, these fragments are multiples of about 10 bases in length when examined by gel electrophoresis under denaturing conditions. The periodicity may reflect the disposition of nuclease-sensitive (or nuclease-recognition) sites on DNA when the double helix lies on the surface of the protein core[15] or may signal the presence in the wrapped DNA of structural discontinuities at regular intervals, for example, kinks.[55,56] The wrapping of DNA is thought to arise from a distribution of basic binding groups for the nucleic acid on the core of the nucleosome (made up of the globular carboxyl-terminal two-third of the histones). For a review of features of histone structure and sequence, see Isenberg.[57]

The occurrence of single-stranded DNA fragments that are integral multiples

356

of 10 bases in a digest of chromatin does not necessarily mean that the nuclease
nicks DNA at every site 10 bases from the ends of the core particle segment.
For example, cuts at 10, 20, 60, 90, and 110 bases (among a group of combinations) from the ends of the DNA will create a series of fragments spanning the
multiples of 10 from 10 to 140. To determine accurately the location and
relative susceptibilities of nuclease cutting sites in the core particle,
one must create a fixed reference point for measurements. This can be done
by using radioactive ATP and polynucleotide kinase[20,22] to label the 5' ends
of the DNA with phosphorus-32.

B. Experimental Methods

Isolated nucleosomal core particles are labeled at the 5' end by using
$[\gamma\text{-}^{32}P]$ATP and polynucleotide kinase. Homogeneity of the particle DNA length is
critical to high-resolution mapping. Particles, at A_{260}= 1-10, are incubated
in 10 mM Tris/Cl (pH 7.5), 5 mM dithiothreitol, 7.5 mM $MgCl_2$,& 0.25 mM EGTA
at 37°C. Labeled ATP, plus carrier if desired, is added with polynucleotide
kinase to a concentration of 40 U/ml. The labeling reaction attains equilibrium within about 20 min. For more nearly quantitative labeling of the DNA
ends (60-80%), carrier ATP is added to bring the total concentration to a value
that is appropriate, on the basis of known equilibrium constants for the labeling reaction and the concentration of 5' DNA ends in the reaction mixture.[20,58]
Higher specific activities and more efficient use of isotope are achieved by
using the radioactive ATP without added carrier; less total labeling occurs,
but the proportion of the isotope incorporated is higher. The reaction is stopped by cooling to 0°C and adding EDTA to 25 mM. Labeled core particles are
separated from unreacted ATP by gel filtration over a small column of Sephacryl
S-200 in 10 mM Tris/Cl (pH 8.0) and 1 mM EDTA. The nucleoprotein peak is collected and stored at 4°C.

To ensure constancy of conditions, we usually digest labeled particles in
trace amount (by mass) with unlabeled carrier particles at a total DNA concentration of 0.5 mg/ml. DNAse I digestions are in 10 mM Tris/Cl (pH 8.0), 10 mM
$MgCl_2$, and 1 mM EGTA. DNAse II digestions are in 5 mM sodium acetate (pH 5.0)
and 0.5 mM EDTA. Because there is no way of predicting what extent of digestion will give an optimal display of cutting sites, it is common to use a set
of enzyme concentrations and/or digestion times when dealing with a new particle. Measurement of acid solubilization of the end label may give an idea
of appropriate conditions; note that this determination measures cutting only

at the site 10 bases from the 5' end, inasmuch as all other cuts will lead to acid-precipitable labeled fragments. If the ratio of cutting at the 10-base site to cutting at others differs between a particular particle and the typical core particle, this measurement will be of little help in selecting digestion conditions. Good starting points are the use of DNAse I at 100 U/ml at 37°C for 1, 2, 4, 8, and 16 min or DNAse II at 500 U/ml for the same times. Interest is in the accessibility of cutting sites in the native core particle, so most information is gained at early phases of the digestion; ideally, one would like to see the distribution of fragments after each strand had been cut just once. Unknown secondary effects of site availability may occur when second and third cuts are made in a given strand of nucleosome DNA.

Reactions are terminated by adjustment of aliquots to 25 mM EDTA and 1% sodium dodecyl sulfate. Sample are deproteinized, precipitated, and dissolved in gel sample buffer that contains 98% formamide. After heating to 100°C to ensure denaturation, samples are electrophoresed on slab gels, usually 12% polyacrylamide and 0.4% (bis)acrylamide with 7 M urea. A recently described gel system markedly reduces sequence effects on mobility, thereby enabling resolution of fragments at the single base level up to nearly the size of the core particle.[59] The critical change is an increase in the proportion of the total acrylamide that is (bis)acrylamide. Long (30-40 cm) gels are generally used, and double loading of samples will enhance resolution of larger fragments. Autoradiographic detection of labeled fragments is by conventional means with intensifier screens at -70°C.

Fragments can be sized approximately by calibration of the gel with the known mobilities of bromphenol blue and xylene cyanol FF in the particular gel system.[59,60] Alternatively, end labeling of a DNAse I digest of chromatin or core particles, after phosphatase treatment, will provide a series of bands of length n X 10.4 bases, where n is an integer from 2 to 14. An elegant calibration uses a DNAse I digest of chromatin formed from inner histones and poly (dA-dT); the bands are seen by stain or label every two bases up to at least 150 bases on a standard gel.[53,61]

The amount of label in a given band can be measured by quantitative densitometry of the autoradiogram[21] with appropriate controls to ensure that image density is linear with radioisotope content, or by cutting out of the bands and counting of the isotope with liquid scintillation.[22] Two alternative approaches to obtaining quantitative rate constants for cutting at different sites have been described.[21,22] In general, semiquantitative analysis

358

of digestion patterns by visual approximation[20] has led to conclusions similar to those derived from more extensive quantitation.

C. Results

The major cutting sites for DNAse I in the core particle are at 10, 20, 40,
50, 90, 100, 120, and 130 bases from the 5' end (for convenience, I specify
sites as multiples of 10.0, rather than of 10.4). Cutting at sites 30 and 110
bases from the end are rare relative to cutting at the others. Cutting at the
site 80 bases from the 5' end is virtually never observed. The center of the
particle, 70 bases from either end, is cut at an intermediate frequency.[20-22]
The site 60 bases from the 5' end is usually cut at a low frequency; however, in
core particles containing poly(dA-dT), the homogeneous DNA is cut at the 60 base
site about as infrequently as at the 80 base site, its symmetric partner.[53]
There is an overall symmetry to the cutting-site pattern about the 70 base site.
Furthermore, assuming that the ends of the DNA are resistant sites, 80 bases
separate sites with like properties toward nucleases - 0 and 80, 30 and 110, 40
and 120, etc. If the path of DNA in the core particle has a helical circumference of 80 bp, this places sites with like properties, in terms of DNAse I cutting, vertically above or below one another.

Several different types of chromatin particles have been studied with this
DNAse I cutting-site mapping technique. Tatchell and Van Holde studied compact
dimeric and trimeric particles; mapping allowed a reasonable estimate of the
interactions of histone cores with DNA in these species, which are created by
nucleosome sliding in 0.6 M NaCl.[62] We studied the structure of the chromatosome,[38] a particle 20 bp longer than the core particle that contains, in
addition to the histone octamer, a molecule of H1 or H5. Mapping data suggested that the particle contains an additional 10 bp DNA at each end of the core
particle segment; if the path of DNA persists for these segments, two full
turns of nucleic acid would be wrapped around the histone octamer and stabilized by H1.[38] Core particles from which the histone amino terminal regions have been removed by limited tryptic proteolysis show increased cutting
at the sites normally resistant to DNAse I, both in the center of the particle
and in the region 20-40 bases, although some unusually sharp bands are seen
in the latter region and have not been explained.[63] Core particles containing hyperacetylated histones (generated in vivo by inhibition of histone deace-
tylase with sodium n-butyrate) are cut more frequently than normal particles
at the site 60 bases from the end.[64] The ends of the DNA in particles with

highly acetylated H3 and H4 are also more accessible to nucleases than those in normal core particles.[64] Core particles from sea urchin sperm contain variant forms of H2A and H2B; in one species, the H2B is longer and contains numerous arginyl residues in an extended amino-terminal region.[57] Digestion of the sperm core particle reveals a generalized decrease in rate of cutting (vs. chicken erythrocyte), with the possible exception of the region 60-70 bases from the 5' end. A selective decrease in cutting is observed at the 20 and 40 base sites; both are cut only about as frequently as the 30 base site in the sperm core particle.[65]

Core particles formed from alternating-sequence synthetic polydeoxyribonucleotides have also been mapped for DNAse I cutting sites. Absence of sequence variability in the DNA of these particles might be expected to lead to enhancement in the resolution of cutting-site experiments. The particle containing poly(dA-dT) is preferentially cut at sites like those seen for native core particles, except that there is virtually no cutting at the 60 base site; this makes the pattern of cutting sites quite symmetric.[53,61] In contrast, the poly(dG-dC)-containing particle is cut at low frequency at 30, 40, 80, and 110 bases from the 5' end; in this particle, the 60 base site is cleaved nearly as frequently as the usual highly susceptible site at 50 bases.[53] One interpretation of such data is that these two core particles represent the ends of a spectrum of possible core particle DNA conformations - a spectrum that has particularly marked variations in the central region of the core particle. Variability in cutting at the 60 base site in native core particles, with random-sequence DNA, might reflect the presence of a mixture of conformations in the native particles.

Although preferential cutting sites are at 10 base intervals along the DNA, both the poly(dA-dT)-containing and native core particles can be cut by DNAse I at every site along the length of the DNA.[53,59,61] This makes it unlikely that static kink sites[55,56] lead to the series of 10 band patterns seen for DNAse I digests of chromatin. Because cutting is observed all along the DNA, the kink (if it exists) must be a migrating one. In conjunction with other data, the nuclease digestion studies suggest that most core particle DNA is smoothly wound around the histone octamer and that the periodicity in DNAse I digests reflects the helical periodicity of the nucleic acid.[15]

What leads to the variable susceptibility of different cutting sites along the length of core particle DNA? One suggestion was that it arose from variations in the orientation of the phosphodiester bond along the length of the DNA;[66] this contrasts with the suggestion that the nuclease-resistant sites arose from protection of the DNA by interaction with the highly basic amino-terminal regions of the inner histones.[20-22,24,63-65] The sum of the evidence from core particles with modified histones, histone variants, and different DNA lengths is certainly in favor of the latter interpretation. Assuming that the protection of various sites arises from interaction of DNA with histone amino-terminal tails, the data suggest that H3 and H4 interact with the ends and the center of the core particle segment, whereas H2A and H2B interact with the region on the opposite face of the cylindric core particle, centered at 30 and 110 bases from the ends of the DNA.

High resolution mapping of DNAse I cutting sites in native core particles, with the Lutter modification for gel analysis[59,67] and with poly(dA-dT) containing particles,[53,61] has demonstrated that the helical repeat for DNA in the core particle (actually the spacing of DNAse I sites) is 10.4-10.5 bp - very similar to that for DNA in solution.[68] This makes unlikely one proposed resolution[69] to the apparent paradox between experimentally observed torsional constraints imposed on closed circular DNA by formation of nucleosomes[16,17] and constraints predicted by a smoothly wrapped, continuously helical DNA path encompassing about two turns of nucleic acid around each histone octamer.

The spacing between cutting sites along the length of core particle DNA is apparently not constant for the length of the particle. Thus, near the ends of the DNA, the periodicity of cutting sites is near 10.0 bases, whereas in the central region of the particle, it is nearer 10.8 bases.[59,61] This suggests that different domains of DNA structure types might exist within the nucleosome core particle. The region with the longer spacing is that for which there are variations in cutting sites for the two synthetic core particles.[53] These data suggest caution in modeling the structure of the core particle as one in which all of the DNA is wrapped smoothly around the histone octamer without any structural discontinuities.

Several other nucleases have been used to map cutting sites in native core particles; in two cases, the results of different groups conflict with respect to cutting sites and their susceptibilities. DNAse II has been reported to yield a pattern of cutting sites very similar to that obtained with DNAse I[70] or to digest core particles with little variability in the susceptibility of the various sites.[54] In both cases, cutting was observed at sites that were multiples of about 10 bases from the 5' ends. Micrococcal nuclease has been reported to cut at sites similar to those for DNAse I[70] or to cut primarily at sites 21, 32, 53, 86, 102, and 115 bases from the 5' end.[71] An endonuclease from Aspergillus oryzae has a cutting pattern similar to that for

DNAse I.[54] In recognition of the sequence preference of different nucleases and the possibility of varied DNA conformations in native core particles, it seems worth while to repeat the digestion studies that have yielded conflicting results, using the poly(dA-dT) core particle[53,61] as a high-resolution model for the nucleosome core particle.

D.   Precautions

Technically, mapping of cutting sites in homogeneous core particles is relatively simple. Three features of the experiment deserve comment. First, the homogeneity of the DNA length is critical to obtaining reasonable data. The presence of longer fragments with randomly extended tails longer than the core particle or of a substantial fraction of DNA of shorter length will obfuscate the experimental results. Second, as a control for this and for internal cutting by contaminating micrococcal nuclease during the labeling or DNAse I digestion, it is mandatory to electrophorese a control sample withdrawn at the conclusion of the digestion period from a sample incubated without added nuclease. Because the data at early times of digestion are of most interest, the presence of small amounts of internally cut and labeled contaminants in the core particle preparation can be important. Third, DNAse I exhibits sequence preferences. Like micrococcal nuclease, DNAse I preferentially digests AT-rich segments of DNA. In fact, when used with magnesium alone, this enzyme does not degrade one strand of poly(dG) poly(dC).[72] Addition of small amounts of calcium allows hydrolysis of all DNA without such a marked sequence variation, as does digestion with manganese as the divalent cation.[72] However, if calcium and magnesium are used, there must be careful controls for residual micrococcal nuclease activity.

IV.   EXCISION OF TRANSCRIPTIONALLY ACTIVE CHROMATIN WITH NUCLEASES

A.   Background

It appears that only a fraction of the total chromatin in cells of a given type is available as a template for transcription; this fraction varies from near zero for some terminally differentiated cells to probably almost 100% for yeast, whereas most typical adult cells are thought to transcribe 10% or less of their DNA. The structural features that differentiate this portion of chromatin, competent for transcription, from the bulk of chromatin are not known. Phenomenologic correlations with increased transcription have been made for the presence of acetylated histones,[73] increased content and phosphorylation of nonhistones,[74,75] and absence of H1.[38,75] All these might be expected to alter the rate of digestion of a segment of chromatin, relative to the bulk, by nucleases.

Three enzymes - DNAse I, DNAse II, and micrococcal nuclease - have been used to investigate the features of chromatin that accompany the transcriptionally competent state. The first is used in a destructive test. Active gene sequences are preferentially degraded by DNAse I;[26,27] this probe is therefore used in experiments that alter the composition of chromatin (by selective extraction or by reassociation) and test the state of sensitivity of the gene in question. The other two enzymes excise longer segments of chromatin that contain active genes more rapidly than the chromatin that is not active for transcription.[28-30] They may thus be used to prepare chromatin that is slightly enriched in competent segments; this allows correlations of their composition and structure with transcriptional activity.

B. Experimental Methods

Digestions are done under conditions described above or with variations described in the literature. The original Weintraub and Groudine DNAse I experiments used a low ionic strength buffer without sucrose.[26] Most of the selective solubilizations of active chromatin with micrococcal nuclease are done at 0°C to allow more facile control of the very limited amount of digestion desired. After digestion, DNAse I treated nuclei are treated with proteinase and/or phenol to prepare DNA for analysis. Micrococcal nuclease treated samples are subjected to nuclear lysis with chelating agents, and the solubilized material is analyzed either directly or after nucleosome fractionation with sucrose gradient centrifugation. DNAse II treated samples are precipitated with 2 mM $MgCl_2$; the soluble fraction contains the putative competent chromatin.[30]

Analysis of gene frequency is by hybridization. Initially, studies used solution hybridization with a labeled cDNA prepared from isolated, characterized mRNA,[26,27] or in some cases polysomal poly(A)+RNA,[76] for assessment of the proportion of gene sequences in a given DNA population. More recently, nick-translated cloned DNA has been used, either for solution hybridization or for detection of gene sequences with Southern blots of restriction endonuclease treated DNA.[77,78] It is important to note that several different features of DNAse I cutting are measured in these various study methods. Bulk DNA digestion is usually measured by acid solubility in 5% perchloric acid. This requires cutting to a size of less than 15 bases to be scored as digested. Solution hybridization will score pieces of DNA considerably longer than this (40-70 bases) as digested, owing to the decrease in hybridization efficiency

with decreasing DNA size. In even greater contrast, blot hybridization of
restriction endonuclease treated DNAse I digests will score the first cut as
digestion of the gene sequence; the restriction fragment disappears with the
first cleavage.

C. Results

The cleanest and most extensive results related to the features that characterize competent chromatin are those derived from the use of DNAse I as a
probe. Weintraub and Groudine[26] and Garel and Axel[27] first demonstrated
that transcriptionally active gene sequences were 5-10 times more rapidly degraded by this nuclease than were bulk genomic sequences. Several systems were
tested; most of the data were obtained on globin sythesis in avian reticulocyte
and ovalbumin synthesis in oviduct. One conflicting result in these two reports
was the retention or lack of retention of the DNAse I sensitivity of active
genes after chromatin was degraded to mononucleosomes; a recent manuscript ad-
dressed this subject. Study of an integrated adenoviral sequence allowed Flint
and Weintraub[79] to define the precision of the altered structure leading to
DNAse I sensitivity; within the limits of detection of the hybridization method,
the altered structure corresponded exactly with the ends of the transcribed segment.

Several recent investigations have used the DNAse I sensitivity of active
genes to ask questions about the composition and structure of this chromatin.
Washing chromatin with 0.35 M NaCl abolishes the sensitivity of the globin gene
to DNAse I.[80] This concentration of salt removes a number of nonhistone
proteins from chromatin, but should not remove any of the histones. Among the
proteins in the 0.35 M NaCl extract are the so-called HMG proteins,[81] small
proteins with both basic and acidic regions that are present in relatively large
amounts in many chromatins. Addition of HMG 14 and/or HMG 17 to salt-washed
chromatin restored the nuclease sensitivity of the competent gene; no tissue
specificity was noted for the nonhistones.[77,80] Thus, it seems that HMG
proteins are necessary but not sufficient for the DNAse I sensitivity of active
gene chromatin. The observation that HMG proteins can restore nuclease sensitivity implies that other features of active chromatin must be altered from
the bulk, allowing recognition of proper binding sites by these small proteins.
In experiments designed to address this question, Weisbrod et al.[77] showed
that washing chromatin with 0.6 M NaCl, removing H1, did not abolish the ability
of the HMGs to bind and regenerate the DNAse I sensitivity of the globin gene.

Addition of small amounts of the various inner histones in a competition experiment suggested that H2A and H2B might bind to the same sites as the HMG
proteins.[77] The first interpretation, that HMGs might replace H2A and H2B
in active chromatin, is difficult to reconcile with a <u>decreased</u> rate of digestion of these gene sequences when the HMG proteins are removed, presumably leaving
a segment of DNA wrapped only with an H3 + H4 tetramer. Obviously, more information is needed to define the structural features of transcriptionally competent chromatin.

A recently developed technique that allows extension of the DNAse I sensitivity method to detection of all active gene sequences and leads to a faci-
litation of the assay will be of help in elucidation of such structural features.
Brief treatment of nuclei with low concentrations of DNAse I selectively nicks
active gene sequences; these can then by labeled by nick translation with DNA
polymerase I[82] to generate a substrate chromatin with active gene sequences
labeled with phosphorus-32. The labeled genes are digested at least 10 times
more rapidly than bulk DNA by DNAse I, the selectivity of digestion is lost
after 0.35 M NaCl extraction of nuclei, and the sensitivity can be restored by
reconstitution of salt-washed nuclei and purified HMG proteins.[83]

Two features of the mapping studies for DNAse I cutting of nucleosomes
are perhaps relevant to the structure of active chromatin. First, acetylated
histones lead to increased cutting by DNAse I in the center and at the ends of
the core particle.[64] Although the rate of digestion of total core particle
DNA to acid solubility is not increased by the presence of hyperacetylated histones, it is possible that the rate of digestion to fragments too small for
efficient hybridization may be increased by the presence of acetylated histones.
Second, as noted above, the most frequently cut site in the nucleosome is the
one 10 bases from the ends of the core particle.[20-22] In the chromatosome,[38]
H1 blocks access of this region to nucleases; part of the differential sensitivity of transcribed and nontranscribed chromatin to DNAse I might arise
from protection of the highly digestible site in the latter by H1.

Results obtained in digestion of active gene sequences by micrococcal
nuclease are less strikingly different from the digestion of bulk chromatin
than are the results of DNAse I digestion. Early studies showed that the
proportion of active gene sequences in the 50% limit digest of chromatin was
similar to that in whole cell DNA[84] and that isolated nucleosomes contained
active genes in the proportion expected from whole cell DNA.[25] More recently, two studies concluded that early in the digestion by micrococcal nuclease there is an enrichment of active gene sequences in the material excised

from the beaded chain.[28,29] This early released material contains increased amounts of some HMG proteins,[85] more highly acetylated histone than the remainder of the chromatin,[86] and much less H1 histone;[85] all these compositional features are consistent with the DNAse I - derived conclusions outlined above.

The nucleosomes with active gene segments appear to be excised first, so the scenario of micrococcal nuclease digestion outlined above would lead one to expect that they might also be internally degraded first. If this is so, it could account for the failure in earlier studies to find enrichment in active gene sequences in monomers and for some of the variability in studies of DNAse I sensitivity of gene sequences in monmeric particles.[26,27] A recent critical study has used both micrococcal nuclease and DNAse I to investigate the state of the ovalbumin gene in oviduct nuclei.[87] Hybridization to globin cDNA was used as an internal control throughout the study; the ratio of ovalbumin (expressed) to globin (nonexpressed) is a sensitive measure of enrichment or depletion of active gene sequences in a given sample. Briefly, micrococcal nuclease did excise nucleosomes with active genes preferentially and then degraded these same nucleosomes, also more rapidly than those which contained nontranscribed gene sequences. For nucleosomes with an increased content of active genes, DNAse I sensitivity was greater than for those without, although the differences were not as marked as for whole chromatin. No differential DNAse I sensitivity was found for active genes in nucleosome core particles.

DNAse II digestion was the first method to suggest that active gene sequences could be preferentially digested in chromatin by a nuclease.[30] Although the initial studies were not analyzed by hybridization, now recognized to be the only rigorous criterion for a competent segment of chromatin, later studies demonstrated increased content of sequences complementary to polysomal poly(A)+RNA and to globin mRNA in the magnesium soluble fraction of a DNAse II digest of nuclei.[88] Others have failed to find such an enrichment;[89] the use of this enzyme for preparation of active chromatin requires further investigation.

D. Precautions

Other than caution in selection of sufficiently pure mRNA, if a cDNA probe is used, and the usual care in hybridization, there are few technical pitfalls in the use of nucleases as probes for active chromatin. One must not forget that the digestion of chromatin by any nuclease is a complex process and that nuclease "sensitivity" is a relative difference, not an absolute one. Hence, experiments at different levels of digestion, preferably with quantitation of content of both a transcribed and a repressed gene sequence, seem appropriate. Finally, the differences in extent of digestion necessary for a particular DNA to be scored as "digested" for the various types of analysis used must be kept in mind.

V.   REFERENCES

1. Clark, R.J. and Felsenfeld, G. (1971) Nature (New. Biol.), 229 101-106.
2. Hewish, D.R. and Burgoyne, L.A. (1973) Biochem. Biophys. Res. Commun., 52, 504-510.
3. Burgoyne, L.A., Hewish, D.R., and Mobbs, J. (1974) Biochem. J., 143, 67-72.
4. Noll, M. (1974) Nature, 251, 249-251.
5. Olins, A.L. and Olins, D.E. (1974) Science, 183, 330-332.
6. Woodcock, C.L.F. (1973) J. Cell Biol., 59, 368a.
7. Cruft, H.J., Mauritzen, C.M., and Stedman, E. (1958) Proc. Roy. Soc. Lond. Ser. B, 149, 21-35.
8. Kornberg, R.D. and Thomas, J.O. (1974) Science, 184, 865-868.
9. Van der Westhuyzen, D.R. and Von Holt, C. (1971) FEBS Letters, 14, 333-337.
10. Roark, D.E., Geohegan, T.E., and Keller, G.H. (1974) Biochem. Biophys. Res. Commun., 59, 542-547.
11. D'Anna, J.A., Jr. and Isenberg, I. (1974) Biochemistry, 13, 4992-4997.
12. Kornberg, R.D. (1974) Science, 184, 868-871.
13. Sahasrabuddhe, C.G. and Van Holde, K.E. (1974) J. Biol. Chem., 248, 1080-1083.
14. Noll, M. (1975) Nucleic Acid Res., 1, 1573-1578.
15. Weintraub, H. and Van Lente, F. (1975) Ciba Found. Symp., 28, 291-307.
16. Griffith, J. (1975) Science, 187, 1202-1203.
17. Germond, J.E., Hirt, B., Oudet, P., Gross-Bellard, M., and Chambon, P. (1975) Proc. Natl. Acad. Sci. USA, 72, 1843-1847.
18. Simpson, R.T. and Whitlock, J.P., Jr. (1976) Nucleic Acids Res., 3, 117-127.
19. Shaw, B.R., Herman, T.M., Kovacic, R.T., Beaudreau, G.S., and Van Holde, K.E. (1976) Proc. Natl. Acad. Sci. USA, 73, 505-509.
20. Simpson, R.T. and Whitlock, J.P., Jr. (1976) Cell, 9, 347-353.
21. Noll, M. (1977) J. Mol. Biol., 116, 49-71.
22. Lutter, L.C. (1978) J. Mol. Biol., 124, 391-420.
23. Maxam, A. and Gilbert, W. (1977) Proc. Natl. Acad. Sci. USA, 74, 560-564.
24. Finch, J.T., Lutter, L.C., Rhodes, D., Brown, R.S., Rushton, B., Levitt, M., and Klug, A. (1977) Nature, 269, 29-36.
25. Lacy, E. and Axel, R. (1975) Proc. Natl. Acad. Sci. USA, 72, 3978-3982.
26. Weintraub, H. and Groudine, M. (1976) Science, 193, 848-856.
27. Garel, A. and Axel, R. (1976) Proc. Natl. Acad. Sci. USA, 73, 3966-3970.
28. Levy, W.B. and Dixon, G.H. (1978) Nucleic Acids Res. 5, 4155-4163.
29. Bloom, K.S. and Anderson, J.N. (1978) Cell, 15, 141-150.
30. Gottesfeld, J.M., Garrard, W.T., Bagi, G., Wilson, R.F., and Bonner, J. (1974) Proc. Natl. Acad. Sci. USA, 71, 2193-2197.
31. Kornberg, R.D. (1977) Annu. Rev. Biochem., 46, 931-954.
32. McGhee, J. and Felsenfeld, G. (1980) Annu. Rev. Biochem., 49, 1115-1156.
33. Oudet, P., Gross-Bellard, M., and Chambon, P. (1975) Cell, 4, 281-300.
34. Varshavsky, A., Bakayev, V.V., and Georgiev, G.P. (1976) Nucleic Acids Res., 3, 477-492.
35. Whitlock, J.P., Jr. and Simpson, R.T. (1976) Biochemistry, 15, 3307-3314.
36. Todd, R.D. and Garrard, W.T. (1977) J. Biol. Chem., 252, 4729-4738.
37. Noll, M. and Kornberg, R.D. (1977) J. Mol. Biol., 109, 393-404.
38. Simpson, R.T. (1978) Biochemistry, 17, 5524-5531.

39. Axel, R., Melchior, W., Sollner-Webb, B., and Felsenfeld, G. (1974) Proc. Natl. Acad. Sci. USA, 71, 4101-4105.
40. Hymer, W.C. and Kuff, E.L. (1964) J. Histochem., Cytochem., 12, 359-363.
41. Whitlock, J.P., Jr. and Simpson, R.T. (1976) Nucleic Acids Res., 3, 2255-2266.
42. Klevan, L. and Crothers, D.M. (1977) Nucleic Acids Res., 4, 4077-4089.
43. Tatchell, K. and Van Holde, K.E. (1978) Proc. Natl. Acad. Sci. USA, 75, 3583-3587.
44. Lawson, G.M. and Cole, R.D. (1979) Biochemistry, 18, 2160-2166.
45. Riley, D. and Weintraub, H. (1978) Cell, 13, 281-293.
46. McCarty, K.S., Jr., Vollmer, R.T., and McCarty, K.S. (1974) Anal. Biochem., 61, 165-183.
47. Olins, A.L., Carlson, R.D., Wright, E.B., and Olins, D.E. (1976) Nucleic Acids Res., 3, 3271-3289.
48. Morris, N.R. (1976) Cell, 8, 357-363.
49. Thomas, J.O. and Furber, V. (1976) FEBS Letters, 66, 274-280.
50. Spadafora, C., Bellard, M., Compton, J.L., and Chambon, P. (1976) FEBS Letters, 69, 281-285.
51. Keichline, L.D. and Wassarman, P.M. (1977) Biochem. Biophys. Acta., 246, 139-151.
52. Albright, S.C., Wiseman, J.M., Lange, R.A., and Garrard, W.T. (1980) J. Biol. Chem., 255, 3673-3684.
53. Simpson, R.T and Kunzler, P. (1979) Nucleic Acids Res., 6, 1387-1415.
54. Whitlock, J.P., Jr., Rushizky, G.W., and Simpson, R.T. (1977) J. Biol. Chem., 252, 3003-3006.
55. Crick, F.H.C. and Klug, A. (1975) Nature, 255, 530-533.
56. Sobell, H.M., Tsai, C., Gilbert, S.G., Hain, S.C., and Sakore, T.D. (1976) Proc. Natl. Acad. Sci. USA, 73, 3068-3072.
57. Isenberg, I. (1979) Annu. Rev. Biochem., 48, 159-191.
58. Lillehaug, J.R., Kleppe, R.K., and Kleppe, K. (1976) Biochemistry, 15, 1858-1865.
59. Lutter, L.C. (1979) Nucleic Acids Res., 6, 41-56.
60. Maniatis, T., Jeffrey, A., and Van de Sande, H. (1975) Biochemistry, 14, 3787-3794.
61. Bryan, P.N., Wright, E.B., and Olins, D.E. (1979) Nucleic Acids Res., 6, 1509-1520.
62. Tatchell, K. and Van Holde, K.E. (1978) Proc. Natl. Acad. Sci. USA, 75, 3583-3587.
63. Whitlock, J.P., Jr. and Simpson, R.T. (1977) J. Biol. Chem., 252, 6515-6520.
64. Simpson, R.T. (1978) Cell, 13, 691-699.
65. Simpson, R.T. and Bergman, L.W. (1980) J. Biol. Chem., in press.
66. Trifonov, E.N. and Bettecken, T. (1979) Biochemistry, 18, 454-456.
67. Prunell, A., Kornberg, R.D., Lutter, L.C., Klug, A., Levitt, M., and Crick, F.H.C. (1979) Science, 204, 855-858.
68. Wang, J.C. (1979) Proc. Natl. Acad. Sci. USA, 76, 200-203.
69. Crick, F.H.C. (1978) Cold Spring Harbor Symp. Quant. Biol., 42, 243.
70. Camerini-Otero, R.D., Sollner-Webb, B., Simon, R.H., Williamson, P., Zasloff, M., and Felsenfeld, G. (1978) Cold Spring Harbor Symp. Quant. Biol., 42, 57-74.
71. Whitlock, J.P., Jr. (1977) J. Biol. Chem., 252, 7635-7639.
72. Bollum, F.J. (1965) J. Biol. Chem., 240, 2599-2601.
73. Ruiz-Carillo, A., Wangh. L.J., and Allfrey, V.G. (1975) Science, 190, 117-128.
74. Kleinsmith, L.J. (1978) in The Cell Nucleus, H. Busch, ed. (Academic Press, New York), Vol. 6, Part C, pp. 221-261.
75. Elgin, S.C.R. and Weintraub, H. (1975) Annu. Rev. Biochem., 44, 725-774.

76. Levy-Wilson, B. and Dixon, G.H. (1977) Nucleic Acids Res., 4, 883-898.
77. Weisbrod, S., Groudine, M., and Weintraub, H. (1980) Cell, 19, 289-301.
78. Zasloff, M. and Camerini-Otero, R.D. (1980) Proc. Natl. Acad. Sci. USA, 77, 1907-1911.
79. Flint, S.J. and Weintraub, H. (1977) Cell, 12, 783-792.
80. Weisbrod, S. and Weintraub, H. (1979) Proc. Natl. Acad. Sci. USA, 76, 631-635.
81. Goodwin, G.H., Sanders, C., and Johns, E.W. (1973) Eur. J. Biochem., 38, 14-19.
82. Levitt, A., Axel, R., and Cedar, H. (1979) Dev. Biol., 69, 496-505.
83. Gazit, B., Panet, A., and Cedar, H. (1980) Proc. Natl. Acad. Sci. USA, 77, 1787-1790.
84. Axel, R., Cedar, H., and Felsenfeld, G. (1975) Biochemistry, 14, 2489-2495.
85. Hutcheon, T., Dixon, G.H., and Levy-Wilson, B. (1980) J. Biol. Chem., 255, 681-685.
86. Levy, W.B., Watson, D.C., and Dixon, G.H. (1979) Nucleic Acids Res., 6, 259-274.
87. Senear, A.W. and Palmiter, R.D. (1980) J. Biol. Chem., in press.
88. Gottesfeld, J.M. and Partington, G.A. (1977) Cell, 12, 953-962.
89. Lau, A.F., Ruddon, R.W., Collett, M.S. and Faras, A.J. (1978) Exptl. Cell Res., 111, 269-276.

TRANSCRIPTION OF MAMMALIAN GENES IN VITRO

James L. Manley[+] and Malcolm L. Gefter[++], [+]Department of
Biological Sciences Columbia University, New York, New York 10027, U.S.;
[++]Massachusetts Institute of Technology, Cambridge, MA 02139, USA

I. INTRODUCTION

It is well established that mammalian cells contain three classes of RNA polymerase[1]. One class, RNA polymerase II, has been implicated as that responsible for mRNA synthesis. This assignment is based primarily on studies with the drug α-amanitin: the same very low concentration of the toxin that selectively inhibits purified RNA polymerase II in vitro inhibits the synthesis of hn RNA, the presumptive mRNA precursor, in isolated nuclei[2]. The enzyme has a very high molecular weight (over 500,000) and may contain as many as nine (or even more) different subunits[3].

In addition to RNA polymerase II, a complex array of enzymes is required to complete the synthesis of mature mRNA. An enzyme (or enzyme) is needed to add the 5' "cap" found on all mammalian mRNAs examined so far[4]. Other enzymes are required to methylate the cap and to carry out internal methylation, such as occurs at the N6 position of some adenosine residues[5]. Another set of enzymes are necessary to create the 3' ends of mature mRNA, including at least one specific endonuclease and poly(A) polymerase[6,7]. Finally, some unknown number of enzymes carry out the "splicing" reactions necessary to remove the intervening sequences found in most mRNA precursors[8,9].

Most of the enzymes that catalyze these reactions have not been identi-fied or characterized. This is due in large part to the lack of suitable substrates with which to assay the enzymes. An ideal substrate would be an RNA species synthesized in vitro that corresponds to the primary transcript of a particular well-characterized gene. However, synthesis of such a mole-cule requires in vitro transcription systems capable of initiating transcrip-tion accurately. Such systems derived from bacterial cells have been well characterized, but systems derived from mammalian cells have only recently begun to become available.

In vitro transcription systems would also be useful to an understanding of the regulation of mRNA synthesis. Regulation of gene expression occurs at the level of transcription, and many examples have shown that control of transcription initiation is of primary importance. To understand how such regulation is mediated, it is first required that we know the sequences of DNA that are necessary to bring about accurate inititation of transcription; i.e., what is a promoter? In bacteria, examination of a number of DNA sequences around transcription starting sites has revealed sequences that are highly conserved[10]. Specifically, a heptanucleotide sequence centered 10 base pairs upstream from the mRNA starting point is highly conserved[11]. Another conserved sequence is found at -35 about base pairs[12]. These sequences appear to constitute the bacterial promoter region. This deduction is supported by experimental evidence that a large number of point mutations that affect promoter function are within the conserved regions. In addition, DNA-protection experiments show that RNA polymerase actually covers this region of DNA during initiation.

Elucidation of the structure of promoters in eukaryotic organisms has been much more difficult, for several reasons. First, until the advent of recombinant-DNA technology, it was impossible to analyze eukaryotic genes,

except for some viral genes. Second, the relative lack of genetics in higher
organisms has made it difficult to obtain mutants with which to define presumptive promoter sequences; however, techniques for selectively mutagenizing DNA
in vitro are rapidly becoming available. Finally, in vitro transcription systems from eukaryotic cells have not been available, making it difficult to .
analyze any presumptive promoter mutations.

We describe here two types of in vitro systems that we have recently developed. One contains isolated nuclei as the source of mRNA synthetic machinery
and is capable of accurately initiating transcription and processing the RNA
synthesized from endogenous templates in the reaction. The other is a soluble
whole-cell extract system that will accurately initiate transcription on exo-
genously added DNA and is, in fact, dependent on it for activity. Also presented
in detail are our methods of purifying and analyzing the in vitro synthesized
RNA.

II. PREPARATION OF EXTRACTS

A. Isolated Nuclei.

Nuclei isolated from myeloma cells or adenovirus-infected cells have been
shown to synthesize nuclear RNA for long periods. In the case of the infected
cells, a substantial fraction of the RNA synthesized is virus-specific and has
many properties expected of authentic nuclear mRNA precursors[13-15]. The rationale for producing such a system is that nuclei should contain all the necessary
marcromolecules and structural features of intact cells. Supplementation of the
nuclei with small molecules and monomeric substrates should provide all the conditions needed for correct biosynthesis of RNA.

Nuclei are obtained from cells broken by the combined action of hypotonic
swelling and mechanical shearing. Nonionic detergent is included to free the
nuclei of adherent cytoplasmic material. Nuclei are separated from other cellular constituents by centrifugation through sucrose. Cells grown in tissue
culture have been used throughout.

Nuclei are isolated by a slight modification of the procedure described
by Marzluff et al.[16] All procedures are carried out at 0-4°C unless otherwise
specified.

1) Cells (in log phase) are collected by sedimentation at 600 g for 10
min and washed three times in 30 mM Tris-HCl(pH, 7.5), 120 mM KCl, 5 mM magnesium acetate, and 7 mM 2-mercaptoethanol.

372

2) The packed cells are resuspended in 5 volumes of 0.3 M sucrose containing 2 mM magnesium acetate, 3 mM CaCl$_2$, 10 mM Tris-HCl(pH, 7.6), 0.1% Triton
X-100, and 0.5 mM 2-mercaptoethanol and homogenized in a B type Dounce homogenizer.

3) The homogenate is mixed with an equal volume of 2 M sucrose containing
5 mM magnesium acetate, 10 mM Tris-HCl(pH, 7.5), and 0.5 mM 2-mercaptoethanol.
The mixture is layered over 2 ml of the 2 M sucrose buffer and centrifuged for
45 min at 20,000 rpm and 5°C in a Beckman SW50.1 rotor.

4) The nuclei are resuspended in 25% glycerol containing 5 mM magnesium
acetate, 0.1 mM EDTA, 5 mM 2-mercaptoethanol, and 50 mM Hepes NaOH(pH, 7.5) at
a concentration of 2-3 X 10^8 nuclei/ml and stored at -80°C. Recovery of nuclei
is about 80%.

Nuclei may be rapidly frozen and stored at -80°C without loss of activity
for at least 4 mon.

Reaction mixtures (100 µl) contain the following: 7.5% glycerol, 19 mM
Hepes NaOH (pH, 7.5), 3.8 mM magnesium acetate, 0.5 mM MnCl$_2$, 84 mM potassium
acetate, 25 mM ammonium acetate, creatine phoshophokinase at 30 µg/ml, 10 mM-
creatine phosphate, 0.5 mM S-adenosyl-L-methionine, each of the 20 amino acids,
at 0-4 mM, 1 mM-ATP, GTP and CTP, at 0.25 mM each 40 µM UTP, (including 10 µCi
of [^{32}P]UTP), and 30 µl of nuclei suspension.

Reaction mixtures are incubated at 25°C, with gentle shaking, for the times
required. RNA synthesis under these conditions is linear for up to 2.5 h.

Nuclei that do not support linear synthesis for at least 90 min are discarded. Such preparations occur about 20% of the time for no apparent reason.

B. Whole Cell Lysate.

The nuclear system is capable of supporting synthesis of virus-specific
RNA. This RNA is initiated, capped, elongated, and terminated properly and
contains the appropriate proportion of polyadenylate on its 3' end. In an
attempt to study these reactions further, we sought a soluble system capable
of supporting RNA synthesis in the same manner. Initial attempts to disrupt
nuclei prepared as described above did not yield active extracts. We therefore
adapted a new procedure for the preparation of a soluble extract. We argued
that perhaps nuclei manipulated in vitro lose essential components by leakage
and that making an extract from whole cells, rather than nuclei, might be pro-
ductive. As described below, such extracts do indeed support proper initiation
and capping of mRNA precursors[17].

Extracts are prepared by Dounce homogenization of hypotonically swelled cells. Nuclei are lysed by the addition of $(NH_4)_2SO_4$. We reasoned that most enzymes associated with nucleic acids would be freed from DNA at the concentration of ammonium sulfate used (10% of saturation). Thus, by addition of $(NH_4)_2SO_4$ and gentle stirring, the proteins are released, but the DNA remains at high molecular weight and can be removed from the extract by high-speed centrifugation (175,000 $\underline{g}$). The extract is later concentrated by $(NH_4)_2SO_4$ precipitation, dialyzed, and used in concentrated form (final protein concentration of incubations, approximately 8-10 mg/ml). Extracts are prepared by a modification of a procedure originally described by Sugden and Keller[18]. HeLa cells are grown in suspension culture in Eagle's minimal essential medium supplemented with 5% horse serum to a density of approximately 4-8 X 10^5 cells/ ml. The cell density appears not to be crucial, although we have recently obtained slightly more active extracts with cells at the lower end of the range indicated. From this point on, all operations are carried out at 0-4°.

1) Cells are washed in PBS containing $MgCl_2$, and the cell pellet is resuspended in 4 packed-cell volumes (PCV) of 0.01 M Tris (pH, 7.9), 0.001 M EDTA, and 0.005 M DTT.

2) After 20 min, the cells are lysed by homogenization in a Dounce homogenizer with 8 strokes with a "B" pestle.

3) Four PCV of 0.05 M Tris (pH, 7.9), 0.01 M $MgCl_2$, 0.002 M DTT, 25% sucrose, and 50% glycerol are then added, and the mixture is stirred $\underline{gently}$, to prevent shearing of the DNA as the nuclei lyse. To this suspension, 1 PCV of saturated $(NH_4)_2SO_4$ is added dropwise. After this addition, the highly viscous lysate is gently stirred for an additional 20 min to ensure that the solution is homogeneous.

4) The extract is carefully poured into centrifuge tubes and centrifuged at 47,500 rpm for 3 h in a Beckman 60 Ti rotor. Because of the high viscosity of the extract, wide-mouth tubes should be used; otherwise, it will be impossible to fill them.

5) The supernatant is decanted so as not to disturb the pellet (the last 1-2 ml is left behind) and precipitated by addition of solid $(NH_4)_2SO_4$ (0.33 g/ml of suspension). After the $(NH_4)_2SO_4$ is dissolved, 1 N NaOH at 0.1 ml/10 gm $(NH_4)_2SO_4$ is added and the suspension is stirred for an additional 30 min.

6) The precipitate is collected by centrifugation at 15,000 $\underline{g}$ for 20 min and resuspended with one-tenth volume of the high-speed supernatant into a buffer containing 20 mM Hepes (pH, 7.9), 100 mM KCl, 12.5 mM $MgCl_2$, 0.1 mM EDTA,

2 mM DTT, and 17% glycerol.

7) This suspension is dialyzed against 100 volumes of the same buffer for 4-8 h, after which the buffer is changed and dialysis is continued for another 4-8 h.

8) The dialysate is centrifuged at 10,000 g for 10 min to remove insoluble material, and the supernatant is quick-frozen in small aliquots in liquid nitrogen and stored at -80°C. Extracts retain full activity for at least 6 mo.

C. In Vitro Incubation.

A standard 50-μl reaction mixture contains 12 mM Hepes (pH, 7.9), 7 mM $MgCl_2$, 60 mM KCl, 0.2 mM EDTA, 1.3 mM DTT, 10% glycerol, 50 μM ATP, 50 μM CTP, 50 μM GTP, 50 μM UTP containing 10 μCi of [α-^{32}P]UTP, 4 mM creatine phosphate disodium salt, 30 μl of extract, and 2.5 μg of DNA. Addition of creatine phosphokinase is not required, as there appears to be a sufficient amount of this activity already present in the lysate. Any of the four ribonucleotide/triphosphates may be added as radioactive precursor. UTP results in slightly less background labeling (see below) than does CTP or ATP, and is very stable. GTP produces the cleanest backgrounds, but is somewhat less stable than the other triphosphates. Incubations are for 60 min at 30°C. When reaction mixtures are incubated at 37°C, the RNA is badly degraded; however, no such degradation occurs at 30°C. We believe that this is the result of the destruction of some endogenous RNAse inhibitor at the higher temperature. The DNA concentration is critical: at suboptimal concentrations of DNA, no transcription is detected; at higher than optimal concentrations, the amount of nonspecific transcription increases, but specific transcription decreases. It appears that this requirement is for total mass of DNA, rather than specific promoter-containing sequences: when the total DNA concentration is held constant with inactive carrier DNA, such as pBR322, and the concentration of promoter-containing DNA is varied, the amount of specific transcription is directly proportional to the amount of promoter-containing DNA in the reaction mixtures. This is unlike the result obtained when the total DNA concentration is varied, in which case a sharp all-or-none curve is often observed (e.g., see Manley et al.[17]). Cloned DNAs, viral DNAs, or restriction fragments derived from them may be used as templates. Linear DNA is much more efficient than supercoiled DNA in initiating specific transcription. We have not determined the reason for this, but we believe that it results from the extra tension in the supercoiled DNA, which may cause increased "breathing" in A-T-rich regions. These sites may then serve as artificial transcription starting-sites. (The amount of nonspecific transcription is indeed higher when supercoiled DNA is used as a template.)

When new DNAs are being tested for their ability to direct accurate transcription in vitro, we suggest that at least the concentrations of DNA, extract, ribonucleotide triphosphates, and monovalent cation be carefully titrated. Changes in the concentrations of all these factors can drastically affect the degree of specific transcription.

As might be expected from the nature of the lysate (a crude whole cell extract), a number of enzyme activities are present which can cause complications in interpreting the results of in vitro transcription experiments. These include end-addition activities, which catalyze the addition of α-^{32}P ribotriphosphates onto the ends of endogenous RNA molecules, such as 28S, 18S, 5S and tRNA's; kinase activities, which preclude the use of γ-^{32}P triphosphates; phosphate and pyrophosphate exchange activities, which makes difficult the use of not only α-^{32}P triphosphates, but also β-^{32}P triphosphates, a DNA ligase activity, wich brings about the ligation of linear DNA molecules; a DNA topoisomerase activity, which results in the concatenation of supercoiled DNA molecules; and a CCA adding activity, which will add these nucleotides to the 3' end of tRNA molecules.

D. Analysis of RNA Synthesized In Vitro.

The usefulness of the transcription systems just described depends on one's ability to extract undegraded RNA from reaction mixtures and then to size and map the transcripts accurately. This section describes our methods for purifying and analyzing RNA synthesized in vitro.

III. EXTRACTION OF RNA FROM AN IN VITRO REACTION MIXTURE

We extract RNA from either type of in vitro reaction mixture described above by similar procedures. The method used to purify RNA from the whole-cell soluble system is outlined below.

1) Terminate transcription reactions (50 μl) by addition of 250 μl of 8 M urea, 0.5% SDS, and 10 mM EDTA (pH, 8.0) containing 25 μg of yeast tRNA.

2) Add an equal volume (300 μl) of a 1:1:0.05 mixture of phenol; chloroform; and isoamyl alcohol; vortex for 30 s; spin in an Eppendorf centrifuge for 5 min; remove the aqueous phase, leaving the interphase.

3) Add to the organic phase 150 µl of 7 M urea, 0.35 M NaCl, 1% SDS, 10 mM EDTA, and 10 mM Tris (pH, 8.0); vortex; spin in an Eppendorf centrifuge for 20 s. A very dense protein precipitate will form at the interphase, making removal of the aqueous phase impossible. Therefore, remove the _organic_ phase with a pasteur pipette, and re-extract the interphase plus aqueous phase with 200 µl of chloroform; after brief vortexing and centrifugation, the aqueous phase can be easily removed (there should be essentially no interphase).

4) Pool the aqueous phases (the volume will now be 450 µl); add an equal volume of a mixture of phenol chloroform; isoamyl alcohol; vortex; centrifuge for 5 min, and remove the aqueous phase, leaving the small interphase behind.

5) Re-extract the aqueous phase two times with equal volumes of chloroform. After the first of these extractions, a small sticky protein precipitate can sometimes be detected; this should be left. After the second extraction, the interface should be completely clear.

6) Add 1 ml of ethanol; mix; hold in a dry ice and ethanol bath for 5 min; (or at least 2 hr at -20°C); spin for 5 min in an Eppendorf centrifuge in a cold room, and decant the supernatant.

7) Resuspend the pellet in 200 µl of 0.2% SDS; add 200 µl of 2 M NH_4Ac; add 1 ml of ethanol; and precipitate the nucleic acid as described in 6.

8) Dissolve the pellet in 400 µl of 0.3 M NaAc (pH, 5.2) and 0.2% SDS; add 1 ml of ethanol, and precipitate the nucleic acid as described in 6.

9) Resuspend the dried pellet in 60 µl of 0.2% Sarkosyl and 1 mM EDTA (pH, 8.0); store at -20°C.

The RNA solution obtained by this method is free of unincorporated ribonucleotide triphosphates, and the amount of phosphorus-32 incorporated into nucleic acid can be simply determined by measuring the Cerenkov radiation. If the RNA is to be analyzed by the S1 technique (see below), DNA must first be removed, as follows:

1) Nucleic acid from step 7 should be dissolved in 0.3 M NaAc without SDS and then precipitated with ethanol.

2) Resuspend the dried pellet in 100 µl of 10 mM Tris (pH, 7.5) and 100 mM NaCl.

3) Add DNAse (Worthington, RNAse-free, which has been treated with iodoacetate as described by Zimmerman and Sandeen[19]) to 50 µg/ml and $MgCl_2$ to 10 mM.

4) Incubate for 5 min at 37°C.

5) Add 100 µl of 10 mM EDTA (pH, 8.0) and 0.2% SDS.

6) Extract with an equal volume of a mixture of phenol, chloroform and isoamyl alcohol, and re-extract the aqueous phase with an equal volume of chloroform.

7) Add NaCl to a final concentration of 150 mM, add 250 µl of ethanol, and precipitate RNA as above.

8) Resuspend the RNA pellet as in 9 above.

A. Sizing and Mapping of RNAs by RNA-DNA Hybridization, Nuclease S1 Digestion, and Agarose-Gel Electrophoresis.

This technique is a modification of one developed by Berk and Sharp[20,21], who used [^{32}P]DNA to size and map steady-state, unlabeled RNA. The method described here allows one to size and map simultaneously steady-state RNA (by staining of DNA-RNA hybrids) and pulse-labeled RNA or RNA synthesized in vitro (by autoradiography of hybrids formed from labeled RNA).

1) Mix RNA and DNA. We routinely use 2.5 µg of adenovirus DNA (35,000 base pairs) or the amount of a purified restriction fragment or cloned fragment that would be obtained from this amount of intact DNA. Two factors determine how much RNA must be added. First, to prevent more than one RNA molecule from hybridizing to a given DNA molecule, hybridization should be carried out with at least a 3- to 4-fold excess of DNA over complementary RNA. This requires that not more than about 0.5 µg of complementary RNA be added to hybridization reactions. Second, for reasons not understood, the RNA concentration during the hybridization reaction must be greater than 750 µg/ml if the S1 nuclease-resistant duplexes are to be analyzed by electrophoresis through neutral-agarose gels. For 15-µl hybridizations, 10 µg of RNA is therefore required. These conditions (i.e., 10 µg of RNA containing less than 0.5 µg of RNA complementary to the DNA probe) are often met with crude RNA preparations, such as total cytoplasmic or nuclear RNA. However, more purified RNAs, such as poly A$^+$ RNA, require the addition of carrier RNA. Total HeLa cell cytoplasmic RNA functions adequately in this respect, although yeast tRNA does not. RNA and DNA are conveniently mixed by adding to 100 µl of 0.2 M NaAc (pH, 5.2) and 0.1% SDS and precipitating with 250 µl of ethanol in Eppendorf tubes. (Precipitations are done by incubating the tubes for 5 min in an alcohol and dry ice bath and centrifuging for 10 min in an Eppendorf centrifuge.)

2) Dissolve RNA and DNA in hybridization buffer. Decant the ethanol carefully, and allow samples to air-dry for about 10 min. It is important that the precipitates not become completely dry, or it will be impossible to dissolve

them. Resuspend the samples in 15 µl of hybridization buffer (80% formamide, 0.4 M NaCl, 0.04 M PIPES (pH, 6.5), and 1 mM EDTA. The formamide should first be deionized by shaking with 6 g of Amberlite mixed-bed resin per 100 ml of formamide for 3 h. The resin is removed by filtration, and the formamide is stored at -20°C. Formamide prepared and stored in this manner shows no sign of breakdown for at least 6 mo.)

3) Denature the DNA and hybridize. Denature and strand-separate the DNA by heating to 68°C for 10 min. Hybridize by quickly transferring the tubes to 60°C and incubating for 45 min. The hybridization time and temperature were determined by A. J. Berk for adenovirus 2 DNA. These numbers can easily be determined for other DNAs as described by Berk and Sharp[20].

4) Digest the unhybridized nucleic acid with nuclease S1. Rapidly dilute the 15 µl hybridization reaction mixture with 200 µl of S1 solution-0.03 M NaAc, (pH, 4.5), 0.25 M NaCl, 1 mM $ZnSO_4$, 5% glycerol, and nuclease S1 at 2×10^3 units/ml (Boehringer Mannheim). The S1 solution should be cooled to 0-4°C, and the samples removed from the 60°C water bath, diluted, and mixed one at a time. This procedure prevents any DNA-DNA reannealing and possible displacement of hybridized RNA. Nuclease digestion is carried out by incubating at 45°C for 30 min.

5) Prepare samples for agarose-gel electrophoresis. Add 200 µl of phenol and chloroform (1:1), and 10 µg of tRNA; mix, and spin for 30 s. Remove the aqueous phase and precipitate nucleic acids with 500 µl of ethanol. Dry the pellets, resuspend them in 25 µl of 20 mM Tris (pH, 8.0) and 2 mM EDTA, and add 5 ml of 50% glycerol saturated with bromophenol blue.

6) Electrophorese RNA-DNA duplexes through an agarose gel. We routinely use 20-cm-long slab gels in a high-salt buffer-40 mM Tris-HCl (pH, 8.3), 50 mM NaAc and 2 mM EDTA-to ensure that short RNA-DNA duplexes do not denature. A 1.2% agarose (Sigma) gel allows one to resolve RNA-DNA duplexes of about 300-10,000 base pairs. Gels should be run until the blue dye reaches the bottom. This takes about 500 V-h. (The high salt concentration results in a high current, which can deplete the buffering capacity. Therefore, buffer should be recirculated.)

7) Visulize the RNA-DNA hybrids. If more than about 10 ng is present in a band 5 X 0.3 cm, it can be easily detected by ethidium bromide staining. (Soak gel in ethidium bromide at 0.5 µg/ml for 20-30 min and visualize hybrids by fluorescence under UV illumination.) For RNA synthesized in vitro, bands should be detected by autoradiography. Unless the RNA-DNA duplexes are to be eluted from the gel, it should be dried down (sharper bands are obtained). In

fact, it is necessry to dry the gel if it is to be exposed with intensifying screeens at -70°C (the gel will shatter otherwise). About 10-counts/min phosphorus-32 in a band will produce a dark band after an overnight exposure with screens.

B. Sizing of RNA by Glyoxalation and Agarose-Gel Electrophoresis

To determine the size of a particular RNA species, we have found the glyoxalation-gel technique of McMasters and Carmichael[22] to be simple and accurate. Molecules 200-5,000 nucleotides long can be readily resolved on one gel, and their mobilities are directly proportional to the log of their molecular weights.

1) Glyoxalate the RNA. RNA stored in the sarkosyl-EDTA solution described above can be directly glyoxalated. The factor that limits how much RNA can be analyzed in one gel slot is the large amount of ribosomal RNA in crude RNA samples. For example, there is about 10 μg of 18 S rRNA in 30 μl of the whole-cell extract. We therefore glyoxalate and electrophorese almost only one-fourth of the RNA obtained from a standard 50 μl reaction mixture, to avoid overloading the gel. A convenient protocol for glyoxalation is as follows: 15 μl (of 60 μl) of RNA solution and 27 μl of a solution that contains 210 μl of DMSO, 4.2 μl of 1 M Na phosphate is mixed with (pH, 6.8) and 60 μl of glyoxal (Eastman). The glyoxal must be first deionized as described above for formamide preparation, except that 25 g of amberlite is used per 100 ml and the glyoxal is stored in 0.5 ml aliquots at -20°C. Glyoxal prepared in this manner is good for at least 3 mo. The samples are mixed and then incubated at 50°C for 1 h.

2) Electrophorese RNA through agarose gel. After addition of bromophenol blue, samples are loaded onto a 1.4% agarose gel made in 10 mM $NaPO_4$ (pH, 6.8), which is also the reservoir buffer. Gels 20 cm long are run at 120 V for 2.5 h. The reservoir buffer must be rapidly recirculated throughout the run to prevent changes in the pH. We have obtained our best results with this voltage. Lower voltages, and the resulting longer running times required, lead to more diffused bands.

3) Visualize the RNA by staining. We routinely stain our gels with ethidium bromide. Although this procedure essentially detects only endogenous rRNA and tRNA carrier, it is a simple and useful method to check both the recovery of RNA from sample to sample and any degradation of RNA which might have occured in any of the samples. Before the RNA can be stained with ethidium bromide, it must be deglyoxalated by treatment with base. Soak the gel in 10

volumes of 50 mM NaOH and ethidium bromide, at 0.5 µg/ml for 20 min; rinse; and soak in 10 volumes of a high-salt buffer, such as the electrophoresis buffer described above, but containing ethidium bromide at 0.5 µg/ml, for 40 min. Visualize the RNA by fluorescence under UV illumination. Dry the gel and expose it to x-ray film as described above.

IV. USES OF IN VITRO SYSTEMS

The two in vitro systems described offer novel opportunities for studying aspects of regulation of gene expression. The isolated-nuclei system has been shown to synthesize mRNA precursors accurately. These nuclei have so far been prepared from adenovirus-infected cells; it has been possible to demonstrate that natural precursors are produced that are stable in vitro. The absence of further processing steps in this system can thus be used to advantage. The soluble system offers opportunitites for analyzing both the exogenous DNA and the protein components needed to promote accurate initiation of mRNA precursor synthesis.

The isolated-nuclei system prepared from adenovirus-infected cells taken at late stages of infection gives rise to virtually all the precursors of late mRNAs. Their stability in vitro allows for their isolation, mapping, and chemical characterization. Nuclei isolated from any virus-infected cell should, in principle, similarly give rise to the virus-specific products and thus be useful in analysis of their structure and composition. The system may also be useful for analyzing RNA products of any active gene. In most cases, our knowledge of primary transcription products of genes is severely limited. One important characteristic of the isolated nuclei is the ease with which "pulse-chase" experiments can be carried out. Such experiments are difficult with whole cells owing to the impermeability of the cell membrane to ribonucleotide triphosphates and the large pool sizes of these molecules within cells. With such a system, we showed that the 3' ends of several adenovirus mRNAs were created by cleavage of larger precursors, and those of others apparently by transcription termination[6].

Isolated nuclei also represent a starting-point for further studies. For example, we first demonstrated transcription initiation from the adenovirus late promoter in isolated nuclei before succeeding with the soluble extract system, which is more amenable to study. However, the soluble system currently does not form mature 3'ends, nor does it splice. We will probably need to use intact nuclei to begin to study these activities.

The soluble transcription system, like the nuclei system, accurately initiates RNA transcripts that _in vivo_ are typically mRNA precursors. Unlike the nuclear system, however, the soluble system does not produce natural 3' ends or polyadenylated transcripts. The main advantage of the soluble system is that any DNA of which one has a sufficient amount may be analyzed for its ability to promote transcription. In virtually every case in which the presence of a "promoter" was known, the soluble system initiated RNA synthesis correctly. The systems that have been used successfully to date include adenovirus and SV40 early and late genes[17,23,24] and human and mouse globin genes[25]. Preliminary success with RNA tumor virus proviral DNA, cloned silk fibroin and mouse insulin genes has been communicated to us. Thus, with a reasonable degree of confidence, an unknown DNA may be examined for its content of promoter sequences. This is especially useful for systems in which genomic clones of active genes have been obtained and the regulatory sequences associated with their transcription need to be determined. Although such sequences hve not been defined with accuracy, experiments that alter "promoter regions" by _in vitro_ mutagenesis should provide information regarding structure-function relationships.

The soluble system described here has only been prepared from uninfected HeLa cells. In semiquantitative terms, all promoters tested are active. At least the protein elements needed for the observed activities may be purified and studied. It is of obvious interest to prepare the soluble system from cells expressing a differentiated function or from cells that have been infected with viruses. Undoubtedly, as these studies become more refined and influences of more subtle regulation are observed, isolation of "factors" from these types of cells will be of interest. This is also true of a similar transcription system described by Weil, Roeder, and co-workers[26,27]. This system has so far generated qualitatively equivalent results.

The descriptions of the systems studied so for are of only a rudimentary nature. We look forward to the elucidation of the biochemistry of gene regulation as these systems become refined and widely used.

REFERENCES

1. Roeder, R. G., (1976) in "RNA Polymerase" (R. Losick and M. Chamberlin, eds.), p 285. Cold Spring Harbor Labs, New York.
2. Zylbert, E. and Penman, S. (1971) Proc. Nat. Acad. Sci. USA 71, 2337.
3. Schwartz, L. and Roeder, R. G. (1975) J. Biol. Chem. 150, 3221.
4. Shatkin, A. J. (1976) Cell, 9, 645.
5. Perry, R. P. and Kelley, D. E., (1974) Cell, 1, 37.

6. Manley, J. L., Sharp, P. A., and Gefter, M. L. (1980) manuscript in preparation.
7. Winters, M. A. and Edmonds, M. (1973) J. Biol. Chem. 248, 4763.
8. Chow, L. T., Gelinas, R. E., Broker, T. R., and Roberts, R. J. (1977) Cell 12, 1.
9. Berget, S. M., Moore, C., and Sharp, P. A., (1977) Proc. Nat. Acad. Sci. USA 74, 3171.
10. Rosenberg, M. and Court, D. (1979). Ann. Rev. Genet. 13, 319-354.
11. Pribnow, D. (1975) J. Mol. Biol. 99, 419-443.
12. Takanami, M., Sugimoto, K., Sugisaki, H., and Okamoto, T. (1976). Nature, 118-121.
13. Mory, Y. Y. and Gefter, M. L. (1977) Nucleic Acids Res. 5, 1739-1757.
14. Manley, J. L., Sharp, P. A. and Gefter, M. L. (1979) Proc. Nat. Acad. Sci. USA 76, 160-164.
15. Manley, J. L., Sharp, P. A. and Gefter, M. L. (1979) J. Mol. Biol. 135, 171-197.
16. Marzluff, W. F., Murphy, E. G. and Huang, R. C. C. (1973) Biochemistry 12, 3440-3446.
17. Manley, J. L., Fire, A., Cano, A., Sharp, P. A. and Gefter, M. L. (1980) Proc. Nat. Acad. Sci. USA 77, 3855-3859.
18. Sugden, B. and Keller, W. (1973) J. Biol. Chem. 1248, 3777-3788.
19. Zimmerman, S. B. and Sandeen, G. (1966) Anal. Biochem. 14, 269-277.
20. Berk, A. J. and Sharp. P. A. (1977) Cell 12, 721-732.
21. Berk, A. J. and Sharp. P. A. (1978) Proc. Nat. Acad. Sci. USA 75, 1274-1278.
22. McMaster, G. K. and Carmichael, G. C. (1977) Proc. Nat. Acad. Sci. 74, 4835-4838.
23. Manley, J. L., Handa, H., Huang, S. Y., Gefter, M. L. and Sharp, P. A. (1980). XVI Miami Winter Symposium. (in press).
24. Handa, H., Kaufmann, R., Manley, J. L., Gefter, M. L. and Sharp, P. A. (1980) Nucleic Acids Res. (in press).
25. Proudfoot, N., Shatner, M., Manley, J. L., Gefter, M. L. and Maniatis, T. (1980) Science (in press).
26. Weil, P. A., Luse, D. S., Segall, J. and Roeder, R. (1979) Cell 18, 469-484.
27. Luse, D. S. and Roeder, R. G. (1980) Cell, 20, 691-700.

A SYSTEM TO STUDY PROMOTER AND TERMINATOR SIGNALS

RECOGNIZED BY ESCHERICHIA COLI RNA POLYMERASE

KEITH McKENNEY, HIROYUKI SHIMATAKE, DONALD COURT,
URSULA SCHMEISSNER, CATHERINE BRADY AND
MARTIN ROSENBERG

Laboratories of Biochemistry and Molecular Biology
National Cancer Institute
National Institutes of Health
Bethesda, Maryland 20205

I. INTRODUCTION

RNA transcription is regulated by specific signals encoded in DNA, some of which (promoters) specify start sites and others (terminators) stop sites for RNA polymerase. More than 60 promoter sites and some 35 terminator sites recognized by E. coli RNA polymerase have been structurally defined[1-4]. Comparison of these regulatory sequences has identified common features in these sites that are thought important for promoter and terminator function. Mutational analyses, transcription studies, and experiments probing the direct interaction of the RNA polymerase with the regulatory site have all contributed to our understanding of promoter and terminator function. These studies have helped to elucidate certain aspects of the RNA polymerase-DNA interaction, but very little is known about how the DNA sequence actually contributes to specifying promoter or terminator function. For example, different promoters and terminators function with a wide spectrum of efficiencies (i.e., strengths), and have varied requirements for different ancillary factors. Clearly, the information that specifies the strength and factor dependence of any site must be encoded in the DNA, and the structural differences among these sites must correlate with their functional differences in some way. We do not yet fully understand these correlations.

One reason for this lack of knowledge has been our inability to make valid comparisons between the efficiencies and factor requirements of various signals. Because each regulatory site is usually studied in context with the particular gene(s) it controls, there is no standard for comparison. Hence, the DNA sequences of the regulatory sites that have been characterized cannot be compared with respect to their relative efficiencies. A second difficulty in acquiring information about these sites is that many promoters and terminators regulate the expression of gene products that cannot be readily monitored. It is difficult to study a regulatory signal that controls a gene with no assayable product or selectable genetic marker. This problem is complicated further if the control signal of interest regulates an operon which is vital to the development of the bacterial or phage system. Moreover, terminator signals that occur at the ends of operons present a particular problem, in that there is usually no downstream function to monitor.

One conventional approach that has proven extremely useful in the study of operon control involves the technique of gene fusion[5-7]. The concept involves genetic fusions of an assayable, selectable gene function (e.g., lac, gal, and trp genes) to the operon of interest. Such fusions place the known gene function under the control of different regulatory elements. Although the fusion systems have allowed us to expand our studies of transcriptional regulation, the methods have several severe limitations: the precise location of the fusion in the operon cannot be preselected, nor can the DNA sequence at the fusion point be known without further characterization; creation of the fusion may result in multiple regulatory sites preceding the fused region, thereby complicating studies of any one regulatory site; vital functions present particular problems since fusions into these operons are lethal; and fusions may introduce polar effects (transcriptional, translational, or both) within the operon. These effects can be quite variable, depending on the position at which the fusion occurs. Apparently, fusion of the same gene at even slightly different locations in the operon can result in substantially different degrees of expression[18-19]. These position effects make it impossible to use conventional fusion systems for comparing different control signals.

The following section describes how gene fusion can be combined with recombinant DNA technology to provide a new approach to the study of transcriptional control signals. The approach involves precise in vitro fusion of DNA fragments containing specific transcription regulatory sites to an assayable, selectable gene function. This approach circumvents many of the limitations inherent in conventional techniques and introduces an entirely new approach to the genetic and biochemical study of regulatory signals.

The system to be described can be used for the isolation, comparison, and complete characterization of almost any promoter or terminator signal, as well as for studies involving various effector molecules that function at these sites in vivo. Moreover, this system has the added flexibility that any in vitro construction can be transferred from plasmid to phage and to the bacterial chromosome (either as a lysogen or as a true bacterial recombinant), thereby allowing study of the regulatory signal in both multiple and single gene copies.

II. CONSIDERATIONS IN USING A RECOMBINANT VECTOR SYSTEM

We outline here a number of important considerations that should be incorporated into the design and construction of any recombinant vector system used to study transcriptional regulatory signals. In the sections that follow, we describe how each of these requirements has been met in our system. It should be noted that several other vector systems have been developed in other laboratories[20-22]. Although each has proved extremely useful in the isolation and study of regulatory sites, none satisfies all of the following requirements.

1. Fusions should be to a gene with a readily assayable product. The assay should be simple, sensitive, and linear over a wide range.

2. Expression of the gene function should be proportional to the amount of transcription, thereby measuring the overall efficiency of the promoter or terminator signal. This requires that the translation efficiency of the mRNA remain nearly constant, irrespective of the regulatory region to which it is fused. To achieve this, all translation except that of the assayable function must be uncoupled from the transcription unit to avoid differential effects on the expression of the gene. In addition, the translation efficiency of the mRNA must be relatively immune to changes in the upstream 5' terminal RNA structure resulting from the fusion.

3. The gene function should be a readily selectable genetic marker. This allows characterization of the regulatory site by a variety of mutational analyses. Gene functions that have both positive and negative selections are the most flexible.

4. A second independent selectable marker other than that used for characterization of the regulatory site should be part of the vector.

5. The existence of several unique cloning sites allows diversity, simplicity, and precision in fusing small DNA fragments that contain the regulatory site of interest to the gene.

6. Transcriptional expression of the gene must depend solely on the inserted regulatory site. Transcription originating elsewhere on the vector must not traverse the gene.

7. The exact DNA sequence should be known between the cloning site and the gene. In addition, extensive restriction information of the gene is extremely useful.

8. The ability to determine accurately and to manipulate the copy number of the cloning vector allows one to relate activity tc copy number and, more importantly, study the regulatory site in both single and multiple copies.

III. USING THE E. COLI GALACTOKINASE GENE TO STUDY PROMOTERS AND TERMINATORS

The galactose operon (gal) of E. coli consists of three genes: epimerase (galE), transferase (galT), and kinase (galK). These genes are normally expressed from a polycistronic mRNA in the order E, T, K[23-24]. We have chosen the promoter distal gene, galK, for fusion to transcriptional regulatory signals. Galactokinase catalyzes the following reaction:

$$\text{galactose} + \text{ATP} \longrightarrow \text{galactose-1-phosphate}$$

This reaction can be monitored with a simple, sensitive, and linear assay (from about 5 to 200,000 molecules/cell, see appendix).

Translation of the galK gene normally initiates in an intercistronic region between galT and galK several thousand nucleotides from the start of its mRNA. Apparently, the structure of this region has developed to ensure proper ribosome recognition and translation initation of galK, despite its location well within the high-molecular-weight gal operon transcript. We reasoned that, if the natural T-K boundary region were left intact, then the efficiency of galK translation might be relatively independent of the RNA structure upstream of this region. That is, the RNA structure that naturally ensures galK expression from the gal mRNA should predominate, even when fused to other RNAs. Results from experiments in which the galK gene was genetically fused at different distances from the phage λ promoter, P_L, support this contention; galK expression remained proportional to the level of P_L transcription[25]. In contrast, it has been well documented that, when promoter proximal genes such as β-galactosidase[10], gal epimerase[18], and the λ gene products cro[26] and repressor[27] are used in similar genetically or in vitro constructed fusions, the translation efficiencies of these genes are highly sensitive to changes in upstream RNA structure. For example, when one of these genes is fused at even slightly different positions and distances from the same promoter signal, the translation efficiency of the gene varies in an unpredictable manner[26-27]. Presumably, because these genes are normally expressed from the 5' ends of their mRNAs, they have not developed the structure necessary to ensure their constant expression from an internal site in a larger RNA transcript.

A. The Plasmid Vector pKO-1

For the reasons discussed above, the galK gene, including 168 base pairs of the region normally preceding the galK coding sequence, was inserted into a plasmid pBR322 derivative from which the entire Tc region of the plasmid was deleted (see Fig. 1). Several unique restriction endonuclease sites were constructed to allow precise cloning of a variety of DNA fragments upstream of galK. The Sma I site at position -185 (Fig. 1) is an extremely versatile blunt-end cloning site. This site is also recognized by Ava I and Xma I. Moreover, any molecular recombination "linker" can be cloned here to add more diversity. For example, we inserted a Bam HI linker at this site, thereby allowing cloning of DNA fragments resulting from restriction with Bam HI, Sau 3A, Bcl I, and Bgl II. Fifteen base pairs upstream of the Sma I site at position -200 (Fig. 1) is a Hind III restriction site that can be used to insert DNA fragments restricted with Hind III, to insert fragments to which Hind III linkers have been attached and to insert in an oriented way any Hind III-blunt ended DNA fragment between the Hind III and Sma I sites. An Eco RI cleavage site is positioned 310 base pairs upstream of the Sma I site and can be used in conjunction with either the Hind III or Sma I site (e.g. see plasmid construction PKG-1800 in Fig. 3). Any of these sites can be used to insert DNA fragments with either defined or potential transcriptional regulatory signals. When fused in the proper orientation, the inserted regulatory signal will control galK expression. The galK gene is not expressed by the starting vector, pKO-1 (shown in Fig 1), and this plasmid can not complement an $E^+T^+K^-$ host. Thus, there is little, if any, transcription of the galK coding region without insertion of a promoter site in the proper orientation at one of the cloning sites.

Many DNA fragments that carry transcription regulatory sites also contain translation start sites. We were concerned that translation starting in the inserted DNA of various constructions might traverse all or part of the leader region preceding galK and thereby exert different effects on galK translation efficiency. This appears to be a common problem with other gene fusion systems in genetic as well as in vitro constructions. To minimize the potiential effects of upstream translation, we constructed translation stop codons in all three reading frames beyond the cloning sites, located at 174, 121 and 74 base pairs before the galK coding sequences. These stop codons uncouple from the galK transcript all translation initiating in any DNA insert. Thus, galK has a relatively uniform untranslated leader region. The purpose of this leader is

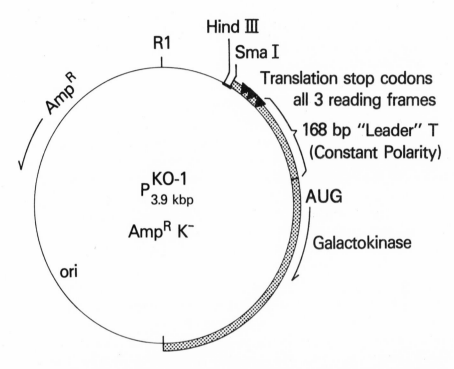

Fig. 1. The pKO-1 plasmid vector. The dotted area represents the galacto-kinase gene (galK) and 168 base pairs of "Leader" galT gene sequences preceding the AUG of galK. Translation stop codons in all three reading frames prevent any translation originating in the inserted DNA fragment from reaching the ribosome binding site and AUG of galK. The R1, Hind III and Sma_RI restriction endonuclease sites can be used to insert DNA fragments. The Amp^R denotes the β-lactamase gene providing a second independent selectable marker, ampicillin resistance. The Ori represents the replication function of this plasmid vector, derived from pBR322.

to ensure constant efficiency for galK translation.

The β-lactamase (amp) gene of pBR322 remains intact in pKO-1 as a second, independent, selectable marker (Fig. 1)[28,29]. Cells transformed with these plasmids can be selected initially on media that contain ampicillin.

The entire nucleotide sequence of the pKO-1 starting vector is known, except for the core region (700 base pairs) of the galK gene. This sequence, in combination with restriction site information for the galK gene, allows rapid and precise determination of the size, position, and orientation of inserted DNA fragments. Moreover, if the DNA sequence of the insert is known, the precise nature of the fusion junction is defined.

B. Positive and Negative Selection for galK Expression

The galK gene provides a readily selectable genetic marker that can be used for the isolation and mutational analysis of DNA fragments that carry transcriptional regulatory sites. galK expression can be made either essential or lethal to cell growth under appropriate conditions[31]. This ability to select either positively or negatively for galK expression is extremely useful.

A cell with the genotype $E^+T^+K^-$ does not metabolize galactose when carrying the pKO-1 vector. It does not grow on minimal media that contain galactose as the sole carbon source and grows as a white colony on MacConkey galactose indicator plates (Mac gal). This is because the galK gene of pKO-1 is not transcribed. However, when a DNA fragment carrying a promoter signal is properly inserted into pKO-1, galK is transcribed and the plasmid complements the galK$^-$ host resulting in growth on minimal media that contain galactose and growth as red colonies on Mac gal plates. This positive selection system has been used to isolate promoters and to obtain promoter "up" mutations that activate promoter function (K. McKenney and M. Rosenberg, unpublished). The gal$^+$ selection has also been used to obtain mutations that reduce transcription termination (K. McKenney and M. Rosenberg, unpublished). Transcription termination signals cloned between a promoter and galK may result in shutoff of galK expression. Mutants can be selected that recover their ability to express galK. Some of these have been characterized for one terminator and found to be point mutations in the terminator signal (see Table 2).

A cell with the genotype $E^-T^-K^+$ is defective in galactose metabolism and accumulates the toxic intermediate galactose-1-phosphate in the cell[41]. Such cells are killed by galactose. Thus, pKO-1 derivatives containing a promoter will kill a gal deleted host ($E^-T^-K^-$) when grown in the presence of galactose. This selection can be used to isolate promoter "down" mutations, screen DNA fragments inserted between the promoter and galK for terminator function, obtain mutations that increase the efficiency of a weak terminator, and obtain host mutations that can survive the toxic effect of galactose-1-phosphate.

C. Manipulation of Copy Number

Although very useful, the plasmid vector system has several limitations, most of which are due to the high copy number. For example, a cloned regulatory signal when present in a large number of copies, may not respond normally in experiments designed to study in vivo regulation of the site. If the signal is affected by various ancillary factors, these may become limiting in the cell. In addition, the cell may respond abnormally to the presence of many copies of the regulatory site, again complicating regulatory studies. A different problem arises from potential variations in plasmid copy number. If the plasmid copy number varies for different constructions, it is not possible to make valid comparisons among them.

These and other potential problems created by the plasmid vector system can be alleviated by transferring the regulatory site-galK fusion to a single copy genome. For example, any in vitro pKO derivative can be transferred precisely into either a phage λ genome or into the bacterial chromosome (see section VI and appendix for procedures).

The resulting bacterial or phage vector carries the identical sequences that had been constructed originally on the plasmid vector. Most importantly, galK expression remains under the control solely of the inserted regulatory information. The phage vector can be integrated into various bacterial hosts for study of the regulatory site in single copy. The bacterial recombinant, which contains neither plasmid nor phage sequences, substitutes a single copy of the in vitro constructed galK fusion for the normal gal operon at the wild type chromosomal location.

This single-gene-copy vector system allows valid comparisons between different cloned regulatory sites, studies that accurately reflect the normal in vivo functioning of the regulatory signal, mutational analysis of any cloned signal by standard phage and bacterial genetic procedures, and examination of dose effects on any regulatory signal by factors provided in trans.

IV. THE PLASMID VECTOR - A MULTICOPY SYSTEM

A. Cloning Promoter Signals

The design and construction of the starting plasmid vector, pKO-1, was described in the previous section. DNA fragments that carry promoter signals can be inserted into this vector, thereby controlling galK expresssion. To demonstrate the function and application of this system, we inserted a small, well characterized DNA fragment derived from phage λ into pKO-1. This fragment contains no known promoter signal and has no promoter activity in the pKO-1 system. The same fragment was also obtained from a phage λ derivative that carries a mutation in this region that is known to create a promoter signal (λc17)[30]. Insertion of this fragment into pKO-1 activates galK expression (Fig. 2). The

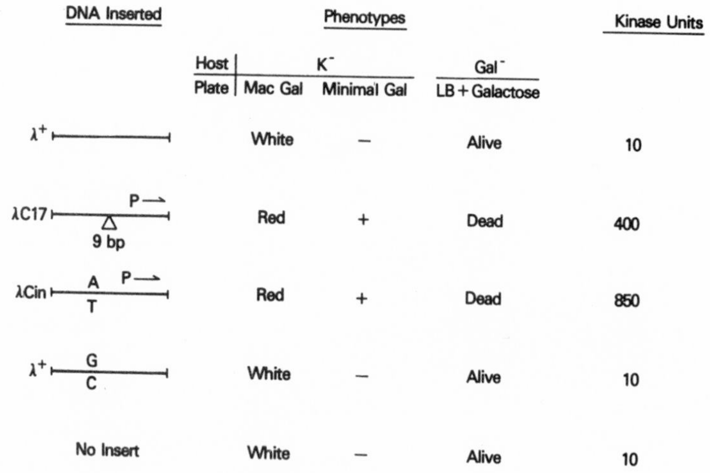

DNA Inserted	Host Plate	Phenotypes K⁻ Mac Gal	Minimal Gal	Gal⁻ LB + Galactose	Kinase Units
λ⁺ ├────────────┤		White	—	Alive	10
λC17 ├───P⟶───┤ △ 9 bp		Red	+	Dead	400
λCin ├─A P⟶──┤ T		Red	+	Dead	850
λ⁺ ├─G───────┤ C		White	—	Alive	10
No Insert		White	—	Alive	10

Fig. 2. Each bar represents a specific λ phage DNA fragment inserted into the pKO-1 plasmid vector. The first pair λ⁺ and λc17 are identical sequences except for the c17 mutation, a 9 base pair insertion that creates a promoter. The second pair (λcin and λ⁺) represents two fragments that differ by a single base pair change, AT to GC, that destroys the λcin promoter. No insert refers to the pKO-1 plasmid vector alone. The phenotypes demonstrate the various selections available using galK, either positive as in a galK⁻ host on minimal media containing galactose as the sole carbon source, or negative as in a gal deleted host in the presence of galactose. The galactokinase units demonstrate that galK expression is dependent on a promoter insert, either P_{c17} or P_{cin}.

dramatic difference in galK expression exhibited by these two essentially identical constructions demonstrates that galK expression depends solely on the promoter activity of the DNA insert.

In a similar way, we demonstrate the usefulness of the system for monitoring promoter inactivation, promoter "down" mutations. A DNA fragment, analogous to the one used above, was isolated from another phage λ derivative (λcin)[30] and inserted into pKO-1. This fragment contains a known promoter signal (H. Shimatake and M. Rosenberg, in preparation), that is oriented toward galK in the plasmid construction. As expected, the PKO-1 derivative exhibits

TABLE I

PROMOTERS INSERTED INTO THE pKO PLASMID VECTOR

PROMOTER	GALACTOKINASE UNITS[*]
E. coli Lac	520
E. coli Gal	680
E. coli Lac i^Q	900
Plasmid TET	550
$\lambda\ P_{482}$**	210
$\lambda\ P_{c17}$	430
$\lambda\ P_{cin}$	600
$\lambda\ P_o$	2250

[*]Galactokinase units expressed as nanomoles of galactose phosphorylated per minute per ml of cells at OD_{650} = 1.0.

** $\lambda\ P_{482}$ was created by in vivo recombination between two structural genes, E. coli gal T and λ N gene, that bring together -35 and -10 sequences for promoter activity (K. McKenney and M. Rosenberg, in preparation).

galK expression (Fig. 2). The same fragment was obtained from a λ phage carrying a point mutation in this promoter which was known to inactivate promoter function. Insertion of this DNA fragment completely eliminated galK expression (Fig. 2). Thus, galK expression is correlated with the degree of promoter activity on the DNA insert.

The pKO system has been used to fuse a variety of promoter signals to galK and to monitor their relative abilities to activate galK expression (see Table 1). Among the various phage and bacterial promoters examined, we observe a ten-fold range in strengths. This is the first time that a set of different promoter sequences have been aligned with respect to their relative in vivo strengths.

B. Termination Signals

The plasmid pKG-1800 (Fig. 3) is a derivative of pKO-1 in which the DNA region between the RI and Hind III sites has been removed and replaced with an RI-Hind III DNA restriction fragment that contains the promoter signal for the E. coli galactose operon (P_G). In pKG-1800, galK expression is controlled by P_G-promoted transcription. As with all pKO derivatives that contain a promoter, pKG-1800 is lethal in an $E^-T^-K^-$ host on media that contain galactose. In an $E^+T^+K^-$ host pKG-1800 permits growth as a red colony on Mac gal plates (Table 2).

DNA fragments can be inserted into pKG-1800 at the Hind III and/or Sma I (Ava I, Xma I) restriction site(s) positioned between P_G and the galK coding sequence (Fig. 3). Although insertion of any fragment usually results in a small polar effect on galK expression, (in the range of 10 to 20%), the insertion of a fragment that carries an authentic terminator oriented properly with respect to P_G transcription results in a drastic reduction in galK expression. Transcription terminators can be readily selected, inasmuch as these pKG-1800 derivatives will no longer cause an $E^-T^-K^-$ host to be sensitive to galactose, and will cause an $E^+T^+K^-$ host to grow as a white colony on Mac gal plates. The effects on galK expression of a variety of different terminators inserted into pKG-1800 are given in Table 2. The extent to which a terminator reduces galK expression is a direct measure of its efficiency. This is confirmed by examining the relative amounts of transcription that stop at or go through the termination site (see the next section).

The choice of the gal promoter to isolate and characterize terminators was arbitrary; any promoter can be used. In fact, by replacing P_G with other promoters, we have begun to examine the effect of varying promoter strength on the

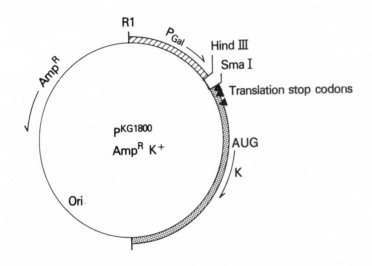

Fig. 3. The pKG-1800 plasmid vector contains the promoter for the gal operon (p$_G$) inserted between the R1 and Hind III sites of pKO. P$_G$ controls galK expression. Transcription termination sites are inserted at either the Hind III or Sma I sites, and their efficiencies determined by assaying the reduction in galK expression.

efficiency of terminator function. Our data suggest that terminators function with the same relative efficiency, regardless of the promoter to which they are fused (Table 2).

Few mutations in termination signals have been isolated with conventional approaches. In most cases no selection exists for such mutations. The pKO system overcomes these problems and allows the selection and analysis of mutations that affect terminator function. For example, a DNA fragment that contained a phage λ terminator (t$_0$) was inserted into pKG-1800; this caused a 95% reduction in galK expression (see Table 2). This plasmid failed to complement the galK$^-$ phenotype of an E$^+$T$^+$K$^-$ host and made this host grow as a white colony on Mac gal plates. Cells that contained this plasmid were mutagenized with nitrosoguanidine by standard procedures. Plasmid DNA was isolated and used to transform an E$^+$T$^+$K$^-$ host. The transformants were plated on Mac gal plates and

396

TABLE II

PROMOTERS AND TERMINATORS INSERTED INTO THE PLASMID VECTOR

INSERTED DNA Promoter	Terminator	PHENOTYPE**	GALACTOKINASE UNITS
P_O	-	Sick red	2,400
P_O	$\lambda\ t_O$	Red	50
P_{gal}	-	Red	680
P_{gal}	$\lambda\ t_O$	White	12
P_{gal}	$\lambda\ t_{o1}^*$	Red	220
P_{gal}	$\lambda\ t_{o2}^*$	Red	400
P_{gal}	$\lambda\ t_{o3}^*$	Red	450
P_{gal}	$\lambda\ t_{int}$	White	15
P_{gal}	$\lambda\ t_{rl}$	Red	185
P_{gal}	t_{IS2}	White	10

* These $\lambda\ t_O$ sites are mutants of the wild type $\lambda\ t_O$

** Phenotype represents the color that an E. coli host SA1943 has on Mac gal plates when carrying different DNA inserts in the pKO vector.

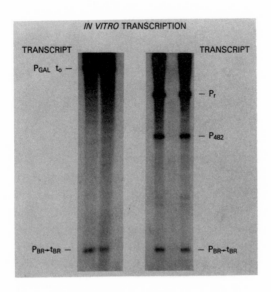

Fig. 4. Polyacrylamide gel electrophoresis of ^{32}P-labeled RNA synthesized <u>in</u>
<u>vitro</u> from several pKO plasmid templates (details in text).

red colonies were obtained at a frequency between 1 in 10^4 and 1 in 10^5.
Plasmid DNA from these red colonies was later characterized and mutations were
found in the t_o termination signal. Moreover, by comparing the levels of <u>galK</u>
expression from the different mutants (see Table 2), the structural changes
found in the terminator site can be correlated directly with the extent to which
terminator function has been impaired.

With the same procedures in combination with the variety of available mu-
tagenic techniques, the pKO system should allow isolation and characterization
of mutations in many different transcription termination sites.

C. Transcription Analysis <u>In Vivo</u>

The pKO-1 plasmid vector functions as an excellent DNA template to examine
transcription <u>in vitro</u> and <u>in vivo</u>. The small size and simple transcription
pattern allow easy monitoring of new transcripts resulting from the insertion

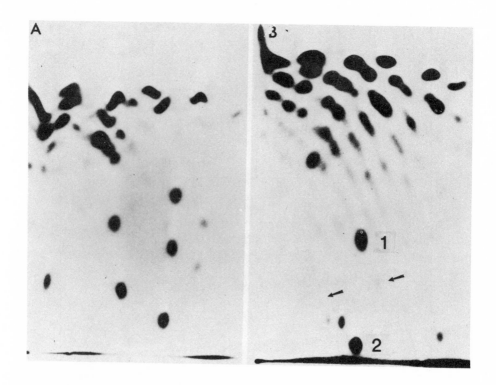

Fig. 5. Two-dimensional fingerprints of ribonuclease T1 oligonucleotide products derived from RNA synthesized _in vivo_ by (A) pKOc17 or (B) pKG-1820 plasmid. Total RNA was hybridized to λ DNA filters to select the λ RNA sequences cloned in the pKO vector system. Horizontal dimension: electrophoresis on Cellogel strips at pH 3.5. Vertical dimension: chromatography on thin layer plates of DEAE-cellulose. In the pKG-1820 fingerprint, oligonucleotides No. 1 and No. 2 derive from sequences immediately preceding the t_0 termination site. The two arrows designate the position of oligonucleotides immediately after the t_0 site. Overexposure of the fingerprint shows that these two oligonucleotides are present at less than 1% of oligonucleotides No. 1 and No. 2.

of a regulatory signal. Transcripts made either from the β-lactamase promoter or the "rep" promoter[42] (see Fig. 4) serve as internal standards and permit direct comparison between different pKO derivatives. Moreover, a cloned regulatory signal can be studied _in vitro_ from either a supercoiled or a relaxed template because the plasmid is isolated as a supercoil and can be readily linearized by restriction. The occurrence of a variety of convenient restriction sites also allows one to monitor discrete "run-off" transcripts originating from within the cloned insert. For example, in Fig. 4 we show a polyacrylamide gel analysis of RNAs synthesized _in vitro_ in a standard transcription reaction[30] using two different pKO derivatives as DNA templates. The RNAs designated P_R and P_{482} are "run off" transcripts generated from the linear template. The RNA designated Pgal $-t_o$ is a transcript initiating at the P_G promoter of PKG-1800 and terminating within the cloned insert at the t_o transcription termination site. The template, in this case, is supercoiled. Further characterization of these transcripts by standard RNA fingerprinting analysis[40] was used to define precisely their _in vitro_ start and stop sites.

The pKO system also provides an excellent means for monitoring RNAs synthesized _in vivo_. As detailed in the appendix, cells containing a pKO derivative can be pulse-labeled with ^{32}P and the labeled RNA isolated. A single-step hybridization procedure is then used to select RNA complementary to the region of interest. This RNA can be characterized by fingerprint analysis. Fig. 5 shows typical T_1 oligonucleotide fingerprints of RNAs prepared _in vivo_ from a pKO derivative and isolated by filter hybridization. The oligonucleotide pattern is specific for the region carrying the cloned regulatory signal. There is no detectable transcription traversing the cloned region which originates from elsewhere on the plasmid. Moreover, because of the multiple number of copies per cell, labeled RNA is obtained in quantities sufficient for several analyses. These techniques have been applied to 1) defining the _in vivo_ 5' initiation site of promoter signals, 2) defining the precise 3' transcription stop site of terminator signals, 3) monitoring the _in vivo_ efficiency of transcription termination at several different terminator signals and 4) monitoring differences in the _in vivo_ efficiency of transcription termination at a specific terminator which has been mutationally altered.

V. THE PHAGE AND BACTERIAL VECTORS-SINGLE COPY SYSTEMS

A. Transfer to the Phage

Genetic recombination of plasmid constructions with phage λ requires that DNA sequences on both sides of the inserted regulatory site be homologous with DNA sequences on phage λ. The procedure for transferring various pKO constructions to phage λ can best be demonstrated with derivatives of the pKG-1800 vector (Fig. 3). DNA fragments are inserted into pKG-1800 between the gal promoter (P_G) and the galK gene. In order to transfer the P_G-insert-galK construction to the phage, a λ derivative, (λ gal8[32]) is used. λ gal8 is a transducing phage which carries the entire gal operon (Fig 6). This phage has two regions which are homologous to sequences carried on all pKG-1800 derivatives:

sequences coding for P_G and galK. Infection of a bacterial strain harboring a pKG-1800 derivative with λ gal8 yields a small percentage (0.1 to 1.0%) of progeny phage which have recombined with pKG-1800 as diagramed in Fig. 6. These λ recombinants are easily detected because they are gal⁻. The ability to select the phage which has recombined by a double, homologous recombination event with the plasmid allows any P_G-insert-galK construction to be transferred to the phage (λ gal1800 derivatives, see Fig. 6).

This same procedure is used to transfer regulatory regions inserted in the pKO-1 vector to the λ gal8 phage. However, an additional manipulation is required to permit the double homologous recombination event. Although all pKO-1 derivatives carry the galK homology with λ gal8, they do not contain the necessary second homologous region upstream of the cloning sites (Fig. 1). To provide this second homologous region, a DNA fragment was obtained by R1 restriction of λ gal8. This fragment contains the region upstream of the gal operon and carries bacterial sequences positioned between the gal operon and the λ phage attachment site, region B in Fig. 7. This fragment is cloned into pKO-1 and pKO-1 derivatives at the R1 site to make pKB-2000 derivatives (see Fig. 7). These pKB-2000 derivatives contain the two homologous regions required to transfer any B-insert-galK construction to λ gal8, as diagramed in Fig. 7. Again, the ability to detect gal⁻ phage allows selection of the appropriate recombinants (see appendix for details). The phage derivatives which carry the B-insert-galK region are called λ gal2000 derivatives (Fig. 7).

The λ gal2000 and λ gal1800 derivatives can be used to lysogenize various bacterial cells. When introduced into a galK⁻ strain as a prophage, these derivatives can be used to study a single copy of any regulatory signal which was originally carried on the multicopy plasmid (see Figs. 8, 9 and appendix). Using this procedure we have transferred several plasmid constructions to single copy and compared the relative galactokinase levels obtained in single and multiple copies. The results in Table 3 indicate that various constructions maintain a constant proportionality in the relative galK levels. We conclude that none of the various regulatory signals which have been cloned into the pKO system cause a significant change in plasmid copy number.

The single copy λ lysogens exhibit the expected growth responses when plated in the presence or absence of galactose on the appropriate indicator plates (Table 4). Moreover, these single copy derivatives more accurately reflect the regulatory functions of the cloned signal within the cellular environment. For example, when the promoters for the gal and lac operons were cloned into pKO-1, they were found to express galK constitutively. Their

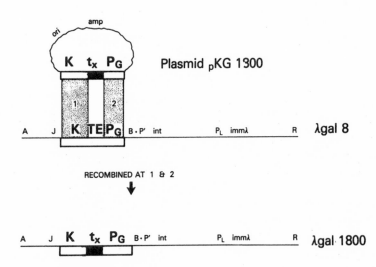

Fig. 6. Transfer of terminator sites from plasmid to λ phage by recombination. Plasmid DNA, wavy lines represent pBR322 DNA; open rectangles represent gal operon DNA, galK gene (K) and gal promoter region (P_G); filled rectangle represents cloned terminator DNA (t_x); in the case of pKG-1800, t_x represents only the cloning sites between galK and galP$_G$. λ gal8 phage DNA, line represents phage DNA with λ genes A-J-int-P_L-immλ-R denoted; bacterial gal operon (KTEP$_G$) is shown as rectangle; B•P' is attachment site of λ gal8 transducing phage. Stippled areas 1 and 2 represent regions of identical sequence carried in phage and plasmid DNA. Recombination in both these regions allows replacement of DNA on phage by plasmid as shown in recombinant λ gal1800.

ability to respond to normal induction had been lost, presumably because the large number of copies of the regulatory signal titrated the host lac and gal repressors. However, upon transferring these constructions to single copy prophages (λ gal2001 and λ gal1800, respectively), normal induction and repression were established. galK expression was now dependent upon galactose induction of the gal promoter and correspondingly, IPTG induction of the lac promoter (Table 4).

402

TABLE III

GALACTOKINASE LEVELS FROM MULTIPLE AND SINGLE GENE COPY

Regulatory Site Terminators*	Multiple Copy K Units **	Single Copy K Units
t_{rl}	210	4
t_o	50	1
t_{IS2}	10	0.3
t_{INT}	50	1

* Each terminator is cloned between P_G and galK;
single copy activity is measured with P_G fully
induced by fucose.

**See Appendix for assay conditions.

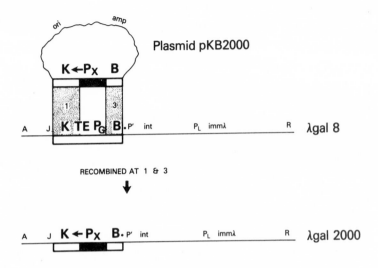

Fig. 7. Transfer of promoters from plasmid to λ phage by recombination. This system is very similar to that described for terminators. The major difference is that homologous sequences are now galK (stippled area 1) and bacterial DNA (B) between gal promoter and attachment site (stippled area 3). B region on plasmid was derived from λ gal8 phage and inserted into pKO-1 in vitro at the Eco R1 site (details in text).

B. Transfer to the Chromosome

It is also possible to transfer the single copy cloned regulatory signal into the bacterial chromosome without accompanying phage λ sequences. In this case, the cloned insert-galK fusion replaces the normal gal operon of the cell, maintaining the control of galK expression by the regulatory signal. The procedure, outlined in Figs. 8 and 9, and detailed in the appendix, involves two steps. First, the phage (either a λ gal1800 or λ gal2000 derivative, Figs. 8 and 9, respectively) is integrated into a gal⁺ strain (N4903) to form a lysogen. The strain is then cured of the phage. The phage occasionally comes out of the bacterial genome by a homologous recombination event occurring between the two copies of galK sequences. The result is the deletion of all phage sequences,◦

TABLE IV

SINGLE COPY PROMOTER DEPENDENT galK EXPRESSION

λ Lysogens in E. coli S165 (Gal E⁻T⁻K⁻)	TB + Galactose		TB - Galactose	
	-IPTG	+IPTG	-IPTG	+IPTG
λ gal2000	+	+	+	+
λ gal2001 (lac)	+	Dead	+	+
λ gal1800 (gal)	Dead	Dead	+	+

λ Lysogens in E coli N100 (Gal E⁺T⁺K⁻)	Maconkey galactose	
	- IPTG	+IPTG
λ gal2000	White	White
λ gal2001 (lac)	White	Red
λ gal1800 (gal)	Red	Red

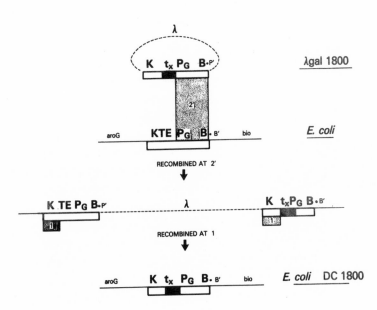

Fig. 8. Transfer from phage to the E. coli chromosome - terminators.
λ gal1800 is the same phage as shown in Fig. 7. The dashed line represents the
circular phage DNA. The rectangular areas are described in the legend of
Fig. 7. The E. coli strain is N4903. The rectangle represents the gal operon
(KTE P$_G$) in the chromosome with bacterial attachment site B B'. Stippled area
2' is identical in phage and chromosome and includes P$_G$ to B B' attachment site.
Recombination within this region of homology by the general recombination pathway or between B·B' of E. coli and B P' of λ gal1800 catalyzed by the phage
site-specific integration system results in formation of the prophage strain
illustrated between the arrows. This stable intermediate can undergo a second
homologous recombination event within galK genes flanking the λ prophage. This
results in loss of phage DNA and the galTE region. The cloned construction is
left in the chromosome. Formally, a reversal of order of the two recombination
events yields the same end product, and only the intermediate has a different
structure.

as well as the original intact gal operon of the cell, leaving the insert-galK

fusion in the bacterial genome. The cells which are cured of phage by this re-

combination event are readily detected because they show a gal⁻ phenotype and

are sensitive to galactose if they produce enough galactokinase. Strains

which are constructed in this way exhibit galK levels similar to the phophage strain containing the same inserted regulatory signal. In addition, we have constructed plasmid derivatives which express the galE and galT genes. When these plasmids are introduced into the bacterial recombinants, they allow selection for galK expression.

C. Potential for Selection of Mutations in Single Gene Copy

The construction of the single copy phage and bacterial vectors also allows the use of standard phage and bacterial genetic techniques for selecting mutations in any cloned regulatory signal. For example, when a terminator is cloned between the gal promoter and galK, galK expression is reduced such that a galK⁻ strain lysogenized by a phage carrying the terminator remains defective for galactose metabolism. Mutations in the terminator site can be generated by direct in vitro mutagenesis of the free phage particle[31]. Terminator mutations carried on the phage are revealed after integration of these phage at the bacterial attachment site of a recA⁻galK⁻ host. These mutants express galK and are able to complement the host galK⁻ defect.

Similar procedures are used for the isolation of promoter mutants. However, instead of galK⁻ hosts, recA⁻galE⁻T⁻K⁻ hosts are lysogenized. In these strains, galK expression in the presence of galactose is lethal. Mutants are selected which fail to kill. These mutations can be either in the galK gene or in the promoter. A simple screening method has been developed to differentiate these two possibilities: we utilize the technique of escape synthesis, that is, transcription of galK from the λ P_L promoter. The λ phage contains a temperature-sensitive repressor that blocks transcription from P_L at 32°C. Shifting the lysogen to 39°C activates the P_L promoter which transcribes through galK. If the mutation is in the promoter insert, then the cell will die at 39°C in the presence of galactose. The ability to select promoter and terminator mutations of both plasmid and phage constructions takes advantage of the different levels of galK produced by the multiple copy plasmid versus the single λ phage. These different levels allow selection of a variety of mutational defects ranging from weak to strong in both terminators and promoters. For example, many weak terminator mutations allow so little galK expression that they are only found when galK is present in plasmid copy number, and are not detected in the single copy prophage. In contrast, weak promoter mutations are selected more readily using the single copy prophage constructions. This is because small changes in promoter efficiency can be detected better from a single copy than from multiple copies.

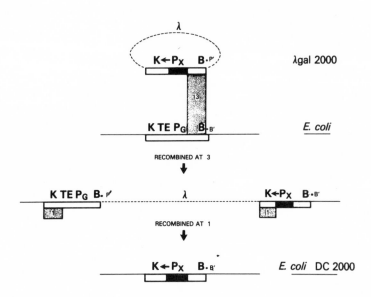

Fig. 9. Transfer from phage to E. coli chromosome - promoters. This system is similar to that described for terminators; the main difference is that the region of homology represented by stippled area 3 is smaller in the promoter constructions.

ACKNOWLEDGMENTS

We thank S. Adhya for providing certain galK⁻ E. coli strains, P. Donoghue and G. Taff for typing the manuscript, and M. Singer and D. Schumperli for helpful comments.

APPENDIX

1. The DNA sequence of pKO-1 from the Hind III site to the AUG of galK

CCAAGCTTACT   CCCCATCCCC   GGGCAATAAG
  Hind III   Sma I

GGCTGCACGC   GCACTTTTAT   CCGCCTCTGC
TGCGCTCCGC   CACCGTACGT   AAATTTATGG
TTGGTTATGA   AATGCTGGCA   GAGACCCAGC
GAGACCTGAC   AGTCAGCGAT   ATCCATTTTC
GCGAATCCGG   AGTGTAAGAAATG
       T stop codon   fmet of galactokinase

2. DNA sequence from R1 to Hind III on pKO-1 (derived from the λ 0 gene)[33]

Eco R1

AATTCTGGCGAATC CTCTGACCAG CCAGAAAACG ACCTTTCTGT GGTGAAACCG GATGCTGCAA TTCAGAGCGC
CAGCAAGTGG GGGACAGCAG AAGACCTGAC CGCCGCAGAG TGGATGTTTG ACATGGTGAA GACTATCGCA
CCATCAGCCA GAAAACCGAA TTTTGCTGGG TGGGCTAACG ATATCCGCCT GATGCGTGAA CGTGACGGAC
GTAACCACCG CGACATGTGT GTGCTGTTCC GCTGGGCATG CCAGGACAAC TTCTGGTCCG GTAACGTGCT
GAGCCCGGCCAAGCTT
        Hind III

3. Restriction Sites from Hind III through galK in pKO-1

	Hind III 0						
Hae III	-5			+290			
Hae II				+270			
Taq I					+440		+1090
Hinf I			+185 +205			+855	
Sma I	+17						
Pvu I						+920	

Sites not on pKO-1; Bam HI, Sal I, Cla I, Hpa I, Kpn I, Sst I and II, Xba I, Xho I and Bst EII.

4. Blunt End Cloning

    For blunt end cloning into the Sma I site of pKO-1, efficient ligation is obtained using a two to four fold molar ratio of insert to plasmid; 0.3-0.6 picomoles of insert to 0.15 picomoles of recipient plasmid, in a 50 μl ligation mix, with 1 Weiss unit of T4 DNA ligase, ligated at 16°C overnight[34].

After ligation, the ligase is inactivated at 65°C for 5 min. The sample is split in two, one half being used directly, and the other half restricted again with Sma I to eliminate those plasmids that have not picked up an insert and simply closed on themselves, recreating the Sma I site. This restriction is done in a volume of 200 µl with an excess of Sma I, 3 units for 3 hours, then DNA is extracted with phenol after addition of 5 µg tRNA carrier and precipitated in ethanol. This DNA is then used for transforming C600 K⁻ cells.

With the Sma I restriction we find that about 80% of the recombinants contain inserts. Note that this procedure cannot be used if the insert contains a Sma I site or regenerates a Sma I site upon insertion.

5. Galactokinase Assay

This assay derives from

Wilson et al., Methods in Enzymology, vol. 8, pp 229.

Adhya and Miller, Nature, vol. 279, No. 5713, pp 492, June 7, 1979.

Cells containing plasmid are grown to log phase (OD_{650} = 0.6) in M-56 medium with fructose as a carbon source. A 1.0 ml sample is removed to a screw cap tube and 40 µl of lysis buffer (Mix No. 3) and a few drops of toluene are added. Samples are vigorously vortexed, then placed in a 37°C shaker to facilitate evaporation of the toluene. At this point the cell lysate is ready to be assayed.

A 20 µl sample is added to 80 µl of a reaction mix consisting of

20 µl mix No. 1

50 µl mix No. 2

10 µl ^{14}C-galactose (Mix No. 4)

Incubate at 32°C for 15-30 min, remove 50 µl to a 2.5 cm diameter DE81 filter and wash with water such that a blank has less then 1000 cpm. Galactokinase units are expressed as nanomoles of galactose phosphorylated per min per ml of cells at OD_{650} of 1.0.

After completion of the assay, 25 µl from two randomly selected sample tubes are transferred to two additional filter discs. These two discs remain unwashed. All of the filter discs are dried under a heat lamp and counted with 5 ml of Econoflour.

Mix No. 1	Mix No. 2	Mix No. 3 Lysis Buffer
5 mM DTT	8 mM $MgCl_2$	100 mM EDTA
16 mM NaF	200 mM Tris-Cl pH 7.9	100 mM DTT
	3.2 mM ATP	50 mM Tris pH 8.0

Mix No. 4 ^{14}C-galactose

We use Amersham D-(1-^{14}C) galactose at 40-60 mCi per µmole, No. CFA 435.
The labeled galactose is diluted to a final specific activity of 4.5 x 10^6 dpm
per µmole. The label is filtered twice through DE81 filters using a Swinnex
filter. Note: depending upon the % of counts converted to galactose phosphate,
the cell sample may have to be concentrated or diluted for the assay. The assay is linear up to conversion of 25% of the galactose to galactose phosphate.
To calculate galactokinase units, the following formula may be used. It assumes a 40% counting efficiency for ^{14}C and that you have used a 20 µl aliquot
for the assay;

$$\text{galactokinase units} = \frac{(\text{cpm - blank}) \times 5200}{\text{average of unwashed filters} \times \text{time of incubation}} \times OD_{650}$$

6. ## Transfer from Plasmid Constructions to Phage λ

Materials: Bacteria: SA1943 with plasmids, S165 (gal KTE deleted)
            Phage: λgal8cI857[32]
Methods: Infection to allow recombination between phage and plasmid.

Cells are grown to saturation in L-broth. Each culture is infected with
λgal8 phage (approximately 10^8 bacteria and 10^2 phage are mixed) at 23°C, and
after 10 min, 2.5 ml of melted 0.6% tryptone agar (at 50°C) are added. The
entire mix is immediately poured on 1% tryptone agar in petri dishes and allowed to harden (∿10 min). Each plate is incubated at 39°C, overnight, to allow
plaque formation. Plaques contain ∿10^6 phage and represent several cycles of
growth of λgal8 upon the plasmid strain. One plaque from each plate is picked
and a mini-lysate of each is made by addition to 5 ml of TM buffer (Tris Cl,
.01 M; MgCl, .01 M pH 7.5).

Screening for Recombinants

λgal8 contains the entire gal operon. When λgal8 infects a strain
deleted for all gal operon genes then only those cells infected are able to
ferment galactose. Thus, when λgal8 forms plaques on such bacteria in the
presence of galactose the lysogenic bacteria in the center of the plaques grow
very dense relative to the uninfected bacteria. λ+ or a λgal8 phage with a
mutation in galK are unable to ferment galactose and make normal turbid plaques
under these conditions. A third type of plaque can also be discerned; λgal8

phage with mutations in galE and T form clear plaques. The cells in these plaques are killed because galactokinase is made and the galactose-1-phosphate formed is toxic to the cells.

Recombination between the plasmid constructions and λgal8 is outlined in Fig. 6 and 7. The recombinant formed has an intact galK gene, but the region galE and T is deleted. This recombinant should make a clear plaque under the conditions discussed above.

The mini-lysates were titered on 1% tryptone agar plates at 32°C with 20 mg per plate of galactose. The cell infected was the bacterial strain S165, deleted for gal genes E, T and K. Approximately 0.1 to 1.0 percent of all phage from the mini-lysates form clear plaques. Clear plaques from each mini-lysate were isolated. Lysates were made on gal deleted hosts to prevent possible gal$^+$ recombinants. They form clear plaques on S165 with galactose and normal turbid plaques at 32°C without galactose. Such phage transduce galK but not galT$^-$ or galE$^-$ strains to gal$^+$. This indicates that they carry only the galK gene on the phage as expected. λgal8 grown on control cells produced fewer clear plaques, < 0.1%, and these were clear also in the absence of galactose.

Phage constructions carrying strong terminators also plated clear on the galactose agar plates; this indicated that kinase was still expressed. One possibility is that the gal promoter was able to read through the terminator; however, a more likely alternative is that the phage P_L promoted transcription reads into the galK region. This has been shown to occur in other experiments where strong terminators are placed between the P_L promoter and galK[35].

Lysogenization

N100 (galK2 recA13) and N4903 (gal+ rec+) were both lysogenized by each phage construction. Single lysogens of N100 were made to measure galacto-kinase expression from one copy of each construction in a prophage. The phenotype of such lysogens was also determined on MacConkey galactose agar. Lysogen of N4903 were made in order to later recombine out the phage to leave the in vitro construction without the prophage in the galactose operon of the E. coli chromosome.

Both strains, N100 and N4903, were grown in L-broth to saturation and 10^8 cells were overlaid on tryptone agar in 2.5 ml of 0.6% agar. Each phage lysate was then spotted on top of the bacterial lawn and incubated at 32°C overnight. Cells from the center of the spot were selected for the presence of the λ

412

prophage. Subsequently monolysogens of N100 were found as low λ phage yielders relative to multiple lysogens after superinfection with a phage of different immunity, λ imm21[36].

Curing of the N4903 Lysogens

Lysogens of N4903 are gal+ and temperature sensitive. N4903 itself is gal+ and each prophage carries the gal-deleted construction. Also, the prophage carries the temperature-sensitive repressor, so that at 39°C the prophage is induced and kills the cell. Recombination (N4903 is rec+) within the homologous re gions of the bacterial gal DNA allow the prophage to delete and the cell to be cured of the phage. Such cells are resistant to heating at 39°C. As out-lined in Figs. 8 and 9, a large portion of these cured recombinants are gal⁻ and carry a single copy of the cloned construction at the position of the galac-tose operon in the host chromosome.

7. In Vivo ^{32}P-labeling and RNA Extraction

Cells containing the appropriate plasmid vector were initially grown over-night in supplemented MOPS medium at 37°C (A_{650} = 4.0). MOPS medium[37] was sup-plemented with 0.2% Difco casamino acids pretreated to remove inorganic phos-phate[38], 25 µg/ml ampicillin and 1.3 mM KH_2PO_4. This culture was diluted 50-fold into MOPS medium containing 0.5 mM $KH_2 PO_4$. Incubation was continued until A_{650} = 0.4, and then cells were pelleted by centrifugation (3000 rpm x 10'). The pellet was resuspended in 0.1 vol MOPS medium without KH_2PO_4 supplement and the suspension was diluted 10-fold into the same medium prewarmed at 37°C. After 5 min of incubation (^{32}P)H_3PO_4 (1-5 mCi, ICN, carrier-free) was added. After a 2 min pulse-labeling, the growth of the culture was stopped by the addition of ice and 0.1 ml sodium azide. The chilled culture was centrifuged and the pellet of cells was resuspended in 2 ml of cold lysis-buffer (10^{-2}M Tris-HCl, pH 7.5; 10^{-3}M EDTA; 10^{-2} M sodium azide, and 0.2 mg/ml egg white lysozyme). The cells were frozen and thawed twice in this buffer. $MgCl_2$ (final conc. 10 mM) and pancreatic DNase (20 µg/ml) were added and the disrupted cells were incubated on ice for at least 30 min. 0.4 ml of acetic acid (0.02 M) and 0.1 ml of sodium dodecyl sulfate (10%) were added and the mixture was incubated at room temperature until the lysate cleared. The lysate was extracted with an equal volume of phenol (pre-equilibrated in RNA buffer: 0.02 M sodium acetate pH 5.2, 0.02 M potassium chloride, 0.01 M magnesium chloride). The aqueous lay-er was reextracted and the RNA precipitated with 2 vol absolute ethanol at -20°C.

The pellet was dried under vacuum and the RNA was dissolved in 0.3 ml water. The RNA solution was passed through a Millipore filter (45 μm) and the filter was washed with 0.2-0.3 ml H_2O.

Hybridization to Select RNA for Fingerprint Analysis

The hybridization procedures are derived from those described in Reference 39. For filter hybridization, the appropriate DNA (ca 80 μg double - stranded λ DNA) was denatured in 0.5 ml NaOH (0.2 M) for 10 min at room temperature, and the reaction was stopped by the addition of 0.1 ml of HCl (1.0 N). The denatured DNA was diluted into 10 ml cold 6x SSC (1xSSC is 0.15 M sodium chloride, 0.015 M sodium citrate, pH 7.0) and collected on nitrocellulose filters (Schleicher and Schuell B6, 25 mm) which were then washed with 100 ml of 6xSSC. The filters were dried overnight at room temperature and then in a vacuum oven at 80°C for two hours. The filters can either be used directly or 4 to 5 smaller discs can be punched out of one large filter prior to drying at 80°C. Each small filter contains sufficient DNA to obtain labeled RNA for fingerprint analysis. The use of these smaller filters significantly improved the quality of the fingerprints.

^{32}P-labeled RNA was denatured at 90°C for 1.5 min and hybridized in the presence of 100 μg of unlabeled carrier tRNA to filter-bound DNA in 2xSSC (1 ml for large filters, 0.6 ml for small filters). Hybridizations were carried out at 65°C for six hours in cases where the 5' nucleoside triphosphate residue of the in vivo-made transcript was to be determined. In all other instances, hybridizations were done for 15-18 hrs.

After hybridization, all filter-bound hybrids were treated with RNase T1 (5 units/ml) in 2xSSC (1 ml) for 20 min at room temperature. The filters were washed with 6xSSC (10 ml), then with 2xSSC (5 ml), and incubated for 3 min at 90°C in distilled water (0.8 ml) to elute the RNA. This elution was repeated with 1/2 volume of water for 1.5 min. The combined eluates were adjusted to 10 mM Tris-HCl, pH 7.5, extracted with an equal volume of phenol and the RNA was precipitated by the addition of 1/10 vol. of 3.0 M sodium acetate and 2.5 vol. of cold ethanol.

The pellet was dissolved in 0.2 ml distilled water and the solution passed through a Millipore filter. The filter was then rinsed with 0.3 ml of water and the RNA was precipitated as described above.

414

Two-dimensional Fingerprint Techniques

The isolated RNA fragments were characterized by standard sequencing techniques,[40]. The RNA was digested with ribonuclease T1 (25 units/100 µg RNA) in 5-7 µl, and the resulting oligonucleotides fractionated in two dimensions. The first dimension was electrophoresis on Cellogel strips at pH 3.5; the second dimension was homochromatography on thin-layer plates of DEAE-cellulose (Analtech) using a mixture of 30 min and 60 min hydrolyzed homochromatography B buffer[40]. The fingerprint was recorded by autoradiography.

REFERENCES

1. Rosenberg, M. and Court, D. 1979. Ann. Rev. Genet. 13, 319-53.
2. Scherer, G.E.F., Walkinshaw, M.D. and Arnott, S. 1978. Nucleic Acids Res. 5, 3759-73.
3. Doi, R.H. 1977. Bacteriol. Rev., 41 (1977) 568-594.
4. Gilbert, W., Starting and stopping sequences for the RNA polymerase, in Losick, R. and Chamberlin, M. (Eds.), RNA Polymerase, Cold Spring Harbor Laboratory, Cold Spring Harbor, NY, 1976.
5. Cuzin, F. and Jacob, F. 1964. C.R. Acad. Sci. 258, 1350.
6. Jacob, F., Ullmann, A. and Monod, J. 1965. J. Mol. Biol. 13, 704.
7. Kessler, D. and Englesberg, E. 1969. J. Bacteriol. 98, 1159.
8. Beckwith, J. 1970. In The Lactose Operon (Zipser, D. and Beckwith, J., Eds.). pp. 5-26, Cold Spring Harbor Laboratory, New York.
9. Miller, J.H., Reznikoff, W.S., Silverstone, A.E., Ippen, K., Singer, E.R., and Beckwith, J.R. 1970. J. Bacteriol. 104, 1273.
10. Reznikoff, W. and Thornton, K. 1972. J. Bacteriol. 109, 526.
11. Franklin, N.C. 1974. J. Mol. Biol. 89, 33.
12. Ketner, G. and Campbell, A. 1974. Proc. Natl. Acad. Sci. 71, 2698.
13. Errington, L., Glass, R.E. Hayward, R.S. and Scaife, J.G. 1974. Nature 249, 519.
14. Mitchell, D.H., Reznikoff, W.S. and Beckwith. 1975. J. Mol Biol. 93, 331.
15. Casadaban, M. 1976. J. Mol. Biol. 104, 541-555.
16. Bassford, P., Beckwith, J., Berman, M., Brickman, E., Casadaban, M., Guarente, L., Saint-Girons, I., Sarthy, A., Schwartz, M., Schuman, H. and Silhavy, T. 1978. Genetic Fusions of the lac operon: A New approach to the Study of Biological Processes in the Operon, Eds. Miller, J. and Reznikoff, W. Cold Spring Harbor Laboratory, Cold Spring Harbor, NY. pp. 245-261.
17. Casadaban, M. and Cohen, S.N. 1979. Proc. Natl. Acad. Sci. 76, 4530-4533.
18. Mercereau-Puijalon, O. and Kourilsky, P. 1976. J. Mol. Biol. 108, 733-752.
19. Merril, C., Gottesman, M., Court, D., and Adhya, S. 1978. J. Mol. Biol. 118, 241-245.
20. West, E.W., Neve, R.L. and Rodriguez, R.L. 1979. Gene 7, 271-288.
21. An, G. and Friesen, J.D. 1979. J. Bacteriol. 140, 400-10.
22. Casadaban, M. and Cohen, S.N. 1980. J. Mol. Biol. 138, 179-208.
23. Michaelis, G. and Starlinger, P. 1967. Mol. Gen. Genet. 100, 210.
24. Adhya, S. and Shapiro, J.A. 1968. Genetics 62, 231.
25. Adhya, S., Gottesman, M., de Crombrugghe, B. and Court, D. 1976. Transcription Termination Regulates Gene Expression. In RNA Polymerase (Losick, R. and Chamberlin, M., Eds.), pp. 719-730, Cold Spring Harbor

Laboratory, NY.
26. Backman, K. and Ptashne, M. 1978. Cell 13, 65-71.
27. Roberts, T.M., Kacich, R. and Ptashne, M. 1979. Proc. Natl.. Acad. Sci. USA 761, 760-764.
28. Rodriguez, R.L., Bolivar, F., Goodman, H.M., Boyer, H.W. and Betlach, M.C. Construction of new cloning vehicles, in Nierlich, D.P., Rutter, W.J. and Fox, C.F., (Eds.), Molecular Mechanisms in the Control of Gene Expression, ICN-UCLA Symposia on Molecular and Cellular Biology, Vol. V, Academic Press, NY, pp. 471-477, 1976.
29. Rodriguez, R.L., Tait, R., Shine, J., Bolivar, F., Heyneker, H., Betlach, M. and Boyer, H.W., Characterization of tetracycline and ampicillian resistant plasmid cloning vehicles, in Ninth Miami Winter Symp., Vol. 13, Academic Press, NY, pp. 73-84, 1977.
30. Rosenberg, M., Court, D., Shimatake, H., Brady, C., and Wulff, D.L. 1978. Nature 272, 414.
31. Miller, J.H. 1972. Experiments in molecular genetics. Cold Spring Harbor Laboratory, Cold Spring Harbor, NY.
32. Feiss, M., Adhya, S., and Court, D. 1972. Genetics 71, 189-206.
33. Scherer, G. 1978. Nucleic Acids Res. 5, 3141-3156.
34. Ullrich, A., Shine, J., Chirgwin, J., Pictet, R., Tischer, E., Rutter, W.J. and Goodman, H.M. 1977. Science 196, 1313-1319.
35. Adhya, S., Gottesman, M. and de Crombrugghe, B. 1974. Proc. Nat. Acad. Sci., USA 71, 2534-2538.
36. Mousset and Thomas, 1968. Dilysogenic excision: an accessory expression of the termination function Cold Spring Harbor Symp. Quant. Biol. 33, 749.
37. Neidhardt, F.C., Bloch, P.L. and Smith, D.F. 1974. J. Bacteriol. 119, 736-747.
38. Summers, W.C., Brunovskis, I. and Hyman, R.W. 1973. J. Mol. Biol. 74, 291-300.
39. Bovre, K. and Szybakski, W. 1971. Multistep DNA-RNA Hybridization Techniques. In Methods in Enzymology (Grossman, K. and Moldave, L., eds.), Vol. 21, part D. pp. 350-383, Academic Press, NY and London.
40. Barrell, B.F. 1971. In Procedures in Nucleic Acid Research (Cantoni, G. and Davies, D., eds), pp. 751-779, Harper and Row, NY.
41. Fukasawa, T. and Nikrido, H. 1961. Biochem. Biophys. Acta 48, 470.
42. Levine, A.D. and Rupp, W.D. 1978. In "Microbiology 1978," Schlessinger, D., eds. pp. 163-166. Amer. Soc. Microbiology.

STRUCTURAL GENE IDENTIFICATION UTILIZING EUKARYOTIC
CELL-FREE TRANSLATIONAL SYSTEMS

BRUCE M. PATERSON* AND BRYAN E. ROBERTS

*National Cancer Institute, Laboratory of Biochemistry
Bethesda, Maryland 20014; Department of Biological Chemistry,
Harvard Medical School, 25 Shattuck Street, Boston, MA. 02115

I. INTRODUCTION

With the advent of recombinant DNA technology it is now possible to isolate, in pure form, a fragment of a gene, or, the entire gene encoding the information for a particular polypeptide. However, the investigator is often
faced with the problem of identifying the recombinant DNA molecules containing
the sequence of interest and must resort to one of a few basic approaches; 1)
recombinants are identified with a specific probe for the sequence of interest,
or, 2) several recombinants must be screened in some fashion to select those
containing the sequence desired. The former situation is rarely the general
case for most researchers, therefore we are going to focus on the application
of methods developed utilizing cell-free translational systems in the identification of specific recombinant molecules. The basis of the approach relies on
the fact each recombinant DNA sequence will hybridize to its corresponding mRNA
and this can be used, therefore, to select the encoded mRNA for subsequent
translation in a cell-free protein synthesizing system to give the final identification of the encoded polypeptide.

We are going to give the details of this approach in its current, most
widely used format with particular note to the following items: 1) methods for
the preparation of the most active mRNA with emphasis on the elimination of nuclease contaminants, 2) preparation of two widely used eukaryotic cell-free
translational systems; the wheat germ extract and the rabbit reticulocyte
lysate; and 3) hybridization mRNA selection procedures in solution and on solid
state supports for the isolation of the mRNA sequences encoded in the recombinant DNA molecules. Each section will contain a "methods" section in order to
maintain continuity.

II. METHODS FOR THE PREPARATION OF mRNA

For years the standard protocol for RNA isolation has utilized phenol
extraction as the basis for the removal of proteins and the inactivation of
nucleases in RNA preparations. However, with the advent of new techniques

TABLE I

RNA EXTRACTION WITH GUANIDINE HYDROCHLORIDE

Stock Solution:
    8 M guandine hydrochloride (BRL cat. no. 5502).
    2 M Sodium acetate pH 5.0: the molarity is based on the
        sodium acetate. Adjust pH with glacial acetic
        acid
    4.5 M Sodium acetate pH 6.0: same as above.
    0.5 M EDTA pH 7.5: mix equal number of moles of NaOH and
        EDTA.

Extraction of RNA
    Extraction is carried out in the cold room on ice.
    Extraction buffer: 8 M guanidine hydrochloride, 25 mM
        sodium acetate pH 5.0, 0.1 M β-mercaptoethanol.

1. For 150 mm cluture plates: drain medium thoroughly, then add 5 ml of extraction buffer to one dish. Scrape cells with spatula and transfer viscous extract to another drained dish. Put the equivalent of five plates into 5 mls of extraction buffer.
For cells in roller bottles: Remove cells with scraping, EDTA or dilute trypsin. In the latter case wash cells with saline to remove trypsin. For approximately 0.1 to 0.2 ml of packed cells add 5 ml of extraction medium while vortexing the cell pellet for immediate disruption of cells.
For whole tissues: Make a 20% w/v homogenate in extraction buffer in a Waring blender. Add tissue while extraction medium is mixing in blender. Blast to top speed for 3 minutes or so until everything is in solution and the DNA is completely sheared (no viscous threads, drops easily from dispo pipette).
Add an equal volume of chloroform and blend for 3 more minutes. Spin the chloroform-guanidine suspension (at 5 kRPM for 5 minutes) and remove upper aqueous phase. Re-extract it in the same manner until the interface is clear, then precipitate the aqueous phase as described below.

The extracts of cultured cells are dounced to shear the DNA. The DNA must be well sheared as otherwise the RNA will not pellet. At this point spin out any particulate material at 5 kRPM for 10 minutes prior to the addition of alcohol.

2. Add <u>one-half</u> volume of ethanol to the extract, vortex and place at -20°C for 24 hours. Prechilling in dry ice speeds the precipitation.

3. Spin at 4-5000 rpm for 15 minutes at 0°C. The low speed is crucial, as otherwise the pellet will not resuspend easily.

4. Pour off ethanol supernatant solution (containing DNA, tRNA, dsDNA, and proteins). For cultured cells dissolve pellet in 1/4 to 1/2 original extraction volume of 8 M guanidine HCl containing 25 mM EDTA pH 7.5. For tissues, redissolve RNA pellet in 1/10 original volume of guanidine hydrochloride. Vortex and heat under tap water to get material into solution. If it is slightly cloudy it is O.K. Add Na acetate, pH 5.0, to 50 mM and add one-half volume of ethanol. Place at -20°C for 2 hours. Repeat this step two more times.

5. Resuspend pellet in 25 mM EDTA pH 7.5 and add 2 volumes of n-butanol: chloroform (1:4 v/v). Vortex vigorously and spin at 4°C 4-5000 rpm to separate phases. Remove upper aqueous phase to a Corex centrifuge tube. Reextract the organic phase 2 more times with 1/2 the first volume of 25 mM EDTA pH 7.5 and pool these extracts with the first. There is often a precipitate of material at the interface. This is to be discarded.

6. To the material in the Corex centrifuge tubes add two volumes of 4.5 M Na acetate pH 6.0 and leave overnight at 4°C. This is a standard high salt precipitation of single-stranded RNA. Spin out RNA at 12-15,000 rpm for 30 minutes at 4°C. Dissolve pellet in water to check concentration and spectra. The 260/230 and 260/280 absorbance ratios should read greater than 2.0 or very close to this. The material is then adjusted to 0.2 M Na acetate, pH 5.0, and precipitated with two volumes of ethanol at -20°C overnight. The final pellet is taken up at 2 mg/ml in water and stored at -70°C.

with which to study the structure and function of RNA it became apparent that phenol extraction was not the panacea once thought. Nuclease inactivation with phenol was not complete, and, depending upon the tissue or culture cells, the RNA yield and template activity varied widely. Cox[1] reported the use of the chaotropic salt guanidinium hydrochloride in the preparation of undegraded RNA from plant tissue, a notoriously rich source of nucleases. The procedure did not involve phenol but relied on the disruptive and solubilizing power of guanidine salts to denature the nucleases and solubilize cellular structure. The approach opened a new era in the methodology of RNA isolation and has been used successfully to prepare RNA from tissues and cells often refractile to phenol isolations. One could now routinely obtain RNA preparations capable of directing the cell-free synthesis of such large polypeptides as myosin (200,000)[2], thyroglobulin (300,000)[3], collagen (180,000)[4] and vitellogenin (220,000)[5]. This step was essential if one was to identify recombinants encoding these large polypeptides with the mRNA selection techniques.

There are currently two different guanidine salts in use; guanidine hydrochloride and guanidine thiocyanate, the later being the most effective denaturant. Procedures for both salts shall be described. The hydrochloride is the most easily available salt with the highest purity so it is the method of choice unless nucleases appear refractile to this salt. The methods are given below in detail.

A. Guanidine Hydrochloride Extraction of RNA

The following protocol in TABLE I is based on the original procedure described by Cox[1] as modified by Chirgwin and Przybyla in W. Rutter's laboratory, and first presented by Strohman et al.[6]

B. Comments on the Guanidine Hydrochloride Extraction Procedure

It is important to maintain the guanidine hydrochloride concentration above 6 M in order to effectively inactivate most nucleases. Enzymatic activity is further retarded if the extraction is carried out in the cold, but the denaturing power of the guanidine is greater at higher temperatures; thus, the optimum procedure must be defined for each particular tissue. In general, it has proven best to do the initial extraction at 4°C. In order to recover the greatest yield of RNA, one should try to maintain a fairly concentrated extract (10-20% homogenate) during the isolation procedure. This is easily done after

TABLE II

GUANIDINE THIOCYANATE RNA EXTRACTION PROCEDURE

Stock solutions:

Same stocks as TABLE I plus,
    4 M Guanidine thiocyanate solution:
    50 g guanidine thiocyanate (Fluka, purum grade, cat. no. 50990)
    0.5 g Sodium-N-Lauroylsarcosine (Sigma, cat. no. L-5125
    0.1 ml β-mercaptoethanol
    5.0 ml 0.5 EDTA pH 7.5
    Water to 100 ml
    Heat slightly to solubilize then filter solution.
    Store in tightly closed bottle at room temperature.

Extraction of RNA

Extractions are done at room temperature
    1. Proceed as in TABLE I for cell disruption and tissue homogenization.
Remove debris with centrifugation prior to the addition of alcohol.

    2. Add 0.75 volumes of alcohol to the guanidine thiocyanate extracts
after adjusting the solutions to 50 mM sodium acetate pH 5.0. Precipitate at
-20°C overnight.

    3. Continue at step 3, TABLE I by dissolving the RNA pellets in 8 M
guanidine hydrochloride, 25 mM EDTA pH 7.5.

the initial extraction. Do not go above a 20% homogenate for tissues or the concentration of guanidine will be near the lower limit. Inclusion of a mercaptan is thought to improve the RNA yield although this has only been documented for pancreatic tissue[7]. It is however routinely included in the extraction solutions.

The greatest problems with the procedure occur during the initial precipitation step. A few key points must be considered: the DNA must be sheared so that the viscosity of the solution does not interfere with the low speed pelleting of the RNA. This is done in a Dounce homogenizer for cultured cells and in the blending step for whole tissues. In addition to the viscosity of the DNA, lipoproteins, glycoproteins, and general tissue matrix material often interfere with the precipitation. We have found that repeated blending of the guanidine homogenate with an equal volume of chloroform removes some undefined viscous or gelatinous material into the chloroform phase. The aqueous phase can now be precipitated as usual after a few chloroform extractions. This was essential for the recovery of thyroglobulin mRNA from rat thyroid glands (R. DiLauro, personal communication). The final RNA preparation appears free of contaminating DNA as judged by the diphenylamine assay[5], depleted of double stranded nucleic acids as judged by the absence of tRNA from the preparation[7], and appears relatively free of proteins. This last contaminant is variable and may cause problems in some applications unless a final phenol extraction is included. We routinely use the RNA without a phenol extraction. However, S1 digestions were irreproducible unless the RNA was first treated with phenol. The RNA is an excellent substrate for protein synthesizing systems and synthesis of cDNA without phenol extraction. Furthermore, electrophoresis of guanidine extracted RNA on fully denaturing gels containing methylmercuric hydroxide demonstrates the RNA to be completely intact.

C. Guanidine Thiocyanate Extraction of RNA

Early reports by the protein biochemists[7] demonstrated that guanidine thiocyanate was about 2.5-fold more effective as a protein denaturant on a per mole basis than guanidine hydrochloride. The former salt is composed of two strongly chaotropic ions while in the later only the guanidine ion is the chaotopic ion. Chirgwin et al.[7] have shown the application of this salt in a similar fashion to that of guanidine hydrochloride in the preparation of RNA from pancreas tissue. The following procedure has been taken from Chirgwin et al.[7] (TABLE II).

A variation of the guanidine thiocyanate procedure has been employed
by N. Davidson's group at Cal. Tech. and involves banding the RNA in CsCl-guan-
idine thiocyanate gradients. It is presented as an alternative procedure in
TABLE III.

D. Comments on the Guanidine Thiocyanate Extraction Procedure

One of the problems inherent in this procedure is the quality of the guanidine thiocyanate. The only reliable source to date has been the Fluka preparation (Purum grade, cat. no. 50990). The Eastman Chemicals product was too
variable in quality and often contained so many impurities that approximately
50% of the initial material would not dissolve. Similar results were obtained
with preparations from K and K Chemicals. If mercaptans are to be included
(essential for pancreatic tissue) one must use only mercaptoethanol with the
guanidine thiocyanate as dithiothreitol reacts with thiocyanate to color the
solution green while releasing hydrogen sulfide gas[7]. The ethanol concentration
is increased in the initial precipitation step to prevent the crystallization of
the guanidine thiocyanate at -20°C.

E. Extraction of RNA with Ribonucleoside Vanadyl Complexes

The following is a modification of the procedures presented by Berger and
Burkenmeier[8]. (TABLE IV).

F. Comments on the Vanadyl Procedure

Unlike the guanidine procedure, this procedure enables one to prepare
cellular fractions enriched for nuclear RNA. In order to obtain active mRNA for
translational systems the vanadyl complex must be completely removed with phenol extraction. Three extractions are usually sufficient. The complex will
inhibit translation.

The potent inhibitory action of the complex on nuclease activity is compatible with cDNA synthesis reactions and allows one to synthesize higher molecular weight single and double stranded cDNAs when concentrations of 1-5 mM
complex are employed in the reaction.

G. Isolation of Polyadenylated mRNA

The total RNA preparations obtained with the guanidine salt procedures are
highly active in cell-free translational systems and in directing cDNA synthesis.

TABLE III

CsCl BANDING OF RNA EXTRACTED WITH GUANIDINE THIOCYANATE

Stock solution:

4 M guanidine thiocyanate solution as in TABLE II.

RNA Extraction Procedure

1. Homogenize tissues or cells in 4 M guanidine thiocyanate solution at room temperature as a 20% homogenate. Shear the DNA completely and centrifuge out any insoluble debris at 5 kRPM for 10 minutes.

2. For each ml of extract add 1 gm of CsCl and put in appropriate centrifuge tubes, depending upon the initial volume. Underlay with 0.3 ml of CsCl solution (p = 1.82) buffered with 50 mM sodium acetate pH 5, 2 mM EDTA. Spin 20-40 hours at 40-50 kRPM at 20°C. RNA will band just below the middle of the tube. Overloading may cause the RNA to form a film which is difficult to redissolve. One can load up to 3 mg of RNA in a Ti. 60 tube containing roughly 35 ml. The band is removed by puncturing the side of the tube. Dilute the CsCl 3-fold with water and precipitate the RNA with 2 volumes of ethanol.

Repeat the precipitation.

TABLE IV

PREPARATION OF RIBONUCLEOSIDE-VANADYL COMPLEX

Stock solutions:

2 M vanadyl sulfate (Fisher, cat. no. V-8); 10 ml in water
250 mM adenosine; boil to solubilize salt.
Mix 24 ml of hot adenosine solution with 3 ml of 2 M vanadyl sulfate.
The solution is a cloudy grey-green.
Adjust to pH 6.0 with 10 N NaOH then to pH 7.0 with 1 N NaOH. When neutrality is reached the solution will turn a clear to dark greenish-black color. Adjust volume to 30 ml with water and store frozen at -20°C in 2-5 ml aliquots. The final stock is 200 mM complex and is used at 10 mM in all solutions. Phenol with hydroxyquinoline will discolor in contact with the complex. Proceed with standard RNA extraction protocols using phenol.

One merely has to increase the concentration of RNA in order to compensate for the ribosomal RNA in the preparation. We assume 5% of the RNA by mass is mRNA. Poly A RNA can be prepared from these preparations by any of the standard protocols using poly-U Sepharose ( ) or poly-U Sephadex[5] ( ) and oligo-dT cellulose chromatography[9]. With regard to the latter, the most satisfactory commercial preparations of oligo-dT cellulose have been obtained from PL Biochemicals. Such preparations routinely bind high molecular weight mRNAs in good yield. Thyroglobulin mRNA (10,000 nucleotides) is efficiently bound only on this commercial preparation.

III. PREPARATION OF CELL-FREE PROTEIN SYNTHESIZING SYSTEMS

We have chosen to present protocols for the preparation of two of the most widely used eukaryotic translational systems, wheat germ extract and the rabbit reticulocyte lysate. Any mRNA-dependent system can be used in the final assay for mRNA selection but these systems are easy to prepare and can be obtained commercially. Their widespread use makes comparisons between researchers much more facile. The following are the procedures we use that have given us reliable performance.

A. Preparation of Wheat Germ Extract - (TABLE V)

Comments on the Wheat Germ Extract

Pool only the most turbid fractions from the G-25 column and keep no more than the original loaded volume for the most active preparation. The nuclease digestion conditions are the same as those described for the reticulocyte lysate in TABLE VI. Freezing and thawing will diminish the activity of the extract, so do not reuse it more than three times. There appears to be a tRNA dependence for the synthesis of high molecular weight polypeptides, as yields can be improved with the addition of exogenous tRNA. This is included in the translation assays.

B. Preparation of mRNA-Dependent Rabbit Reticulocyte Lysate - (TABLE VI)

Comments on the Lysate Preparation

It is very important to use young rabbits 4-6 weeks of age. Older rabbits do not yield active lysates. Weigh the rabbits to be sure they are within the weight range as the dose of acetylphenylhydrazine is adjusted for body weight. Note the protocol does not say to remove the "buffy coat". This layer is just

TABLE V

PREPARATION OF WHEAT GERM EXTRACTS

---

1. Obtain good wheat germ (not toasted or otherwise abused by health stores). W. C. Mailnot, General Mills, Inc., Way Zata Blvd., Minneapolis, Minn. 05426, supplies excellent material free of charge. Otherwise, get stocks from someone already doing protein synthesis.

2. Store over $CaCl_2$ under vacuum at 4°C. The germ is hygroscopic and will pick up moisture and lose viability.

3. Prepare S30's by the following modifications of the protocol of Roberts and Paterson (PNAS, 70: 2330-2334, 1973).

The 30,000 x g supernatant (S-30) of wheat germ was prepared by modifications of published procedures. Wheat germ (12 g) was ground in a chilled mortar with an equal weight of preignited sand (Baker) and 28 ml of a solution containing: 20 mM HEPES (pH 7.6) (adjusted with KOH), 100 mM KCl, 1 mM magnesium acetate, 2 mM $CaCl_2$, and 6 mM 2-mercaptoethanol. The homogenate was centrifuged at 30,000 x g for 10 minutes at 0-2°, and the supernatant was removed, avoiding both the surface layer of fat and the pellet. The S-30 fraction was made up to 3.5 mM magnesium acetate. Finally, 10-12 ml of the S-30 fraction was passed through a 1.5 x 50 Bio Rad disposable column of Sephadex G-25 (coarse), equilibrated with 20 mM HEPES (pH 7.6), 120 mM KCl, 5 mM magnesium acetate, and 6 mM 2-mercaptoethanol at a flow rate of 1.3-1.4 ml/min. The peak of the turbid fraction was pooled and dispensed through a sterile syringe into liquid nitrogen. The frozen spheres were stored thus for 6 months without any detectable loss of activity. Spheres can be stored at -80°C in the Revco.

4. Remember to add the creatine phosphokinase (Sigma) to the extract to a final concentration of 40 μg/ml. The stock is 40 mg/ml in 50% glycerol/H_2O solution.

5. Unpreincubated S30's can be digested with micrococcal nuclease to remove any residual endogenous mRNA in the extract (see TABLE VI). N.B. one obvious endogenous protein which labels with ^{35}S-methionine, at an apparent MW of 94,000 is untouched by preincubation or micrococcal nuclease digestion. This labeling is probably not directed by ribosomes and may be end addition of the methionine. Occasionally, older extracts are less capable of making large proteins (>150,000) than are freshly prepared ones.

6. A number of workers have found a stimulation of synthesis on the addition of polyamines. Subsequently we found that the addition of 40-80 μM of spermine free base or 400-800 μM of spermidine free base not only increased incorporation but also dramatically reduced "early quitters" or prematurely terminated proteins. We find a much higher and broader potassium optimum when potassium acetate is used instead of potassium chloride in the assay. Remember, this is plant material; incubate at 22°C, not at 37°C, for 2 to 3 hrs.

TABLE VI

PREPARATION OF MICROCOCCAL NUCLEASE-TREATED RETICULOCYTE LYSATE

Reagents:
Saline: 0.140 M NaCl; 1.5 mM $MgAc_2$ 5 mM KCl
Acetylphenylhydrazine (Sigma) 1.2%²solution neutralized to pH 7.0 with 1 M
HEPES pH 7.5 Lysate preparation as per Lodish, Methods in Enzymology. Vol.
XXXX, Part F.

Six New Zealand White rabbits weighing 4-6 pounds are made anemic by
subcutaneous injection of 1.2% acetylphenylhydrazine according to the following schedule: 2 ml on day 1, 1.6 ml on day 2, 1.2 ml on day 3, 1.6 ml on day
4, and 2 ml on day 5. On days 7, 8 and 9, the rabbits are bled: One ear is
swabbed with cotton saturated with xylene and a single incision using a new
razor blade is made in the posterior ear vein about midway along the length
of the ear. Each rabbit should yield 50-60 ml of blood, which is collected
into 50 ml of chilled saline containing 0.001% heparin. The blood is filtered
through cheesecloth and then centrifuged at 3500 rpm for 5 minutes. The cells
are washed by centrifugation three times, with the last centrifugation at
7000 rpm. Packed cells are lysed at 0° with an equal volume of cold distilled
water 40 µg/ml hemin. Lysate is centrifuged at 15,000 rpm for 20 minutes.
Aliquots (usually 0.5 ml) of the supernatant are frozen at -80°, at which
temperature template activity is stable for several months. Procedure for
microccal nuclease treatment of retic (as Pelham and Jackson Eur. J. Biochem. (1976) 67: 247.

Stock solutions:
      Hemin-4 mgs/ml in ethylene glycol per normal
      50 mM $CaCl_2$
      100 mM EGTA (Kodak 8276) pH 7.5
      Micrococcal nuclease concluded stored in 50 mM glycine pH 9.2
      150,000 units/ml, 5 mM/CaCl
      CPK Type I Sigma (No. 3755) conc. at 40 mg/ml in 50% glycerol/H_2O

Hemin Stock:
      20 mg Hemin (Kodak)
      0.4 ml 0.2 M KOH (vortex between each addition)
      0.6 ml H_2O
      0.1 ml 1 M Tris-HCl pH 7.8
      (about 7.8 with HCl and adjust if needed)
      Add 4 ml ethylene glycol
      Spin 5 KRPM for 5 minutes; discard insoluble pellet

Digestion Conditions:
      for 100 parts reticulocyte lysate add:
      2 parts $CaCl_2$
      Mix
      Add 0.05 parts micrococcal nuclease
      Mix
      Incubate 15 min 20°C
      Add 2 parts EGTA, mix
      0.4 parts CPK
      Aliquot and freeze

above the majority of the immature reticulocytes, the latter supplying the active components in the translational system. Do not disturb the "buffy coat". The washing cycles will remove the serum components. Lysis is now carried out in the presence of hemin in order to maintain the activity of the initiation factors. The cells in the buffy coat will not lyse and are removed in the high speed spin after lysis of the reticulocytes. For the nuclease digestion, it is a good idea to maximize the surface-to-volume ratio in order to facilitate temperature equilibration for large volumes of lysate. Do not go above 20° during the incubation to avoid activation of translational components in the lysate. Aliquot the lysate into smaller fractions of approximately one ml, as continued freezing and thawing will diminish the lysate's translational activity. Like the wheat germ extracts, synthesis of high molecular weight polypeptides is greatly improved with the addition of exogenous tRNA. The lysate has already been shown to have a tRNA requirement after nuclease digestion, but this improved synthesis of large polypeptides requires a concentration of tRNA on the order of 400 µg/ml, which is approximately 10-fold higher than needed after nuclease digestion of the lysate. The best commercial source for tRNA is Boehringer-Mannheim calf liver tRNA (cat. no. 109576). It can be used directly in the required high concentrations with no inhibitory effects.

C. Assembly of the Translational Assay

Reaction conditions and components for the reticulocyte lysate and wheat germ extract systems are given in TABLE VII. Proper treatment of the components for the translational assay is just as important as the preparation of the extracts. All components of the system are kept on dry ice prior to use, at which time they are thawed on ice, immediately refrozen after use and stored at -80°C. The cocktail described below is equally compatible with the reticulocyte lysate and the wheat germ extract. Polyamines are included to enhance translation[10,28]. 10 ml batches of cocktail are prepared amd then aliquoted into 0.5 ml fractions and stored at -80°C. Cocktail can be frozen and thawed several time without loss of activity. The isotope employed in the reaction depends upon the amino acid composition of the protein under study. ^{35}S-methionine is the most commonly used amino acid because of its high specific activity, ease of detection and its presence in most proteins.

TABLE VII

TRANSLATION PROTOCOLS

Reticulocyte Lysate (treated with micrococcal nuclease ala Pelham Jackson)

25 μl reaction:
Order of addition - on ice
1. H_2O and/or RNA 2-3 μl RNA in H_2O w/o EDTA or phenol.
                              5-10^{2}μg of total RNA or
                              0.5-1μg of A$^+$ RNA.
2. tRNA   1 μl 1.5 mg/ml calf liver tRNA from
                              Boehringer-Mannheim in H_2O.
3. -Met AA   1 μl 2 mM amino acids in water without methionine
                              or labeled amino acid.
4. -AA cocktail   5 μl translation salts without amino acids.
                              Recipe below.
5.* ^{35}S-Met   6 μl aliquot and store at -80°C. Thaw just prior
                              to use and refreeze immediately.
6. Lysate   10 μl (can refreeze 4-5 times). Store at -80°C

Incubate 30°C 1-2 hrs. 1 hour. is sufficient for small polypeptides.
^{35}S-Met at 500-1000 Ci/mmole - Amersham is best

Wheat Germ

25 μl reaction: order of addition - on ice

1. H_2O and/or RNA 9 μl
2. -Met AA   1 μl - 2 mM without methionine or labled amino acid.
3. -AA cocktail   5 μl - Recipe below
4. ^{35}S-Met   5 μl - thaw just prior to use; refreeze immediately on
                              dry ice.
5. S-30   5 μl - can refreeze 4-5 times. Store at -80°C or liquid
                              nitrogen.

Incubate at 23-25°C 2-3 hours.

*
    The wheat germ S-30 has to be treated with micrococcal nuclease prior to use. For every 100 μl S-30 add, on ice; 1 μl of 100 mM $CaCl_2$, and 2 μl of nuclease (1 mg/ml and 8000 units/mg made up in H_2O). Incubate 15' at 20°C. Stop on ice with 2.5 μl of 100 mM EGTA pH 7.5. Ready to use immediately.

To analyze: spot 1 μl from each reaction on 1/2 of a 2.5 cm Whatman 3 MM filter disc. Toss into 10% TCA for 5'. Boil 15' in fresh 5% TCA. Rinse 2X with 5% TCA, 1X with ethanol, 1X with acetone. Dry in oven and count in toluene scintillation fluid with wide ^{14}C window.

-AA coctail contains:

Stocks	µl
1 M KCl	600
100 mM magnesium acetate	100
*100 mM ATP	100
50 mM GTP	40
800 mM Creatine phosphate	100
1 M DTT	20
1 M Hepes pH 7.6 (with KOH)	240
100 mM Spermidine (free base)	60
H_2O	740

* Neutralize ATP in unbuffered Tris°-200 mM Tris° for 100 mM ATP. Aliquot in 200 µl aliquots and store at -80°C.

The ^{35}S-methionine is the most labile component in the reaction and must be handled properly. The material is thawed on ice after puncturing the top of the vial with a syringe needle to release pressure inside the bottle. The methionine is immediately aliquoted into 50 µl samples in Eppendorf tubes and frozen on dry ice. Prolonged exposure to air or warmer temperatures results in the oxidation of the methionine to the sulfone which is inactive for translation. Since methionine is the most commonly used label in translational systems, this is an important precaution. Other amino acids are handled according to their particular requirements and the unlabeled amino acid mixture is adjusted accordingly.

The assembly of the reaction is carried out on ice. Components are added in order according to their stability; RNA and water first with methionine and extract last. The reticulocyte lysate is incubated 2 hr at 30°C instead of 37°C as one can achieve an increased synthesis of high molecular weight polypeptides at this temperature. Wheat germ extract is incubated at 23-25°C as plant translational components give optimal synthesis at these lower temperatures. After the incubation period, hot TCA precipitable counts are measured by standard procedures. The assays can be frozen at -20°C until they can be conveniently analyzed.

IV. IDENTIFICATION OF THE STRUCTURAL GENES IN RECOMBINANT DNA MOLECULES BY
     mRNA HYBRIDIZATION-SELECTION AND CELL-FREE TRANSLATION

The techniques used to identify a particular recombinant molecule rely upon the fact that, under the appropriate conditions, mRNA will form a stable hybrid with its complementary DNA sequence in the recombinant which can subse-

quently be purified away from all of the unhybridized mRNA sequences. The mRNA
is then isolated from the hybrid and translated in a cell-free system to identify the encoded sequence in the recombinant molecule. This specific hybridization is utilized in a number of different methods in which either 1) the DNA
is bound to a solid phase support for the hybridization or 2) the hybrid is
formed in solution prior to its separation from the unhybridized material.
The methodology currently most reliable in the view of the authors shall be
presented in detail for each approach. Direct visualization techniques will
not be described in this chapter. The reader is referred to a review[11] for
in depth coverage of this approach.

V. SOLID STATE mRNA-HYBRIDIZATION SELECTION

Hybridization selection was first used to purify mRNA coding for the major
capsid protein VPI of SV40 by hybridization of total RNA isolated from infected
cells to viral DNA immobilized on nitrocellulose filters[12]. Subsequently,
selections have been carried out with DNA immobilized on nitrocellulose[13,14,15,16]
cellulose[17,18], Sepharose[19], oligo(dT)-cellulose[20] and diazobenzyloxymethyl (DBM)
paper[21].

This approach defines those specific mRNAs whose sequences are encoded
within the DNA. To locate the position of an mRNA sequence more precisely
within the DNA, the DNA is digested with the appropriate restriction enzyme(s),
and the resulting fragments are fractionated by electrophoresis in agarose gels
and purified from the agarose by various methods. The isolated DNA fragments
are immobilized on a support and used to select mRNA. Alternatively, a more
convenient approach utilizes the blotting method of Southern[22], in which the
DNA fragments are directly transferred from the agarose gel onto nitrocellulose
sheets[23] or DBM paper[21]. The portion of the support containing the fragment
is then cut out and used in hybridization selections to identify the mRNA sequences complementary to specific DNA fragments as determined by cell-free
translation of the mRNA. Of all of the solid state supports used, we have found
nitrocellulose to be the most convenient and reliable material due to its relatively low cost, uniform quality, ease of handling, high capacity for nucleic
acids, availability and applicability to other methods used in recombinant DNA
research. There is no need to use the more expensive and unpredictable derivatized cellulose supports since nitrocellulose can be used to bind DNA, RNA, and
protein very effectively[24].

The hybridization conditions are the most controversial part of the procedure and are constantly changing; thus, we can only give the conditions that we find the most reliable at the present time. Formamide is often a component in the hybridization reaction and is likely to be the culprit if things do not work. If it is to be used, the material must be deionized either with resins or by recrystallization[25]. The latter method is preferred since there is no material introduced into the formamide as there is in the case of the deionizing resins. Store the clean formamide at -70°C in 1 ml aliquots. The inclusion of formamide en ables one to hybridize at lower temperatures, thus preserving mRNA, and to block DNA reassociation while encouraging DNA-RNA hybridization. This is not so important in the solid-state hybridization conditions but is essential in the liquid hybridization selection procedures. We have recently been using the same hybridization conditions as those presented in TABLE VIII at slightly higher temperatures, 50°C, without formamide, for shorter periods of hybridization (2-4 hrs). The rates of hybridization are faster without formamide[25] so the time of hybridization can be decreased. Background varies with RNA preps and most probably reflects the protein contaminating the RNA: more protein helps to nonspecifically bind RNA to the filter which is then eluted with the boiling step. Our typical procedure for solid-state selection is given in TABLE VIII.

VI. LIQUID PHASE HYBRIDIZATION SELECTION OF mRNA

The hybridizations are carried out in formamide under conditions allowing only DNA-RNA hybrids to form[25]; these are the standard R-loop conditions.

In this procedure, DNA-mRNA hybrids are separated from the unhybridized RNA and single-stranded DNA by column chromatography[26]. The advantage of performing the initial reaction in the liquid phase is that hybridization occurs faster than when the DNA is immobilized. The DNA molecules containing R-looped RNA are fractionated from the unhybridized RNA by gel filtration chromatography on agarose A-150 M equilibrated in 0.8 M NaCl. The fractionation is based on the fact that, in high concentrations of sodium ion, double-stranded DNA is in an extended conformation whereas unhybridized RNA exists in a compact collapsed structure. Therefore, upon gel filtration in agarose A-150 M, the DNA and DNA containing R-loops are excluded from the matrix, whereas the unhybridized RNA and single-stranded DNA are included. Both fractions are collected, heat-dissociated and translated in a cell-free system to identify the polypeptide encoded by the mRNA that hybridized to the DNA. A disadvantage

TABLE VIII

SCREENING OF GENOMIC OR cDNA CLONES BY mRNA SELECTION

I. Application of DNA to nitrocellulose

   A. Use linearized DNA (mini preps of banded phage/plasmids)
       1. If plasmid DNA, first linearize by restriction digestion
       2. Phenol/chloroform extract to remove enzyme
       3. Suspend EtOH-precipitated. DNA in TE (10 mM Tris, pH 7.5; 2 mM Na_2 EDTA)
   B. Dose of DNA
       1. 1-5 $\mu g/cm^2$
       2. Higher concentrations of DNA come off of the paper during hybridization and hybridize mRNA in solution
   C. Spotting of DNA
       1. Mark off a 1 cm square on nitrocellulose paper spot DNA i TE (or suitable buffer) onto dry nitrocellulose paper using a 20 μl micropet. Try to keep the spot small by letting it air dry between applications
       3. Let air dry 30'-1 hr
       4. Alkali denature DNA on paper by laying nitrocellulose over What-man 1 paper (3 layers) soaked in:
           0.5 N NaOH
           1.5 M NaCl
           Repeat 3 times for 1 min each, letting area blot lightly between applications on a sheet of 3 MM paper
       5. Neutralize in same manner with:
           2 M Tris-HCl (pH 7.4)
           2 X SSC
       6. Wash in 2 X SSC in the same manner. Then immerse in 2 X SSC for 30' with shaking to wash
       7. Air dry
       8. Bake at 80° for 2 hours

II. Hybridization, Washing, and Elution Conditions

   A. Cut the 1 cm square containing the DNA into 16 squares with a razor blade
       1. Can wet with TE to aid slicing
       2. Place pieces in microfuge tube with a syringe needle
   B. Pre-hybridize for 30 min in 100 μl hybridization buffer
       Remove buffer
   C. Hybridize overnight at 37° in 30 μl or less of hybridization buffer

   Hybridization Buffer:
       50% formamide (Fluka, puriss.)
       0.1 M Tris-HCl (pH 7.5) or 40 mM PIPES (pH 6.4)
       0.75 M NaCl
       0.002 M Na_2EDTA (pH 7.5)
       0.4% SDS
       Add 10 μg tRNA as carrier (200 μg/ml single-stranded E. coli DNA, 500 μg/ml tRNA)
       Add 10 μg Poly A + mRNA for selection

D. Wash Conditions (Jet Washes)
   1. 5 times with hybridization buffer at 37°
   2. 3 times with 5 mM Na pyrophosphate pH 7.5 at room temperature (optimal)
   3. 5 times with 10 mM Tris pH 7.5 2 mM EDTA at 52°C
E. Elute selected mRNA in 300 μl of:
       10 mM Tris-HCl pH 7.5 or 10 mM PIPES pH 6.4
       2 mM Na_2EDTA
       5 to 10^2 μg carrier tRNA
   by placing in boiling water bath for 1-1.5 minutes; quick quench.
F. EtOH ppt. mRNA with 3 volumes of EtOH after adjusting salt to 0.2 M NH_4Ac - 10 mM $MgCl_2$
   1. At lower tRNA concentrations you can quantitatively precipitate mRNA by adding $MgCl_2$ to a final concentration of 10 mM
G. Wash precipitated RNA 2 times with 70% EtOH containing 50 mM NH_4 Acetate
H. Dry pellet
I. Resuspend in 5 μl water for translation or directly in the incubation mixture

III. Comments
   A. Can hybridize for shorter time (i.e., 4 hrs) but overnight gives a better selection
   B. We have tried denaturing and neutralizing DNA prior to spotting on nitrocellulose paper with poor results. (Procedure of Kafatos, Jones and Efstratiadis, Nucleic Acids Research 7:1541-1552 (1979)
   C. We have also restriction-digested phage λ-containing cloned DNA prior to spotting on nitrocellulose and found this step not necessary.

of this approach, particularly for refined mapping of mRNAs, is that isolated DNA fragments are required. In addition, the separation on agarose of the included RNA and excluded small DNA fragments is less pronounced. This latter difficulty could be overcome by ligating the small DNA to a larger carrier DNA.

The same rationale for liquid hybridization was employed by Persson[27] utilizing Sepharose as the matrix for the gel filtration. Hydroxylapatite has also been used to separate single-stranded, unhybridized RNA from the duplex DNA-mRNA complex[28].

Alternatively, the hybrids can be selected by filtration through nitrocellulose in moderate salt solutions. The hybrids are retained on the filter by virtue of the long single stranded DNA tails in the hybrid. Single-stranded DNA will also bind but unhybridized RNA will wash through the filter in moderate salt solutions (greater than 0.3 M NaCl). The hybridized RNA is then removed by melting the hybrid as in TABLE VIII.

VII. HYBRID-ARRESTED TRANSLATION

In contrast to hybridization selection which identifies DNA molecules which encode particular mRNA sequences, hybrid-arrested translation (HART) identifies those sequences within the DNA containing coding information for the polypeptide of interest. This approach was based upon the observation that mRNA in hybrid form with its complementary DNA is not translated in eukaryotic cell-free systems, while heat dissociation of the hybrid returns complete trans-lational activity[13,30].

HART is accomplished by hybridizing mRNA to specific DNA fragments under conditions of high salt and formamide that favor DNA-mRNA hybrid formation over DNA/DNA reannealing. Half of each completed reaction is maintained in the hybrid form, while the other half is disrupted by heat melting and serves as a control. Both reactions are introduced into an eukaryotic cell-free protein synthesizing system and the ^{35}S-methionine-labeled polypeptides are analyzed on SDS-polyacrylamide gels. (TABLE IX).

A number of different DNA-mRNA interactions may occur and the location of this duplex relative to the mRNA coding sequences has characteristic effects on its translation. A lack of complementarity between the mRNAs and the DNA used in the hybridization results in no duplex formation and does not alter the translation of the mRNA. Duplex formation involving the entire translated sequence or that portion of it corresponding to the N-terminus of the polypep-tide abolishes its translation. Interestingly, duplex formation involving only the C-terminal portion of the coding sequence may result in the synthesis of a truncated polypeptide[13]. In all of these examples, the full translational acti-vity of the hybridized mRNA is restored upon boiling the reaction.

Duplex formation on the 3' terminal side of the coding sequence of an mRNA has no detectable effect on translation in vitro[30]. Duplex formation on the 5'-terminal side of the coding sequence sometimes results in a reduction of translational activity, but this effect is variable and has never completely eliminated messenger translation[29].

Hybrid-arrested translation has been successfully applied in numerous in-stances to identify DNAs containing sequences complementary to specific mRNAs[29]. However, since it is a subtractive method, it is limited to mRNAs whose encoded polypeptide is both obvious and clearly resolved from other polypeptides on the final SDS-polyacrylamide gel.

Another important application of HART is in defining the location within an mRNA of those sequences necessary for its translation[29]. This information, combined with that from hybridization selection will provide a complete picture of the non-coding sequences and those sequences necessary for the translation of the mRNA. It should be noted that arrest of translation could be the result of either hybridization to coding sequences in the mRNA or to controlling regions required for translation of the coding sequences.

The coding sequences can only be located precisely if a number of truncated polypeptides are produced by duplexes formed at different positions within the carboxyterminal portions of the coding region. The sizes of the truncated polypeptides can be directly correlated to the position of the duplexes on the DNA and thus accurately define the location of the coding sequence. The precise location of the coding region can ultimately be achieved by determining the DNA sequence of the appropriate DNA fragment.

TABLE IX

HYBRID-ARRESTED TRANSLATION
_____

Stock Solutions:
    10 X Hybridization buffer: 4 M NaCl
                                     100 mM Pipes, 6.4
                                     20 mM EDTA
    Deionized Formamide   Fluka, puriss.

Reaction Conditions:
    Very clean DNA. Linearize plasmids with insert, or phage DNA with restriction enzymes. Phenol/chloroform extract after digestion and precipitate with ethanol.

Mix RNA and DNA (5 x DNA sequence excess) in minimum volume of 10 mM PIPES pH 6.4 and boil for one minute. Chill quickly on ice. Add formamide first, then 10 x hybridization salts to 1 x.
Place at 52°C for 2-4 hrs.

Split reaction in equal portions and dilute 20 fold with water. Boil one portion for 30 seconds and chill quickly.

Add 10 µg of tRNA and precipitate both reactions. Translate as per usual.

REFERENCES

1. Cox, R. A. (1968) Methods Enzymol. 12B, 120-129.
2. Paterson, B. M. and Bishop, J. O. (1977) Cell 12, 751-765.
3. DiLauro, R., personal communication.
4. Adams, S. L., Alwine, J. C., Pastan, I., and deCrombrugghe, B. (1979) J. Biol. Chem. 254, 4935.
5. Gordon, J. I., Deeley, R. G., Burns, A. T. H., Paterson, B. M., Christmann, J. L., and Goldberger, R. F. (1977) J. Biol. Chem., 252, 8320-8327.
6. Strohman, R. C., Moss, P. S., Micou-Eastwood, J., Spector, D., Przybayla, A. E. and Paterson, B. M. (1977) Cell, 10, 265-273.
7. Chirgwin, J. M., Przybyla, A. E., MacDonald, R. J. and Rutter, W. J. (1979) Biochem., 18, 5294-5299.
8. Berger, S. L. and Birkenmeier, C. S. (1979) Biochem., 18, 5143-5149.
9. Aviv, H. and Leder, P. (1972) Proc. Natl. Acad. Sci. USA, 69, 1408-1412.
10. Hunt, T. personal communication.
11. Davis, R. W., Simon, M. N. and Davidson, N. (1971) Methods Enzymol., 21, 413-428.
12. Prives, C. L., Aviv, H., Paterson, B. M., Roberts, B. E., Rozenblatt, S., Revel, M. and Winocour, E. (1974) Proc. Natl. Acad. Sci. USA, 71, 302.
13. Paterson, B. M., Roberts, B. E. and Kuff, E. L. (1977) Proc. Natl. Acad. Sci. USA, 74, 4370.
14. Buttner, W., Veres-Molnar, Z. and Green, M. (1974) Proc. Natl. Acad. Sci. USA, 71, 2951.
15. Harpold, M. M., Dobner, P. R., Evons, R. M. and Bancroft, F. C. (1978) Nucleic Acids Res., 5, 2039.
16. McGrogan, M., Spector, D. J., Goldenberg, C. J., Halbert, D. and Raskas, H. J. (1979) Nucleic Acids Res., 6, 593.
17. Noyes, B. E. and Stark, G. R. (1975) Cell, 5, 301.
18. Shih, T. Y. and Martin, M. A. (1974) Biochemistry, 13, 3411.
19. Gilboa, E., Prives, C. L. and Aviv, H. (1975) Biochemistry, 14, 4215.
20. Venetianer, P. and Leder, P. (1974) Proc. Natl. Acad. Sci. USA, 71, 3892.
21. Goldberg, M. L., Lifton, R. P., Stark, G. R. and Williams, J. G. (1979) Methods Enzymol., 68, 206 .
22. Southern, E. M. (1975) J. Mol. Biol., 98, 503.
23. Ricciardi, R. P., Miller, J. S. and Roberts, B. E. (1979) Proc. Natl. Acad. Sci. USA, 76, 4927.
24. Bowen, B., Steinberg, J., Laemmli, U. K. and Weintraub, H. (1980) Nucleic Acids Res., 8, 1-20.
25. Casey, J. and Davidson, N. (1977) Nucleic Acids Res., 4, 1539-1552.
26. Rosbash, M., Blank, D., Fahrner, K., Hereford, L., Ricciardi, R. P., Roberts, B. E., Ruby, S. and Woolford, J. (1979) Methods Enzymol., 68, 454 (Academic Press).
27. Persson, H. (1979) Dissertation (Uppsala, Sweden).
28. Lewis, J. B., Atkins, J. F., Anderson, C. W., Baum, P. R. and Gesteland, R. F. (1975) Proc. Natl. Acad. Sci. USA, 72, 1344.
29. Miller, J. S., Ricciardi, R. P., Roberts, B. E., Paterson, B. M. and Mathews, M. B., J. Mol. Biol., in press.
30. Kronenberg, H. M., Roberts, B. E. and Efstratiadis, A. (1979) Nucleic Acids Res., 6, 153.

IN VITRO TRANSLATION OF EUKARYOTIC MESSENGER RNA

DON HENDRICK

Bethesda Research Laboratories, Inc.
411 N. Stonestreet Avenue
P.O. Box 6010
Rockville, Maryland 20850

I. INTRODUCTION

In the eleven years since the identity of mouse globin messenger RNA was
formally established by the demonstration of its biological activity in a rabbit
reticulocyte lysate[1], in vitro translation systems have become indispensible
tools for the molecular biologist and have provided exciting and important
areas of research into the mechanisms and regulatory events of protein
synthesis. It is the aim of this review to describe the relative merits of the
currently available cell-free translation systems, to consider some of the
practical aspects of cell-free mRNA translation and identification of the
translation products, and to briefly describe some of the current applications
of cell-free translation.

II. SYSTEMS FOR mRNA TRANSLATION

The earliest used, and still perhaps the most active cell-free system,
is the reticulocyte lysate. This lysate is prepared from immature red blood
cells, or reticulocytes, of a number of different animal species after induction of anaemia with phenylhydrazine. Such lysates, when supplemented with
hemin and a system for energy regeneration, continue to synthesize globin
chains at rates close to those of intact cells[2]. Addition of exogenous mRNA
results in its efficient translation, and such lysates, usually made from
rabbit reticulocytes, are used to translate a number of eukaryotic mRNAs. The
details of the preparation and use of such reticulocyte lysates are reported
in detail[2-7]. The major disadvantage of this original lysate, however, is in
its high endogenous globin mRNA activity. Hence, the biological activity of an
added mRNA cannot be monitored simply by the stimulation of incorporation of a
radioactive amino acid precursor into acid-soluble products and sophisticated
techniques of product analysis are required for each translational assay.

This problem led to the extensive use of two mRNA-dependent in vitro
translation systems derived from wheat germ[8-11] and from mouse ascites
cells[12,13]. A supernatant fraction capable of supporting mRNA-dependent protein synthesis can be made from both of these sources. Wheat germ extracts
generally have low endogenous mRNA activity and ascites cell extracts can be
preincubated to reduce their endogenous activity. Therefore, both systems are
mRNA dependent and the activity of added mRNA can be easily and conveniently
monitored. In addition, both extracts can be passed over a Sephadex-G-50
column to effect a very significant reduction of the endogenous amino acid
pools. This result leads to efficient labelling of translation products

which are of high specific activity. However, the translational activity of both the wheat germ and ascites cell-free systems are substantially less than that of the reticulocyte lysate. Thus, mRNAs are probably only translated 1-5 times in wheat germ or ascites extracts, compared with up to 50 times in the reticulocyte lysate[14]. In addition, there may be lower efficiencies of the production of complete translation products from very large mRNAs in wheat germ and ascites extracts. This deficiency is probably related to both premature peptide chain termination, as well as to significant levels of ribonuclease and protease activities in the extract. Nevertheless, due to its relative ease of preparation, the wheat germ cell-free system in particular is still widely used for the effective and efficient translation of many eukaryotic mRNAs.

A recent procedure, developed by Pelham and Jackson[15] resulted in a major return to using the reticulocyte cell-free system. This modification involved the treatment of the lysate with a calcium-dependent micrococcal nuclease to digest the endogenous globin mRNA. The nuclease activity is then removed by binding of Ca^{++} ions with the chelating agent (EGTA) to leave the system totally mRNA-dependent but retaining up to 70% of its original protein synthetic activity. These features, together with the virtual absence of endogenous nuclease or protease activities[15] and its recent commercial availability, make the rabbit reticulocyte lysate the system of choice for cell-free translation, particularly for large mRNAs. The only disadvantage that remains is the presence of substantial endogenous pools of some amino acids. It has been reported[16,17] that the reticulocyte lysate can be desalted by passage over Sephadex G-50, but it is not yet clear what effect this has on the biological activity of the lysate or on its storage properties.

The general principle of nuclease-treatment has now been extended to other cell-free systems, most recently to develop mRNA-dependent translation systems from Drosphila tissue culture cells and embryos[18] and yeast spheroplasts[19]. Such systems will doubtless be of use in analyzing detailed mechanisms of protein synthesis, since both of these organisms are readily amenable to genetic manipulation.

The final system to be considered is not a "cell-free" system, though it has been used to translate a variety of eukaryotic mRNAs. This system is the microinjected amphibian oocyte, pioneered as a system for mRNA translation by Gurdon and his collaborators[20-22]. While this system requires some sophisticated equipment for micromanipulation and microinjection and a source of oocytes, its major advantages are its sensitivity (as little as 10 ng. of mRNA can be readily detected) and the stability of the mRNA in the oocyte after injection. It has been reported that injected globin mRNA can continue to be translated for as many as 100,000 times over a period of as long as two weeks[21]. In addition, the transcription of sequences from injected DNA or whole nuclei have been detected[22] and such transcripts may be processed to mature mRNA molecules within the oocyte[23].

III. TECHNICAL ASPECTS OF IN VITRO MESSENGER RNA TRANSLATION

A. Total RNA Isolation

The mRNA to be translated must be intact and free of any other components that can inhibit cell-free translation. Provided these criteria can be met, total cell or tissue RNA can be added to translation systems and the mRNA present successfully translated, even though mRNA normally represents only 1-2% of the total RNA of a cell. Great care must invariably be exercised, however, in the isolation and handling of RNA required for cell-free translation. Ribonucleases which will very rapidly destroy the biological activity of mRNA are ubiquitous and very stable. It is therefore important to maintain very vigorous standards of cleanliness and often sterility in reagents and glassware. The use of diethyl pyrocarbonate as an inhibitor of nucleases and as a sterilant for materials and reagents which cannot be autoclaved has been described in detail[24]. A list of other commonly used RNase inhibitors can be found in a recent review[25].

There are a variety of procedures currently used to isolate RNA[14,25-27]. Most of these methods involve the deproteinization of cells or tissues with phenol as a denaturant in the presence of a detergent to dissolve ribonucleoprotein complexes and to inhibit ribonuclease activity. Often the most critical step in RNA isolation is the initial homogenization, particularly in tissues where nuclease levels are high. The recent modifications[28,29] of a procedure initially developed by Cox[30] involve the direct homogenization of cells or tissues with high concentrations (6M) of the powerful denaturant guanidinium hydrochloride, followed by the differential precipitation of RNA with ethanol. The very rapid inactivation of endogenous nucleases is probably the reason why this procedure has yielded RNA preparations with high biological activities of mRNAs for very large proteins, such as vitellogenin[29], myosin heavy chain[28,31], chick type I procollagen and fibronectin[32], chick chondroitin sulfate proteoglycan core protein, and type II procollagen[33]. A further variation involving an even more powerful denaturing agent, guanidinium thiocyanate[34], has allowed for successful isolation of biologically active α-amalyse mRNA from such nuclease-rich tissues as the canine pancreas.

B. Messenger RNA Isolation

Most eukaryotic mRNAs have at their 3' terminus a tract of polyadenylic acid. This poly A "tail" allows the ready separation of poly A^+ mRNAs from the bulk of the cellular ribosomal RNA by affinity chromatographic procedures. A number of affinity matrices have been used for this purpose. The most commonly used matrix is oligo-dT cellulose[35], but others include poly U-agarose[36,37], nitrocellulose filters[38,39], unmodified cellulose[40-43] and, recently, poly U bound to Sephadex G-10 beads[29]. The principles and procedures for using these matrices were reviewed in detail[25]. With poly U-agarose the contamination of the poly A^+ mRNA fraction with rRNA is much less than for oligo-dT cellulose[44]. However, it was recently reported[29] that very large mRNAs, such as those for chick vitellogenin, can be excluded from the agarose beads of poly U-agarose and thus be unable to bind with poly U attached inside. Poly U-Sephadex G-10 was used to overcome this problem and to obtain high yields of vitellogenin mRNA[29]. The matrices containing immobilized poly U also seem to be more efficient at binding mRNAs with short poly A tracts than is oligo-dT cellulose, but the lability of the poly U, which can elute bound to the poly A tail, can cause problems, particularly if the mRNA is required for reverse transcription. Further purification of individual mRNAs from total poly A^+ mRNA is usually performed on the basis of the mRNA size, i.e., by sucrose density gradient centrifugation or gel electrophoresis.

C. Messenger Ribonucleoprotein Complexes

It is likely that inside cells mRNA exists as a complex of RNA and protein, rather than as free RNA molecules. Two lines of evidence lead to this conclusion. First, treatment of isolated polysomes (the complex of mRNA and ribosomes actually engaged in protein biosynthesis) with chelating agents such as EDTA causes the release of mRNA-containing particles consisting of about 60% protein and 40% RNA[45]. In the case of reticulocyte polysomes, globin mRNA was released as a 15S ribonucleoprotein (RNP) particle before the demonstration of the 3' poly A tract on eukaryotic mRNAs. This method was used as a means of purifying globin mRNA[46-48]. Other messenger RNAs have also been purified by the isolation of specific mRNP particles, including those for chick keratin[49], calf lens crystallin[50,51], trout testes protamine[52] and chick myosin[53]. In addition, other procedures have been developed to allow the purification of mRNAs directly from dissociated polysomes using oligo-dT cellulose[54,55]. This binding of the poly A tract to oligo-dT suggests that at least a part of the poly A in mRNA is not covered by protein in mRNPs and is accessible for interaction with the oligo-dT.

Second, a substantial portion of cellular mRNA is often not associated with polysomes[26,56-58]. In duck reticulocytes, globin mRNA exists as free 20S mRNPs in addition to the 15S mRNP that can be recovered by EDTA treatment of polysomes[59]. In unfertilized sea urchin eggs, mRNA is sequestered as mRNP particles that are unable to be utilized for protein synthesis[60-61].

Despite a large number of studies (see 56-58), the biological significance of most mRNP particles remains unclear. This is due, at least in part, to the difficulty in preparing and handling highly purified mRNPs. The rabbit globin mRNP contains two major protein components[48], one of which may be involved in binding the poly A segment of the mRNA[62].

Isolated globin mRNPs can be translated in cell-free systems. Rabbit globin mRNP is able to be translated 30-40% more efficiently in cell-free systems derived from ascites cells and from chick embryo brain than is the purified mRNA[63,64]. Other reports have suggested, however, that globin mRNPs were no more efficient as templates in cell-free systems than 9S mRNA[65-67], although definitive product analyses were not done in these experiments.

In any event, the above observations, together with the fact that naked mRNA functions efficiently as a template in cell-free protein synthesis, suggest that the proteins of mRNPs may not be absolutely required for protein synthesis and that they are unlikely to confer any tissue specificity to the translation of particular mRNAs[64].

D. Optimization of Cell-Free Translation

Translation of mRNA is generally performed to identify the biological activity of a particular mRNA species. The cell-free translation system chosen must be optimized for the translation of that particular mRNA since the optima for a number of parameters, such as time, temperature and RNA concentration, may be solely a function of the translation system and may be optimized with standard mRNAs, usually globin mRNA. However, others, particularly the optimal concentration of monovalent cations such as potassium may vary with different mRNAs. Thus, rat liver albumin mRNA, when translated in the wheat

germ cell-free system, requires higher levels of potassium ions than does total liver mRNA[68,69]. When optimizing cell-free translation systems for K^+ ion
concentration, it is important to use the acetate salt of this cation (with the
pH of the solution appropriately adjusted) since high levels of chloride ions
can inhibit cell-free translation by interfering with the initiation process[70].

The addition of polyamines such as spermine or spermidine to cell-free
systems to stimulate translation has been reported for the wheat germ cell-
free system[71], and these polyanions have also been used in the reticulocyte lysate[17]. The addition of polyamines seems to increase the yield of full-length
translation products and to lower the Mg^{++} concentration required for optimal
cell-free synthesis.

The translation of any one particular mRNA among a mixture of mRNAs may
also be influenced by competition for some rate-limiting components of the
translational machinery. Such competition forms the basis of a model of translational regulation proposed by Lodish[72]. Some experimental evidence for such
phenomena in cell-free systems exists[73]. The various parameters which need to
be considered when attempting to determine the levels of translatable mRNA for
a particular protein, especially if that protein is large, are well illustrated
in a study of the translation of myosin heavy chain mRNA[31].

Finally, it may be important to denature the mRNA before translation. The
usual procedure is to subject the RNA to a brief heat treatment, followed by
rapid chilling. Recently, it was reported that treatment with the denaturant
methyl mercury hydroxide led to the more efficient translation of conalbumin
mRNA, as well as to more efficient transcription with reverse transcriptase[74].
However, adequate safety precautions must be used when dealing with this very
toxic compound.

E. Identification of the Cell-Free Translation Product

There are a number of methods used to establish the identity of a cell-
free translation product. A general first step is to determine the size of
the product by SDS polyacrylamide gel electrophoresis, followed by fluorography. However, a great many proteins, particularly those destined for export
from the cell, are synthesized as precursor forms (pro, or pre-pro forms - see
below). In such cases the size of the primary translation product is often
larger than that of the final mature protein.

A second common means of identification involves the precipitation of the
cell-free product with a specific antibody to the protein in question. The ab-

solute amounts of translation products made by cell-free systems are almost always very small. Thus, the "sandwich" technique, whereby the first antibody
with the specific translation product bound to it is precipitated either by a
second antibody or by such reagents as protein A of Staphylococcus aureus, is
commonly used. The resulting precipitate is then washed, solubilized and analyzed by SDS-polyacrylamide gel electrophoresis. Such procedures are used to
identify the majority of the vast number of eukaryotic mRNAs that are translated in vitro.

Additional criteria for the identity of cell-free translation products
which have been used are their chromatographic behavior on ion exchangers or
molecular sieves, and most rigorously, their peptide maps or their amino acid
sequence. In a very few cases, where assays of sufficient sensitivity exist,
translation products have been identified by their biological activity; examples of this include interferon[75,76] and mouse B-glucuronidase translated in
Xenopus oocytes[77].

IV. USES OF CELL-FREE TRANSLATION SYSTEMS

A. Aid to mRNA Purification and Identification of Cloned DNA Sequences

One of the most common current uses of cell-free translation systems is in
the monitoring of the purification of a particular mRNA through a series of
procedures to yield a preparation of sufficient purity to prepare a complementary DNA copy. With the powerful techniques of DNA cloning[78,79], a very high
degree of purity is no longer required. However, the particular mRNA species
in question must be a major component of the mRNA preparation. Complementary
DNA can then be prepared using the mRNA as template and then converted to
double-stranded form by synthesis of the DNA strand complementary to the first
DNA strand synthesized. Insertion into a bacterial plasmid follows. The unam-
biguous identity of a particular recombinant DNA sequence can be achieved by
the use of translation systems. A number of procedures, such as hybrid arrest-
ed[80,81] or hybrid selected[82,83] translations were developed for such identification. The procedures are discussed in detail elsewhere in this volume[84].
Clearly, however, cell-free translational systems are of major importance in
the current studies of the structural organization and regulation of expression
of eukaryotic genes.

B. Quantitation of Specific mRNA Levels

In studies on developmental and hormonal regulation of the expression of specific proteins, cell-free translation systems are commonly used to monitor translatable levels of particular mRNA species. Most frequently the proportion of their specific translation products are measured relative to total translation products. Such studies are best performed together with experiments quantitating the same mRNA sequence by molecular hybridization procedures since, as discussed above, a number of the parameters of cell-free translation may markedly influence the amount of a particular translation product. There are innumerable examples of such investigations, including the now-classic demonstration of the reduced levels or absence of particular globin mRNAs in the thalassemia syndromes[85] and in studies of the effects of hormones on the production of such proteins as ovalbumin[86,87], casein[88] and vitellogenin[89-91] and on developmental changes in mRNAs for myosin[28,31] and procollagen[17].

C. Investigation of Regulatory Mechanisms at the Translational Level

Probably the best understood example of the translational regulation of a particular protein is that of globin synthesis in reticulocytes by hemin. In the absence of hemin, an inhibitor of polypeptide chain initiation, the hemin-controlled repressor, is activated. This repressor, a cyclic AMP-independent kinase, inhibits chain initiation through phosphorylation of one of the three subunits of the initiation factor, eIF-2. The details of our current understanding of this mechanism have been the subjects of recent detailed reviews[92,93]. It is likely that similar mechanisms operate to varying extents in different cells as part of normal regulatory processes, as a part of the mechanism of action of the antiviral glycoprotein interferon[92-94], in the shut-off of cellular protein synthesis in virus-infected cells[92], and in the regulation of myosin mRNA utilization during myogenesis[95].

There are a number of other examples where potential translational regulatory mechanisms have been examined using cell-free translational systems. One such mechanism concerns low molecular weight RNAs[96-98] which appear to inhibit translation of mRNAs. Another concerns the increased efficiency of the rat pancreas after treatment with glucose[99]. The transfer of anchorage-dependent mouse 3Tb fibroblasts to suspension culture is accompanied by an 85% decline in protein synthesis with no change in the cytoplasmic poly A$^+$ mRNA levels[100,101]. This change is apparently mediated by reversible changes in mRNA molecules, causing them to be inefficiently translated in *vitro* as well as _in vivo_. Finally, the roles of the 5' cap structure in the translation process have been intensively investigated with cell-free translation systems[102-105] and it has recently been shown that prokaryotic mRNAs are inefficiently translated in eukaryotic cell-free systems due to the lack of a cap structure[106].

D. Studies on the Mechanism of Protein Secretion

An area of investigation where cell-free systems have been particularly important is in studies of the mechanisms by which extracellular proteins are transported outside the cell. The "signal peptide" concept, first proposed by Blobel and Sabitani[107], has now been experimentally verified by a number of studies which have demonstrated the presence of a short, hydrophobic peptide extension of 12-30 amino acids at the N-terminus of secreted proteins. Such a peptide, first found in immunoglobulins[108], functions to attach a nascent polypeptide chain and hence the ribosomes synthesizing it to membranes and allows the newly made chain to be extruded directly into the lumen of the rough endoplasmic recticulum. Such signal peptides have been identified, and in many cases have had their amino acid sequences determined, by the analysis of cell-free translation products. Several examples exist, including those for immunoglobulin[108-112], lysozyme[113], ovomucoid[114], serum transferrin and egg-white conalbumin[115]. This signal peptide is, in most cases, cleaved off after transfer across the membrane. By combining cell-free translational systems with preparations of dog pancreas membranes, the entire process of secretion, followed by cleavage, has been achieved _in vitro_ for an immunoglobin[116] and for fish proinsulin[117]. An interesting variation on this theme is ovalbumin which, despite being the major protein secreted by the hen oviduct, does not contain an N-terminal signal peptide[118,119], nor is its N-terminus highly hydrophobic. However, it is likely that a hydrophobic segment in the middle of the molecule may serve as the signal[120,121].

A similar signal peptide appears to be used by proteins which are constituent parts of, and span across, biological membranes[122-124], and in at least some cases the N-terminal signal peptide is not cleaved[125,126]. Recently, signal peptides and analogous secretory and membrane assembly systems have been found in bacteria. These studies are the subjects of a recent detailed review[127].

E. Coupled Transcription - Translation Systems

Numerous attempts have been made to establish a eukaryotic cell-free system in which transcription and translation could be linked or coupled. Early systems[128] used E. coli RNA-polymerase, naked DNA fragments, and the wheat germ cell-free system. Subsequently, a wheat germ RNA polymerase was used[129] and such systems are still used in some cases[130,131]. However, in view of the recent discovery of discontinuous eukaryotic genes and RNA splicing, the biological relevance of such systems is not clear.

Alternative systems, where transcription and translation have been coupled, involve the use of virus cores from viruses such as vaccina[132-134], influenza[135], or vesicular stomatitis virus[136] and cell-free systems such as those from reticulocytes, wheat germ or L-cells. In these cases, the endogenous polymerases of the virus perform the transcription and authentic translation products corresponding to viral protein have been detected.

Recent developments in transcriptional systems have focused on the use of crude cell extracts for transcription and these are considered elsewhere in this volume[137]. However, the transcription of adenovirus[138,139] and conalbumin genes[139] has been reported. It will be intriguing to see if such systems are able to perform the correct splicing of transcripts to produce functional mRNA which can be translated in a coupled or linked translation system. Such a system would be useful for advancing our understanding of regulatory processes of gene expression.

V. CONCLUSIONS

The major contributions of cell-free systems to our current knowledge of molecular events of gene expression have been outlined. While cell-free translation is now in many laboratories a routine and straight-forward undertaking, new and novel ways to use such systems are continually being developed. This trend will doubtless continue, and cell-free translation systems will in the future continue to play a pivotal role in studies of the process of gene expression and the mechanisms by which this process is regulated.

REFERENCES
1. Lockard, R. E., and Lingrel, J. B. (1969) Biochem. Biophys. Res. Commun. 37, 204-212.
2. Woodard, W. R., Ivey, J. L., and Herbert, E. (1974) Methods in Enzymology 30F, 724-731.

3. Schimke, R. T., Rhoads, R. E., and McKnight, G. S. (1974) Methods in Enzymology 30F, 694-701.
4. Villa-Komaroff, L., McDowell, M. J., Baltimore, D., and Lodish, H. F. (1974) Methods in Enzymology 30F, 709-723.
5. Palmiter, R. D. (1973) J. Biol. Chem. 248, 2095-2106.
6. Lingrel, J. B. (1972) Methods in Molecular Biology, (Last, J., and Laskin, A., editors) 2, 231-263.
7. Hunt, T., and Jackson, R. J. (1974) Modern Trends in Human Leukemia (Neth, R., Gallo, R. C., Spiegelman, S. and Stohlmann, F., editors) pp. 300-307, J. F. Lehmanns Verlag, Munich.
8. Marcus, A. (1970) J. Biol. Chem. 245, 955-961.
9. Marcus, A., Efron, D., and Weeks, D. P. (1974) Methods in Enzymolgogy 30F, 749-754.
10. Shih, D. S., and Kaesberg, P. (1973) Proc. Natl. Acad. Sci. USA 70, 1799-1803.
11. Roberts, B. E., and Paterson, B. M. (1973) Proc. Natl. Acad. Sci. USA 70, 2230-2234.
12. Mathews, M. B., and Korner, A. (1970) Eur. J. Biochem 17, 328-338.
13. Aviv, M., Boime, I., and Leder, P. (1971) Proc. Natl. Acad. Sci. USA 68, 2303-2307.
14. Mathews, M. B. (1973) Essays in Biochemistry 9, 59-102.
15. Pelham, H. R. B., and Jackson, R. J. (1976) Eur. J., Biochem. 67, 247-256.
16. Palmiter, R. D., Gagnon, J., Ericsson, L. H., and Walsh, K. A. (1977) J. Biol. Chem. 252, 6386-6393.
17. Rowe, D. W., Moen, R. C., Davidson, J. M., Byers, P. H., Bornstein, P., and Palmiter, R. D. (1978) Biochemistry 17, 1581-1590.
18. Scott, M. P., Storti, R. V., Pardue, M. L., and Rich, A. (1979) Biochemistry 18, 1588-1593.
19. Gasior, E., Herrera, F., Sadnik, I., McLaughlin, C. A., and Moldave, K. (1979) J. Biol. Chem. 254, 3965-3969.
20. Gurdon, J. B., Lane, C. D., Woodland, H. R., and Marbaix, G. (1971) Nature 233, 177-182.
21. Gurdon, J. B., Lingrel, J. B., and Marbaix, G. (1973) J. Mol. Biol. 80, 539-551.
22. Gurdon, J. B., De Robertis, E. M., and Partington, G. (1976) Nature 260, 116-120.
23. De Robertis, E. M., and Olsen, M. V. (1979) Nature 278, 137-143.
24. Ehrenberg, L., Fedorcsak, I., and Solymosy, F. (1976) Prog. Nucleic Acid Res. Mol. Biol. 16, 189-262.
25. Taylor, J. M. (1979) Ann. Rev. Biochem. 48, 681-717.
26. Brawerman, G. (1974) Ann. Rev. Biochem. 43, 621-642.
27. Parish, J. H. (1972) Principles and Practice of Experiments with Nucleic Acids, Wiley, New York, pp. 104-125.
28, Strohman, R. C., Moss, P. S., Micou-Eastwood, J., Spector, D., Przybyla, A., and Paterson, B. (1977) Cell 10, 265-275.
29. Deeley, R. G., Gordon, J. I., Burns, A. T. M., Mullinix, K. P., Bina-Stein, M., and Goldberger, R. F. (1977) J. Biol. Chem. 252, 8310-8319.
30. Cox, R. A. (1968) Methods in Enzymology 12B, 120-129.
31. Benoff, S., and Nidal-Ginard, B. (1979) Biochemistry 18, 494-500.
32. Adams, S. L., Sobel, M. E., Moward, B. J., Olden K., Yamada, K. M., deCrombrugghe, B., and Pastan, I. (1977) Proc. Natl. Acad. Sci. USA 74, 3399-3403.
33. Upholt, W. B., Vertel, B. M., and Dorfman, A. (1979) Proc. Natl. Acad. Sci. USA 76, 4847-4851.

34. Chirgwin, J. M., Przybyla, A. E., MacDonald, R. J., and Rutter, W. J., (1979) Biochemistry 18, 5294-5299.
35. Aviv, M., and Leder, P. (1972) Proc. Natl. Acad. Sci. USA 69, 1408-1412.
36. Lindberg, U., and Persson, T. (1972) Eur. J. Biochem. 31, 246-254.
37. Adesnik, M., Salditt, M., Thomas, W., and Darnell, J. E. (1972) J. Mol. Biol. 71, 21-30.
38. Lee, S. Y., Mendecki, J., and Brawerman, G. (1971) Proc. Nat. Acad. Sci. USA 68, 1331-1336.
39. Brawerman, G., Mendecki, J., and Lee, S. Y. (1972) Biochemistry 11, 637-641.
40. Kitos, P. A., Saxon, G., and Amos, H. (1972) Biochem. Biophys. Res. Commun. 47, 1426-1433.
41. Schutz, G., Beato, M., and Feigelson, P. (1972) Biochem. Biophys. Res. Commun. 49, 680-689.
42. DeLarco, J., and Guroff, G. (1973) Biochem. Biophys. Res. Commun. 50, 486-493.
43. Sullivan, N., and Roberts, W. K. (1973) Biochemistry 12, 2395-2403.
44. Shapiro, D. J., and Schimke, R. T. (1975) J. Biol Chem. 250, 1759-1764.
45. Burny, A., Huez, G., Marbiax, G., and Chantrenne, H. (1969) Biochem. Biophys. Acta, 190, 228.
46. Lebleu, B., Marbiax, G., Huez, G., Temmerman, J., Burny, A., and Chantrenne, H. (1971) Eur. J. Biochem. 19, 264-269.
47. Morel, C., Kayibanda, B., and Scherrer, K. (1971) FEBS Letters 18, 84-88.
48. Blobel, G. (1972) Biochem. Biophys. Res. Commun. 47, 88-95.
49. Kemp, D. J., Partington, G. A., and Rogers, G. R. (1974) Biochem. Biophys. Res. Commun. 60, 1006-1014.
50. Chen, J. H., Lavers, G. C., and Spector, A. (1976) Biochem. Biophys. Acta. 418, 39-51.
51. Berns, A. J. M., and Bloemendal, H. (1974) Methods in Enzymology 30, 375-394.
52. Gedamu, L., Iatrou, K., and Dixon, G. H. (1978) Biochem. J. 171, 589-599.
53. Robbins, J., and Heywood, S. M. (1978) Eur. J. Biochem. 82, 601-608.
54. Linberg, U., and Sundquist, B. (1974) J. Mol. Biol. 86, 451-468.
55. Burns, A. T. H., and Williamson, R. (1975) Nucleic Acids Res. 2, 2251-2256.
56. Spirin, A. S. (1969) Eur. J. Biochem. 10, 20.
57. Georgiev, G. P., and Samarina, O. P. (1971) Advan. Cell Biol. 2, 47.
58. Williamson, R. (1973) FEBS Letters 37, 1-6.
59. Civelli, O., Vincent, A., Buri, J. F. and Scherrer, K. (1976) FEBS Letters 77, 281-286.
60. Kaumeyer, J. F., Jenkins, N. A., and Raff, R. (1978) Develop. Biol. 63, 266-278.
61. Young, E. M., and Raff, R. A. (1970) Develop. Biol. 72, 24-40.
62. Blobel, G. (1973) Proc. Natl. Acad. Sci. USA 70, 924-928.
63. Hendrick, D., Knochel, W., Schwarz, W., Pitzel, S. and Tiedemann, H. (1974) Develop. Biol. 36, 299-310.
64. Hendrick, D., Schwarz, W., Pitzel, S., and Tiedemann, H. (1974) Biochemica Biophysica Acta 340, 278-284.
65. Sampson, J., Matthews, M. B., Osborne, M. and Borghetti, A. (1972) Biochemistry 11, 3636-3640.
66. Nudel, V., Lebleu, B., Zehavi-Willner, T., and Reud, M. (1973) Eur. J. Biochem. 33, 314-322.
67. Sampson, J., and Borghetti, A. (1972) Nature New Biol. 238, 200-202.
68. Tse, P. T. H., and Taylor, J. M. (1977) J. Biol. Chem. 252, 1272-1278.
69. Sonenshein, G. E., and Brawerman, G. (1977) Biochemistry 16, 5445-5448.
70. Weber, L. A., Hickey, E. D., Maroney, P. A., and Baglioni, C. (1977) J. Biol. Chem. 252, 4007-4010.

452

71. Hunter, A. R., Farrell, P. J., Jackson, R. J., and Hunt, T. (1977) Eur. J. Biochem. 75, 149-157.
72. Lodish, H. F. (1976) Ann. Rev. Biochem. 45, 39-72.
73. Herson, D., Schmidt, A., Seal, S., Marcus, A., and van Vloten-Doting, L. (1979) J. Biol. Chem. 254, 8245-8249.
74. Payvar, F., and Schimke, R. T., (1979) J. Biol. Chem. 254, 7636-7642.
75. Babu Kishan Raj, N., and Pitha, P. M. (1977) Proc. Natl. Acad. Sci. USA 74, 1483-1487.
76. Cavalieri, R. L., Havell, E. A., Vilcek, J., and Pestka, S. (1977) Proc. Natl. Acad. Sci. USA 74, 3287-3291.
77. Labarca, C., and Paigen, K. (1977) Proc. Natl. Acad. Sci. USA 74, 4462-4465.
78. Maniatis, T., Kee, S. C., Efstratiadis, A., and Kafatos, F. (1976) Cell 8, 163-182.
79. Higuchi, R., Paddock, C. J., Wall, R. and Salzer, W. (1976) Proc. Natl. Acad. Sci. USA 73, 3146-3150.
80. Paterson, B. M., Roberts, B. E., and Kuff, E. L. (1977) Proc. Natl. Acad. Sci. USA 74, 4370-4374.
81. Hastie, N. D., and Held, W. A. (1978) Proc. Natl. Acad. Sci. USA 75, 1217-1221.
82. Woolford, J. L., Jr., and Rosbash, M. (1979) Nucleic Acids, Res. 6, 2483-2497.
83. Ricciardi, R. P., Miller, J. S., and Roberts, B. E. (1979) Proc. Natl. Acad. Sci. USA 76, 4927-4931.
84. Patterson, B., and Roberts, B. (1980). This volume.
85. Bunn, M. F., Forget, B. G., and Ranney, H. M. (1977) Human Hemoglobins, Saunders, W. B., Philadelphia, pp. 140-192.
86. Cox, R. F. (1977) Biochemistry 16, 3433-3443.
87. Pennequin, P., Robins, D. M., and Schimke, R. T. (1978) Eur. J. Biochem. 90, 51-58.
88. Rosen, J. M., and Barker, S. W. (1976) Biochemistry 15, 5272-5280.
89. Skipper, J. K., and Hamilton, T. H. (1977) Proc. Natl. Acad. Sci. USA 74, 2384-2388.
90. Gordon, J. I., Deeley, R. C., Burns, A. T. M., Paterson, B. M., Christmann, J. L., and Goldberger, R. F. (1977) J. Biol. Chem. 252, 8320-8327.
91. Burns, A. T. H., Deeley, R. G., Gordon, J. I., Udell, D. S., Mullinix, K. P., and Goldberger, R. F., (1978) Proc. Natl. Acad. Sci. USA 75, 1815-1819.
92. Revel, M., and Groner, Y. (1978) Ann. Rev. Biochem. 47, 1079-1126.
93. Ochoa, S., and deHaro, C. (1979) Ann. Rev. Biochem. 48, 549-580.
94. Baglioni, C. (1979) Cell 17, 255-264.
95. Gette, W. R., and Heywood, S. M. (1979) J. Biol. Chem. 254, 9879-9885.
96. Bogdanovsky, D., Hermann, W., and Schapiro, G. (1973) Biochem. Biophys. Res. Commun. 54, 25-32.
97. Heywood, S. M., Kennedy, D. S., and Bester, A. (1974) Proc. Natl. Acad. Sci. USA 71, 2428-2431.
98. Lee-Huang, S., Sierra, J. N., Naranjo, R., Filipowicz, W., and Ochoa, S. (1977) Arch. Biochem. Biophys. 180, 276-287.
99. Itoh, N., and Okamoto, M. (1980) Nature 283, 100-102.
100. Benecke, B. J., Ben-Ze'ev, A., and Penman, S. (1978) Cell 14, 931-939.
101. Farmer, S. R., Ben-Ze'ev, A., Benecke, B. J., and Penman, S. (1978) Cell 15, 627-637.
102. Shatkin, A. (1976) Cell 9, 645-653.
103. Filipowicz, W., (1978) FEBS Letters 96, 1-11.

453

104. Weber, L. A., Hickey, E. D., and Baglioni, C. (1978) J. Biol. Chem. 253, 178-183.
105. Sonenberg, N., Trachsel, H., Hecht, S., and Shatkin, A. J. (1980) Nature 285, 331-333.
106. Paterson, B. M., and Rosenberg, M. (1979) Nature 279, 692-696.
107. Blobel, G., and Sabitani, D. D. (1971) Biomembranes, (Manson, L. A., editor,) Plenum, New York, Vol. 2, 193-195.
108. Milstein, C. Brownlee, G. G., Marrison, T. M., and Mathews, M. B. (1972) Nature New Biol. 239, 117-120.
109. Swan, D., Aviv, H., and Leder, P. (1972) Proc. Natl. Acad. Sci. USA 69, 1967-1971.
110. Mach, B., Faust, C., and Vassalli, P. (1973) Proc. Natl. Acad. Sci. USA 70, 451-455.
111. Schechter, I. (1973) Proc. Natl. Acad. Sci. USA 70, 2256-2260.
112. Tonegawa, S., and Baldi, I. (1973) Biochem. Biophys. Res. Commun. 51, 81-87.
113. Palmiter, R. D., Gagnon, J., Ericsson, L. H., and Walsh, K. A. (1977) J. Biol. Chem. 252, 6386-6393.
114. Thibodeau, S. N., Palmiter, R. D., and Walsh, K. A. (1978) J. Biol. Chem. 253, 9018-9023.
115. Thibodeau, S. N., Lee, D. C., and Palmiter, R. D. (1978) J. Biol. Chem. 253, 3771-3774.
116. Dobberstein, B., and Blobel, G. (1977) Biochem. Biophys. Res. Commun. 74, 1675-1682.
117. Shields, D., and Blobel, G. (1977) Proc. Natl. Acad. Sci. USA 74, 2059-2063.
118. Palmiter, R. D., Gagnon, J., and Walsh, K. A. (1978) Proc. Natl. Acad. Sci. USA 75, 94-98.
119. Gagnon, J., Palmiter, R. D., and Walsh, K. A. (1978) J. Biol. Chem. 253, 7464-7478.
120. Lingappa, V. R., Shields, D., Woo, S. L. C., and Blobel, G. (1978) J. Cell Biol. 79, 567-572.
121. Lingappa, V. R., Lingappa, J. R., and Blobel, G. (1979) Nature 281, 117-121.
122. Katz, F. N., Rothman, J. E., Lingappa, V. R., Blobel, G., and Lodish, H. F. (1977) Proc. Natl. Acad. Sci. USA 74, 3278-3282.
123. Lingappa, V. R., Katz, F. N., Lodish, H. F., and Blobel, G. (1978) J. Biol. Chem. 253, 8667-8670.
124. Dobberstein, B., Garoff, M., Warren, G., and Robinson, P. J. (1979) Cell 17, 759-769.
125. Bonatti, S., and Blobel, G. (1979) J. Biol. Chem. 254, 12261-12264.
126. Bar-Nun, S., Kreibich, G., Adesnik, M., Alterman, L., Negishi, M., and Sabitini, D. D. (1980) Proc. Natl. Acad. Sci. USA 77, 965-969.
127. Davis, B. D., and Tai, P. C. (1980) Nature 283, 433-437.
128. Roberts, B. E., Goreki, M., Mulligan, R. C., Danna, K. J., Rozenblatt, S. and Rich, A. (1975) Proc. Natl. Acad. Sci. USA 72, 1922-1926.
129. Rozenblatt, S., Mulligan, R. C., Gorecki, M., Roberts, B. E., and Rich, A. (1976) Proc. Natl. Acad. Sci. USA 73, 2747-2751.
130. Kronenberg, H. M., Roberts, B. E., Habener, J. F., Potts, J. T., and Rich, A. (1977) Nature 267, 804-807.
131. Kronenberg, H. M., Roberts, B. E., and Efstratiadis, A. (1979) Nucleic Acids Res. 6, 153-166.
132. Pelham, H. R. B., Sykes, J. M. M., and Hunt, T. (1978) Eur. J. Biochem. 82, 199-209.
133. Cooper, J. A., and Moss, B. (1978) Virology 88, 149-165.
134. Bossart, W., Nuss, D. L., and Paloetti, E. (1978) J. Virol. 26, 673-680.

454

135. Content, J., deWit, L., and Horisberger, M. A. (1978) J. Virol. 26, 817-821.
136. Ball, L. A., and White, C. N. (1978) Virology 84, 479-495.
137. Manley, J., (1980). This volume.
138. Weil, P. A., Luse, D. S., Segall, J. and Roeder, R. G. (1979) Cell 18, 469-484.
139. Wasylyk, B., Kedinger, C., Colden, J., Brison, O., and Chambon, P. (1980) Nature 285, 367-373.

DNA TOPOISOMERASES

JAMES C. WANG AND KARLA KIRKEGAARD

DEPARTMENT OF BIOCHEMISTRY AND MOLECULAR BIOLOGY, HARVARD UNIVERSITY
CAMBRIDGE, MASSACHUSETTS 02138

I. INTRODUCTION

DNA topoisomerases are enzymes that catalyze the breakage of DNA backbone bonds and their subsequent rejoining. Breakage and rejoining usually occur in a concerted way, in that the breakage of a DNA backbone is followed efficiently by the rejoining of the bond. Thus, these enzymes are almost invariably detected and monitored by their promotion of interconversions between different topological isomers, or topoisomers, of DNA - hence the name "topoisomerases" (for a review, see Wang and Liu[1]).

Although the breakage of a DNA backbone bond by a topoisomerase is usually followed by the reformation of the same bond, it is plausible that under some conditions two different backbone bonds will be broken in concert and pairwise switching will occur before the reformation of the bonds. If the bonds involved are on different DNA strands, such a sequence will lead to strand switching. The most studied example involves the phage λ integrase, which has a topoisomerase activity, as shown by its relaxation of supercoiled DNA[2]. If a particular E. coli protein is present in addition to the phage integrase, however, the occurrence of two pairwise strand switches can be readily demonstrated[3]. Both intrastrand and interstrand modes of reactions are presumably catalyzed by the same catalytic site(s) on the integrase. In most of this discussion, we do not consider reactions involving strand switching.

Mechanistically, the known DNA topoisomerases are in two catagories. Type I enzymes catalyze topoisomerization reactions that involve transient breaks in one DNA strand at a time. Type II enzymes catalyze topoisomerization reactions that involve transient breaks made in a double-stranded fashion. Fig. 1 illustrates diagrammatically four types of reactions that are catalyzed by E. coli DNA topoisomerase I[4] and M. luteus topoisomerase I[5], two of the most extensively studied Type I enzymes. The relaxation of negatively supercoiled DNA (A) and the intertwining of single-stranded DNA rings of complementary sequences (C) are topologically the same kind of reaction. In both cases, the topological quantity that is changed is the linking number between the complementary single-stranded rings, and reaction C can be viewed as a special case of reaction A. All four of the reactions are probably also catalyzed by the Type I eukaryotic topoisomerases[23]. The eukaryotic enzymes can catalyze the relaxation of both positively and negatively supercoiled DNAs[6]; the bacterial enzymes can relax negatively superhelical DNAs, but have little effect on positively supercoiled ones[4,5].

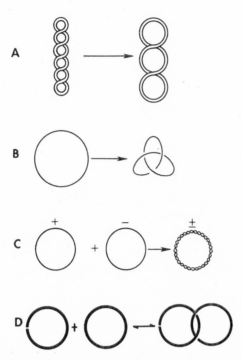

Fig. 1. Topoisomerizations that are known to be catalyzed by Type I topoisomerases. **A**, removal of superhelical turns from covalently closed duplex DNA[4,6]. **B**, introduction and removal of topologic knots in single-stranded DNA rings[7]. **C**, intertwining of complementary single-stranded DNA rings to form a closed duplex ring[8,9]. **D**, catenation and decatenation of double-stranded rings, provided that one of the rings contains a single-strand break[10].

Among the Type II enzymes, the bacterial gyrases[11-13] are unique in their catalysis of the negative supercoiling of DNA in the presence of ATP. Other Type II enzymes--such as the ATP-dependent topoisomerases from phage T4[14,15], Xenopus laevis[16], and Drosophila melanogaster[17]--can catalyze the relaxation of positively or negatively supercoiled DNAs, but not the supercoiling reaction. In addition to the supercoiling and relaxation reactions, the catenation and decatenation and the knotting and unknotting of covalently closed or nicked double-stranded DNA rings by the Type II enzymes have been observed. These interconversions are illustrated diagrammatically in Fig. 2.

458

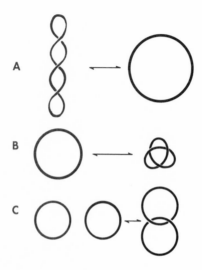

Fig. 2. Types of reactions catalyzed by Type II topoisomerases. **A**, introduction or removal of superhelical turns in closed duplex DNA rings (see text
for specific examples)[11-17]. **B**, interconversion of closed double-stranded DNA
rings and without topological knots[19,20]. **C**, formation or resolution of catenane
of closed duplex DNA rings[16,17,21].

Although all known DNA topoisomerases can relax negatively supercoiled
DNAs, the reactions catalyzed by Type I and Type II enzymes differ in the
quantization of linking-number changes: the Type II enzymes catalyze changes
in linking numbers by integral multiples of 2; while no such restriction has
been observed for the Type I enzymes. This dyadic linking-number change is a
hallmark of catalysis by the Type II enzymes[14,16,20,22].

This paper is intended to provide a guide to the isolation of DNA topoisomerases and their use in DNA research. Comprehensive reviews of the abundance,
physical properties, mechanistic characteristics and _in vivo_ functions of these
enzymes may be found elsewhere[1,23-26].

II. ENZYMATIC ASSAYS

General Considerations

In general, the assay of an enzyme is based on the reaction(s) that it catalyzes, and the choice of assay is usually determined by convenience, rapidity, reliability, and the nature of interfering activities. For the enzymatic assays of the DNA topoisomerases, the most problematic interfering activities are nucleases. With prokaryotic organisms, the use of endonuclease-deficient strains as the source of the topoisomerase is helpful. The addition of tRNA, which reduces the degradation of the DNA substrate by E. coli endonuclease I and similar bacterial endonucleases, is also advisable. With eukaryotic organisms, the major Type I topoisomerase activity can be assayed in the absence of the divalent metal ions that are required by most nucleases; omission of the divalent ions from the assay buffer usually alleviates the nuclease problem. If these approaches are impractical, preliminary fractionation of the topoisomerase activity from the interfering nucleases is mandatory, unless a nonenzymatic assay can be devised. For the ATP-dependent topoisomerases, other ATPases could conceivably interfere with the topoisomerase assay, although there is no documented example of such interference. ATP is usually present in excess in the assay mixture; if necessary, an ATP-regenerating enzyme system[27] can be included.

The relaxation of supercoiling of a covalently closed double-stranded circular DNA continues to be the reaction that is usually chosen for the assay of topoisomerases. In assaying a Type II enzyme in the presence of an excess Type I topoisomerase activity, however, several other reactions may be more advantageous. The catenation and decatenation and the knotting and unknotting of covalently closed duplex rings are unique reactions catalyzed by the Type II, but not the Type I, enzymes. These reactions can therefore be used to assay a Type II enzyme in the presence of a Type I enzyme.

When a Type I enzyme is to be assayed in the presence of a Type II enzyme, several options are available. The omission of ATP (if ATP is absolutely required by the interfering activity) and the use of specific inhibitors or mutants deficient in the Type II enzyme are two obvious approaches. For the known topoisomerases, the linking of single-stranded rings of complementary sequences and the knotting or unknotting of single-stranded DNA rings appear to be reactions that are catalyzed by Type I, but not Type II, enzymes. Another method of assay is to use one topoisomer of a covalently closed

double-stranded DNA ring, which can be prepared by first resolving electrophoretically the topoisomers of different linking numbers in a covalently closed
DNA sample[28]. Because Type II activities can change the linking number of the
substrate DNA only by an even number, the topoisomer bands that differ in linking number from the original species by an odd number must be generated by a
Type I activity.

A. Monitoring the Topoisomerization Reactions

During the purification of the first DNA topoisomerase a decade ago, more
than 1,000 assays were performed by sedimentation velocity measurements in the
ultracentrifuge. Such tedious and laborious assays have long since been replaced by more rapid and convenient ones. The method of choice now appears
to be agarose gel electrophoresis[29]. Although this method is not as rapid or
convenient as some fluorescence assays[30,31], it has the advantage of reliability. The electrophoretic resolution of various DNA species allows one to examine both the disappearance of the input DNA substrate and the nature of the
product formed. Spectroscopic methods, on the other hand, usually provide only
one parameter and are therefore more prone to interference from side reactions.
If a long horizontal gel with several separate rows of sample is used, as many
as 100 samples can be assayed on a single gel.

In the event that thousands of samples are to be assayed, the more rapid
fluorometric methods deserve consideration. One method takes advantage of
the greater sensitivity of a negatively supercoiled DNA than of a relaxed
DNA to single-strand-specific nucleases[32]. Cleavage of the supercoiled DNA
by the nuclease, followed by a cycle of heating and cooling at high pH[33],
yields denatured DNA. If ethidium is added, it binds to the DNA, but its
fluorescence is not greatly increased. However, if the negatively supercoiled
DNA is first relaxed by a topoisomerase, it becomes resistant to the single-
strand-specific nuclease and remains double-stranded after a cycle of heating
and cooling. When ethidium is added, it binds to the duplex DNA, causing a
fluorescence increase of two orders of magnitude. These assays can be performed on microtiter plates with the mass-processing techniques described by
Weiss and Milcarek[34]. After completion of the assays, the plastic microtiter
plates are placed on a long-wavelength ultraviolet source. Wells that contain
topoisomerase activity are displayed as red dots because of the increased
ethidium fluorescence, whereas wells that contain no topoisomerase activity
appear as dark dots. This assay was used by R. Depew and J. C. Wang in their

search for E. coli topoisomerase mutants (unpublished results). A similar approach has been independently devised and used to assay a bovine topoisomerase[35].

B. Specific Examples

The most commonly used assay for both Type I and Type II enzymes is the relaxation of negatively supercoiled DNA. Phage PM2 is a convenient substrate because of the high yield of the phage and the predominance of the monomeric species in the supercoiled form. With the availability of small plasmids that can be isolated rapidly and in high yields, however, it is more convenient to use a plasmid, such as pBR322[36], as the source of DNA. Some plasmid DNA preparations from rec[+] hosts contain multimeric species, and in these cases DNA preparations that contain a higher proportion of monomers can be obtained with rec A strains[37,38].

The buffer is usually selected by trial and error. For the Type I enzymes from E. coli, M. luteus, B. megaterium, and S. typhimurium, as well as those from eukaryotic organisms, a dilute buffer (10 mM Tris at a pH of 8, for example) containing 100 mM Na^+ is adequate. The bacterial enzymes require a concentration of several mM of Mg^{2+} in addition. Selection of 37°C as the assay temperature should not be automatic; for the Drosophila[39] and sea urchin[40] enzymes, for example, relaxation activity is detectable only at lower temperatures. The temperature also affects the assay in a different way. With the eukaryotic enzymes, the complete relaxation of a supercoiled DNA at 37°C yields a population of topoisomers that does not resolve well from the nicked circular form by agarose gel electrophoresis if the gel is run at room temperature in an electrophoresis buffer containing no magnesium. But if the assays are carried out at a lower temperature, such as 20°C, the electrophoretic mobilities of the relaxed topoisomers are quite distinct from that of the nicked species.

Fig. 3 illustrates the assay of topoisomerases by their relaxation of negatively supercoiled DNA. The lanes left to right denote fractions collected during the chromatography of a Drosophila extract. Assays shown in the upper frame were carried out in the absence of ATP. An activity centering around fraction 140 is seen. Assays shown in the lower frame were carried out in the presence of ATP and reveal an ATP-dependent activity in fractions 40 and 50.

Another convenient assay for the Type II topoisomerases is the decatenation of networks of catenated duplex DNA rings. Giant catenated networks can be formed in vitro with Type II topoisomerases[16,17,21] or obtained from

462

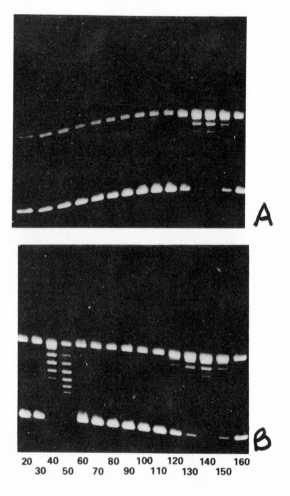

Fig. 3. The assay of topoisomerases by the relaxation of negative supercoils. Details are given in text. Reprinted with permission from Hsieh and Briutlag[17].

kinetoplasts (for a recent review, see Englund[41]). Such giant networks hardly enter agarose gels and are detected as a fluorescent band on top of the gel after electrophoresis in the presence of ethidium. Decatenation of the rings by a topoisomerase yields monomeric rings that migrate much faster on electrophoresis[42]. If the networks consist of covalently closed duplex rings, decatenation to closed monomers can be catalyzed only by a Type II topoisomerase.

This assay has been used in the purification of an ATP-dependent Type II enzyme from HeLa cells (L. Liu, personal communication).

III. NONENZYMATIC ASSAYS

Recent developments have made nonenzymatic assays feasible and more convenient in some instances. In the cases of the E. coli and M. luteus DNA gyrases, for example, the two subunits dissociate readily. Because the individual subunits have no catalytic activity[12], the combination of fractions by trial and error is necessary for the assay of enzymatic activities. The application of the radioactive-tracer method is therefore advantageous (W. D. Rupp, personal communication). This method requires the cloning of the genes coding for the enzymes. An appropriate strain carrying a plasmid that contains the gene coding for the desired enzyme or enzyme subunit is first grown and then subjected to the maxi-cell treatment[45]. Synthesis of the host proteins is abolished, thus permitting the specific labeling of the plasmid-coded proteins. These small numbers of labeled plasmid-coded species are resolved by electrophoresis through a nondenaturing gel, and the band that contains the desired protein is sliced out. Elution of the protein from the gel slice, either electrophoretically or by soaking of the meshed gel, provides the radioactively labeled tracer, which can be mixed with the cell extract before fractionation. The E. coli gyrase genes have been cloned by Mizuuchi et al.[20] and by P. Martens and S. Swanberg (unpublished work). The gene (top) that codes for E. coli DNA topoisomerase I has also been cloned (Wang, unpublished).

If antibodies directed against the desired enzyme are already available, radioimmunological assays can be carried out. One such method is the antibody-sandwich technique[46]. A drop containing about 50 µl or less of each fraction is pipeted onto the surface of a petri dish that contains agarose gel. If the salt concentration in the fractions is higher than about 0.2 M, dilution of the fraction with a dilute buffer that contains bovine plasma albumin at 100 µg/ml before their delivery to the gel surface is advisable. One standard-size petri dish can easily accommodate 100 drops. A plastic sheet coated with the specific antibodies is then placed over the gel surface as described by Broone and Gilbert[46]. After standing for several hours in the refrigerator to allow binding of antigens by the antibodies adhered to the plastic sheet, the sheet is lifted off, washed, and stained with ^{125}I-labeled antibodies[47]. Autoradiography of the stained sheet gives a semiquantitative measurement of the amounts of the protein in the various fractions. The radioimmunologic methods detect antigenic determinants that were recognized in the production of the antibodies. Thus, the presence of the intact protein and of its subfragments can be monitored. Antibodies raised against proteins recovered from dodecyl sulfate gels can usually recognize the native proteins as well. Because methods like dodecyl sulfate gel electrophoresis and isoelectric focusing are very powerful separation procedures, initial isolation of a protein by these methods for the purpose of raising antibodies is advantageous in many cases. The use of nonenzymatic assays is generally applicable to the identification and purification of all proteins, and specific applications to topoisomerases are not described here.

IV. PURIFICATION

A number of DNA topoisomerases have been purified to homogeneity or near-homogeneity. A large-scale purification of E. coli DNA topoisomerase I to near-homogeneity with a 16% recovery has been reported[48]. The procedure followed the fairly common practices of cell breakage, removal of nucleic acids by streptomycin, fractionation by $(NH_4)_2SO_4$ precipitation, and column chromatography through phosphocellulose, hydroxyapatite, and DNA cellulose. Similar procedures were used in the purification of M. luteus[5] and B. megaterium topoisomerase I[49].

The use of polymin P[50] is effective as an initial fractionation step in the purification of bacterial gyrases[11,12,51]. In the earlier procedures, the supercoiling activity was followed during fractionation[11,51], and enzyme purified to near-homogeneity was obtained[51]. The dissociation of the subunits of bacterial gyrases during some steps, however, favors the purification of the subunits individually[12,43,44]. This is true especially when the genes coding for the two subunits are cloned on separate plasmids[20]. Reconstitution of the purified subunits, followed by chromatography on DNA cellulose, has yielded homogeneous M. luteus gyrase (see Klevan and Wang[52] and Klevan and Wang, unpublished results).

The type II topoisomerase from T4 phage[14,15] can be prepared in high yield from strains of phage-infected E. coli cells that accumulate large amounts of the T4 enzyme. The three subunits comprising the topoisomerase remain associated throughout purification, and as much as 30 mg of homogeneous enzyme can be obtained from 200 g of cells (K. Kreuzer and B. Alberts, personal communication).

Early work on a major eukaryotic Type I topoisomerase yielded a nearly homogeneous protein with weight of 66,000 daltons[53], and all reported eukaryotic topoisomerase activities were found to be coincident with proteins of similar weight[1,23]. More recently, the adoption of procedures using high concentrations of proteinase inhibitors and rapid fractionation steps has yielded enzymes of higher specific activity with weights of about 100,000 daltons. Purification of the HeLa enzyme by the isolation of nuclei from fresh cells, fractionation by polyethylene glycol extraction, and successive chromatography on hydroxyapatite, phosphocellulose, and DNA cellulose gives a homogenous preparation of a 100,000-dalton protein. When frozen cells are used as the starting material, however, a 67,000-dalton active enzyme is obtained (L. Liu, personal communication). Active Type I topoisomerases with weights around 100,000 daltons have also been purified from wheat germ[5] and Drosophila embryos (K. Javaherian and J. C. Wang, unpublished results).

Type II topoisomerase activities have been detected in extracts prepared from Drosophila melanogaster[16] and Xenopus laevis[17] cells. Recently, a topoisomerase with a weight of 170,000 daltons has been prepared from HeLa cells to a purity of about 80%. The enzymatic properties of all these eukaryotic Type II topoisomerases closely resemble those of the ATP-dependent activity from phage T4 (L. Liu, personal communication).

V. USES OF DNA TOPOISOMERASES

This section does not discuss specific in vitro systems in which topoisomerases are required for cellular processes involving DNA, such as gyrase-dependent replication of Col E1 or λ integrase-promoted site-specific recombination, but rather summarizes the general use of the DNA topoisomerases.

A. Relaxation of Supercoiled DNA

The removal of superhelical turns from DNA is most easily accomplished with a Type I topoisomerase from a eukaryotic source. When the reaction is carried to completion, the distribution in linking number of the product is indistinguishable from that of the same DNA after a cycle of nicking by a nuclease and resealing by a ligase[56]. Prokaryotic Type I enzymes are efficient in relaxing moderately negatively supercoiled DNAs, but are sluggish with DNAs that are only slightly negatively supercoiled or are positively supercoiled.

B. Probing Structural Changes in DNA

Relaxation of a covalently closed DNA under a given set of conditions yields a Boltzmann population of DNA molecules differing only in linking numbers; these topoisomers are resolvable by gel electrophoresis. If two samples of the same DNA are relaxed under different sets of conditions, the variation of the average linking numbers of the two populations can be measured. Such measurements yield information on the changes in the DNA structure brought about by changes in the conditions of relaxation.

The complete relaxation of a supercoiled DNA by a topoisomerase is formally equivalent to the sequential actions of a nuclease that introduces single-chain scissions in the DNA and a ligase that reseals them. The nuclease-ligase approach has the advantage that the complete relaxation of the DNA is ensured. The use of the topoisomerase is simpler experimentally and has the further advantage of concerted breakage and rejoining of the DNA, inasmuch as proteins or other reagents that bind specifically to single-chain nicks can render the nuclease-ligase approach unfeasible. In using a topoisomerase, however, one must take precautions to ensure the complete relaxation of the DNA, especially when the reaction conditions are not optimal for the enzyme. The classical way of determining whether equilibrium has been established is to start from different sides of the equilibrium position and ask whether the same final state is reached. In this particular case, two DNA samples with average linking numbers above and below that of the expected average can be used to see whether the same distribution of topoisomers results after treatment with the topoisomerase.

There have been many instances of the deduction of structural changes in DNA with the approaches just described. Examples are the unwinding of the DNA helix by RNA polymerase[57], and the wrapping of DNA around histones[58] and around gyrase[12]. In all these cases, the linking-number changes caused by the binding of the proteins and then the relaxation of the DNA are measured after the removal of the proteins. The family of topoisomers in a DNA sample relaxed in the presence of RNA polymerase, for example, is displaced on gel electrophoresis from that in a sample relaxed in the absence of RNA polymerase (Fig. 4); that suggests that a change in the helical structure of the DNA has been produced by the binding of the polymerase.

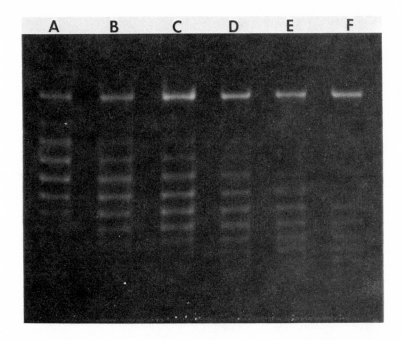

Fig. 4. Unwinding of the DNA helix by RNA polymerase. Samples of duplex fd
DNA that have been relaxed at 5°C in the absence (A) and presence in increasing amounts (B-F) of RNA polymerase holoenzyme. Electrophoresis through a
0.7% agarose gel in a buffer containing 40 mM Tris (pH, 8), 5 mM magnesium
acetate, and 0.1 mM Na_3EDTA clearly resolves the covalently closed topoisomers
from the more slowly migrating nicked double-stranded rings. The increase in
negative superhelicity of the closed DNA after the removal of RNA polymerase
bound at the time of covalent closure has been interpreted as resulting from
the disruption of base-pairing by the enzyme.

As can be seen in Fig. 4, DNA molecules that have been relaxed in the
presence and absence of RNA polymerase differ only in linking number on removal
of the protein. In the region where the two families of topoisomers overlap,
the bands of topoisomers in one family align exactly with those in the other
family: bands of identical electrophoretic mobilities are in fact indistinguishable if mixed. In contrast, if relaxed DNA to which particular agents have
been irreversibly bound or that has undergone irreversible structural changes
is compared with relaxed DNA without these modifications, shifts of the bands
of one family of topoisomers relative to those of the other may occur. A

special case that has been studied in some detail is the addition or deletion of a given number of base pairs (Fig. 5; see Wang[59,60] for discussion). Another example is the photo-cross linking of DNA by psoralen[61].

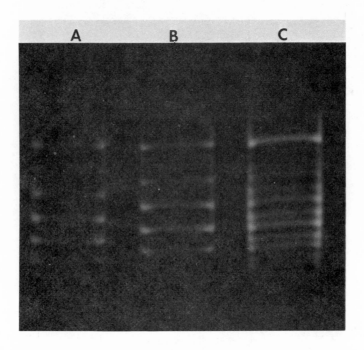

Fig. 5. The "band-shift" method of observing changes in DNA structure. Lane A, topoisomers of pBR322 DNA. Lane B, the same DNA, differing only by the insertion of four base pairs. Lane C, a mixture of the two samples. All DNA samples were relaxed at 0°C under identical conditions; electrophoresis was performed at room temperature with 0.7% vertical agarose slab gels and a buffer containing 90 mM Tris-borate (pH, 7.9) and 20 mM Na$_3$EDTA. Photograph courtesy of Lawrence Peck.

C. Negative Supercoiling of DNA

Plasmid and phage DNAs isolated from cells typically have superhelical densities or specific linking differences (for definition, see Wang[60]) between -0.05 and -0.08, corresponding to five to eight negative superhelical turns for every 1,000 base pairs[62]. In vitro, DNA gyrase can supercoil λ DNA to a negative superhelical density about 50% greater than that of the intracellularly supercoiled λ DNA[11]. A higher negative superhelicity can be achieved with gyrase in the presence of ethidium followed by removal of the intercalating dye after the reaction.

Often, however, it is desirable to prepare negatively twisted DNA with a defined specific linking difference. The most convenient method is the relaxation of a DNA with a eukaryotic topoisomerase or the sealing of a nicked DNA with ligase in the presence of ethidium. Each bound ethidium molecule unwinds the DNA helix by about $26°$[63], so the linking number is reduced by 1 with the binding of about 14 ethidium molecules. If the DNA concentration is 50 µg/ml or higher, the binding of ethidium is nearly quantitative in media that contain a salt at up to about 0.2 M[64]. The topoisomerase reaction can be terminated by rapid extraction with phenol, which also removes ethidium. DNA of a negative superhelicity up to about 0.2 can be so prepared. Smaller changes in linking numbers can also be effected by changing the temperature of relaxation. Each 1°C increment in temperature unwinds the helix by about 0.012° per base pair[65].

D. Preparation of Positively Supercoiled DNAs

Aside from the relaxation of a DNA at a low temperature so that it is slightly positively supercoiled at higher temperatures, the only known method of yielding moderately positively supercoiled DNA is by relaxation in the presence of stoichiometrically bound gyrase molecules[12]. Omission of ATP is mandatory to avoid the negative supercoiling of DNA by gyrase.

E. Formation of Novel Species of DNA

The reactions depicted diagrammatically in Figs. 1 and 2 illustrate the diversity of DNA topological isomers that can be obtained by the use of the topoisomerases. Detailed descriptions of the preparations of these species can be found in the references cited in the captions of these figures. Another unique system for the formation of double-stranded knots and catenanes is the λ integrase-host factor combination[2,3]. In contrast with the formation of duplex knots and catenanes by the Type II topoisomerases, which involves no strand-switching, knotting and catenation by the int system involve pairwise strand exchanges, as mentioned earlier.

On the basis of the known reactions depicted in Figs. 1 and 2, it is anti-
cipated that a plethora of related species can be prepared. For example, the
intertwining of two single-stranded rings with a dimeric single-stranded ring
that contains two tandem copies of sequences complementary to the monomeric
single-stranded rings should yield a Holliday structure[66]. The assimilation
of a homologous single-stranded DNA fragment by a negatively supercoiled duplex
ring is energetically favorable and occurs readily[67,68]. If the single strand
is in the form of a ring, however, its base-pairing with one of the strands of
the duplex ring is hindered by the topological constraint, even in the presence
of the rec A protein and ATP, which normally accelerate the process[69,70]. The
addition of a topoisomerase removes the topological constraint and thus allows
the assimilation of the single-stranded ring (C. Radding, personal communication).

Intertwined duplex rings have been constructed recently by using the same
principle as the intertwining between single- and double-stranded rings.
Closed circular duplexes that contain mismatched single-stranded loops were
prepared by annealing complementary strands that contained regions mismatched
in such a way that the mismatched single-stranded loops in half the heteroduplexes produced complemented the mismatched single-stranded loops in the other
half. As expected, the addition of a topoisomerase readily allows the cross-
renaturation of those molecules via the intertwining of the complementary loops
in the heteroduplex regions (unpublished results).

The fusion of two negative supercoils by intermolecular homologous pairing
is thermodynamically favorable. Such a process can be visualized as first involving the disruption of intramolecular base-pairing in one segment of each
intramolecular molecule, and then pairwise intertwining of the single-stranded
segments of one molecule with their complements in the other molecule. Base-
pairing is therefore restored to a large extent, and the net gain in free energy
comes from a reduction in the negative superhelicity of the molecules. This
reaction is plausible if gyrase is present in excess to keep the molecules in
the negatively supercoiled state and another topoisomerase is present to allow
the cross-intertwining of complementary loops.

Another novel species that can be formed from a closed duplex ring that
contains a giant palindromic region is the hairpin structure[71]. Negative supercoiling by gyrase of a DNA with a large palindromic region facilitates intrastrand base-pairing at the expense of interstrand pairing, thus yielding a
molecule with two long hairpins.

VI. CONCLUDING REMARKS

In recent years, the field of DNA topoisomerases has expanded dramatically. New enzymes and new reactions have been discovered rapidly, and it now appears that, in the presence of these enzymes, no reaction involving DNA is forbidden topologically. The importance of the biological roles of the bacterial gyrases and phage topoisomerases is well established[24], and preliminary studies with E. coli DNA topoisomerase I mutants have shown pleiotropic effects of the mutations on the transcription of some operons and transposition of several transposons[72]. Although eukaryotic mutants with defective topoisomerases have yet to be found, the rapid advancement of in vitro recombinant DNA methods has provided tools for the genetic analysis of complex genomes that were unavailable a few years ago. It is likely that substantial progress will be made in the near future in elucidating the biological functions of the eukaryotic DNA topoisomerases. They are likely to be involved in a multitude of processes, judging from the roles played by their counterparts in the prokaryotes.

At the same time, methods for the assay of various DNA topoisomerases have been greatly streamlined, and the availability of bacterial topoisomerase I mutants will make the search for other topoisomerases easier. The existence of RNA topoisomerases has not yet received serious consideration, but it appears feasible to adapt some of the assays developed for the DNA topoisomerases to the assay of possible RNA enzymes.

ACKNOWLEDGMENTS

This work was supported in part by grants from the U.S. Public Health Service (GM 24544) and the National Science Foundation (PCN-78-05892). We wish to express our gratitude to all those who communicated their results to us before publication.

REFERENCES

1. Wang, J. C., and Liu, L. F. (1979) In Molecular Genetics (J. M. Taylor, Ed.), part 3, pp. 65-88. Academic Press, New York.
2. Kikuchi, Y., and Nash, M. A. (1979) Proc. Natl. Acad. Sci. USA 76, 3760-3764.
3. Kikuchi, Y., and Nash, M. A. (1978) J. Biol. Chem. 253, 7149-7157.
4. Wang, J. C. (1971) J. Mol. Biol. 55, 523-540.
5. Kung, V. T., and Wang, J. C. (1977) J. Biol. Chem. 252, 5398-5402.
6. Champoux, J. J. (1978) Annu. Rev. Biochem. 47, 449-463.

7. Liu, L.F., Depew, R.E., and Wang, J.C. (1976) J. Biol. 106, 439-452.
8. Champoux, J.J. (1977) Proc. Natl. Acad. Sci. USA 74, 5328-5332.
9. Kirkegaard, K., and Wang, J.C. (1978) Nucl. Acids. Res. 5, 3811-3820.
10. Tse, Y.-C., and Wang, J.C. (1980) Cell 22, 269-276.
11. Gellert, M., Mizuuchi, K., O'Dea, M. N., and Nash, N. A. (1976) PNAS 73, 3872-3876.
12. Liu, L. F., and Wang, J. C. (1978) Proc. Natl. Acad. Sci. USA 75, 2098-2102.
13. Sugino, A., and Bott, K. F. (1980) J. Bacteriology, 141, 1331-1339.
14. Liu, L. F., Liu, C.-C., and Alberts, B. M. (1979) Nature 281, 456-461.
15. Stetler, G. L., King, G. J., and Huang, W. N. (1979) PNAS 76, 3737-3741.
16. Baldi, M. I., Benedetti, P., Mottoccia, E., and Tocchini-Valentini, G. P. (1980) Cell 20, 461-467.
17. Hsieh, T.-S. and Brutlag, D. (1980) Cell 21, 115-125.
18. Brown, P.O., Peebles, C.L., and Cozzarelli, N.R. (1979) Proc. Natl. Acad. Sci. USA 76, 6110-6114.
19. Liu, L.F., Lir, C.-C., and Alberts, B.M. (1980) Cell 19, 697-707.
20. Mizuuchi, K., Fisher, L.M., O'Dea, M.H., and Gellert, M. (1980) Proc. Natl. Acad. Sci. USA 77, 1847-1851.
21. Kreuzer, K.N., and Cozzarelli, N.R. (1980) Cell 20, 245-254.
22. Brown, P.O., and Cozzarelli, N.R. (1979) Science 206, 1081-1083.
23. Champoux, J.J. (1978) Annu. Rev. Biochem. 47, 449-463.
24. Cozzarelli, N. R. (1980) Science 207, 953-960.
25. Wang, J. C. (1981) in The Enzymes (Boyer, P. D., ed.). In press.
26. Gellert, M. (1981) in the Enzymes (Boyer, P. D., ed). In press.
27. Barnett, R. E. (1970) Biochemistry 9, 4644-4648.
28. Pulleyblank, D. E., Shure, M., Tang, D., Vinograd, J., and Vosberg, M. (1975) Proc. Natl. Acad. Sci. USA 72, 4280-4284.
29. McDonell, M. W., Simon, M. N., and Studier, F. W. (1977) J. Mol. Biol. 110, 119-146.
30. Paoletti, C., LePecq, J.-B., and Lehman, I. R. (1971) J. Mol. Biol. 55, 77-100.
31. Morgan, A. R., Evans, D. N., Lee, J. S., and Pulleyblank, D. E. (1979) Nucleic Acids, Res. 7, 571-594.
32. Kato, A. C., Bartok, K., Fraser, M. J., and Denhardt, D. T. (1973) Biochim. Biophys. Acta. 308, 68-78.
33. Morgan, A. R., and Pulleyblank, D. E. (1974) Biochem. Biophys. Res. Comm. 61, 396-403.
34. Weiss, B., and Milcarek, C., (1975) in Methods in Enzymology 29, 180-193.
35. Kowalski, D. (1980) Analyt. Biochem. 107, 311-313.
36. Bolivar, F. R. L., Rodriguez, P. J., Green, M. C. Betlach, M. L., Heyneker, H., Boyer, H. W., Croser, J. H., and Falkow, S. (1977) Gene 2, 95-113.
37. Hobom, G., and Hogness, D. S. (1974) J. Mol. Biol. 88, 65-87.
38. Potter, H., and Dressler, D. (1977) Proc. Natl. Acad. Sci. USA 74, 4168-4172.
39. Baase, W. A., and Wang, J. C. (1974) Biochemistry 13, 4299-4303.
40. Poccia, D. L., LeVine, D., and Wang, J. C. (1978) Dev. Biol. 64, 273-283.
41. Englund, P. T. (1980) in Biochemistry and Physiology of Protozoa (Levandowsky, M., and Hutner, S. H., eds.) 2nd Ed., Vol. 4, Acad. Press, New York.
42. Marini, J. C., Miller, K. G., and Englund, P. T. (1980) J. Biol. Chem. 255, 4976-4979.
43. Higgins, N.P., Pebbles, C.L., and Englund, P.T. (1980) J. Biol. Chem. 255, 4976-4979.

44. Mizuuchi, K., O'Dea, M.H., and Gellert, M. (1978) Proc. Natl. Acad. Sci. USA 75, 960-5963.
45. Sabcar, A., Hack, A. M., and Rupp, W. D. (1979) J. Bacteriology 137, 692-693.
46. Broome, S., and Gilbert, W. (1978). Proc. Natl. Acad. Sci. USA 75, 2746-2749.
47. Hunter, W. M., and Greenwood, F. C. (1964) Biochem. J. 91, 43-46.
48. Depew, R. E., Liu, L. F., and Wang, J. C. (1978) J. Biol. Chem. 253, 511-518.
49. Burrington, M. G., and Morgan, A. R. (1976) Can. J. Biochem. 54, 301-306.
50. Burgess, R. R., and Jendrisak, J. J. (1975) Biochemistry 14, 4634-4638.
51. Sugino, A., Peebles, C. L., Kreuzer, K. N. and Cozzarelli, N. R. (1977) Proc. Natl. Acad. Sci. USA 74, 4767-4771.
52. Klevan, L., and Wang, J.C. (1980) Biochemistry, in press.
53. Champoux, J.J., and McConaughy, B.L. (1976) Biochemistry 15, 4638-4643.
54. Banks, G.R., Boezi, J.A., and Lehman, I.R. (1979) J. Biol. Chem. 254, 9886-9892.
55. Dynan, W.S., Jendrisak, J.J., Hager, D.A., and Burgess, R.R. (1980). Manuscript submitted.
56. Pulleyblank, D. E., Shure, M., Tang, D., Vinograd, J., and Vosberg, H. (1975) Proc. Natl. Acad. Sci. USA 72, 4280-4284.
57. Wang, J. C., Jacobsen, J. H., and Saucier, J.-M. (1977) Nucleic Acids Res. 4, 1225-1241.
58. Germond, J. E., Hirt, B., Oudet, P., Gross-Bellard, M., and Chambon, P. (1975) Proc. Natl. Acad. Sci. USA 72, 1843.
59. Wang, J. C. (1979) Proc. Natl. Acad. Sci. USA 76, 200-203.
60. Wang, J. C. (1980) Trends in Biochem., Sci. 5, 219-221.
61. Wisehahn, G., and Hearst, J. E. (1978) Proc. Natl. Acad. Sci. USA 75, 2703-2707.
62. Bauer, W. R. (1978) Ann. Rev. Biophys. Bioeng, 7, 287-313.
63. Wang, J. C. (1974) J. Mol. Biol. 89, 783-801.
64. LePecq, J.-B. and Paoletti, C. (1967) J. Mol. Biol. 27, 87-106.
65. Depew, R. E., and Wang, J. C. (1975) Proc. Natl. Acad. Sci. USA 72, 280-4284.
66. Holliday, R. (1967) Genet. Res. 5, 282-304.
67. Liu, L. F., and Wang, J. C. (1975) Biochem. Biophys. Acta 395, 405-412.
68. Holloman, W. K., Wiegand, R., Hoessli, C., and Radding, C. M. (1975) Proc. Natl. Acad. Sci. USA 72, 2394-2398.
69. Shibata, T., DasGupta, C., Cunningham, R. P., and Radding, C. M. (1979) Proc. Natl. Acad. Sci. USA 76, 1638-1642.
70. DasGupta, C., Shibata, T., Cunningham, R. P., and Radding, C. M. (1980) manuscript submitted.
71. Gellert, M., Mizuuchi, K., O'Dea, M. H., Ohmori, H., and Tomizawa, J. (1978) Cold Spring Harbor Symp. Quant. Biol. 43 35-40.
72. Sternglanz, R., D. Nardo, S., Voelker, K. A., Nishimura, Y., Hirota, Y., Becherer, K., Zumstein, L., and Wang, J. C. (1980) Proc. Natl. Acad. Sci. USA, in press.

DNA HELIX-DESTABILIZING PROTEINS

KENNETH R. WILLIAMS AND WILLIAM KONIGSBERG
Department of Molecular Biophysics and Biochemistry, Yale University
P.O. Box 3333 New Haven, Connecticut 06510

I. INTRODUCTION

The introduction of DNA-cellulose-affinity chromatography in 1968[1,2] has facilitated the isolation of a wide variety of proteins from prokaryotic and eukaryotic sources. Besides binding to single-stranded DNA (ssDNA) or double-stranded DNA (dsDNA), these proteins seem to have a number of diverse functions in vivo.[+] This review is restricted to a few of the well-characterized DNA binding proteins that alter the secondary and higher-order structure of DNA in such a way as to make it a more suitable substrate for other proteins involved in DNA metabolism. Included among these so-called Class I DNA binding proteins[7] are those involved in unwinding (rep), untwisting (topoisomerase), twisting (gyrase), and condensing (histones) DNA, as well as a subgroup of proteins that have been referred to as helix-destabilizing[8] or ssDNA binding proteins. The proteins in this subgroup bind preferentially to ssDNA, thereby lowering the melting temperature of dsDNA or partially double-stranded polynucleotides such as poly d(A-T)[8]. Under conditions in which dsDNA is stable, such as those found in vivo, many of these helix-destabilizing proteins also catalyze renaturation of ssDNA. The ability of some of these proteins to catalyze renaturation suggests that the nucleotide bases are exposed and are oriented in a regular array in these DNA:protein complexes. Because of their tight binding and relative abundance in vivo, the helix-destabilizing proteins would be expected to completely cover the transient single-stranded regions of DNA that normally arise in vivo as a result of DNA replication, repair, and recombination. In addition, binding of these proteins to DNA results in the removal of the hairpin-like, base-paired loops often found in ssDNA and further constrains the ssDNA in an extended conformation, making it a more suitable substrate for other proteins involved in DNA metabolism in vivo. Another important property of most helix-destabilizing proteins (HDP's) is their ability to stimulate in vitro the homologous DNA polymerase. Thus the T4 gene 32 protein stimulates T4 DNA polymerase but not any of the host DNA polymerases. The specificity of this stimulation probably arises from direct protein: protein interactions (which can only occur between a DNA polymerase and its homologous HDP) as well as from subtle differences in the conformation of the ssDNA in these various HDP:ssDNA complexes.

The helix-destabilizing proteins coded for by E. coli (single-strand binding protein, SSB), and by the bacteriophages fd (gene 5 protein, 5P) and T4 (gene 32 protein, 32P) have been extensively studied. Among these three proteins, 5P is the best characterized and also the most highly specialized.

[+]Some care needs to be exercised in ascribing in vivo functions on the basis of binding to DNA cellulose. For instance, proteins that apparently have no role in DNA metabolism such as the P-1 protein (protocollagen precursor from fibroblasts[3]) and three serum proteins involved in complement activation (C3DP[4], factor B[5], and β1H[6]) nevertheless bind to ssDNA cellulose.

Its major role in vivo is to shut off the synthesis of the double-stranded replicative form of fd DNA, thereby promoting the synthesis of the ssDNA found
in the fd virion. The gene 5 protein appears to bind ssDNA as it is synthesized, thus preventing complementary strand synthesis by host enzymes[9]. Both
the gene 32 protein from bacteriophage T4 and the single-strand binding protein
from E. coli are essential for DNA replication, recombination, and repair and
should serve as good models for proteins of similar function that are undoubtedly present in eukaryotes.

While several ssDNA binding proteins have been found in higher organisms,
(see Coleman and Oakley[9], and Falaschi et al.[10] for recent reviews), only a few
of these proteins have been characterized well enough to determine if they are
functionally homologous to the prokaryotic HDP's. The helix-destabilizing proteins isolated from mouse myeloma[11], rat liver[12,13], and calf thymus[14,15,16]
all stimulate their homologous DNA polymerase and may be analogous to the SSB
and gene 32 proteins. In addition, a ssDNA binding protein has been found in
Drosophila that cross reacts with antibodies to highly purified rat liver
HDP[17]. This Drosophila ssDNA binding protein has been localized within chromosome puffs that are known to be transcriptionally active and suggests that some
HDP's may function in transcription as well as in DNA replication[17]. In contrast to HDP's, the DNA binding protein from adenovirus infected cells, although
it is required for DNA replication, actually stabilizes dsDNA against thermal
denaturation and is therefore not a HDP[18-22]. Since the adenovirus protein
binds preferentially to the ends of dsDNA[23], it is probably not functionally
homologous to the prokaryotic HDP's. This review will concentrate primarily on
the helix-destabilizing proteins from bacteriophage T4 (32P) and E. coli (SSB),
with particular emphasis on results obtained within the last year.

II. GENE 32 PROTEIN FROM BACTERIOPHAGE T4

A. Multifunctional Properties

The availability of temperature-sensitive and amber mutations in gene 32
has greatly facilitated studies on the function of the gene 32 protein in vivo.
For instance, in bacteriophage T4 containing the P7 temperature-sensitive mutation in gene 32, all DNA synthesis stops within one minute of a temperature
shift from 25°C to 42°C[24]. In vitro, the presence of the gene 32 protein (32P)
increases the rate at which T4 DNA polymerase copies primed ssDNA templates by
5-10 fold, and this stimulation is temperature-sensitive for the P7 gene 32

protein[25,26]. Maximum rate enhancement by 32P is observed at low temperature, high ionic strength, and with sufficient 32P to bind all the ssDNA present; all of which suggest that 32P is stimulating the T4 DNA polymerase by removing inhibitory secondary structure from the template DNA[25]. More direct evidence for this effect of 32P on the template DNA has been obtained by Huang and Hearst[27] who observed that the addition of 32P results in a 7.5 fold increase in the rate of progression of T4 polymerase through one of the most stable helical region in single-stranded fd DNA. In the presence of the three T4 polymerase accessory proteins (44P-62P, 45P), the addition of 32P leads to a tremendous increase in the processivity[28] and a rate of DNA synthesis[29]. Gene 32 protein also plays a role in strand-displacement DNA synthesis. Although T4 DNA polymerase (43P) cannot use nicked dsDNA as a template-primer, the addition of 32P does allow limited strand-displacement DNA synthesis to occur at low ionic strength[26]. At physiological ionic strength, strand displacement DNA synthesis requires a minimum of five proteins (32P, 43P, 44P-62P, 45P) and ATP hydrolysis[30,31]. The addition of 32P to this system results in a 200 fold increase in the rate at which the T4 DNA polymerase copies nicked dsDNA[31,32]. Since the rate of fork movement in this system is proportional to the free 32P concentration (that is 32P is always present in vast excess over that required to cover all of the existing ssDNA), it appears that 32P may exert "pressure" at the fork and hence play an active role in helix unwinding ahead of the leading strand DNA polymerase molecule[32]. It is evident from these results that 32P has a number of functions in T4 DNA replication. Because of the inability of the E. coli HDP to substitute for 32P in these reactions[33,34] some of these effects are undoubtedly due to direct protein: protein interactions between 32P and the other T4 DNA replication proteins. In support of this idea there is in vitro evidence that 32P binds to T4 DNA polymerase[25,35] to the polymerase accessory proteins, 44P-62P[36], and to the RNA priming protein, 61P[35].

In addition to DNA replication, 32P is also essential for T4 DNA recombination[37] and the repair of ultraviolet light-damaged DNA[38]. One of the probable functions of 32P in DNA repair is to bind single-stranded gaps and prevent them from becoming double-stranded cuts that cannot be repaired[38]. In support of this idea, it has been observed that in cells infected with T4 producing the temperature-sensitive P7 gene 32 protein, the normal 800-1,000S intracellular T4 DNA is completely converted to small DNA segments ranging from one-fourth to two genomes in length (39-80S) within 2 min of a shift to high temperature[39]. Gene 32 protein has also been shown to protect ssDNA from nuclease digestion in vitro[39]. The role of 32P in T4 DNA recombination is likely to be complex.

The ability of 32P to catalyze renaturation of complementary strands[40], to protect single-strands from nuclease digestion[38,39], and to promote strand displacement[25] may all be relevant to recombination. Genetic studies suggest that gene 32 protein also interacts directly with two enzymes, T4 DNA ligase[41] and T4 gene 46/47-controlled nuclease[42], as well as membrane proteins[43,44] that are thought to be involved in T4 DNA recombination. Thus, it has been proposed that 32P functions in recombination by forming a membrane-bound complex containing DNA and recombination enzymes[43,44]. Because 32P plays such a central role in DNA metabolism, a detailed analysis of its structural and functional properties should lead to a better understanding of the mechanisms involved in DNA replication, recombination, and repair.

While cooperative binding of 32P to ssDNA is undoubtedly crucial to most of its functions in DNA metabolism, the ability of this protein to also bind to ssRNA allows 32P to self regulate its own rate of synthesis. As 32P is synthesized _in vivo_ it binds first to all of the ssDNA sequences that are present. At this point the "free" 32P concentration rises until it reaches a particular threshold level necessary for 32P to bind specifically to its own mRNA and prevent further 32P synthesis[45-48]. The 10^1 to 10^4 higher affinity (depending on the particular homopolymer tested) of 32P for ssDNA over ssRNA assures that all of the intracellular ssDNA will be saturated with 32P before binding occurs on mRNA[49]. As predicted by this model, mutations that either increase the amount of ssDNA (nonsense mutations in gene 30, DNA ligase; or in genes 41 and 61, RNA priming proteins) or decrease the amount of functional 32P present _in vivo_ (temperature sensitive mutants in gene 32 grown at restrictive temperatures or nonsense mutations in gene 32) result in overproduction of 32P[45-47]. This autogenous regulation of 32P synthesis has been verified _in vitro_ by adding purified 32P to a cell-free translation system. Large amounts of 32P can be synthesized _in vitro_ when unfractionated T4 RNA is added to E. coli extracts (ribosomes and supernatant fractions). If exogenous 32P (final concentration about 8 μM) is added to this system, the rate of 32P synthesis is decreased by 75%. In contrast, this concentration of 32P has no effect on the rate of translation of other T4 mRNA's[48]. These _in vitro_ results require that 32P binds preferentially to a control site on its own mRNA before binding to other T4 mRNA's. This specificity could be accounted for if gene 32 mRNA contains an unusually long sequence (if all other aspects of binding affinity are equal, longer single-stranded sequences will be saturated with 32P prior to shorter sequences[49]) near the initiation site for ribosome binding which is devoid of double-stranded hairpin loops too stable to be melted out by gene 32

TABLE I

CALORIMETRIC AND THERMODYNAMIC PARAMETERS[a] FOR 32P, 32P*-A, AND 32P*-B'

Protein	Additions	T_d, °C	ΔH_{cal}, kJ/mol	N, $\dfrac{\Delta H_{vH}}{\Delta H_{cal}}$	ΔG, KJ/mol
32P	--	56.3	813	0.99	0
32P	poly(dT)	58.9	811	2.45	6.3
32P*-A	--	51.0	961	1.26	-16
32P*-A	poly(dT)	60.9	881	3.04	12
32P*-B'	--	54.1	771	1.03	-5.2
32P*-B'	poly(dT)	54.8	737	1.05	-3.4

Source: K.R. Williams et al. (1979) J. Biol. Chem. 254, 6426-6432 and unpublished observations of L. Sillerud and K.R. Williams.

[a] T_d, thermal denaturation temperature; N, number of molecules undergoing the observed transition as a unit; ΔG, GIBBS free-energy change relative to native 32P.

protein[47]. While the recent studies of von Hippel and his colleagues[49-50] demonstrate that just such a model is qualitatively consistent with the known binding properties of 32P, a specific quantitative model for this system will require additional information on the secondary structure and base composition of the initiation sequences of the relevant mRNA's (see Krisch et al., ref. 51, for preliminary DNA sequencing studies on gene 32).

B. Physical Chemistry of Binding to DNA

Several physicochemical approaches have been used to explore the molecular details involved in 32P binding to ssDNA. Formation of the 32P:DNA complex results in changes in the thermal denaturation and intrinsic fluorescence of 32P, as well as alterations in the ultraviolet absorbance and circular dichroism spectra of the DNA. Differential scanning microcalorimetry[52] reveals that native 32P undergoes a single endothermic transition at 56.3°C, which is lowered to 54.1°C by removal of the NH_2-terminal B' region (residues 1-9)[+] and to 51.0°C by removal of the COOH-terminal A region (residues 254-301) (Figure 1). The calculated cooperative unit size is very close to 1.0 for each of these transitions; implying that even though both 32P and 32P*-A (32P lacking the A region[++]) are present as high-molecular weight aggregates, in each case the monomer is the species undergoing the thermal transition. In the presence of DNA, all three forms of 32P have increased thermal stability (Table I), but only intact 32P and 32P*-A undergo narrow thermal transitions indicative of cooperative protein:protein interactions (Figure 1). In the presence of poly (dT), the cooperative unit size increases to at least 4 for both 32P and 32P*-A, whereas it remains at 1.0 for 32P*-B' or 32P*-(A+B)[52]. Oligo $d(pT)_8$, which is too short to permit cooperative binding, has no effect on the cooperativity of these thermal transitions, although it does result in increased thermostability[52]. These data indicate that the NH_2-terminal B region, in particular the first 9 amino acids, is essential for cooperative protein:protein interactions between adjacent 32P monomers bound to DNA.

[+]L. Sillerud and K. R. Williams, unpublished observations

[++]The intact 32P contains 301 amino acids[53,54]. Partial proteolysis with trypsin can be used to remove either the carboxy-terminal "A" region (residues) 254-301) or both the "A" region and the amino terminal "B" region (residues 1-21) to give 32P*-A and 32P*-(A+B) respectively. While it is difficult to limit trypsin digestion to only the "B" region, half of this region (residues 1-9) can, however, be removed with Staphylococcal Protease to give 32P*-B'.

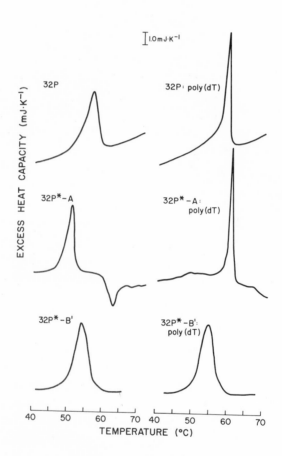

FIG. 1. Effect of poly(dT) on thermal denaturation of 32P and its partial proteolysis products. 32P (29 μM), 32P*-A (22 μM), and 32P -B' (31 μM) were scanned at 1°/min, as described by Williams et al. (1979) J. Biol. Chem. 254, 6426-6432. Data on 32P*-B' are from unpublished observations of L. Sillerud and K.R. Williams.

Because the cooperativity parameter, ω (defined as the equilibrium constant for translocation of a 32P molecule from an isolated to a contiguous binding site), for 32P binding to ssDNA is approximately 1,000[55], removal of the B region from 32P should result in a substantial decrease in the association constant of the 32P:ssDNA, but not of the 32P:d(pT)$_8$ complex. These bindings constants can be readily determined by taking advantage of the intrinsic fluorescence of 32P. While 32P contains 8 tyrosines and 5 tryptophans per monomer[53,54], the intrinsic fluorescence appears to result entirely from the tryptophan residues[56,57]. The tyrosine residues of 32P either are totally quenched or fluorescence only via quantitative energy transfer to tryptophan[56,57]. Binding to nucleotide ligands quenches the intrinsic fluoresence of 32P from approximately 10% for dinucleotides to 20-57% for oligonucleotides and polynucleotides longer than 7 residues[58]. Maximal quenching occurs at a nucleotide-to-protein ratio of approximately 5-6[59,60], which is close to the binding-site size of about 7, as estimated by other methods[55]. On the basis of fluorescence quenching, 32P, 32P*-A, and 32P*-(A+B) all have equilibrium binding constants between 6 X 10^5 M^{-1} and 1.2 X 10^6 M^{-1} for d(pT)$_8$[59] thus indicating that the oligonucleotide binding site is within the trypsin-resistant core of 32P. Both 32P and 32P*-A bind cooperatively to poly(dT) as evidenced by a 400- to 1,000-fold increase in association constant for poly(dT) compared with d(pT)$_8$. In contrast, the binding constant for the 32P*-(A+B):poly(dT) complex is only 3 times higher than that for the 32P*-(A+B):d(pT)$_8$ complex[59], which confirms the proposal based on scanning microcalorimetry that the first nine amino acids are essential for cooperative 32P:32P protein interactions.[+]

These early studies suggested that the approximately 10^8-10^9 M^{-1} binding constant for a 32P:ssDNA complex might be viewed as simply the product of the binding constant for an oligonucleotide ($\sim 10^5$-10^6 M^{-1}) and the cooperativity parameter ($\omega = 10^3$). However, the recent discovery of the very different effects of ionic strength and base composition on the affinity of 32P for oligonucleotides versus polynucleotides suggests there are important differences in the DNA:protein interactions involved in noncooperative versus cooperative binding[49,58]. The binding affinity of 32P for oligonucleotides (2 to 8 nucleotides long) is approximately 10^5 M^{-1} and is essentially independent

---

[+]Our preliminary results indicate that removal of half of the B region (residues 1-9) has a similar effect on the DNA binding properties of 32P as the removal of the entire B region (residues 1-21).

of base composition and sugar type[58]. The binding also has little dependence on ionic strength (a tenfold increase in salt concentration results in only about a 30% decrease in association constant), suggesting that only a single ionic interaction is involved in the oligonucleotide:32P complex[58]. In contrast, 32P has a high degree of binding specificity for polynucleotides, with net binding affinities ($K\omega$) ranging from 10^6 M^{-1} for poly(C) to 10^{11} M^{-1} for poly(dT) in 0.2 M NaCl[49,61]. In general, the binding affinity for a particular homopolyribonucleotide is less (by a factor of $10-10^4$) than the affinity for the corresponding homopolydeoxyribonucleotide. In addition, cooperative binding of 32P to polynucleotides depends heavily on ionic strength, with a tenfold increase in salt concentration resulting in a decrease in binding constant by a factor of about 10^7. Because most of the salt dependency of cooperative 32P binding is due to anion (rather than cation) displacement effects, the actual 32P:polynucleotide complex probably involves about three ionic interactions[58]. For all polynucleotides, virtually the entire base and sugar dependence of binding resides in the association constant (K) for 32P binding to an isolated site on a polynucleotide, rather than in the cooperativity parameter (ω). Thus, cooperative 32P:32P interactions are independent of ionic strength, as well as of the sugar and base composition of the polynucleotide[49,58]. A molecular interpretation of these results requires that 32P exist in two distinct conformations: an oligonucleotide, and a polynucleotide binding mode[58]. The conformational change that accompanies cooperative binding of 32P to ssDNA (see next section) apparently allows additional electrostatic contacts to occur between 32P and the functional groups in the nucleotide bases, which accounts for the increased binding specificity in the polynucleotide binding mode[49].

Fluorescence studies on proteolysis fragments of 32P suggest that the COOH-terminal A region of 32P has relatively little effect on oligonucleotide or polynucleotide binding[59,62]. The apparent association constants for oligonucleotides, as well as the polynucleotide binding specificity and cooperativity, are similar for both 32P and 32P*-A[62]. The removal of the A region does, however, permit the formation of about two additional ionic interactions between 32P*-A and a bound oligonucleotide, thus increasing the salt sensitivity of this interaction[62]. These two positively charged sites in 32P appear to be masked by the negatively charged A region in free or oligomer-bound 32P. The conformational change that results from cooperative DNA binding appears to unmask these two sites, so that the ionic-strength dependence of 32P and 32P*-A binding to polynucleotides is very similar. 32P*-A appears to bind polynucleotides approximately 2-4 times more tightly than native 32P[62]; a finding which is consistent with the increased thermostability of the 32P*-A:poly(dT) complex compared with the 32P:poly(dT) complex[52]. The additional removal of the B region, to form 32P*-(A+B), reduces the cooperativity parameter (ω) from about 10^3 to 1, the binding-site size from 7 to 6 nucleotides, and also increases the salt sensitivity of oligonucleotide binding[62].

The small difference in binding affinity between 32P and 32P*-A for ssDNA[59,62] is not sufficient to account for the increase in helix-destabilizing ability that occurs on removal of the A region. Both 32P and 32P*-A decrease the melting temperature of poly[d(A-T)] [d(A-T)] by at least 40°C[63,64], but only 32P*-A can melt native dsDNA. Under conditions in which 32P*-A destabilizes T4 DNA by 56°C, 32P has no significant effect on the T_m for this dsDNA. Because 32P and 32P*-A have similar equilibrium binding constants for ssDNA, there must be a kinetic, rather than thermodynamic, explanation for the inability of 32P to melt native DNA[59,62]. Most helix-destabilizing proteins melt dsDNA by "trapping" ssDNA loops that open spontaneously as a result of dsDNA "breathing," rather than by binding to dsDNA and forcing the strands apart[66]. Therefore, it has been suggested that 32P is "kinetically blocked" from denaturing dsDNA, because native DNA does not contain single-stranded sequences long enough to permit effective initiation of cooperative 32P binding[55]. Apparently, synthetic polynucleotides, such as poly[d(A-T)] [d(A-T)], do contain single-stranded loops that are long enough for 32P to nucleate cooperative binding[55]. Removal of the A region from 32P somehow eliminates this "kinetic block" to denaturing native DNA[65].

When polynucleotides, such as poly[d(A-T)], are melted by 32P or its proteolysis products, 32P*-A and 32P*-(A+B), there is a large hyperchromic shift which is greater than that produced by temperature-induced melting[55,62]. This increase in absorbance results from loss of base-stacking interactions in double-helical, interchain, hydrogen-bonded sequences, as well as from alterations in the residual stacking interactions between the bases along the single strand[55]. The binding of 32P to single-stranded polynucleotides, such as poly (A) or poly(dA), which are about 50% base-stacked at room temperature[67], results in about a 15% increase in the absorbance at 260 nm[49]; hence, 32P binding may lead to a loss of residual intrachain base-stacking. A similar conclusion was reached from studies on the effect of 32P on poly(1-N⁶-ethenoadenylic acid) [poly (ϵA)] fluorescence. The extended ring system of the ribo(1-N⁶-ethenoadenylic acid) results in a marked fluorescence at 410 nM, which is largely quenched in the polymer as a result of base-stacking[68]. As expected,

32P binding to poly(ϵA) leads to a 2-5 fold increase in the fluorescence of the polynucleotide[58,69], which is similar to that induced by thermal or solvent denaturation of poly(ϵA)[68,70].

In addition to changing the absorbance of polynucleotides at 260 nm, 32P binding also alters the ultraviolet circular dichroism of polynucleotides. In general, there is a decrease in the positive ellipticity between 260 and 280 nm, as well as changes in the negative ellipticity between 240 and 260 nm[55,63,71]. Similar, though not identical, effects on the circular dichroism spectra of polynucleotides are seen with 32P*-A[62,64] and 32P*-(A+B)[62]. Although the circular dichroism changes in the 260-280 nm region have been interpreted in terms of uncoupling of the base chromophores as a result of decreased intrastrand base-stacking in the 32P:polynucleotide complex[9,55] (which is consistent with 32P-induced hyperchromicity in single-stranded polynucleotides), these changes in the 260-280 nm bands are not the same as those observed upon thermal denaturation of synthetic polymers[55,63,64]. This latter observation suggests that the polynucleotide is held in a particular conformation by 32P. In line with this, one circular-dichroism study concluded that 32P binding increases intrastrand base-stacking and that the nucleotide base planes in the ssDNA: 32P complex are actually rigidly fixed relative to each other[63]. In this regard, it is of interest that 32P binding to poly(dT) (which is the only polynucleotide that does not seem to contain significant intrachain base-stacking) results in a small hypochromism[49,63], which at least in this instance, suggests increased base-stacking as a result of 32P binding. While the changes in the circular dichroism spectra probably reflect alterations in intrachain base-stacking, possibly due in part to tyrosine[71] or tryptophan[57] intercalation, it is as yet difficult to interpret the circular dichroism of polynucleotides in terms of a precise conformation[9].

Although 32P melts double-stranded polymers, it also catalyzes renaturation of denatured DNA under conditions in which dsDNA is stable[40]. 32P facilitates DNA renaturation by accelerating the nucleation rate for complementary pairing between two single strands; which is the first and rate-limiting step in DNA renaturation[40]. It probably does so by holding the ssDNA in a highly extended form, thus leaving the bases available for pairing. Sucrose gradient centrifugation reveals that the fd DNA:32P complex sediments only 1.3 times faster than free fd DNA, even though its mass is approximately 16 times greater[40]. This implies that the frictional coefficient of fd DNA increases by a factor of at least 6 in the complex, which is consistent with the DNA being

held in an extended conformation. The expansion of the fd DNA can be observed
directly by electron microscopy, which shows extended, free fd DNA to have a
contour length under denaturing conditions of 1.92 μm (3.0 Å/nucleotide, assuming fd DNA contains 6,408 nucleotides[72]) and fd DNA complexed with 32P to have
a contour length of 3.02 μm (4.7 Å/nucleotide[73]). In contrast to intact 32P,
32P*-A does not catalyze the nucleation step in renaturation; hence, 32P*-A has
only a slight effect, compared to 32P, on the renaturation of alkaline-denatured
T4 DNA[65]. 32P*-A does, however, catalyze the normally rapid second step in renaturation, that is the snapping back of denatured strands already joined by
short stretches of dsDNA. The inability of 32P*-A to catalyze renaturation may
be due to small differences in the conformation of the ssDNA in 32P:ssDNA versus
32P*-A:ssDNA complexes, as was suggested by circular dichroism[63,64].

The data obtained on the physicochemical properties of 32P:DNA complexes
have provided a promising start toward an understanding of the mechanisms involved in the cooperative binding of 32P to ssDNA. Additional studies are
needed to clarify the role of the NH_2-terminal B region in cooperative DNA
binding and the effect of removal of the COOH-terminal A region on the kinetics
of 32P induced melting of dsDNA, as well as to identify particular amino acids
involved in 32P:DNA interactions.

C. Protein Chemistry

The primary structure of 32P has recently been determined[53,54], which
should allow as detailed a physicochemical analysis of this protein:DNA complex
as was used to identify the amino acids involved in fd gene 5 protein:DNA and
protein:protein interactions[9,74]. Krisch et al.[+] have determined the DNA sequence of the first 204 nucleotides in gene 32 and their results are in complete
agreement with the amino acid sequence given in Figure 2 for residues 1-68 in
32P. However, there is some disagreement in the literature concerning the amide
content of 32P (several of the glutamine and asparagine residues in Figure 2
were identified as aspartic acid and glutamic acid respectively in the partial
32P sequence given in reference 75) and the amino acid sequence of the carboxy
terminal A region (compare residues 254-301 in Figure 2 with the sequence of
the A region given in reference 35). As shown in Figure 2 and Table II, 32P
contains 301 amino acids, giving a total molecular weight of 33,487 and a net

[+]Personal communication and reference 51.

488

TABLE II

AMINO ACID COMPOSITIONS OF DNA HELIX-DESTABILIZING PROTEINS

Amino Acid	fd 5P[a]	T4 32P[b]	E. coli SSB[c]
Cysteine	1	4	0
Aspartic acid/asparagine	5	51	16
Threonine	4	14	9
Serine	7	25	11
Glutamic acid/glutamine	10	28	28
Proline	6	8	12
Glycine	7	18	28
Alanine	4	26	13
Valine	8	19	14
Methionine	2	9	5
Isoleucine	4	10	5
Leucine	10	19	8
Tyrosine	5	8	4
Phenylalanine	3	18	4
Histidine	1	2	1
Lysine	6	33	6
Arginine	4	4	10
Tryptophan	0	5	4
TOTAL	87	301	178

[a] Calculated from the amino acid sequence of 5P, Y. Nakashima, et al. (1974) FEBS LETT. 43, 125.

[b] Calculated from the amino acid sequence of 32P, K.R. Williams et al. (1980) Proc. Nat. Acad. Sci., 77, 4614-4617.

[c] Calculated from the amino acid sequence of SSB, A. Sancar et al. (1981) Proc. Nat. Acad. Sci., in press.

```
                        10                                    20
Met-Phe-Lys-Arg-Lys-Ser-Thr-Ala-Glu-Leu-Ala-Ala-Gln-Met-Ala-Lys-Leu-Asn-Gly-Asn-

                        30                                    40
Lys-Gly-Phe-Ser-Ser-Glu-Asp-Lys-Gly-Glu-Trp-Lys-Leu-Lys-Leu-Asp-Asn-Ala-Gly-Asn-

                        50                                    60
Gly-Gln-Ala-Val-Ile-Arg-Phe-Leu-Pro-Ser-Lys-Asn-Asp-Glu-Gln-Ala-Pro-Phe-Ala-Ile-

                        70                                    80
Leu-Val-Asn-His-Gly-Phe-Lys-Lys-Asn-Gly-Lys-Trp-Tyr-Ile-Glu-Thr-Cys-Ser-Ser-Thr-

                        90                                   100
His-Gly-Asp-Tyr-Asp-Ala-Cys-Pro-Val-Cys-Glu-Tyr-Ile-Ser-Lys-Asn-Asp-Leu-Tyr-Asn-

                       110                                   120
Thr-Asp-Asn-Lys-Glu-Tyr-Ser-Leu-Val-Lys-Arg-Lys-Thr-Ser-Tyr-Trp-Ala-Asn-Ile-Leu-

                       130                                   140
Val-Val-Lys-Asn-Pro-Ala-Ala-Pro-Glu-Asn-Glu-Gly-Lys-Val-Phe-Lys-Tyr-Arg-Phe-Gly-

                       150                                   160
Lys-Lys-Ile-Trp-Asp-Lys-Ile-Asn-Ala-Met-Ile-Ala-Val-Asp-Val-Gln-Met-Gly-Glu-Thr-

                       170                                   180
Pro-Val-Asp-Val-Thr-Cys-Pro-Trp-Glu-Gly-Ala-Asn-Phe-Val-Leu-Lys-Val-Lys-Gln-Val-

                       190                                   200
Ser-Gly-Phe-Ser-Asn-Tyr-Asp-Glu-Ser-Lys-Phe-Leu-Asn-Glu-Ser-Ala-Ile-Pro-Asn-Ile-

                       210                                   220
Asp-Asn-Glu-Ser-Phe-Gln-Lys-Glu-Leu-Phe-Glu-Glu-Met-Val-Asp-Leu-Ser-Glu-Met-Thr-

                       230                                   240
Ser-Lys-Asn-Lys-Phe-Lys-Ser-Phe-Glu-Glu-Leu-Asn-Thr-Lys-Phe-Gly-Gln-Val-Met-Gly-

                       250                                   260
Thr-Ala-Val-Met-Gly-Gly-Ala-Ala-Ala-Thr-Ala-Ala-Lys-Lys-Ala-Asp-Lys-Val-Ala-Asp-

                       270                                   280
Asp-Leu-Asp-Ala-Phe-Asn-Val-Asp-Asp-Phe-Asn-Thr-Lys-Thr-Glu-Asn-Asn-Phe-Met-Ser-

                       290                                   300
Ser-Ser-Ser-Gly-Ser-Ser-Ser-Ser-Ala-Asp-Asp-Thr-Asp-Leu-Asp-Asp-Leu-Leu-Asn-Asp-Leu
```

FIG. 2. Amino acid sequence of the bacteriphage T4 gene 32 protein. The B region (residues 1-21) and the A region (residues 254-301) are italicized. (From Williams et al. (1980) Proc. Nat. Acad. Sci. USA, 77, 4614-4617.

charge of approximately -9 at a pH of 7. These values agree with the molecular weight of 35,000, as determined by sodium dodecyl sulfate polyacrylamide gel electrophoresis[40], and the isoelectric point of 5.0, as obtained from isoelectric focussing in urea-containing polyacrylamide gels[65]. The charge distribution within the 32P primary sequence is asymmetric, with the NH_2-terminal half having a net charge of +8 and the COOH-terminal half having a net charge of -17. The secondary structure of 32P predicted from the amino acid sequence by the method of Chou and Fasman[76] is shown in Figure 3. On the basis of this analysis, 32P contains 36% α-helix, 18% β-sheet, and 46% random coil. Circular dichroism spectra have previously indicated that 32P contains 22% α-helix, 26% β-sheet, and 52% random coil[63], thus, the Chou and Fasman analysis may be overestimating the amount of α-helix and underestimating the amount of β-sheet. As shown in Figure 3, 32P appears to contain three domains with respect to secondary structure; the NH_2-terminal region (residues 1-35) and the COOH-terminal region (residues 187-301) are primarily α-helical, while the central region (residues 36-186) contains most of the β-sheet and 11 of the 15 predicted β-turns.

DNA binding results in alterations in the structure of 32P that can be detected by using partial trypsin digestion as a conformational probe. The native 32P contains two regions, the NH_2-terminal B region and the COOH-terminal A region, that can be removed by partial proteolysis with several proteases[65,77,78]. When 32P is subjected to limited trypsin digestion, cleavage occurs first at lysine 253 to produce 32P*-A and then at lysine 21 to produce 32P*-(A+B)[35,53,78]. Cooperative binding of 32P to ssDNA increases the rate at which trypsin removes the COOH-terminal A region[78] and decreases the susceptibility of the NH_2-terminal B region to tryptic hydrolysis[65,78]. In contrast, oligonucleotides that are too short to allow cooperative binding of 32P have no effect on the rate of removal of either the A or the B region by trypsin[78]. A model which accounts for these findings requires that cooperative binding of 32P to ssDNA induces a conformational change that results in increased exposure of lysine 253 and decreased exposure of lysine 21 to trypsin[78]. While trypsin preferentially removes the A region from 32P, Staphylococcal protease cleaves first at glutamic acid 9,[+] thus removing half the B region and producing 32P*-B'. Because removal of the A or B region changes several characteristic properties of 32P, these proteolysis fragments have been valuable in determining the location of functional domains in 32P.

[+]Reference 35 and unpublished observations of L. Sillerud and K. Williams.

Neither the cooperativity of 32P binding to DNA nor the self-association of 32P in solution appears to be affected by removal of the COOH-terminal A region. Both native 32P[79] and 32P*-A undergo such extensive aggregation in solution that, at high protein concentrations, they both elute in the void volume of a Bio-Gel A-5M column.[+] The association constant for indefinite aggregation of 32P is estimated as 3.3×10^5 M^{-1} and is very sensitive to ionic strength[80]. Since the cooperativity parameter for DNA binding by 32P is about 1,000[55], it appears that 32P:32P interactions are less favored when they occur on a DNA lattice. Above a salt concentration of 1.0 M, no aggregation is seen beyond the dimer; and this "high-salt dimer" is apparently a homologous dimer unrelated to the heterologous association of 32P at moderate ionic strengths[80]. In contrast to intact 32P, 32P*-B', which contains residues 10-301 and therefore lacks only half of the B region, elutes from a Sephadex G-100 column run in low salt at a position corresponding to that of a monomer.[+] These findings suggest that 32P self-aggregation may arise from ionic interactions between the Lys-Arg-Lys sequence in the B region (residues 3-5, Figure 2) of one 32P molecule and an anionic site on a neighboring 32P molecule.

Chemical modification experiments have indicated that the binding of 32P to ssDNA may involve tyrosine intercalation between bases[71]. Treatment of 32P with tetranitromethane nitrates five of the eight tyrosine residues in the protein and abolishes the tight binding to ssDNA. In the presence of ssDNA all the tyrosine residues are protected from reacting with tetranitromethane[71]. During the course of the nitration reaction, extensive 32P-32P cross-linking occurs which is not seen in the presence of ssDNA. These results suggest that the tyrosine residues involved in DNA binding give rise to tetranitromethane-induced 32P-32P cross-linking in the absence of DNA. In view of the possible involvement of tyrosine residues in base intercalation, it is tempting to speculate that the region of 32P between residues 72 and 116 (Figure 2) is involved in DNA binding. This region is unusual in that it contains six of the eight tyrosine residues in 32P. Five of the six tyrosine residues in this region are separated by six to eight amino acids and are therefore almost equally spaced in the primary sequence. Because tyrosine intercalation is known to be involved in the binding of the fd ssDNA binding protein (5P) to DNA[9], it is of

---

[+]Reference 35 and unpublished observations of L. Sillerud and K. Williams.

interest to compare the structure between residues 72 and 116 of 32P with those
regions of 5P involved in DNA binding. Although there is no extensive similarity between the primary structures of the ssDNA binding proteins from fd and T4,
we have noticed the following limited homology:

32P	Glu-Tyr-Ser	-		Leu-Val-Lys	
	(105)			(110)	
5P	Glu-Tyr-Pro	-	Val	-	Leu-Val-Lys
	(40)				(46)

The cited region of 5P contains tyrosine 41, which has been shown by a number
of criteria to be directly involved in DNA binding[9]. There is also some similarity between the predicted secondary structure of the 32P region containing
residues 72-116 and that of the 5P region involved in DNA binding. Most of
the amino acids involved in DNA binding in 5P are in a three-stranded antiparallel β-sheet arising from residues 12-49[74]. As shown in Figure 3, the
tyrosine-rich region of 32P (residues 72-116) is also predicted to contain
three short regions of β-sheet, as well as several β-turns.

Although we have not noticed any extensive homology between 32P and other
DNA binding proteins, the spatial arrangement of cysteine residues in 32P is
very similar to that found in RNA binding proteins from Type C viruses.[+]
The RNA binding protein from Rauscher murine leukemia virus (R-MuLV) has been
sequenced, and it contains 53 amino acids[81]. Partial amino acid sequences of
the homologous proteins from AKR, Moloney, and Friend murine leukemia viruses
indicate that these structures differ from the R-MuLV protein by only one or
two amino acid substitutions.[+] As shown here,

R-MuLV[+]   Cys-Asp-Lys-Ala-Trp-His-Gly-Lys-Glu-Lys-Cys-Tyr-Ala-Cys
   (39)   (35)   (30) (26)

32P   Cys-Ser-Ser-Thr-His-Gly-Asp-Tyr-Asp-Ala-Cys-Pro-Val-Cys
 (77)   (81)   (86)   (90)

if the region spanning residues 26-39 in the R-MuLV protein is turned around so
as to read from the COOH-terminal to the NH_2-terminal end, it can be aligned

---

[+]L. Henderson, Frederick Cancer Research Center, personal communication.

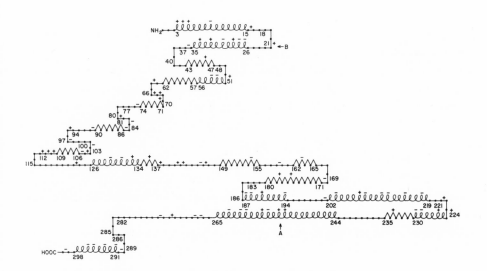

FIG. 3. Schematic diagram of the predicted secondary structure of the gene 32 protein. The secondary structure predictions are from Williams et al. (1981) J. Biol. Chem., in press. Residues are represented in helical (ℓ), β-sheet ($\wedge$), and coil (-) conformations. β-turns are denoted by chain reversals. The positions of charged residues are indicated, and conformational boundary residues are numbered. A and B indicate the trypsin-sensitive bonds in the native protein that, when cleaved, give rise to 32P*-A and 32P*-(A+B).

with residues 77-90 in 32P so that there is a histidine-glycine sequence in
approximately the same position in both sequences and the three cysteines
occur at identical positions. This cluster of cysteine residues is within the
tyrosine-rich region of 32P that we suggest is involved in DNA binding. In addition, ssDNA has been shown to protect two cysteines in 32P from reacting with
5, 5'-dithiobis(2-nitrobenzoic acid),[+] which further suggests that the homology
cited above may be of functional importance.

Ultraviolet light has been used extensively to covalently cross-link
proteins to nucleic acids. In the few instances (including cross-linking of
the fd 5P to DNA) where the reaction products have been carefully characterized
and there is crystallographic data on the three-dimensional structure of the
protein:nucleic acid complex[82,83,84], the cross-linking has been shown to
occur only at the protein:nucleic acid interface. Thus, photoinduced cross-
linking of 32P to ssDNA is an alternative approach to chemical modification
to localize regions of 32P involved in DNA binding. Figure 4 demonstrates
that intact 32P and its proteolysis products 32P*-A and 32P*-(A+B) can all be
cross-linked to [^{32}P]-labeled fd DNA by ultraviolet light. This finding is
consistent with previous results that showed that the DNA binding site is within the trypsin-resistant core of 32P[52,59]. Inasmuch as no cross-linking
is observed with denatured 32P or in the presence of salt concentrations that
are high enough to disrupt the 32P:ssDNA complex, the cross-linking appears
to be specific and should facilitate the identification of 32P amino acids involved in DNA binding.

D. Functional Domains within the Gene 32 Protein

In vitro studies suggest that 32P has at least three separate functional
domains:the NH$_2$-terminal B' region (residues 1-9), a tyrosine-rich region
(residues 72-116), and the COOH-terminal A region (residues 254-301). While
the NH$_2$-terminal B' region has been shown to be essential for both 32P self-
aggregation and cooperative binding to ssDNA, there may be important differences in the actual protein:protein contacts involved in these two processes.
Hence, the 32P:32P interaction has been estimated to be about 300 fold stronger
in the absence of ssDNA than in the presence of ssDNA[55,80]. In addition,
indefinite 32P self-association has no effect on the width of the 32P thermal

[+]R. Mittle, M. LoPresti, and K. Williams, unpublished observations.

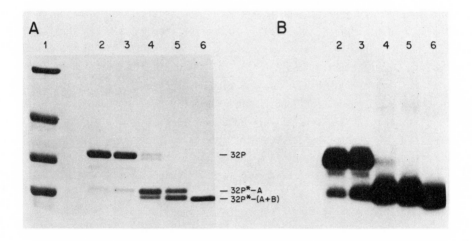

FIG. 4. Gel electrophoresis of partial tryptic digests of the gene 32 protein cross-linked to fd DNA. A, 360 pmole 32P was digested for 60 min at 22°C at the following weight ratios of 32P to trypsin: 1080, 540, 216, 21.6 (lanes 3-6, respectively), then mixed with fd [^{32}P]ssDNA. The samples were then photo-cross-linked by exposure to a germicidal lamp, digested with nuclease, and subjected to polyacrylamide gel electrophoresis. The 32P in lane 2 was cross-linked to fd [^{32}P]DNA without prior trypsin digestion, and lane 1 contains the following standard proteins:bovine serum albumin, ovalbumin, pancreatic deoxyribonuclease, and chymotrypsinogen. B, autoradiogram of the polyacrylamide gel in Fig. 4A.

transition whereas the cooperative 32P:32P protein interactions involved in
binding ssDNA considerably sharpen this transition[52]. Finally, the cooperative
32P:32P protein interaction is not affected by increasing ionic strength whereas
32P self-aggregation in solution is curtailed by increasing concentrations of
NaCl[58,80]. These data suggest that the B' region may be directly involved in
32P self-aggregation but only indirectly involved in cooperative 32P:32P interactions. In view of the insensitivity of these cooperative 32P:32P interactions
to ionic strength, it seems unlikely that the positively charged B' region is
directly involved in these protein:protein contacts. An alternative possibility
is that the conformational change from the oligonucleotide to the polynucleotide
binding mode might permit the B' region to contact the DNA lattice (thus account-
ing for the additional 2-4 basic amino acids in contact with DNA in the cooperative polynucleotide binding mode[58]) and this DNA:protein interaction helps to
drive the conformational change. In the absence of the B' region, this conformational change might be energetically unfavorable and hence account for the inability of 32P*-B to bind cooperatively to ssDNA. Protein:protein cross-linking
reagents, such as a water-soluble carbodiimide which forms covalent bonds between
residues in elecrostatic contact[85], might be useful to determine whether the
B' region is directly or only indirectly involved in cooperative 32P:32P interactions.

Both chemical-modification experiments and the finding of homologous
sequences with other DNA binding proteins have implicated the involvement of
the tyrosine-rich region of 32P (residues 72-116) in DNA binding. In addition
to six tyrosine residues, some of which may be involved in intercalation between bases in DNA[71], this region contains two tryptophan residues and a Lys-
Arg-Lys sequence (residues 110-112). Previous studies indicated that there is
at least one tryptophan residue at the DNA binding site[56,57,69], and this Lys-
Arg-Lys sequence may be important for ionic interactions with the phosphodiester backbone of DNA. These general conclusions concerning the location of the
DNA binding site in 32P are in agreement with the in vivo studies of Breschkin
and Mosig[86]. On the basis of an examination of the in vivo properties of bacteriophage T4 containing temperature-sensitive and amber mutations in gene 32,
they concluded that the NH$_2$-terminal half of 32P contains all the interaction
sites that are essential for DNA binding and replication[86]. More detailed in
vitro chemical-modification and 32P-DNA cross-linking studies should help to
define further the amino acids involved in DNA binding.

Removal of the COOH-terminal A region from 32P eliminates the kinetic block that prevents 32P from denaturing native dsDNA. The small hyperchromicity observed when 32P is mixed with dsDNA[65] suggests that some limited denaturation, probably in A-T-rich regions, is occurring. In line with this idea, if 32P is mixed with dsDNA, fixed with glutaraldehyde, and then subjected to electron microscopy, a denaturation map of the dsDNA is obtained which is similar to that produced by mild alkaline denaturation[73]. That is, the A-T-rich regions have been melted. This suggests that a small number of 32P molecules normally bind to transient single-stranded DNA loops produced by "breathing" of the DNA. Whereas these 32P molecules usually dissociate before cooperative binding is established, the addition of glutaraldehyde irreversibly cross-links the 32P to the ssDNA, thus driving the reaction towards denaturation. One possible explanation of the inability of 32P to melt DNA is that the conformational change required for tight, cooperative binding to DNA provides the "kinetic block" that prevents 32P from denaturing native dsDNA. Hence, 32P dissociates from the transient ssDNA loops in dsDNA before it can undergo the conformational change from the oligonucleotide binding mode to the cooperative, polynucleotide binding mode. Fluorescence studies have established that removal of the A region from 32P exposes two additional positively charged sites that can interact with ssDNA and are usually exposed only when 32P is in the cooperative ssDNA binding conformation[58]. If, as these data suggest, 32P*-A is already in the conformation that favors polynucleotide binding (even before binding to ssDNA), then one might expect 32P*-A to be a much more effective helix-destablizing protein than intact 32P. While further kinetic studies are needed to establish the exact nature of the influence of the A region on 32P binding, possible in vivo consequences of a 32P lacking the A region are readily apparent in the in vitro T4 DNA replication system. Studies with this system have demonstrated that, although 32P*-A can substitute for 32P in leading strand synthesis, there is complete absence of lagging strand synthesis in the presence of 32P*-A.[35] The 32P*-A inhibition of RNA primer extension (lagging strand synthesis) appears to be due to destabilization of the 3'-hydroxy chain terminus by 32P*-A. Because 32P, but not 32P*-A, binds to the suspected T4 RNA primase (61P), the primase may interact with intact 32P to block melting of the newly replicated duplex region - a control exerted through (or at least requiring) an intact A domain[35].

III. SINGLE-STRAND BINDING PROTEIN FROM E. COLI

A. Multifunctional Properties

In vitro studies reveal that the E. coli helix-destabilizing protein is required both for the conversion of phage ssDNA to the duplex replicative form[87] and for the synthesis of single-stranded viral DNA on this dsDNA template[88]. Following bacteriophage fd infection, E. coli single-strand binding protein (SSB) coats the single-stranded fd DNA, preventing nonspecific binding by RNA polymerase, and thus directing the RNA polymerase to the small hairpin duplex in fd DNA that contains the initiation point for RNA primer synthesis[89]. The RNA primer is then extended by DNA polymerase III holoenzyme in a reaction that is stimulated by the E. coli SSB.[87] This stimulation of polymerase III holoenzyme is specific; the helix-destabilizing proteins from bacteriophages fd and T4 can not substitute for SSB[87]. In the elongation reaction SSB is probably acting in a manner analogous to that proposed for 32P, that is binding to the template and holding it in an extended conformation favorable to the DNA polymerase. SSB addition also results in a 10-fold increase in the activity of E. coli DNA polymerase II on extensively gapped templates[33]. The SSB stimulation of DNA polymerase II activity again appears to be mediated via the DNA template, since maximal stimulation is achieved at a fixed DNA-to-protein ratio and is independent of polymerase concentration[90]. The SSB protein increases the processivity of DNA polymerase II on primed ssDNA templates by removing sites of secondary structure in the ssDNA, such as hairpin loops, that otherwise act as "barriers" to further primer extension[91]. Some of the stimulation may, however, result from direct SSB:polymerase II interactions, since SSB forms a complex with DNA polymerase II, but not polymerase I or III, that can be isolated by glycerol gradient sedimentation[92].

Despite the dependence on SSB for phage replication in vitro, it was not certain that replication in vivo had a similar requirement until an E. coli mutant was found with a temperature-sensitive SSB. In E. coli containing the temperature-sensitive SSB-1 mutation, DNA synthesis stops immediately when the temperature is raised from 30°C to 42°C[93]. Similarly, phage ssDNA replication is defective both in vivo and in vitro in this E. coli mutant at 42°C[93]. The SSB-1 mutant is also extremely sensitive to ultraviolet irradiation and only about one-fifth as active as wild-type E. coli in recombination[94]. Thus, the E. coli helix-destabilizing protein is implicated in repair and recombination, as well as in replication. As found with the T4 32P, stoichiometric addition of SSB protects ssDNA from a wide variety of nucleases[92,95-97], thus preventing the formation of dsDNA cuts that cannot be repaired. The finding that lex C, a gene proposed as a regulator of inducible DNA repair[98], is probably alle-lic with SSB suggests that SSB is also involved in regulating the ultraviolet-induced repair system[94]. _In vitro_ studies[99,100] indicate that one of the func-tions of SSB in DNA recombination is to promote the ssDNA assimilation reaction catalyzed by the recA protein. SSB increases the rate and extent of strand assimilation into homologous duplex DNA (D-loop formation), as well as reducing the high concentration of recA protein usually required for this reaction[99,100]. In contrast to the stimulation of T4 DNA polymerase, _E. coli_ DNA polymerase II and the polymerase III holoenzyme, which all require the presence of the homo-logous ssDNA binding protein[33,87], an identical stimulation of the reaction catalyzed by _E. coli_ recA protein was observed with either the _E. coli_ SSB or bacteriophage T4 gene 32 protein[100]. The SSB protein purified from _E. coli_ containing the lex C mutation was, however, considerably less effective than wild-type SSB in stimulating strand assimilation[99]. Inasmuch as the SSB puri-fied from a lex C mutant retains the ability to bind cooperatively to ssDNA[99], the lex C mutation may be in a particular region of SSB that is required for strand assimilation, but not for DNA binding. This implies that SSB, like T4 32P, may have discrete functional domains for the various reactions in which it participates.

B. Physical Chemistry of Binding to DNA

The relative difficulty in obtaining large amounts of SSB, compared with the availability of T4 32P, accounts for the scarcity of physical and chemical information on SSB. Preliminary differential scanning calorimetry studies indicate that SSB undergoes an irreversible thermal transition at 70.9°C.[+] The enthalpy of denaturation of SSB is only about 320 kJ/mol (monomer of mole-cular weight 19,500), compared with 813 kJ/mol for 32P. The cooperativity of the transition ($\Delta H_{vH}/\Delta H_{cal}$ = 2.3 at a nominal scanning rate of 1°/min) suggests that the tetramer is the species actually undergoing the transition. [On the basis of the effect of scanning rate on the 32P transition[52], the cooperative unit size ($\Delta H_{vH}/\Delta H_{cal}$) for SSB will probably increase from 2.3 to approximate-ly 4 as the scanning rate is decreased.] The irreversibility of the transition was surprising at first, in view of the reported stability of SSB to boiling[87].

[+]L. Sillerud, J. Chase, and K. R. Williams, unpublished observation.

This apparent irreversibility is probably a kinetic effect caused by the extended exposure of the protein (at least 90 min in this particular experiment) to
temperatures above 45°C. Previous experiments have demonstrated that long
exposure of SSB to temperatures as low as 45°C leads to irreversible denaturation[96]. This characteristic property of SSB (stability to short periods of
boiling) is shared even by the temperature-sensitive SSB-1 protein, which,
although it is functionally inactive above 43°C, can be boiled and then returned
to a permissive temperature (37°C) with complete recovery of activity[101].

   Oligonucleotide binding to SSB has been studied with equilibrium dialysis
and fluorescence quenching. The equilibrium-dialysis experiments used [^{125}I]
labeled oligonucleotides of varying length and base composition to probe the
non-cooperative binding properties of SSB. This study[96] demonstrated that the
oligonucleotide:SSB interaction is not markedly affected by iodination of the
oligonucleotide or by the base composition or length of the oligonucleotide in
the range of 6-18 bases. The increase in the binding constant from about
5×10^3 M^{-1} for d(pCpT)$_2$ to about 5×10^4 M^{-1} for d(pCpT)$_3$[96] suggests that the
SSB-oligonucleotide interaction may involve as many as six nucleotides. A
Scatchard plot of the binding data indicated that there are two identical
oligonucleotide binding sites per SSB tetramer and that probably only one of
these sites is occupied when SSB binds cooperatively to ssDNA[96]. Although the
oligonucleotide binding constant is not effected by increasing the NaCl concentration from 0.04 to 0.20, a precipitous decrease in association constant (to
less than 10^3 M^{-1}) is observed when the NaCl concentration is increased above
0.2 M[96]. Because sedimentation analysis reveals that the tetrameric structure
of SSB is not altered by increasing the NaCl concentration to 1.0 M[96], this
abrupt decrease in binding affinity (above 0.2 M NaCl) may result from a cooperative conformational change within the SSB tetramer that occurs at high
ionic strength[96].

   SSB has an intrinsic fluorescence which, like that found in 32P, appears
to result entirely from the five tryptophan residues in SSB (Table II). This
fluorescence is partially quenched by nucleic acids[102,103]. The quenching is
proportional to the length of the oligonucleotide, with d(pT)$_{10}$, d(pT)$_{16}$, and
fd DNA reducing the quantum yield of SSB by 10%, 50%, and 80% respectively[103].
This finding indicates that some of the quenching with polynucleotides may be
brought about by changes in the protein conformation and that the final SSB
conformation depends on the length of the oligonucleotide[103]. On the basis of
fluorescence quenching, SSB was found to have association constants of about

2×10^6 M^{-1} and 4×10^8 M^{-1} for d(pT)$_8$ and d(pT)$_{16}$, respectively. The binding constant for d(pT)$_{16}$ is much higher than that observed with equilibrium dialysis[96] and suggests that, under the conditions used in the fluorescence study[102] (low ionic strength and very dilute solutions of SSB, about 0.5 μg/ml), SSB binds cooperatively to d(pT)$_{16}$. In contrast, at higher ionic strengths and protein concentrations, SSB binds noncooperatively, with an association constant of about 10^5 M^{-1}, to oligonucleotides containing less than 18 bases[96]. Gradient centrifugation studies indicate that, at very low protein concentrations and particularly in the presence of d(pT)$_{16}$, the usually stable SSB tetramer dissociates to a dimer[102] Apparently, these conditions allow cooperative protein: protein and protein:DNA interactions to occur that are not found at more physiological ionic strengths and at higher SSB concentrations[96].

Gel filtration[87], density gradient centrifugation[33] electron microscopy[104], and fluorescence quenching[102] all indicate that SSB binds cooperatively to ssDNA with a stoichiometry of about eight nucleotides per SSB monomer of molecular weight 20,000, and with an apparent association constant (K) of 10^9-10^{10} M^{-1}. Both the intrinsic binding constant (K)[96] and the cooperativity parameter ( )[104] are substantially decreased by increasing the NaCl concentration above 0.2 M. The cooperativity parameter (which apparently reflects the strength of the SSB tetramer:tetramer interaction on a DNA lattice[104]) is estimated as about 10^5 M^{-1}. This value is approximately 100 times larger than that seen for 32P binding to ssDNA[55]. SSB binds polynucleotides in the order poly(dT) > poly(dA) > poly(dC) > poly(rA) > poly (rC), with an association constant decreasing from about 10^9 M^{-1} for poly(dT) to about 8×10^6 M^{-1} for poly(rC)[102]. This order of preferential binding to polynucleotides is almost identical with that for 32P and may reflect differences in the secondary structures of these polynucleotides, rather than individual base specificities[87]. Neither SSB[102] or 32P[55] has an appreciable affinity (<10^5 M^{-1}) for dsDNA.

The helix-destabilizing "activity", renaturing ability, and alterations in the circular dichroism of the DNA induced by SSB are all more similar to those observed with 32P*-A than to those observed with 32P. Both 32P*-A[65] and SSB[33] can denature native T4 DNA, whereas 32P appears to be "kinetically blocked" from denaturing dsDNA[55]. With both 32P*-A and SSB, the "melted" DNA is still held together by short stretches of dsDNA, so that, when the Mg^{2+} concentration is increased, the partially separated single strands rapidly reanneal[33,65]. Neither SSB[33] nor 32P*-A[65] catalyzes renaturation of alkaline-denatured (strand-separated) dsDNA under conditions (10mM NaCl, 60 mM Mg^{2+}) that are optimal for 32P-catalyzed renaturation[65]. Hence, under these conditions, neither SSB nor

$32P^*$-A catalyzes the nucleation reaction that is a prerequisite for renaturation of dsDNA. Catalysis of DNA renaturation by SSB does occur at low pH in the presence of divalent cations, but at pH 7 SSB does not catalyze DNA renaturation unless polyamines are present[97]. All three proteins (32P, $32P^*$-A, and SSB) induce similar ellipticity changes in the base chromophores of DNA[64,71]. However, the magnitude of the negative ellipticity change in the region between 270 and 290 nm is greater for $32P^*$-A[64] and SSB[71] than for 32P. While density gradient centrifugation indicates that both SSB[33] and 32P[40] hold ssDNA in a conformation which is greatly expanded compared to that of ssDNA at physiological salt concentrations, electron microscopy indicates that these two proteins have opposite effects on the internucleotide spacing along fully denatured ssDNA. Whereas 32P binding increases the distance between bases by 57%[73] SSB actually decreases this distance by 35%[33], compared with that observed in the absence of added protein.

These preliminary studies have posed several questions about the DNA binding properties of SSB. Even though this protein appears to be a stable tetramer containing four identical subunits, it contains only two oligonucleotide binding sites, and apparently only one can be occupied when SSB binds cooperatively to ssDNA[96]. Although NaCl concentrations below 0.2 M have little effect on DNA binding, higher salt concentrations result in a dramatic decrease in cooperativity (due to disrupting the interactions between neighboring tetramers), as well as the intrinsic binding of the protein to oligonucleotides[96,104]. The similarity of the DNA binding properties of SSB and $32P^*$-A is intriguing and may help to explain the role of the COOH-terminal A region of 32P. Apparently, E. coli DNA replication can tolerate a stronger helix-destabilizing protein than is compatible with T4 DNA replication.

C. Protein Chemistry

SSB can be purified to homogeneity by taking advantage of its tight binding to ssDNA cellulose[33] or its stability to boiling[87]. A convenient filter binding assay is available to measure the amount of SSB even in crude extracts[105]. Crude extracts prepared from bacteria carrying the SSB gene on multicopy plasimds[106] contain about 13 times as much SSB as those prepared from wild-type bacteria[107]. Hence, relatively large amounts of homogeneous SSB (about 20-25 mg/100 g of cells) can now be obtained[107]; which should facilitate more detailed physical and structural analysis of SSB.

Native SSB is a tetramer containing four identical subunits of molecular weight 18,984[108]. The tetramer is stable within the range of 75-750 μg/ml of SSB at NaCl concentration of 0.1[90]. There is no evidence for higher-molecular-weight aggregates beyond the tetramer. Both SSB and 32P are acidic (pI for SSB, 6.0[87]), and they have similar secondary structures (about 20% α-helix, 20% β-structure, and 60% random coil), as determined by circular dichroism[71]. The amino acid sequence of SSB has recently been established and it is shown in Figure 5[108]. Aside from the observation that there seems to be some positional

```
                              10                          20
Ala-Ser-Arg-Gly-Val-Asn-Lys-Val-Ile-Leu-Val-Gly-Asn-Leu-Gly-Gln-Asp-Pro-Glu-Val-

                              30                          40
Arg-Tyr-Met-Pro-Asn-Gly-Gly-Ala-Val-Ala-Asn-Ile-Thr-Leu-Ala-Thr-Ser-Glu-Ser-Trp-

                              50                          60
Arg-Asp-Lys-Ala-Thr-Gly-Glu-Met-Lys-Glu-Gln-Thr-Glu-Trp-His-Arg-Val-Val-Leu-Phe-

                              70                          80
Gly-Lys-Leu-Ala-Glu-Val-Ala-Ser-Glu-Tyr-Leu-Arg-Lys-Gly-Ser-Gln-Val-Tyr-Ile-Glu-

                              90                         100
Gly-Glu-Leu-Arg-Thr-Val-Arg-Lys-Trp-Thr-Asp-Gln-Ser-Gly-Gln-Asp-Arg-Tyr-Thr-Thr-

                             110                         120
Glu-Val-Val-Val-Asn-Val-Gly-Gly-Thr-Met-Gln-Met-Leu-Gly-Gly-Arg-Gln-Gly-Gly-Gly-

                             130                         140
Ala-Pro-Ala-Gly-Gly-Asn-Ile-Gly-Gly-Gly-Gln-Pro-Gln-Ser-Gly-Trp-Gly-Gln-Pro-Gln-

                             150                         160
Gln-Pro-Gln-Gly-Gly-Asn-Gln-Phe-Ser-Gly-Gly-Ala-Gln-Ser-Arg-Pro-Gln-Gln-Ser-Ala

                             170
Pro-Ala-Ala-Pro-Ser-Asn-Glu-Pro-Pro-Met-Asp-Phe-Asp-Asp-Asp-Ile-Pro-Phe
```

FIG. 5. Amino acid sequence of the E. coli single strand binding protein (SSB). (From reference 108).

preference for the location of positively charged residues near the NH_2-terminus, there is no extensive homology between this sequence and the sequences of the helix-destablilizing proteins from T4 (Fig. 2)[53,54] or fd[109-111]. All three of these proteins (5P, 32P, and SSB) have positively charged residues at positions 3 and 21, while 5P and SSB have lysines at position 7, and 5P has an arginine and 32P a lysine at position 16. 32P also has positively charged residues at positions 4 and 5, which do not correspond to lysines or arginines in SSB or 5P. It is premature to attempt to draw any conclusions from these locations of lysine and arginine residues. Although there is no apparent sequence homology between 5P and SSB, antibodies to SSB cross-react with 5P, but do not cross-react with 32P[87]. Chemical-modification studies have indicated that, in contrast to both 32P and 5P, surface tyrosine residues are not involved in SSB binding to ssDNA[71]. Incubation of SSB with N-bromosuccinimide or acetic anhydride destroys the ability of SSB to bind DNA[103], which suggests that tryptophan and lysine residues are essential either for interactions with DNA or for the maintenance of the native SSB conformation. Partial proteolysis can be used to generate two stable cleavage products from SSB (about 16,000 and 12,500 daltons, respectively), that retain the ability to bind ssDNA and may be helpful in locating functional domains within this protein.[+]

IV. CONCLUSIONS

Proteins that bind ssDNA cellulose have now been isolated from a large number of diverse prokaryotic and eukaryotic sources (see Coleman and Oakley[9] and Falaschi et al.[10] for recent reviews). These proteins fulfill a wide variety of functions related to DNA metabolism in vivo. The single-stranded DNA binding proteins from bacteriophage T4 and E. coli belong to a class of proteins whose primary function seems to be to impose an ordered, extended structure onto ssDNA, thus making it a more suitable substrate for other proteins involved in DNA replication, repair and recombination. When normalized to the number of replication forks, SSB and gene 32 proteins are present in nearly equivalent amounts in vivo:there is sufficient SSB and 32P to cover about 1,600[87] and 1,200[40] nucleotides per replication fork, repectively. Although these two proteins appear to be functionally homologous, there is no evidence that they are structurally homologous. More detailed analysis of the SSB and gene 32 proteins should lead to a better understanding of the molecular details involved in DNA: protein interactions, as well as of the mechanisms of DNA replication, repair, and recombination. Of particular interest are the contribution of these ssDNA

[+]K. Williams and J. Chase, unpublished observations.

binding proteins to the structural integrity of replication and recombination complexes, their role in unwinding dsDNA, and their direct interactions with DNA polymerases, ligases, nucleases, and membrane proteins. These studies will undoubtedly facilitate the identification of functionally homologous single-stranded DNA binding proteins in higher organisms.

ACKNOWLEDGEMENTS

We thank several of our colleagues for sending us copies of manuscripts that are in press, and R. L. Burke and B. Alberts for their helpful comments on this manuscript.

REFERENCES

1. Alberts, B., Amodio, F., Jenkins, M., Gutman, E., and Ferris, F. (1968) Cold Spring Harbor Symp. Quant. Biol., 33, 289-305.
2. Litman, R. (1968) J. Biol. Chem. 243, 622-6233.
3. Tsai, R. and Green, H. (1972) Nature New Biol., 237, 171-173.
4. Parsons, R. and Hoch, J. (1976) Eur. J. Biochem. 71, 1-8.
5. Gardner, W., Haselby, J., and Hoch, S. (1980) J. Immunology, 124, 2800-2806.
6. Gardner, W., White, P., and Hoch S. (1980) Biochem. Biophys. Res. Commun., 94, 61-67.
7. Kornberg, A. (1980) in DNA Replication, W. H. Freeman and Company, San Francisco, pp. 277-291.
8. Alberts, B. and Sternglanz, R. (1977) Nature (London), 269, 655-661.
9. Coleman, J. and Oakley, J. (1979) Crit. Rev. Bioch. 3, 247-289.
10. Falaschi, A., Cobianchi, F., and Riva, S. (1980) Trends in Biochem. Sci., 154-157.
11. Planck, S., and Wilson, S. (1980) J. Biol. Chem. 255, in press.
12. Duguet, M., and de Recondo, A. (1978) J. Biol. Chem. 253, 1660-1666.
13. Patel, G. (1980) Biochemistry, in press.
14. Herrick, G., and Alberts, B. (1976) J. Biol. Chem. 251, 2124-2132.
15. Herrick, G., and Alberts, B. (1976) J. Biol. Chem. 251, 2133-2141.
16. Herrick, G., Delius, H., and Alberts, B. (1976) J. Biol. Chem. 251, 2142-2146.
17. Patel, G., and Thompson, P. (1980) Proc. Nat. Acad. Sci. USA 77, in press.
18. van der Vliet, P., Levine, A., Ensinger, M., and Ginsberg, H. (1975) J. Virology 15, 348-354.
19. Ginsberg, H., Lundholm, U. and Linne, T. (1977) J. Virology 23, 142-151.
20. Horwitz, M. (1978) Proc. Nat. Acad. Sci. USA 75, 4291-4295.
21. Klein, H., Maltzman, W., and Levine, A. (1979) J. Biol. Chem. 254, 11051-11060.
22. Schechter, N., Davies, W. and Anderson, C. (1980) Biochem. 19, 2802-2810.
23. Fowlkes, D., Lord, S., Linne, T., Petterson, U., and Philipson, L. (1979) J. Mol. Biol. 132, 163-180.
24. Riva, S., Cascino, A., and Geiduschek, E. (1970) J. Mol. Biol., 54, 85-102.
25. Huberman, J., Kornberg, A., and Alberts, B. (1971) J. Mol. Biol., 62, 39-52.
26. Nossal, N. (1974) J. Biol. Chem., 249, 5668-5676.

506

27. Huang, C., and Hearst, J. (1980) Anal. Biochem. 103, 127-139.
28. Newport, J., Kowalcyzkowski, S., Lonberg, N., Paul, L., and von Hippel, P. (1980) in Mechanistic Studies of DNA Replication and Genetic Recombination. ICN-UCLA Symposium on Molecular and Cellular Biology (Alberts, B. and Fox, C. eds.), Academic Press, New York, in press.
29. Morris, C., Sinha, N., and Alberts, B. (1975) Proc. Nat. Acad. Sci. USA 72, 4800-4804.
30. Liu, C., Burke, R., Hibner, U., Barry, J., and Alberts, B. (1979) Cold Spring Harbor Symp. Quant. Biol. 43, 469-487.
31. Nossal, N., and Peterlin, B. (1979) J. Biol. Chem. 254, 6032-6037.
32. Alberts, B., Barry, J., Bedinger, P., Burke, R., Hibner, U., Liu, C., and Sheridan, R. (1980) in Mechanistic Studies of DNA Replication and Genetic Recombination. ICN-UCLA Symposium on Molecular and Cellular Biology (Alberts, B. and Fox, C. eds.), Academic Press, New York, in press.
33. Sigal, N., Delius, H., Kornberg, T., Gefter, M., and Alberts, B. (1972) Proc. Nat. Acad, Sci. USA, 69, 3537-3541.
34. Burke, R., Alberts, B., and Hosoda, J. (1980) J. Biol. Chem. 255, in press.
35. Hosoda, J., Burke, R., Moise, H., Kubota, I., and Tsugita, A. (1980) in Mechanistic Studies of DNA Replication and Genetic Recombination. ICN-UCLA Symposium on Molecular and Cellular Biology (Alberts, B. and Fox, C. eds.) Academic Press, New York, in press.
36. Alberts, B., Barry, J., Bittner, M., Davies, M., Hama-Inaba, H., Liu, C., Mace, D., Morgan, L., Morris, C., Piperno, J., and Sinha, N. (1977) in Nucleic-Acid Protein Recognition, Vogel, H. ed., Academic Press, New York.
37. Tomizawa, J., Anraku, N., and Iwama, Y. (1966) J. Mol. Biol. 21, 247-253.
38. Wu, J., and Yeh, Y. (1973) J. Virology, 12, 758-765.
39. Curtis, M., and Alberts, B. (1976) J. Mol. Biol., 102, 793-816.
40. Alberts, B., and Frey, L. (1970) Nature, 227, 1313-1318.
41. Mosig, G. and Breschkin, A. (1975) Proc. Nat. Acad. Sci., 72, 1226-1230.
42. Mosig, G., and Bock, S. (1976) J. Virology, 17, 756-761.
43. Mosig, G., Berquist, W., and Bock, S. (1977) Genetics, 86, 5-23.
44. Mosig, G., Luder, A., Garcia, G., Dannenberg, R., and Bock, S. (1979) Cold Spring Harbor Symp. Quant. Biol., 43, 501-515.
45. Krisch, H., Bolle, A., and Epstein, R. (1974) J. Mol. Biol. 88, 89-104.
46. Gold, L., O'Farrell, P., and Russel, M. (1976) J. Biol. Chem. 251, 7251-7262.
47. Russel, M., Gold, L., Morrissett, H., and O'Farrell, P. (1976) J. Biol. Chem. 251, 7263-7270.
48. Lemaire, G., Gold, L., and Yarus, M. (1978) J. Mol. Biol. 126, 73-90.
49. Newport, J., Lonberg, N., Kowalczykowski, S., and von Hippel, P. (1980) J. Mol. Biol., in press.
50. Kowalczykowski, S., Lonberg, N., Newport, J., Paul, L., and von Hippel, P. 1980, Biophysical J., in press.
51. Krisch, H., Duvoisin, R., Allet, B., and Epstein, R. (1980) in Mechanistic Studies of DNA Replication and Genetic Recombination. ICN-UCLA Symposium on Molecular and Cellular Biology (Alberts, B. and Fox, C. eds.) Academic Press, New York, in press.
52. Williams, K.R., Sillerud, L., Schafer, D., and Konigsberg, W. (1979) J. Biol. Chem. 254, 6426-6432.
53. Williams, K.R., LoPresti, M., Setoguchi, M., and Konigsberg, W. (1980) Proc. Nat. Acad. Sci., 77, 4614-4617.
54. Williams, K., LoPresti, M., and Setoguchi, M. (1981) J. Biol. Chem. 256, in press.
55. Jensen, D., Kelly, R., and von Hippel, P. (1976) J. Biol. Chem., 251, 7215-7228.

56. Kelly, R., and von Hippel, P. (1976) J. Biol. Chem. 251, 7229-7239.
57. Helene, C., Toulme, F., Charlier, M., and Yaniv, M. (1976) Biochem. Biophys. Res. Commun. 71, 91-98.
58. Kowalczykowski, S., Lonberg, N., Newport, J., and von Hippel, P. (1980) J. Mol. Biol., in press.
59. Spicer, E., Williams, K. R., and Konigsberg, W. (1979) J. Biol. Chem., 254, 6433-6436.
60. Kelly, R., Jensen, D., and von Hippel, P. (1976) J. Biol. Chem., 251, 7240-7250.
61. Bobst, A., and Pan, Y. (1975) Biochem. Biophys, Res. Commun., 67, 562-570.
62. Lonberg, N., Kowalczykowski, S., Paul, L., and von Hippel, P. (1980) J. Mol. Biol., in press.
63. Greve, J., Maestre, M., Moise, H., and Hosoda, J. (1978) Biochemistry, 17, 887-893.
64. Greve, J., Maestre, M., Moise, H., and Hosoda, J. (1978) Biochemistry, 17, 893-898.
65. Hosoda, J., and Moise, H. (1978) J. Biol. Chem., 253, 7547-7555.
66. von Hippel, P., Jensen, D., Kelly, R., and McGhee, J., (1977) in Nucleic Acid-Protein Recognition, Vogel, H., ed., Academic Press, New York, pp. 65-90.
67. Leng, M., and Felsenfeld, G., (1966) J. Mol. Biol., 15, 455-466.
68. Lehrach, H., and Scheit, K. (1973) Biochem. Biophys. Acta, 308, 28-34.
69. Toulme, J., and Helene, C. (1980) Biochem. Biophys. Acta, 606, 95-104.
70. Steiner, R., Kinnier, W., Lunacin, A., and Delac, J. (1973) Biochem. Biophys. Acta, 294, 24-37.
71. Anderson, R., and Coleman, J. (1975) Biochemistry, 14, 5485-5491.
72. Beck, E., Sommer, R., Auerswald, E., Kurz, C., Zink, B., Osterburg, G., and Schaller, H. (1978) Nucleic Acids Research, 5, 4495-4503.
73. Delius, H., Mantell, N., and Alberts, B. (1972), 67, 341-350.
74. McPherson, A., Jurnak, F., Wang, A., Molineux, I., and Rich, A. (1979) J. Mol. Biol., 134, 379-400.
75. Pan, Y., Nakashima, Y., Sharief, F., and Li, S. (1980) Hoppe-Seyler's Z. Physiol. Chem. 361, 1139-1153.
76. Chou, P., and Fasman, G. (1978) Advances in Enzymology, 47, 45-148.
77. Moise, H., and Hosoda, J. (1976) Nature, 259, 455-458.
78. Williams, K. R., and Konigsberg, W. (1978) J. Biol. Chem., 253, 2463-2470.
79. Bittner, M., Burke, R., and Alberts, B., (1979) J. Biol. Chem., 254, 9565-9572.
80. Carroll, R., Neet, K., and Goldthwait, D. (1975) J. Mol. Biol., 91, 275-291.
81. Henderson, L., Long, C., and Oroszlan, S. (1980) Fed. Proc., 39, 1606.
82. Schoemaker, H., Budzik, G., Giege, R., and Schimmel, P. (1975) J. Biol. Chem., 250, 4440-4444.
83. Havron, A. and Sperling, J. (1977) Biochemistry, 16, 5631-5635.
84. Paradiso, P., Nakashima, Y., and Konigsberg, W. (1979) J. Biol. Chem., 254, 4739-4744.
85. Boulikas, T., Wiseman, J., and Garrard, W. (1980) Proc. Nat. Acad. Sci. USA, 77, 127-131.
86. Breschkin, A. and Mosig, G. (1977) J. Mol. Biol., 112, 279-294.
87. Weiner, J., Bertsch, L., and Kornberg, A. (1977) Proc. Nat. Acad. Sci. USA, 74, 193-197.
88. Scott, J., Eisenberg, S., Bertsch, L., and Kornberg, A. (1977) Proc. Nat. Acad. Sci. USA, 74, 193-197.
89. Geider, K. and Kornberg (1974) J. Biol. Chem., 249, 3999-4005.

508

90. Molineux, I., Friedman, S., and Gefter, M. (1974) J. Biol. Chem., 249, 6090-6098.
91. Sherman, M., and Gefter, M. (1976) J. Mol. Biol., 103, 61-76.
92. Molineux, I. and Gefter, M. (1974) Proc. Nat. Acad. Sci. USA, 76, 1702-1705.
93. Meyer, R., Glassberg, J., and Kornberg, A. (1979) Proc. Nat. Acad. Sci. USA, 76, 1702-1705.
94. Glassberg, J., Meyer, R., and Kornberg, A. (1979) J. Bacteriol., 140, 14-19.
95. Molinex, I. and Gefter, M. (1975) J. Mol. Biol., 98, 811-825.
96. Ruyechan, W. and Wetmur, J. (1976) Biochemistry, 15, 5057-5063.
97. Christiansen, C. and Baldwin, R. (1977) J. Mol. Biol., 155, 441-454.
98. Johnson, B. (1977) Mol. Gen. Genet., 157, 91-97.
99. McEntee, K., Weinstock, G., and Lehman, I. (1980) Proc. Natl. Acad. Sci. USA, 77, 857-861.
100. Shibata, T., DasGupta, C., Cunningham, R., and Radding, C. (1980) Proc. Nat. Acad. Sci. USA, 77, 2606-2610.
101. Meyer, R., Glassberg, J., Scott, J., and Kornberg, A. (1980) J. Biol. Chem., 255, 2897-2901.
102. Molineux, I., Pauli, A., and Gefter, M. (1975) Nucleic Acids Research, 2, 1821-1837.
103. Bandyopadhyay, P., and Wu, C. (1978) Biochemistry, 17, 4078-4085.
104. Ruyechan, W. and Wetmur, J. (1975) Biochemistry, 14, 5529-5533.
105. Whittier, R. and Chase, J. (1980) Anal. Biochemistry, in press.
106. Sancar, A. and Rupp, W.D. (1979) Biochem. Biophys. Res. Commun., 90, 123-129.
107. Chase, J., Whittier, R., Auerbach, J., Sancar, A., and Rupp, W.D. (1980) Nucleic Acids Research, 8, 3215-3227.
108. Sancar, A., Williams, K.R., Chase, J., and Rupp, W.D. (1981) Proc. Nat. Acad. Sci. USA, in press.
109. Nakashima, Y., Dunker, A., Marvin, D., and Konigsberg, W. (1974) FEBS Lett., 40, 290-292.
110. Nakashima, Y., Dunker, A., Marvin, D., and Konigsberg, W. (1974) FEBS Lett., 43, 125.
111. Cuypers, T., vanOuderaa, F., and deJong, W., (1974) Biochem. Biophys. Res. Commun., 59, 557-563.

RNA HELIX-DESTABILIZING PROTEINS

RICHARD L. KARPEL

Department of Chemistry

University of Maryland Baltimore County

Catonsville, Maryland 21228

I. INTRODUCTION

In the 10 yr since the publication of Alberts and Frey's seminal paper on T4 bacteriophage gene 32 protein[1], much work has been done on this and other DNA helix-destabilizing proteins, or HDPs[2-4A]. Considerable less is known about proteins that lower the melting temperature of RNA helices. This paper reviews recent studies on proteins that have such activity _in vitro_ and explores their established and suspected physiologic roles.

Like DNA HDPs, RNA helix-destabilizing (or melting) proteins bring about the denaturation of helices, and thus a lowering of melting temperature, by selectively binding to single strands. Several of these proteins display little selectivity toward binding RNA or DNA single strands, and their involvement in _in vivo_ RNA or DNA function remains to be determined. But RNA HDPs could play a critical role in a number of aspects of RNA physiology, and these are delineated here.

Single-stranded ribonucleic acids, even of random sequence, have a natural tendency to assume conformations of high secondary structure[5,6]. Because there are a number of physiologic processes in which RNA molecules undergo conformational changes that result in the temporary removal or alteration of secondary and tertiary structure, the existence of factors that modulate the extent and kinetics of these changes is likely. Thus, at the level of transcription, in addition to melting of double-stranded DNA, dissociation of the nascent RNA from the DNA template appears to be required. There is good evidence the E. coli RNA polymerase itself can achieve this[7], and the reactivity of DNA bases with dimethyl sulfate indicates that about 11 base pairs are unwound by the enzyme[8]. The activity of the E. coli transcription termination protein, ρ, is influenced by changes in RNA secondary structure, and it has been suggested that this factor has duplex RNA or DNA-RNA hybrid melting activity[9]. Such activity might be coupled to the observed RNA-dependent ATPase activity of ρ, in analogy with the DNA "unwinding enzymes" or "helicases," which derive energy from ATP for their catalysis of strand separation[3].

The transport of RNA from nucleus to cytoplasm in eukaryotes, which for large molecules is believed to take place through the nuclear pore[10], requires local denaturation of helical structure. The size of the nuclear pore would restrict passage of a large globular macromolecule; an estimate of the patent pore diameter is 90Å[11]. It seems reasonable, therefore, that passage of a large RNA molecule through the membrane would be most efficient if the RNA assumed a structure of minimal cross-sectional area (and frictional drag), i.e., a linear conformation. Indeed, there is some electron microscopic evidence suggesting this[10,12]. Single strand-specific RNA binding proteins could thus fulfill an important function in RNA transport by denaturing secondary (and tertiary) structure and forming a linear ribonucleoprotein (RNP) complex that would be easily transportable through the nuclear pore complex. This possible requirement for HDPs may not exist in prokaryotes, in which protein synthesis can be coupled directly with RNA synthesis and newly made RNA molecules may not have the chance to fold into higher-ordered structures.

Another process in which helix destabilization may occur is the reverse transcription of DNA from RNA tumor viral templates, which in avian cells is intiated by a tRNA^{Trp} primer[13]. About 20 residues at the 3' end of the tRNA primer are hydrogen-bonded to the template, corresponding to the loss of secondary structure in the acceptor and TψC stems of the tRNA[14]. In vitro reverse transcription, with a naked template and tRNA, is preceded by the hybridization of the primer to the template, which generally must be annealed at high temperatures for long durations[15], if there is to be sufficient destruction of the tRNA secondary structure. Such heating is not required for reverse transcription in disrupted virions, and at least some tRNA in virus particles is found hybridized to the template[16]. A recent study reported that preincubation of the tRNA and RNA template in the presence of the reverse transcriptase at a moderate temperature (35°C) produces a tRNA-viral RNA complex that is active in initiating DNA synthesis[17]. The enzyme is known to bind the tRNA[18], and this association requires that the tRNA have its native tertiary and most of its primary structure before interaction[19,20]. These results suggest that reverse transcriptase melts part of the secondary structure of tRNA, allowing for hybridization of its 3' end to the template; but this has not been directly demonstrated. An "unwinding-like" activity in reverse transcriptase was demonstrated, through S1 nuclease (single-strand-specific) activity for RNA-DNA hybrids and DNA duplexes, but not for RNA duplexes[21].

RNA tumor viruses contain other RNA binding proteins with specificity for single strands. These include p10 from Rauscher murine oncovirus[22,23] and p12 from avian tumor viruses[24]. Proteins with binding specificity toward single or double strands could regulate the activities and specificities of nucleases involved in RNA processing[25], in which HDPs might play a role (p10 and p12 are treated more fully below).

The temporal destruction of mRNA secondary and tertiary structure at the ribosome during protein synthesis argues for the presence of a helix-destabilizing factor(s). Such a function has been demonstrated in the multiprotein initiation factor eIF3 from rabbit reticulocytes, which stimulates globin synthesis and lowers the melting temperature of RNA polymers[26]. In E. coli, ribosomal protein S1, which is involved in the binding of mRNA to the ribosome[27] and appears to be essential for the translation of natural messages[28], has demonstrable helix-destabilizing activity in vitro[29]. Although its physiologic role has not been established, it has been suggested that S1 functions in conjunction with initiation factor IF3 by recognizing and unfolding elements of the mRNA[30], or possibly 16S rRNA, to facilitate complementary pairing between the two macromolecules[31,32]. Circular dichroism experiments suggest that IF3 also has RNA helix-destabilizing activity[33]. The involvement of HDPs in protein synthesis is detailed later.

A number of tRNA molecules and E. coli 5S RNA take up biologically inactive conformations in vitro when the concentration of counterions is low[34-36]. When physiologic counterion concentration is restored - i.e., 0.01 M Mg^{2+} and 0.1 M Na^+ or K^+ - the conversion of these inactive conformers to functional, active molecules is often quite slow, even at 37°C. Although there is no evidence that these inactive RNAs exist in vivo, they serve as accessible model systems for the study of the involvement of HDPs in RNA conformational interchange. The structural basis of the slow renaturation kinetics is the presence of "incorrect" base pairs in the inactive conformer and possibly other tertiary structural interactions not present in the active form (37; Lindahl et al., in preparation). These incorrect pairs must be broken for the RNA molecule to fold into its native conformational state correctly. The rate-determining step for renaturation, characterized by a high activation barrier (38-40; Lindahl et al., in preparation), is therefore a denaturation event. Factors that encourage helical denaturation, such as HDPs, should accelerate the inactive ⟶ active RNA conformational change.

II. RNA HELIX-DESTABILIZING AND SINGLE-STRAND - SPECIFIC PROTEINS IMPLICATED IN PROTEIN SYNTHESIS

A. E. coli Ribosomal Protein S1

A number of recent studies have concentrated on the nucleic acid helix-destabilizing and single-strand binding activity of S1, a protein isolated from E. coli 30S ribosomal particles. S1 is a single polypeptide of 65,000 daltons, with a very high (10:1) axial ratio[41]. On the basis of sedimentation equilibrium and low-angle x-ray scattering experiments, S1 appears to be a prolate ellipsoid with approximate dimensions of 220 Å x 25-30 Å[41]. Circular dichroism measurements indicate a high degree of β-sheet structure (35-40%), but only 10-15% α-helix[41]. Although there are indications that the nucleic acid binding properties of S1 depend somewhat on the method of protein preparation, these physical properties were found to be independent of the method of preparation.

When stoichiometric amounts of S1 were mixed with single-stranded polyribonucleotides that have helical structure, such as poly(C) and poly(A), an ultraviolet hyperchromic effect was observed at 18°C equivalent to thermal melting at 70°C and 50°C, respectively[29]. The circular dichroism spectrum of poly(C), treated analogously, was substantially diminished, indicating a distortion in the polynucleotide backbone coincident with base unstacking[42]. The circular dichroism spectrum of poly(A) was unaffected by S1, although the salt concentration in these experiments (0.1 M Na^+, 0.01 M Mg^{2+}) was much higher that that in the UV studies (10 mM NaCl, 5 mM Tris-HCl, at a pH of 7.4) and thus less favorable for binding if electrostatic interactions are important. Nitrocellulose filter binding and other experiments have indicated that poly(U) and poly(I) are bound by S1 even more tightly than poly(C) and poly(A)[43].

S1 also lowers the melting temperature of the acidic double helices of poly(C) and poly(A), as well as of the spermine-stabilized double helix of poly(U), poly(U·U)[29,42]. When S1 is added to a solution of the poly(U) random coil, the poly(U·U) helix is not formed on addition of spermine. Likewise, at low sodium ion concentration (<0.025 M), S1 prevents the formation of the poly(A·U) helix (1:1 poly(A) + poly(U)), but not at 0.1 M[29]. At 0.01 M, 75% of the helical structure of coliphage MS2 was seen to melt on addition of S1, whereas at higher sodium ion concentration there was less of an effect, and no denaturation was seen in 0.01 M Mg^{2+} [29]. Similar results were obtained with tRNA and with 16S and 23S E. coli ribosomal RNA. These results indicate that, as in the case of stacked single strands, the ability of S1 to denature or prevent the formation of double helices (presumably by binding tightly to the individual strands) reflects an important electrostatic component in the interaction.

In contrast with the disruption of RNA secondary structure by S1, the optical properties of natural and synthetic DNAs are generally unaffected by this protein. Although heat-denatured E. coli DNA shows a hyperchromic effect on addition of S1, the ultraviolet absorptions of native E. coli DNA, polyd(A-T) (alternating sequence), poly(dA), and poly(dT) are unaffected by the protein, even at very low ionic strength[29]. S1 does slow the formation of the 1:1 complex of poly(dA) and poly(dT), which was interpreted as protein interaction with the fully hyperchromic poly(dT), but not with poly(dA), inasmuch as a hyperchromic change analogous to the effect with poly(A) was expected for interaction with poly(dA). However, fluorescence-quenching experiments clearly showed that S1 binds to poly(dA), and an alternative explanation has been offered (see below). The circular dichroism of a number of synthetic DNA and DNA RNA double helical complexes was unaffected by S1, although that of poly (A dT) was perturbed in the absence of any hyperchromic effect[42].

The interactions and binding selectivity of S1 with polynucleotides have also been investigated, with the protein's intrinsic tryptophan fluorescence as a probe[44-46]. These studies led to the conclusion that there are two nucleic acid binding sites on S1: site I has no specificity for RNA vs. DNA, and site II binds only RNA chains. The evidence for this model is based on the different fluorescence-quenching effects of polydeoxyribonucleotides and polyribonucleotides. Poly(dC) and poly(dA) bring about a substantial degree of quenching in 0.1 M NaCl - 35% and 31%, respectively. Analysis of fluorescence titration and sucrose-gradient sedimentation data, via the overlap method of McGhee and von Hippel[47], indicated that binding of single-stranded DNA to S1 (site I) is noncooperative, with each S1 occluding 5 ± 1 nucleotide residues[44,45]. If S1 is first fully titrated with the oligodeoxynucleotide $(dA)_4$, additional fluorescence quenching of 15-20% occurs when $(rA)_{19}$ or $(rC)_{20}$ is then added (Figure 1). The additional quenching by $(rC)_{20}$ has a stoichiometry of 2 S1 molecules per oligomer, or an occluded site size, n, of about 10 nucleotide residues (site II). Fluorescence quenching with $(rC)_{20}$ in the absence of $(dA)_4$ shows a sharp break indicative of two binding sites (figure 1). Analysis of the initial steep portion of the titration plot yielded a binding affinity and site size correlated with site II; at higher [oligonucleotide]:[S1], the (lower) affinity and site size were indicative of site I.

The characteristics of binding of oligonucleotides and polydeoxynucleotides to site I were studied in detail, mainly by fluorescence-quenching techniques[45]. In 0.12 M Na^+, the association constants of poly(dA) and poly(dC) were similar (1.7×10^6 M^{-1} and 6.0×10^6 M^{-1}, respectively), with no evidence of binding cooperativity (cooperativity parameter, $\backsim 1$). Analysis of the [Na^+] dependence of the association constant by the method of Record et al.[48] indicated that two ion pairs are involved in the interaction at site I (five nucleotide residues are occluded at this site). This analysis also indicated that there

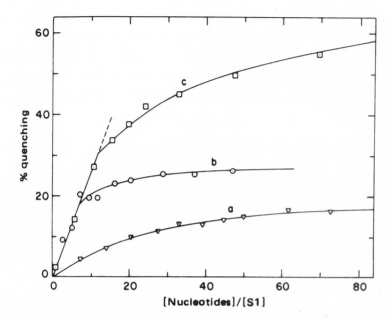

Fig. 1. Fluorescence titrations of S1 with RNA and DNA oligomers. All titrations are in 10 mM phosphate (Na$^+$) (pH, 7.7), 1 mM Na$_2$EDTA, 1 mM 2-mercaptoethanol, 10% v/v glycerol, and 0.1 M NaCl. Concentrations of S1: curve a, 0.65 μM ; curve b, 0.74 μM; and curve c, 0.76 μM. Titrations a and b are in the presence of 16 μM (dA)$_4$; the quenching shown is in addition to the 43% quenching from (dA)$_4$. Titration a is with (rA)$_{\overline{19}}$; titrations b and c are with (rC)$_{\overline{20}}$. From Draper et al.[44].

was a significant nonelectrostatic contribution to the binding free energy, -6.7 kcal/mole. Because there appeared to be little base specificity at this site, it was surmised that interactions between the protein and nucleic acid sugar residues account for this energy. The binding of a number of oligodeoxynucleotides was determined; the extent of S1 fluorescence quenching was similar for all oligomers tested.

The affinity of S1 for double-stranded DNA (determined by sucrose-gradient band sedimentation) was about one-thousandth of the association constant for single-stranded DNA. In line with the lack of base specificity at site I observed with polydeoxynucleotides and oligodeoxynucleotides, the binding of 2'-deoxyribose 5'-phosphate was shown to be comparable with that of dAMP and dCMP. Two d(ApA) molecules were found to bind to two subsites of site I, with rather different affinities - $>10^7$ M^{-1} and 2×10^5 M^{-1}. Iodine quenching experiments confirmed the existence of two subsites, each presumably capable of forming one of the two ion pairs expected to occur with DNA phosphates. Although sites I and II display very different characteristics, they are not independent. Substrates - e.g., (rC)$_{\underline{}}$ - bound to site II diminish the affinity of oligodeoxy-nucleotides for site I. In line with the absence of ultraviolet absorbance or circular dichroism changes in single-stranded DNAs on interaction with S1[29,42], there was no change in the fluorescence, and hence no change in the base stacking of etheno-derivatives of d(pA)$_4$, d(pA)$_8$, or denatured calf thymus DNA.

The interaction of oligo- and polyribonucleotides with site II was studied in the presence of excess d(pA)$_4$, so that site I was fully occupied and inaccessible to RNA[46]. Analysis of the binding of poly(A) to site II indicated a noncooperative association ($\omega \sim 1$), with K = 3.6 x 10^5 M^{-1} and an occluded site size, $\underline{n}$, of 10 residues. One molecule of A(pA)$_9$ binds to site II, with an affinity of 2.1 x 10^5 M^{-1}, about half that of poly(A). With the assumption that it is the base residues of substrate that interact with site II (see below), statistical considerations suggest that this lower association constant reflects an interactive site size $\underline{m}$ (the minimal length of residues interacting with the protein) of 9 residues. The binding of oligo(C) and poly(C) to site II is qualitatively different from that of oligo(A) and poly(A). Two S1 molecules bind cooperatively ($\omega \sim 31$) to oligo(C)$_{\underline{20}}$ site II with an (intrinsic) association constant of 1.0 x 10^6 M^{-1} and an occluded site size of 9.6 ± 2.0 nucleotide bases. K,ω, and $\underline{n}$ were found to be the same for poly(C). So, unlike the association at site I, binding of RNA oligomers and polymers at site II is base specific and cooperative for oligo(C) and poly(C).

The association of RNA with site II is apparently nonelectrostatic and thus does not involve backbone phosphates, in as much as the affinity of oligo(A) or (C) for this site actually _increases_ with increasing NaCl concentration. This does not, however, explain S1-induced poly(A) hyperchromicity seen at low ionic strengths[29] but the absence of observable backbone distortion at higher salt concentration[42]. In fluorescence titration experiments at constant S1 concentration, addition of poly(C) first titrates site II, then site I; this is consistent with the higher affinity of this polymer (taking cooperativity into account) for site II. However, treatment of poly(C) with formaldehyde to form a hydroxymethyl adduct of the exocyclic amino group yields a polymer that titrates site I first, with the affinity for site II relative to that of unmodified poly(C) diminished. These results are consistent with the observed base specificity of site II and the nonspecificity of site I.

The fluorescence-quenching technique was used to determine the affinity of S1 for a dodecameric, pyrimidine-rich fragment from the 3' terminus of _E. coli_

16S rRNA that was generated by the action of RNase T1[49] and contained a sequence that had been proposed to bind S1 specifically[50]. This fragment was shown to have an affinity comparable with that expected for oligo(C)$_{12}$, with an intrinsic binding constant of 1.6×10^6 M^{-1} [46]. There was thus no evidence of site-specific recognition, and any apparent high affinity of this fragment for S1 was probably due to its base composition, rather than its sequence.

Because S1 molecules do not appear to interact with each other[41], it was suggested that the cooperativity seen on binding oligodeoxynucleotides and poly(C) to site II is a consequence of the distortion of the polynucleotide lattice by S1[46]. The binding of an initial S1 molecule results in lattice distortion at that site, so less free energy is required to distort the poly-nucleotide at adjacent sites. In this manner, additional S1 molecules are more likely to bind contiguously, rather than noncooperatively. This interpretation, however, does not explain the lack of cooperativity seen with poly(A) in 0.12 M Na$^+$ [46], which at low ionic strength (about 0.01 M) is probably distorted (a hyperchromic change occurs) on binding S1[29]. In 0.12 M Na$^+$, the affinity of poly(A) is slightly greater for site I than for site II[46]; in 0.01 M Na$^+$, the very different salt dependences of the two sites indicate that site I binding should be even more dominant. However, because there appears to be no single-stranded nucleic acid distortion at site I, as judged by the circular dichro-ism[42] and UV[29] effects of S1 on polydeoxynucleotides, the S1-induced hyper-chromicity of poly(A) at low sodium ion concentration cannot be explained by a greater (site I) affinity. Conceivably, the cooperativity parameter, ω, is itself dependent on sodium ion concentration. Furthermore, the method of pre-paration of S1, which differed among the various studies, may affect its bind-ing properties[51]. Aside from the oligo- and poly(A) results, the hypothesis of two binding sites and the characteristics enumerated above are generally consis-tent with the various physical studies on S1-substrate interactions.

Several recent chemical modification studies have increased the understand-ing of the structural nature of these interactions. Reaction of S1 with N-ethylmaleimide (NEM), a cysteine-modifying reagent, produces a monosubstituted derivative that has largely lost its effect on poly(A), poly(C), or poly(U·U) hyperchromicity[52]. However, poly(C) and MS2 RNA produce substantial quenching of NEM-S1 intrinsic fluorescence, although by less than with unmodified S1. Poly(U) (random coil) quenching of NEM-S1 fluorescence is the same as that seen with unmodified protein. It thus appears that S1-induced hyperchromicity, associated with site II, depends on an essential cysteine residue or some other residue blocked by the modification. The NEM derivative did incorporate into

30S S1-depleted ribosomal subunits, but did not bind MS2 RNA nor form an initiation complex with this natural mRNA. These reconstituted subunits retain their AUG-directed initiator FMet-tRNA^{FMet} binding, as well as poly(U)-directed Phe-tRNA^{Phe} binding. These results are consistent with the loss of helix-destabilization activity of NEM-S1 and with the possible requirement of (at least partial) melting of structure on the message to expose 16S rRNA-binding[31] or tRNA-binding sequences.

In an electron microscopic study[53], S1 and NEM-S1 bound MS2 RNA and ØX174 DNA with a stoichiometry of one S1 per 10-15 nucleotides. The interaction was not highly cooperative. Underivatized S1 melted most of, but not all, the RNA or DNA secondary structure, whereas NEM-S1 did not show this activity. Neither electron microscopic nor sedimentation techniques showed any evidence of S1-effected RNA-RNA or RNA-DNA cross-linking[53], such as might result as a consequence of the two proposed binding sites[44-46]. However, the conditions used in the electron microscopic study were necessarily different from those of the fluorescence titration experiments, which reported only on polymer-oligomer or oligomer-oligomer binding to S1, but not on the simultaneous binding of RNA polymer and DNA polymer.

The involvement of lysine residues in interactions of S1 and single-stranded nucleic acid was explored in a study on the consequences of reductive methylation of the protein[54]. An average of 6 of the 30 lysine residues of S1 were found to be methylated with formaldehyde and NaB^{3}H$_4$, although this figure may be twice the actual number in that it was apparently based on the assumption that monomethylated derivatives were formed, whereas the principle product of this reaction is known to be ϵ-N,N-dimethyllysine[55,56]. Methylated S1, prepared with unlabeled NaBH$_4$, displayed a poly(^{3}H-U) binding activity (on nitrocellulose filters) only about 20% of that of unmodified protein. When reductive methylation was performed in the presence of excess poly(U), the extent of reaction was reduced by 80%. The observation that 2 ion pairs are involved in the interaction of site I with single-stranded substrates[45] leads to the prediction that the number of protected lysine residues should be no more than 2, and possibly less, because only a fraction of the lysine residues can be modified. Thus, if only an average of about three lysine groups are modified under nonprotecting conditions (about three dimethyl groups), then about two lysines are protected by poly(U) against methylation by poly(U); that is consistent with m' = 2. The 20% residual poly(U)-binding activity of methylated S1 may represent interaction at site II, which is nonelectrostatic[46] and thus likely to be less sensitive to lysine modification. Although unmodified S1 can exchange with 30S subunits,

methylated S1 did not[53]. 30S subunits depleted of S1, however, did bind the
derivatized protein. S1-depleted subunits bind poly(U) as well as nondepleted
particles, but subunits reconstituted with methylated S1 had reduced poly(U)-
directed binding of Phe-tRNA[Phe 53].

Limited tryptic digestion of S1 produces a 48,500-dalton fragment that
binds to poly(U) and S1-depleted 30S subunits, but does not function in protein
synthesis[57]. This fragment, which also bound denatured SV40 DNA (to nitrocellulose filters), failed to bind MS2 RNA[58]. Unlike noncleaved S1, the fragment had
no destablilizing effect on the secondary structures of the poly(U•U) helix, MS2
RNA, or ØX174 DNA, as judged by optical and circular dichroic experiments[58].
This cleavage product's properties are similar to those of NEM-S1[53,54] and further support the apparent correlation between S1 helix-destabilizing activity
and its function in protein synthesis. In this regard, a mutant form of S1
lacking about 140 amino acids from the C terminus (about 23% of the total polypeptide) is active in protein synthesis and retains poly(U), MS2 RNA, and denatured SV40 DNA binding activity, as well as poly(U•U), MS2 RNA, and ØX174 DNA
helix-destabilizing activity comparable with that of wild-type protein[58,59].

Although the experiments reported above support a relationship between the
in vitro helix-destabilizing activity of S1 and its involvement in protein synthesis, the actual physiologic function of S1 is not yet fully understood. In
an attempt to increase understanding of its function in the translation process,
several recent studies have concentrated on the interaction of S1 with the 30S
ribosomal subunit and the concomitant effect on polynucleotide binding. Sucrose-
gradient centrifugation experiments indicated that 30S particles have 2 potential S1 binding sites[41,60]. The sites differ by a factor of 10-20 in affinity
for S1 and appear to function independently of each other[60]. Heat-reactivated
30S subunits (with increased 50S- and tRNA-binding properties) have only one
high-affinity site. When 70S ribosomes are treated with colicin E3, a 49-
nucleotide fragment is cleaved from the 3' terminus of 16S RNA[61]. Such cleavage
has no effect on the high-affinity S1 site, although the 3' end of 16S RNA
appears to be required for binding the protein to the low-affinity site[60].
NEM-S1 - which, as we have noted, has no helix-destabilizing activity[52] - binds
to both high- and low-affinity sites on 30S subunits[60]. Thus, it might be expected that site II on S1, which brings about single-stranded RNA distortion, is
not involved in binding this protein to the ribosome. However, it was shown,
with a sucrose-gradient band sedimentation technique, that single-stranded polyribonucleotides, but not polydeoxynucleotides, are effective competitive inhibitors of the interaction of S1 with S1-depleted 30S subunits; so site II, but

520

not site I, seems to be involved[51]. It is possible that NEM modification deri-
vatizes a cysteine or sterically blocks another amino acid at site II that is
essential for helix destabilization, but is not involved in binding the protein
to the ribosome. Furthermore, recent experiments have indicated that S1, although part of the 30S subunit, can bind polynucleotides via site II; so this
site is not completely buried (D. Draper, personal communication). The association constant of the 1:1 complex of S1 with the 30S particle is 2 X 10^8 M^{-1},
which corresponds to a free-energy change of about -10.9 kcal/mole[51]. A major
part of this energy, about -8.0 kcal/mole, can be accounted for by S1-16S RNA
interaction, leaving about -3 kcal/mole for interactions between S1 and other
components of the 30S subunit[51]. The affinity of S1 for 70S ribosomes is about
the same as that for 30S particles; it is much less for the 50S subunit, and
this indicates that the larger subunit is probably not involved in the interaction with S1[51].

Although the 3'-terminus of 16S rRNA does not appear to be involved in the
binding of S1 to active ribosomes, 30S-bound S1 cound still function as a 16S
RNA helix-destabilizer. S1 forms a strong 1:1 complex with the 49-nucleotide
colicin fragment[63]. Temperature-jump experiments on the fragment indicate the
existence of two hairpin double helices of nine and four base pairs[63]. Some
(3.5%) RNA hyperchromicity is seen in the 1:1 complex, corresponding to S1 binding to and melting of the weaker, four-base-pair pyrimidine-rich hairpin. Higher concentrations of S1 yield increased, but weaker, binding and melt the remaining secondary structure, for a total absorbance increase of 30%. The affinity of S1 for S1-depleted 30S subunits was found to be 40 times higher than
that for the colicin fragment-confirmation that this part of the 16S RNA is not
involved in S1-30S binding.

If S1 is involved in the formation of mRNA-16S rRNA and mRNA-tRNA pairings
as part of a stable initiation complex, then its single-stranded-RNA-specific
binding and helix-destabilizing properties suggest that its role is to disrupt
internal RNA secondary structure, rather than to stabilize the RNA-RNA double
helices formed. Intramolecular 16S RNA or mRNA secondary structure may have
to be disrupted in order to form intermolecular base-pairing. In the colicin
fragment of 16S RNA, part of the pyrimidine-rich Shine-Delgarno sequence, which
pairs with complementary residues on mRNA[31,32], is found in the four-base-pair
helix. Although this helix was found to melt below room temperature[63], it is
conceivable that the Shine-Delgarno residues are not fully accessible for pairing when part of the 30S particle. The complementary mRNA residues may also be
internally paired, necessitating S1-effected melting. The stimulatory effect

of S1 on the formation of initiation complexes of the polycistronic R17 RNA was significant for the coat protein and replicase initiator regions, where theoretically stable hairpin loops can form[64]. However, only minimal stimulation was seen for the A-protein region, where theoretically no such structures can form[64,65]. S1 may not be the only macromolecular helix-destabilizer at the E. coli ribosome; there is evidence that initiator factor IF3 also has helix-destabilizing activity.

S1 may also have roles in several aspects of RNA function other than the initiation of translation, e.g., in the elongation cycle. When S1 was added to ribosomes from which it had been depleted, it stimulated poly(U)-dependent polyphenylalanine synthesis and poly(A)-dependent polylysine synthesis[66]. Although S1 has no effect on the binding of poly(U) to 30S subunits, Phe-tRNA binding to subunits in the presence of oligonucleotide or poly(U) was found to be S1-dependent, whereas the analogous interaction with Lys-tRNA and oligo(A) was insensitive to S1[66]. High S1-to-ribosome ratios inhibit protein synthesis, known as interference factor (iα) activity. The inhibition is due to S1-mRNA binding[67] and is strongest on pyrimidine-rich templates[68]. Finally, it should be noted that S1 is the α-subunit of the coliphage Qβ replicase, an RNA-dependent RNA polymerase[69,70].

B. E. coli Initiation Factor 3

E. coli initiation factor 3 (IF3) is essential for the initiation of natural mRNA translation[71-73] and stimulates initiation directed by synthetic polynucleotide messengers[74]. The degree of stimulation of both natural and synthetic mRNA appears to be correlated with the extent of secondary structure in the initiation region of the messenger[73]. Thus, as in the case of S1, there is reason to suspect that IF3 is an RNA helix-destabilizer, although the particular RNAs destabilized (16S or mRNA) remain to be definitively identified[75].

Recent experiments have concentrated on the binding to[75] and helix destabilization of[33] polynucleotides by IF3 using nitrocellulose-filter binding and circular dichroism techniques. In a buffer of low ionic strength (about 0.02 M), IF3 was found to bind poly(^{3}H-A), poly(^{3}H-C), and poly(^{3}H-U) tightly to nitrocellulose filters. In all three cases, the endpoint of titrations run with these polynucleotides was found to be 14 ± 1 nucleotide residues per IF3[75]. IF3 disrupts the circular dichroic spectra and thus the polynucleotide backbone of single-stranded helical poly(A) and poly(C). Titration of these polynucleotides with S1 shows diminishing ellipticity up to an endpoint of 13 ± 1 nucleotides per IF3. In contrast, the oligonucleotides $(A)_{10-20}$ and $(C)_{10-20}$ give, at neutral pH, endpoints of 26 ± 4 nucleotides. A circular dichroism study of the hairpin A_8UGU_6 shows an endpoint of 56 ± 3. These results suggest an occluded-site size of about 13 residues for the polynucleotides and overlapping of two single-stranded oligomers at this site, and/or an additional site that is available for oligomer binding[33]. The circular dichroic spectra of double helical poly(U U), poly(I C), and poly(A U) are somewhat perturbed by IF3, although it appears that the integrity of the A + U helix is preserved. The circular dichroic spectra of E. coli tRNA, MS2 RNA, ØX174 DNA, and sonicated calf thymus DNA were unaffected by IF3. IF3 displays less base specificity than S1; the titration with poly(A) was run under conditions (0.01 M Mg^{2+}) in which S1 has no effect on the circular dichroism of poly(A)[42,76]. In the presence of S1, there was little diminution in the poly(A) circular dichroism until the concentration of IF3 equaled that of S1, which suggested competition between the two proteins[33].

C. Eukaryotic Initiation Factor 3

Eukaryotic initiation factor 3 (eIF3) is a multiprotein complex that contains as many as 10 polypeptide chains and has a weight of about 500,000 daltons[73,26]. In a rabbit reticulocyte cell-free translation system, eIF3 is not needed for efficient $AUG(U)_{\overline{100}}$ - directed polyphenylalanine synthesis in 3 mM Mg^{2+}, but translation of globin mRNA requires this factor[26]. When poly(A) is complexed to $AUG(U)_{\overline{100}}$, no translation occurs with or without eIF3. However, when $AUG(U)_{\overline{100}}$ was complexed to a poly(A,U) copolymer - 90% A + 10% U - efficient (80% of control) translation was achieved in the presence of eIF3, whereas no polyphenylalanine synthesis was observed in its absence[26]. These results suggest that the eIF3 preparation contained a component with helix-destabilizing activity toward short RNA double helices, as would be found in mRNA and poly $(A,U) \cdot AUG(U)_{\overline{100}}$. Indeed, in 0.1 M NaCl, excess eIF3 (57:1 weight ratio) decreased globin mRNA melting temperature by 31°C and that of poly(A·U) by 40°C[26]. The melting observed with poly(A·U) in the face of inability to translate poly(A)·AUG $(U)_{\overline{100}}$ might be related to the much lower eIF3:RNA weight ratio of the latter experiment (about 1). eIF3 also prevents the formation of the poly(U·U) helix in the presence of spermine[77]. The component of the eIF3 preparation responsible for helix destabilization remains to be elucidated. It could conceivably play a role in elongation, rather than in the initiation reaction, inasmuch as the secondary structural impediment to translation of poly(A,U)·AUG(U)$_{\overline{100}}$ occurs during the elongation cycle.

D. A. salina Helix-Destabilizing Protein 40

A high-salt wash of ribosomes from cryptobiotic embryos of the brine shrimp Artemia salina, when subjected to a preparative procedure that included chromatography on denatured DNA agarose, yielded a helix-destabilizing protein, termed HD40[78]. This protein has a molecular weight of 40,000, is monomeric and, on the basis of circular dichroism data, has an α-helix content of 15%. HD40 effectively bound MS2 RNA, as well as single-stranded polyribonucleotides and polydeoxyribonucleotides, to nitrocellulose filters, but did not bind double-stranded RNA (poly(A•U)) or DNA (from SV40). Competition studies with [^{3}H]MS2 RNA and unlabeled nucleic acids indicated a preference for single-stranded RNA over DNA, with poly(U) and poly(dT) the most effective competitors. poly(U)- and A. salina poly(A$^+$) mRNA-directed in vitro protein synthesis in A. salina and wheat germ systems, respectively, was inhibited by HD40. The inhibition could be overcome by an excess of mRNA. HD40 protected a number of single-stranded RNAs and DNAs from digestion by several nucleases, although it had little or no effect on the action of micrococcal nuclease. Although the in vivo function of HD40 is not known, its relative abundance in the cytoplasm (about two molecules per 80S ribosome), inhibition of protein synthesis, and protection of single-stranded RNA against nuclease digestion suggests that it plays a regulatory role in translation[78].

The interaction of HD40 with nucleic acids was studied further with optical and electron microscopic methods[79]. In analogy with the effect of a number of HDPs on single-stranded polynucleotide conformation, the circular dichroic spectra of neutral poly(A) and poly(C) in 0.05 M Na$^+$ are greatly diminished by HD40, with no further effect on ellipticity seen beyond one protein to 12-15 nucleotide residues. Similar results were obtained for polyd(A), polyd(C), MS2 RNA, ØX174 DNA, and poly(U•U). The increase in ultraviolet absorbance seen with poly(U•U) effected by HD40 (about 50%) corresponds to complete melting of the helix. Double helical poly(A•U), poly(I•C), and poly(dA•dT) duplexes, as well as triple-helical poly(A 2U), were unaffected by HD40, even at very low (0.1 mM) NaCl concentration. Likewise, no effect was seen on the circular dichroic spectra of native calf thymus or E. coli DNA, or (alternating sequence) poly d(A-T). Although HD40 did not lower the melting temperature of poly(A•U), it did prevent its formation from poly(A) and poly(U) at sodium ion concentrations of 0.05 M or lower at 20°C. Sedimentation of MS2 RNA or ØX174 DNA with subsaturating concentrations of protein showed only single symmetric bands - an indication that the interaction is not highly cooperative. If it were, two bands would be seen: one corresponding to fully saturated nucleic acid, the other to nucleic acid free of protein. When HD40 was added to MS2 RNA or ØX174 DNA at ratios greater than 1 protein per 12-15 nucleotides, distinct bead-like structures were seen with the electron microscope. This change was not accompanied by any further perturbation of circular dichroic or UV spectra. The beaded structures bear a resemblance to those seen with eukaryotic hnRNA-protein complexes[80] and, with the similarities in amino acid composition, suggest a relationship between HD40 and hnRNP and mRNP proteins (recent studies confirm this; W. Szer, personal communication). The optical and sedimentation studies on HD40 show similarities to the properties of UP1, an extensively studied HDP from calf thymus.

III. RNA HELIX-DESTABILIZING AND SINGLE-STRAND-SPECIFIC PROTEINS WITH UNDETERMINED PHYSIOLOGIC ROLES.

A. Calf Thymus UP1

Several proteins from calf thymus pass through native DNA-cellulose in 0.05 M Na^+, but bind tightly to and are resistant to dextran sulfate elution from denatured DNA-cellulose[81]. After dextran sulfate treatment, the remaining proteins on the denatured DNA-cellulose column can be separated by a sodium chloride gradient. The fraction eluting at 0.3-0.5 M NaCl has been termed UP1, is relatively pure (greater than about 90%), monomeric, heterogeneous in weight (21,000 ± 2000 daltons[82]), and has an isoelectric point of about 7[81]. At stoichiometric concentrations of protein, the melting temperature of the alternating-sequence polymers poly d(A-T) and poly(A-U) is lowered, with no obvious preference for DNA versus RNA[83]. The melting-temperature decrease was large at low ionic strength (over 30°C in 5 mM monovalent cation), but was reduced with increasing ionic strength. UP1 also lowered the melting temperature of double-stranded C., perfringens DNA and unfolded hairpin helices in single-stranded fd DNA or denatured T7 DNA. As seen with the electron microscope, UP1 holds fd DNA in a rigid, extended conformation that has a contour length 20% greater than free DNA[84]. In analogy with the results with A. salina HD40, UP1-fd DNA complexes sedimented as single peaks[84]. The S value of the complexes increased with increasing protein concentration up to a saturation endpoint of about 7 nucleotides per UP1 molecule.

The electron microscopic experiments suggested that UP1 distorts the conformation of single-stranded nucleic acids, and this was directly demonstrated by the effect of the protein on the circular dichroic spectra of poly(A) (Figure 2) and poly(C)[82]. UP1 induces about a 20% increase in the ultraviolet

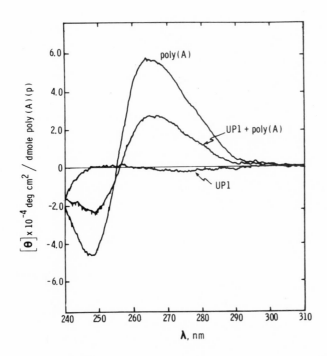

<u>Fig. 2.</u> Effect of UP1 on the circular dichroic spectrum of poly(A); 3.0 mM
Tris-HCl, 0.55 mM Na₂EDTA, 0.10 mM DTT, and 10% v/v glycerol (pH, 8.25), at
20°C. For the poly(A) and poly(A) + UP1 spectra, [poly(A)] = 25.0 μM(p); for
the UP1 and poly(A) + UP1 spectra, [UP1] = 3.6 μM. From Karpel and Burchard[82].

absorbance of poly(A) at 20°C, which is comparable with the effect of heating
this polynucleotide to 75°C in the absence of protein and indicates extensive
disruption of base-stacking[82]. Titration of poly(A) ellipticity and absorbance
by UP1 at low ionic strength (about 0.004 M) gave an occluded-site size of 7
nucleotide residues per protein molecule (Figure 3), which confirmed the result
from the sedimentation studies. The effect on poly(A) was very salt-sensitive -
an indication of a substantial electrostatic component in the interaction - with
the diminution in ellipticity reduced by half at 0.05 M Na⁺. Both the poly(U·U)
and polyd(A-T) double helices are disrupted by UP1. The circular dichroic spectrum of protein-denatured polyd(A-T) is qualitatively different from that of
heat-denatured material, which indicates that UP1 distorts the conformation of
single-stranded DNA, as well as RNA[82].

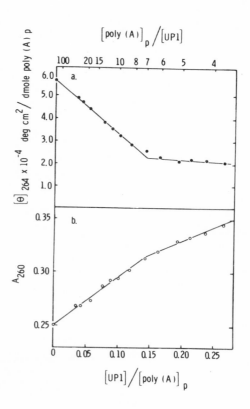

Fig. 3. Determination of the length of single-stranded polynucleotide chain occluded by UP1, at 20°C. Buffer and $[poly(A)]_p$ were the same as in Figure 2. [UP1] varied between 0 and 6.7 μM. a, variation of poly(A) molar (p) ellipticity ($[\emptyset]_{264}$) with $[UP1]/[poly(A)]_p$. b, A_{260} of the data points shown in a. From Karpel and Burchard (82).

A striking property of UP1 is its effect on tRNA and E. coli 5S RNA conformation changes[85,86]. As explained earlier, several small RNAs take up inactive conformations in the absence of Mg^{2+}. On addition of Mg^{2+}, renaturation is very slow and reflects the disruption of "incorrect" secondary structure in the inactive conformer. For example, the half-life of yeast $tRNA_3^{Leu}$ renaturation is about 11 h at 25°C and about 1 yr at 0°C, with an activation energy of about 60 kcal/mol (Lindahl et al., unpublished results; 38). When stoichiometric amounts of UP1 are mixed with inactive $tRNA_3^{Leu}$ in the absence of Mg^{2+} at 0°C, renaturation appears to be virtually instantaneous, as indicated by aminoacylation assay performed in 0.01 M Mg^{2+} at 25°C[85,87].

The acceleration of tRNA renaturation by UP1 is related to the protein's helix-destabilizing properties[82]. UP1 induces a hyperchromic change in inactive $tRNA_3^{Leu}$ of 10% at 258 nm; this corresponds to a disruption of about one-third of the RNA secondary structure. The magnitude of the RNA circular dichroic maximum is reduced by half on stoichiometric addition of UP1. When Mg^{2+} is restored, the circular dichroic spectrum indicates the conversion to the native conformer[82]. Thus, UP1 removes the secondary structural impediment to $tRNA_3^{Leu}$ renaturation on forming a complex with the inactive conformer; with addition of Mg^{2+}, the complex dissociates, and the tRNA refolds readily to its native form. This is reflected in the kinetics of aminoacylation of UP1-inactive $tRNA_3^{Leu}$ mixtures at 0°C, which are considerably slower than the charging of native tRNA[86,87]. However, if UP1-inactive tRNA mixtures are incubated with Mg^{2+} before aminoacylation, the kinetics are identical with those of native tRNA. It is likely that UP1 blocks recognition of the tRNA by the cognate aminoacyl tRNA synthetase, and the complex must be dissociated by Mg^{2+} before charging. The slow aminoacylation kinetics of complexes in the Mg^{2+} - containing charging medium likely represents this dissociation.

The selective affinity of UP1 for single-stranded nucleic acids is reflected in the ability of polynucleotides to inhibit UP1-induced tRNA renaturation[86,87]. Random-coiled poly(U) is the most effective inhibitor, followed by helical (neutral) poly(A) and poly(C). Double helical poly(A-U) is considerably less effective. Because polynucleotides are distorted on interaction with UP1, the energy expended for poly(U) to achieve the particular conformation required at the surface of the protein should be less than that for the base-stacked polymers, poly(A) or poly(C). This would explain the greater affinity of poly(U) for UP1. An alternative explanation, that the protein interacts more strongly with U than with A or C, is less likely, because a polynucleotide-derived polymer devoid of heterocyclic base residues is an effective inhibitor of UP1-induced renaturation[87]. When poly(C) reacts with $KMnO_4$, the product is nonabsorbing (above 240 nm), nondialyzable, and orcinol-sensitive[87]. On the basis of a comparison with the literature of $KMnO_4$ oxidation of cytosine derivatives[88] and chemical analysis, this material probably has urea or biuret residues at the 1'-ribose position, and it has been termed poly(ribosylureaphosphate), or PRUP. PRUP inhibits UP1- induced $tRNA_3^{Leu}$ renaturation much more effectively than poly (C) and nearly as well as poly(U). PRUP has recently been shown to bind tightly to a number of other HDPs (Karpel and Yrttimaa, unpublished observations). Additional evidence that UP1 has little or no direct interaction with the bases was provided by a study of the effect of UP1 on the reactivity of heat-denatured calf thymus DNA toward chloroacetaldehyde[89], which takes place at positions 1 (nitrogen) and 6 (exocyclic amino group) of adenine. A 10-fold weight excess of UP1 had no effect on reactivity. Likewise, the ^{1}H-^{2}H exchange rate of denatured DNA bases was unaffected by UP1[89]. Thus, inasmuch as UP1 has little or no base or ribose (deoxyribose) specificity, the predominant sites of interaction on single-stranded substrates are the phosphate residues. Oligouridylates with as few as 3 phosphates inhibit UP1-effected tRNA renaturation, so the minimal length of residues actually interacting with the protein, $\underline{m}$, is 3 (phosphates)[86,87].

These various results have led to a proposal for the molecular basis of the selective binding of UP1 to single strands[82]. It was proposed that the interaction is predominantly between basic amino acids and phosphate residues. The distances between these amino acids are sufficiently varied and different from the phosphate-phosphate distances found in helices that single-stranded helices would have to distort and double helices would have to denature for the binding of basic amino acids and phosphates to be maximal.

Stoichiometric concentrations of UP1 also accelerate the renaturation of the inactive conformers of $\underline{E}$. $\underline{coli}$ tRNA^{Trp} and 5S rRNA[85]. Renaturation of 5S RNA was followed by a polyacrylamide-gel electrophoresis procedure, which yielded direct evidence of a UP1-5S RNA complex[85]. A UP1-tRNA$_3^{Leu}$ complex was also observed on electrophoresed polyacrylamide gels[86]. Although not explored in detail, the phenomenon of protein-facilitated RNA renaturation is not restricted to calf thymus UP1; several other proteins that were tested have this activity (85; Karpel $\underline{et}$ $\underline{al}$., unpublished results). There is no evidence that the inactive RNA conformers exist $\underline{in}$ $\underline{vivo}$; however, the process by which HDPs partly denature the RNA to bring about the formation of new $\underline{intra}$molecular secondary structure is analogous to the manner by which S1 and/or IF3 may melt internal structure to bring about $\underline{inter}$molecular base-pairing. The $\underline{in}$ $\underline{vitro}$ studies of UP1 and tRNA$_3^{Leu}$ used various magnesium ion concentrations to modulate helix denaturation and reformation. Exactly how much modulation of helix-destabilizing activity is achieved $\underline{in}$ $\underline{vivo}$ remains to be determined.

The RNA helix-destabilizing and renaturing activity of UP1 and the similarities to the activity of S1 and IF3 suggest the possibility of a role in protein synthesis. No UP1 was found in sucrose-isolated calf thymus nuclei[86]; however, neither purified ribosomes nor polysomes contain the protein (Karpel $\underline{et}$ $\underline{al}$., unpublished observations). It should be noted that calf thymus UP1 stimulates the homologous DNA polymerase-α[84]; thus, its $\underline{in}$ $\underline{vitro}$ DNA helix-destabilizing activity could reflect involvement in DNA function.

B. Proteins from RNA Tumor Viruses

As noted above, tRNA-primed reverse transcription of viral RNA templates requires the disruption of about half the tRNA secondary structure (in the acceptor and TψC stem) if an H-bonded complex of tRNA and viral RNA is to be formed[14]. Secondary structure in the viral RNA may also have to be disrupted. This process is formally similar to the formation of mRNA-rRNA and mRNA-tRNA pairings in translation-initiation complexes, where disruption of intramolecular base-pairing may also be required. The disruption of primer tRNA secondary structure must occur well below the melting temperature of the RNA, so some factor(s) must be present in the virus or infected cell that will effect (temporary) helix destabilization. A likely candidate is the reverse transcriptase enzyme itself, which can form a complex with the tRNA[18]. Several studies have indicated that most of the structural elements required to maintain the secondary and tertiary structure of the tRNA are needed for the interaction with the enzyme[19,20]. A 67-nucleotide fragment of bovine liver tRNA^{Trp} - essentially identical with avian tRNA^{Trp} [90] - that lacks 8 residues from the 3' terminus has a high binding affinity for the avian myeloblastosis virus (AMV) reverse transcriptase[20]. However, tRNA fragments with the 3' terminus intact, but lacking as few as 18 residues from the 5' end, do not form stable complexes with the enzyme[19]. If reverse transcriptase brings about partial melting of the tRNA, it would probably not be bound to the 3' terminus, which must pair with complementary residues on the template RNA.

The 5' terminus of tRNA^{Trp}, which is normally resistant to RNase T1 digestion, becomes accessible to attack in tRNA-reverse transcriptase complexes[17]. This result, which could indicate protein-induced melting of the acceptor stem, was also obtained for bovine tRNA^{Val}, and thus is not specific for primer RNA. Other evidence of enzyme-facilitated melting of tRNA is the formation of functional (in DNA synthesis) tRNA^{Trp}-AMV RNA complexes in the presence of reverse transcriptase at 35°C, when such hybridization would not occur in the absence of enzyme[17]. Avian reverse transcriptase also causes RNA-DNA and DNA-DNA, but not RNA-RNA, duplexes to be susceptible to (single-strand-specific) S1 nuclease digestion[21].

A number of structural proteins of RNA tumor viruses bind tightly to nucleic acids, and several appear to be selective for single strands[22-24]. The interactions of AMV p12 (a 12,000-dalton protein) with viral RNA and heterologous single-stranded DNA were studied by nitrocellulose-filter binding and by the effect of complex formation on fluorescamine reaction with the protein[24].

Both substrates had comparable affinities for p12: 1-5 x 10^6 M^{-1} at low ionic
strength (about 1 mM). The magnitude of the association constant was salt-
sensitive. The occluded-site size was about four nucleotide residues per protein molecule. Single-stranded DNA apparently binds p12 more tightly to nitrocellulose filters than does double-stranded DNA; however, because the site size
of the latter substrate was not determined, no definitive conclusions regarding
the specificity of p12 can be drawn[24]. The interaction of Rauscher murine leukemia virus p10 (a 10,000-dalton protein) with nucleic acids was studied by
nitrocellulose-filter binding[22] and formaldehyde crosslinking[23] techniques.
Competition studies demonstrated that heterologous and homologous viral RNA and
heat-denatured DNA had substantially greater affinity for p10 than did double-
stranded DNA[22]. The apparent stoichiometry of p10 binding to a number of single-
stranded substrates was determined by protein-nucleic acid cross-linking followed by velocity sedimentation and density-gradient centrifugation[23]. The
apparent site size varied with the substrate and was as large as about 140
nucleotides (for single-stranded fd DNA). Although no information is available
on the helix-destabilizing activity, if any, of these viral proteins, p10 can
facilitate $tRNA_3^{Leu}$ renaturation (J.R. Fresco, personal communication). Another
viral protein, p12 from mammalian viruses, exhibits specificity for unique sites
on the viral RNA[91].

The host cell may contain HDPs that facilitate reverse transcription. The
activity of reverse transcriptase on Rous sarcoma virus (RSV) RNA-dependent and
native and denatured DNA-dependent DNA synthesis can be stimulated by a protein
isolated from RSV-transformed and normal chicken cells[92,93]. The protein,
isolated on single-stranded DNA cellulose, brings about a time-dependent hyperchromic change in polyd(A-T) of about 25% at 24°C (10:1 weight ratio of protein
to DNA[92]). Under the same conditions, an approximate 20% increase in the A_{260}
of viral RNA occurs on interaction with the protein[93]. DNA products synthesized
in the presence of this protein were considerably longer than those produced in
its absence[93]. The binding and helix-destabilizing effects were sensitive to
NaCl concentration. The stimulation of reverse transcription, as well as the
increased size of the DNA transcripts, may be a consequence of the removal of
secondary structural impediments to the action on the enzyme of the viral RNA
template.

C. Other Proteins

P8 protein, isolated from mammalian cells on single-stranded-DNA-cellulose,
was found to bind selectively to single-stranded DNA[94] and facilitated $tRNA_3^{Leu}$

renaturation[85]. Like a similar protein from yeast, it lowers the melting temperature of synthetic RNA and DNA[94A]. However, the demonstration that P8 is glyceraldehyde-3-phosphate dehydrogenase (or one of several isozymes of the enzyme) makes questionable the physiologic relevance of the protein's nucleic acid-interactive properties[95,96].

The interaction of the T4 bacteriophage HDP (gene 32 protein) with a variety of DNA and RNA substrates has been the subject of extensive study[3,4A]. T4 HDP may regulate its own synthesis by reversibly binding its own mRNA after first titrating the available single-stranded DNA sites in the cell[97-99]. The affinity of the HDP for single-stranded DNA must be greater than that for RNA, and recent experiments have shown that even marginal differences in the association constants for DNA and RNA oligonucleotides, when coupled to the high cooperativity ($\omega \sim 1,000$) characteristic of T4 HDP binding to polynucleotides, can quantitatively account for the autoregulatory properties of this system[99A]. An obvious requirement of this model is the rapid dissociation of HDP-nucleic acid complexes. At low ionic strength (about 0.02 M), T4 HDP denatures about 90% of MS2 RNA, and the hyperchromic effect can be rapidly reversed at higher ionic strengths[100]. The kinetics of salt-induced dissociation, monitored by the loss of hyperchromicity, is of a magnitude ($\omega = 7$ sec) comparable with that of the dissociation kinetics of complexes of T4 HDP and single-stranded DNA, where there is little reformation of secondary structure[100].

IV. A GENERAL APPROACH TOWARD ISOLATING RNA HDPs

DNA-cellulose affinity chromatography has been widely used for the preparation of DNA helix-destabilizing proteins[101]. Because many proteins without specificity for single-stranded DNA still bind tightly to denatured DNA-cellulose, some preparative procedures have used native DNA-cellulose to remove such proteins initially from crude extracts, so that only proteins with specific affinity for single strands are allowed to pass through the denatured DNA cellulose[81]. Although several proteins, such as calf thymus UP1, with high affinity for denatured DNA-cellulose also bind tightly to single-stranded RNA, a procedure aimed primarily at isolating RNA-binding proteins would be useful. RNA-cellulose[102] and RNA-agarose[103] have been used for the purification of a number of proteins, including E. coli ribosomal S1 on poly(C)-cellulose[102] and poly(A)-cellulose[104]. We have developed a procedure, using both RNA- and DNA-affinity columns, that maximizes the effect of the selective affinities of these matrices[94A],[105]. This procedure, performed at 0-4°C, should be most useful for cells that contain a complex set of nucleic acid-interacting proteins, for which separation might be difficult.

The cells are initially disrupted in a Tris buffer at near - neutral pH (7.5), containing DTT, glycerol, EDTA (to sequester heavy metal ions, which could degrade the RNA on the affinity columns[106]), PMSF (a protease inhibitor), and moderate NaCl concentration (0.05 M, to reduce nonspecific binding). Cell debris and crude chromatin are removed by centrifugation, and remaining nucleic acids are precipitated with 10% polyethylene glycol 6000 (PEG) in 2 M NaCl[81],[101]. We have found that, with yeast, virtually all the DNA and 90-95% of the RNA are removed by this procedure[94A]. The supernatant is dialyzed against the cell disruption buffer (without PMSF) and chromatographed over 5 tandem columns in the following order: native DNA-cellulose, aminophenyl-phosphoryl-UMP-agarose (APUP-agarose), poly(I C)-agarose, poly(U)-cellulose, and denatured DNA-cellulose.

Our major interest is in proteins that bind poly(U)-cellulose, because the random coiled poly(U) is among the best substrates for the various RNA HDPs described above. The denatured DNA-cellulose column connected downstream from the poly(U)-cellulose binds proteins with selective affinity for single-stranded DNA (but not RNA). The poly(I C)agarose column (prepared from poly(I)-agarose annealed to poly(C)[103], and commercially available) is connected upstream of the poly(U)-cellulose, to discourage proteins with comparable affinities for single- and double-stranded RNA from binding to the poly(U)-cellulose column. APUP-agarose[107], which is also commercially available, is used to bind any endogenous pyrimidine-specific ribonucleases in the crude extract before elution over the RNA matrices. The first column in the tandem array, native DNA-cellulose, discourages proteins with nonspecific affinities for DNA, RNA, or simply negatively charged material from binding to the poly(U)- cellulose or denatured DNA-cellulose[81].

After the crude extract is run through the tandem columns, the columns are uncoupled and separately eluted with buffer containing 2 M NaCl. The poly(U)-cellulose eluant can then be tested for RNA-binding, RNA-helix-destabilizing, or RNA-renaturing activity with nitrocellulose filter binding, optical, or aminoacylation assays, respectively. The poly(U)-cellulose eluant will, it is hoped, show a relatively simple pattern on SDS-polyacrylamide gels. Further purification may require recycling on some or all of the columns, or the use of other chromatographic techniques. NaCl concentration gradients and elution with polyanions (such as dextran sulfate) may also be required[81]. The use of nucleic acid matrices does not guarantee the isolation of physiologically important HDPs, as the isolation and identification of P8 protein clearly show. However, this approach is both rational and rapid and, when combined with a suitable functional assay, should prove useful.

V. CONCLUSION

Helix-destabilizing proteins may be required in a variety of cellular processes where RNA molecules must undergo conformational changes or otherwise be subjected to temporary denaturation of secondary structure. This review has detailed studies on several proteins, such as E. coli S1 and IF3, for which the connection between in vitro helix destabilization and in vivo function is highly probable. The physiologic roles of other RNA HDPs, such as calf thymus UP1, are not known, but there has been progress in understanding their mechanisms of action in vitro. Clearly, there is good reason to continue the study of RNA helix-destabilizing protein structure, function, and mechanism.

ACKNOWLEDGMENTS

I thank David Draper for useful discussions and Wlodzimierz Szer and Eric Wickstrom for making available papers before publication. Research in my laboratory was supported by NIH grant CA 21374.

REFERENCES

1. Alberts, B.M. and Frey, L. (1970) Nature 227, 113.
2. Alberts, B.M. and Sternglanz, R. (1977) Nature 269, 655.
3. Champoux, J.J. (1978) Ann. Rev. Biochem. 47, 449.
4. Coleman, J.E. and Oakley, J.L. (1980) CRC Crit. Rev. Biochem. 7, 247.
4A. Williams, K.R. and Konigsberg, W. (1981) This Volume 23.
5. Fresco, J.R., Alberts, B.M. and Doty, P. (1960) Nature 188, 98.
6. Gralla, J. and DeLisi, C. (1974) Nature 248, 330.
7. Richardson, J.P. (1975) J. Mol. Biol. 98, 565.
8. Siebenlist, U. (1979) Nature 279, 651.
9. Adhya, S., Sarkar, P., Valenzuela, D. and Maitra, U. (1979) Proc. Natl. Acad. Sci. USA 76, 1613.
10. Franke, W.W. and Scheer, U. (1974) Symp. Soc. Exp. Biol. 28, 249.
11. Paine, P.L., Moore, L.C. and Horowitz, S.H. (1975) Nature 254, 109.
12. Franke, W.W. and Scheer, U. in The Cell Nucleus, Vol. I, H. Busch, Ed., pp. 278, 282, 296, 298, Academic Press, 1974.
13. Harada, F., Sawyer, R.C. and Dahlberg, J.E. (1975) J. Biol.Chem. 250, 3487.
14. Eiden, J.J., Quade, K. and Nichols, J.L. (1976) Nature 259, 245.
15. Waters, L.C., Mullin, B.C., Ho, T. and Yang, W.K. (1975) Proc. Natl. Acad. Sci. USA 72, 2155.
16. Green, M. and Gerard, G.F. (1974) Prog. Nuc. Acid Res. Mol. Biol. 14, 187.
17. Araya, A., Sarih, L. and Litvak, S. (1979) Nuc. Acid Res. 6, 3831.
18. Haseltine, W.A., Panet, A., Smoler, D., Baltimore D., Peters, G., Harada, F. and Dahlberg, J.E. (1977) Biochemistry 16, 3625.

534

19. Cordell, B., Swanstrom, R., Goodman H.M. and Bishop, J.M. (1979) J. Biol. Chem. 254, 1866.
20. Baroudy, B.M. and Chirikjian, J.G. (1980) Nuc. Acids Res. 8, 57.
21. Collett, M.S., Leis, J.P., Smith, M.S., and Faras, A.J. (1978) J. Virology 26, 498.
22. Davis, J., Scherer, M., Tsai, W.O. and Long, C. (1976) J. Virology 18, 709.
23. Schulein, M., Burnette, W.N. and August, J.T. (1978) J. Virology 26, 54.
24. Smith, B.J. and Bailey, J.M. (1979) Nuc. Acids Res. 7, 2055.
25. Leis, J.P., McGinnis, J. and Green, R.W. (1978) Virology 84, 87.
26. Ilan, J. and Ilan, J. (1977) Proc. Natl. Acad. Sci. USA 74, 2325.
27. Szer, W. and Leffler, S. (1975) Proc. Natl. Acad. Sci. USA 71, 3611.
28. Van Dieijen, G., Van der Laken, C.J., Van Knippenberg, P.H. and Van Duin, J. (1975) J. Mol. Biol. 93, 351.
29. Szer, W., Hermoso, J.M. and Boublik, M. (1976) Biochem. Biophys. Res. Comm. 70, 957.
30. Van Diefen, G., Van Knippenberg, P.H. and Van Duin, J. (1976) Eur. J. Biochem. 64, 511.
31. Shine, J. and Delgarno, L. (1974) Proc. Natl. Acad. Sci. USA, 71, 1342.
32. Steitz, J.A. and Jakes, K. (1975) Proc. Natl. Acad. Sci. USA, 72, 4734.
33. Schleich, T., Wickstrom, E., Twombly, K., Schmidt, B. and Tyson, R.W. (1980) Biochemistry 19, 4486.
34. Fresco, J.R., Adams, A., Ascione, R., Henley, D., and Lindahl, T. (1966) Cold Spring Harbor Symp. Quant. Biol. 31, 527.
35. Sueoka, N., Kano-Sueoka, T. and Gartland, W.J. (1966) Cold Spring Harbor Symp. Quant. Biol. 31, 539.
36. Aubert, M., Scott, J., Reynier, M. and Monier, R. (1968) Proc. Natl. Acad. Sci. USA 61, 292.
37. Uhlenbeck, O.C., Chirikjian, J.G. and Fresco, J.R. (1974) J. Mol. Biol. 89, 495.
38. Hawkins, E.R., Chang, S.H. and Mattice, W.L. (1977) Biopolymers 16, 1557.
39. Ishida, T. and Sueoka, N. (1968) J. Biol. Chem. 243, 5329.
40. Richards, E.G., Lecanidou, R. and Geroch, M.E. (1973) Eur. J. Biochem. 34, 262.
41. Laughrea, M. and Moore, P.B. (1977) J. Mol. Biol. 112, 399.
42. Bear, D.G., Ng, R., VanDerveer, D., Johnson, N.P., Thomas, G., Schleich, T. and Noller, H.F. (1976) Proc. Natl. Acad. Sci. USA 73, 1824.
43. Lipecky, R., Vohlschein, J. and Gassen, H.G. (1977) Nuc. Acids Res. 4, 3627.
44. Draper, D.E., Pratt, C.W. and von Hippel, P.H. (1977) Proc. Natl. Acad. Sci. USA 74, 4786.
45. Draper, D.E. and von Hippel, P.H. (1978) J. Mol. Biol. 122, 321.
46. Draper, D.E. and von Hippel, P.H. (1978) J. Mol. Biol. 122, 339.
47. McGhee, J.D. and von Hippel, P.H. (1974) J. Mol. Biol. 86, 469.
48. Record, M.T., Jr., Lohman, T.M. and deHaseth, P. (1976) J. Mol. Biol. 107, 145.
49. Dahlberg, A. and Dahlberg, J. (1975) Proc. Natl. Acad. Sci. USA 72, 2940.
50. Senear, A. and Steitz, J.A. (1976) J. Biol. Chem. 251, 1902.
51. Draper, D.E. and von Hippel, P.H. (1979) Proc. Natl. Acad. Sci. USA 76, 1040.
52. Kolb, A., Hermoso, J.M., Thomas, J.O. and Szer, W. (1977) Proc. Natl. Acad. Sci. USA 74, 2379.
53. Thomas, J.O., Kolb, A. and Szer, W. (1978) J. Mol. Biol. 123, 163.
54. Khanh, N.Q., Lipecky, R. and Gassen, H.G. (1978) Biochem. Biophys. Acta. 521, 476.
55. Means, G.E. and Feeney, R.E. (1968) Biochemistry 7, 2192.
56. Fretheim, K., Iwai, S. and Feeney, R.E. (1979) Int. J. Peptide Protein Res. 14, 451.

57. Suryanarayana, T. and Subramanian, A.R. (1979) J. Mol. Biol. 127, 41.
58. Thomas, J.O., Boublik, M., Szer, W. and Subramanian, A.R. (1979) Eur. J. Biochem. 102, 309.
59. Subramanian, A.R. and Mizushima, S. (1979) J. Biol. Chem. 254, 4309.
60. Laughrea, M. and Moore, P.B. (1978) J. Mol. Biol. 121, 411.
61. Bowman, C.M., Dahlberg, J.E., Ikemura, T., Konisky, J. and Nomura, M. (1971) Proc. Natl. Acad. Sci. USA 68, 964.
63. Yuan, R.C., Steitz, J.A., Moore, P.B. and Crothers, D.M. (1979) Nuc. Acids Res. 7, 2399.
64. Steitz, J.A., Wahba, A.J., Laughrea, M. and Moore, P.B. (1977) Nuc. Acids Res. 4, 1.
65. Steitz, J.A. (1973) Proc. Natl. Acad. Sci. USA 70, 2505.
66. Linde, R., Quoc Khanh, N., Lipecky, R. and Gassen, H.G. (1979) Eur. J. Biochem. 93, 565.
67. Jay, G. and Kaempfer, R. (1974) J. Mol. Biol. 82, 193.
68. Miller, M.J. and Wahba, A.J. (1974) J. Biol. Chem. 249, 3808.
69. Groner, Y., Scheps, R., Kamen, R., Kolakofsky, D. and Revel, M. (1972) Nature New Biol. 239, 19.
70. Kamen, R., Kondo, M., Romer, W. and Weissman, C. (1972) Eur. J. Biochem. 31, 44.
71. Revel, M., Brawerman, G., Lelong, J.C. and Gros, F. (1968) Nature 219, 1016.
72. Iwasaki, K., Sabol, S., Wahba, A.J. and Ochoa, S. (1968) Arch. Biochem. Biophys. 125, 542.
73. Revel, M., "Molecular Mechanisms of Protein Biosynthesis", Weissbach, H. and Pestka, S., Eds., Academic Press, New York, 1977, p. 245.
74. Grunbert-Manago, M. and Gros, F. (1977) Prog. Nuc. Acid Res. Mol. Biol. 20, 209.
75. Wickstrom, E., Tyson, R.W., Newton, G., Obert, R. and Williams, E.E. (1980) Arch. Biochem. Biophys. 200, 296.
77. Szer, W., Thomas, J.O., Freinenstein, C. and Kolb, A. "Proceedings of the International Symposium on Translation", Legocki, A.B., Ed., University of Poznan Press, Poznan, Poland, 1978, p. 70.
78. Marvil, D.K., Nowak, L. and Szer, W. (1980) J. Biol. Chem. 225, 6466.
79. Nowak, L., Marvil, D.K., Thomas, J.O., Boublik, M. and Szer, W. (1980) J. Biol. Chem. 255, 6473.
80. McKnight, S.L. and Miller, O.L., Jr. (1976) Cell 8, 305.
81. Herrick, G. and Alberts, B. (1976) J. Biol. Chem. 251, 2124.
82. Karpel, R.L. and Burchard, A.C. (1980) Biochemistry 19, 4674.
83. Herrick, G and Alberts, B. (1976) J. Biol. Chem. 251, 2133.
84. Herrick, G., Delius, H. and Alberts, B. (1976) J. Biol. Chem. 251, 2142.
85. Karpel, R.L., Swistel, D.G., Miller, N.S., Geroch, M.E., Lu, C. and Fresco, J.R. (1974) Brookhaven Symp. Bio. 26, 165.
86. Karpel, R.L., Miller, N.S. and Fresco, J.R. "Molecular Mechanisms in the Control of Gene Expression", Nierlich, D.P., Rutter, W.J. and Fox, C.F., Eds., Academic Press, New York, 1976, p. 411.
87. Karpel, R.L., Miller, N.S. and Fresco, J.R., (1981) Biochemistry, in press.
88. Chatamra, B. and Jones, A.S. (1963) J. Chem. Soc., 1963, 811.
89. Kohwi-Shigematsu, T., Enomoto, T., Yamada, M.A., Nakanishi, M. and Tsuboi, M. (1978) Proc. Natl. Acad. Sci. USA 75, 4689.
90. Baroudy, B.M., Fournier, M., Labouesse, J., Papas, T.S. and Chirikjian, J.G. (1977) Proc. Natl. Acad. Sci. USA 74, 1889.
91. Sen, A.C., Sherr, C.J. and Todaro, G.J. (1976) Cell 7, 21.
92. Hung, P.P. and Lee, S.G. (1976) Nature 259, 499.
93. Lee, S.G. and Hung, P.P. (1977) Nature 270, 366.
94. Tsai, R.L. and Green, H. (1973) J. Mol. Biol. 73, 307.
94A. Karpel, R.L. and Burchard, A.C., (1981) Biochim. Biophys. Acta, (in press).

536

95. Perucho, M., Salas, J. and Salas, M. (1977) Eur. J. Biochem. 81, 557.
96. Perucho, M., Salas, J. and Salas, M. (1980) Biochem. Biophys. Acta 606, 181.
97. Gold, L., O'Farrell, P.Z. and Russel, M. (1976) J. Biol. Chem. 251, 7251.
98. Russel, M., Gold, L., Morrissett, H. and O'Farrell, P.Z. (1976) J. Biol. Chem. 251, 7263.
99. Lemaire, G., Gold, L. and Yarus, M. (1978) J. Mol. Biol. 126, 73.
99A. Newport, J.W., Lonberg, N., Kowdezykowski, S.C. and von Hippel, P.H. (1981) J. Mol. Biol. 145, 105.
100. Suau, P., Toulme, J.J. and Helene, C. (1980) Nuc. Acids Res. 8, 1357.
101. Alberts, B. and Herrick, G. (1971) Meth. Enz. 21D, 198.
102. Carmichael, G.G. (1975) J. Biol. Chem. 250, 6160.
103. Wagner, A.F., Bugianesi, R.L. and Shen, T.Y. (1971) Biochem. Biophys. Res. Comm. 45, 184.
104. Carmichael, G.G., Weber, K., Niveleau, A. and Wahba, A.J. (1975) J. Biol. Chem. 250, 3607.
105. Karpel, R.L. and Burchard, A.C. (1979) Fed. Proc. 38, 485.
106. Butzow, J.J. and Eichhorn, G.L. (1965) Biopolymers 3, 95.
107. Wilchek, M. and Gorecki, M. (1969) Eur. J. Biochem. 11, 491.

DOT HYBRIDIZATION AND HYBRID-SELECTED TRANSLATION: METHODS FOR DETERMINING NUCLEIC ACID CONCENTRATIONS AND SEQUENCE HOMOLOGIES

Fotis C. Kafatos[*], George Thireos, C. Weldon Jones,
Sonia G. Tsitilou[*] and Kostas Iatrou

The Biological Laboratories
Harvard University
16 Divinity Avenue, Cambridge, MA   02138

[*]Also: Department of Biology
University of Athens
Panepistimiopolis, Kouponia, Athens, Greece 621

I. Introduction

Nucleic acid reassociation or hybridization reactions can yield a wide variety of information about the reacting species. With one of the reactants immobilized on a solid support, such as nitrocellulose or DBM paper, it is possible to obtain information both on the concentration of the other reactant, which is in solution, and on sequence homology between the reactants. In this paper we review some methods for obtaining these types of information.

For both purposes, we routinely use multiple cloned DNAs, individually spotted on the same nitrocellulose filter, in "dots" of uniform diameter and DNA content. In the procedure which we call "dot hybridization"[1,2], such filters are hybridized to an appropriate radioactive probe (DNA or RNA), under conditions where the filter-bound DNA is in excess; before the reaction reaches completion, non-hybridized probe is washed thoroughly away, the filter is autoradiographed, and the extent of hybridization of the probe with each cloned DNA is evaluated from the dot intensities in the autoradiogram. In the hybrid-selected translation procedure, the multiple DNA dots are hybridized with a non-radioactive mRNA preparation, basically as described by Ricciardi et al[3]. After the non-hybridized RNA is washed off, the dots are collected individually, the RNA hybridized to each is recovered by melting and translated in a cell-free system, and the products are analyzed on adjacent slots of an SDS-polyacrylamide slab gel; the final autoradiogram reveals which polypeptide(s) is(are) encoded by mRNAs hybridizing to each of the cloned DNAs[4].

We have used these procedures extensively in our analysis of evolutionarily homologous gene families. For both dot hybridization and hybrid-selected translation, the choice of stringency of the hybridization criterion is crucial. If a single criterion is used, it defines a threshold of maximum mismatch below which related sequences are assayed as a group. If several criteria are used, we can discriminate between individual sequences or subfamilies within the family.

We will first outline the general experimental design and discuss typical results obtained by these procedures. Although the emphasis will be on the procedures we use, we shall refer briefly to possible alternatives, some of which have been used by others. Finally, we shall present our detailed protocols.

II. Results and Discussion

A. Estimates of Sequence Concentration by Dot Hybridization

This application is of wide utility, since it is substantially less laborious than estimates using hybridization in solution. Figure 1 shows results from a typical experiment. DNAs from several clones of interest are linearized with a restriction enzyme, denatured, and bound to a filter in dots of uniform diameter and DNA content. Replicate filters are then hybridized with RNA mixtures, in this case <u>in vitro</u> labeled poly(A)$^+$ RNA preparations from silkmoth follicles of specific developmental stages of choriogenesis. The radioactivity of each dot is estimated by autoradiography and reflects the relative concentration of the respective sequence in the RNA preparation. A reconstruction experiment using plasmid probes of known sequence concentration validated

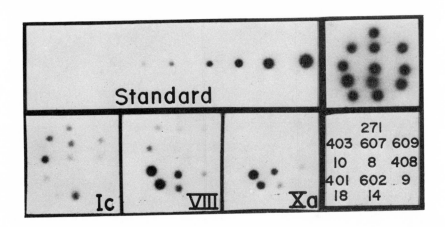

Figure 1: Dot hybridization analysis of the abundance of chorion RNA sequences during development. Twelve chorion cDNA clones were spotted on multiple filters as indicated by the code numbers on the right, bottom. The dots of one such filter were visualized by hybridization with nick-translated ^{32}P-DNA of the plasmid vector, pML-21 (upper right). To determine the abundance of specific RNA sequences during development, poly(A)$^+$ cytoplasmic RNAs were prepared from staged follicles, fragmented, ^{32}P-end labelled, and used as hybridization probes. Results are shown from 3 of 17 stages of choriogenesis. It can be seen for example, that clone pc271 is "early", pc403 and pc10 "middle", and pc401 and pc18 "late". Semi-quantitative estimates can be obtained by comparing dot insensities to a standard autoradiogram of twofold dilution series of ^{32}P-DNA, directly spotted on a filter. For example, the pc18/pc271 RNA abundance ratio is approximately 0.3 at stage Ic, and >50 at stage VIII$_2$ Hybridization conditions as in Figure 4, at 64°C. Modified from Sim <u>et al</u>[2].

540

this interpretation (Table 1). Figure 1 shows, for example, that clone pc271
is developmentally "early" (its sequence is present in follicular RNA primarily
during the early period of choriogenesis), pc10 is "middle", and pc401 is "late".
In such experiments, relative sequence concentrations can be estimated semi-
quantitatively, by excising the dots and counting their radioactivity by liquid
scintillation, or alternatively by visually comparing them to a standard, i.e.
a series of dots containing a dilution series of radioactive DNA (Figure 1).
We prefer the latter alternative because of its reasonable accuracy and superior
sensitivity (spots containing fewer than 5 cpm can be easily quantified visually).

TABLE I

DETERMINATION OF SEQUENCE ABUNDANCE BY DOT HYBRIDIZATION[*]

% of Sequence in Probe Mixture		cpm Hybridized to DNA Dots	
RSF1030	pML-21	RSF1030	pML-21
0	100	0 -	639 (100%)
20	80	136 (20%)	514 (80%)
50	50	319 (47%)	326 (51%)
80	20	526 (78%)	122 (19%)
100	0	677 (100%)	0 -

[*]cRNAs were prepared separately from pML-21 and RSF1030 plasmids, mixed
in various proportions, and hybridized with replicate filters containing pML-
21 and RSF1030 dots. For both plasmids, the extent of hybridization using
pure probe (639 or 677 cmp; 100% hybridization) corresponded to approximately
0.3% of saturation. From (1).

Dot hybridization may be simplified[5] by deriving the DNA dots directly
by lysis on the nitrocellulose filter of bacterial colonies containing the
respective plasmid clones, in an extension of the Grunstein-Hogness colony hybridization procedure[6,7]. In our experience, spotting purified DNA gives more
reproducible results, but careful colony hybridizations can certainly yield
qualitative information on relative RNA concentrations. Figure 2 shows typical
results from an experiment in which Grunstein-Hogness replicates of a follicular
cell cDNA library were screened with probes from wild-type and mutant silkmoth
follicles[4]: certain sequences can be seen to be represented more or less equally in follicles of both genotypes, whereas others are grossly deficient in the
mutant. Similarly, differential screenings of cDNA libraries with different
RNA preparations have permitted recovery of sequences that are rapidly induced

by ecdysone in Drosophila cell cultures[8], and of salivary gland sequences[9] which are suppressed by ecdysone (intermolt of glue protein mRNAs), or which are induced by ecdysone with a long delay (late puff mRNAs). Furthermore, changes in the prevalence of RNA sequences during development of <u>Xenopus</u> embryos[10] have been evaluated qualitatively by this procedure.

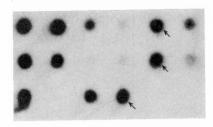 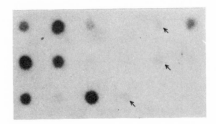

Figure 2: Qualitative dot hybridization by the Grunstein-Hogness procedure. A chorion cDNA library was screened with ^{32}P-cDNA probes, derived from RNA of wild-type <u>Bombyx</u> <u>mori</u> follicles (left) or from follicles of the homozygous Gr^B mutant (right). With these exposures, it can be seen that the mutant is grossly deficient for three sequences which are very abundant in the wild type (arrows); other sequences are either moderately reduced or of reasonably normal abundance. See Iatrou <u>et</u> al[4].

Williams and Lloyd[11] have used "dot hybridization by Grunstein-Hogness" in a semi-quantitative manner, and Lasky et al[12] have further refined this approach. The latter authors first estimate the absolute concentration of a few sequences in the RNA preparation to be used as probe, by a titration method in solution. Then they construct an internal standard curve by including the respective cDNA clones in the dot hybridization experiment: the radioactivities of all other dots in the same experiment are similarly counted and converted to absolute sequence concentrations by reference to the standard curve. Concentrations down to 10^{-4} of the RNA can be estimated, probably to within a factor of two, and even rarer sequences are detectable (10^{-5} of the RNA; one copy per sea urchin gastrula cell). The standard curve is linear on a log-log scale, over three orders of magnitude in sequence concentration (Figure 3).

It is possible to estimate sequence concentrations by dot hybridization in a converse manner: by spotting on a filter a series of DNA or RNA preparations containing a sequence of interest in unknown proportions, and then hybridizing the filter with a pure probe of that sequence (e.g. a nick-translated cDNA clone). Weisrod and Weintraub[13] have used this approach to show that nucleosomes of active and inactive genes can be fractionated on an agarose-HMG 14-17 column. With appropriate standards, the same approach should be applicable to estimating gene copy numbers in a genome, or RNA prevalence. Although such estimates are routinely obtained from Southern and Northern blots, the dot hybridization method may be more convenient for assaying multiple samples, and also more accurate if the sequence in question is distributed over several bands in the Southern and Northern profiles. The accuracy and sensitivity of dot hybridization in this context remains to be established quantitatively, however.

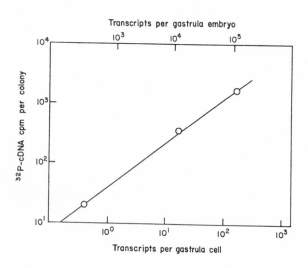

Figure 3: Quantitative dot hybridization by the Grunstein-Hogness procedure. The abundance of transcripts corresponding to three cDNA clones was estimated by a titration method, using sea urchin gastrula RNA. The amount of radioactivity hybridized by the DNA dots of these three clones was determined, in a Grunstein-Hogness screening using gastrula ^{32}P-cDNA as probe. The slope of the line is 0.64, indicating that, over this range, the number of counts hybridized is not directly proportional to transcript prevalence[12].

B. Estimates of Sequence Homology by Dot Hybridization

Mismatching reduces both the rate of hybridization and the stability of hybrids. Therefore, if a series of related sequences are spotted together and hybridized, under an appropriate criterion of stringency, with a pure probe consisting of one of these sequences, the intensities of the dots will reflect the respective sequence homologies to the probe. This is a convenient and powerful method for evaluating homology relationships, especially if the experiment is repeated at several different, progressively more stringent criteria. Figure 4 shows how this approach can be used to work out homology relationships within a multigene family.

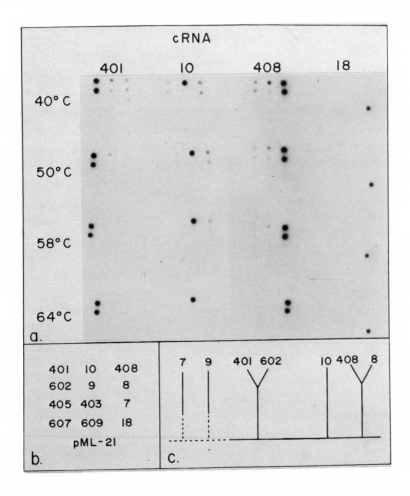

Figure 4: Determination of sequence homologies by dot hybridization. DNAs from twelve chorion cDNA clones, and the plasmid vector pML-21 as a control were spotted on replicate filters, as shown in panel b. Filters were hybridized at different temperatures with four ^{32}P-cRNA probes, synthesized using as templates the chorion sequences of the indicated clones (pc401, pc10, pc408, pc18; panel a). Clone pc18 is a member of the A family, and does not detectably cross-hybridize with any of the other clones. The seven B family clones crosshybridize to variable extents, permitting us to construct a diagram of homology relationships (panel c). From (2).

A series of clones (chorion cDNA clones in this case) are spotted on replicate filters, and hybridized at various criteria with pure probes, consisting of the same cloned sequences (cRNAs or nick-translated DNAs). Distantly related members of a family cross-hybridize only at low criterion, whereas progressively more related sequences cross-hybridize even at progressively higher criteria. The results can be checked by reciprocal cross-hybridizations, and quantified visually or by liquid scintillation. This approach has been used to determine homology relationships in the vitellogenin[14] and chorion[2] gene families. When the DNAs being studied are sufficiently long, however, it may be profitable to perform Southern hybridizations at progressively higher criteria, rather than dot hybridizations: this approach has been used to identify the area of greatest homology between viral genomes[15], and should be a general method for documenting the spatial distribution of mismatching across large genes (e.g. introns vs. exons) or complex chromosomal segments, provided that restriction maps and the distribution of repetitive DNA elements, if any, are known.

We have also obtained satisfactory results with variations of the basic experimental design exemplified by Figure 4. Filters can be hybridized first at high criterion, autoradiographed, and then rehybridized at lower criterion and re-autoradiographed: this variation minimizes the effort of spotting the filters, but increases the chances of failure due to faulty spotting of a single dot, or due to cumulative background problems. Alternatively, a single hybrization at low criterion can be followed by melting and autoradiography at progressively higher temperatures: here the danger is that some of the hybridized probe may become irreversibly bound to the filter, especially if the filter dries out between the melting steps.

Melting experiments of the type just described can be monitored by liquid scintillation[2,14]; the resulting melting curves should reflect quantitatively the sequence homologies (Figure 5). The data base available is as yet insufficient to allow accurate estimates of percent divergence from the ΔTm values, although preliminary information suggests that under the conditions of Figure 5 one degree of ΔTm corresponds to more than one percent mismatch[2,16]. The ΔTm is undoubtedly influenced by the spatial distribution of mismatches as well as by the length of the probe. An as yet not attempted modification which may prove valuable would be to use cRNA as probe and digest with RNase before determining the melting curve: this should both increase the sensitivity of the method (i.e. maximize ΔTm per one percent mismatch) and reduce the influence of possible limited areas of exceptionally high sequence homology, thus improving quantitation.

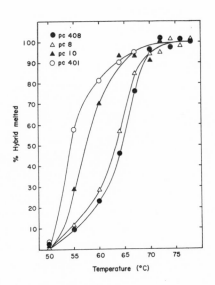

<u>Figure 5</u>: Determination of sequence homologies from the melting profiles of dot hybrids. After 50°C hybridizations with pc408 ^{32}P-cRNA, as in Figure 4, melting curves were determined for dots of pc408, pc8, pc10 and pc401 DNAs. The results are consistent with those of Figure 4. From (2).

C. Hybrid-Selected Translation and Estimates of Homologies Between Proteins

A powerful general method for identifying the protein encoded by a cloned DNA segment is hybridization selection or, as we call it, hybrid-selected translation[3,17]: an mRNA preparation which includes the sequence in question is hybridized to the cloned DNA, previously immobilized on nitrocellulose or DBM paper, and after thorough washing the hybridized RNA is melted off and translated in a cell-free system. The product is identified by SDS-polyacrylamide gel electrophoresis and characterized by standard methods (e.g. immunochemically, or by peptide analysis). An elegant version of this method, as initially described[3], is to restrict the DNA, subject it to electrophoresis, and transfer it to solid support by Southern blotting, so that the polypeptide(s) encoded by each restric-

546

tion fragment can be separately identified. For many purposes, however, especially when multiple DNA samples are involved, it is sufficient to spot the cloned DNAs on a filter, as in the dot hybridization procedure; we routinely use this method.

In the study of multigene families, it is often important to evaluate the evolutionary relatedness of nucleotide sequences encoding distinct polypeptides. Although such evaluations are best performed by cloning the respective sequences and characterizing them by dot hybridization, as discussed above, or by sequence analysis, valuable information can be obtained much more quickly by hybrid-selected translation: in our studies of the chorion system we have found that, by comparing the electrophoretic patterns of the translation products of total and hybrid-selected mRNAs, we could identify both the polypeptide encoded by a clone, and the corresponding family of evolutionarily related polypeptides[17,18]. A powerful extension of this approach is to perform hybrid-selected translations at several criteria of stringency: the "true" product can be identified more definitively, by its increasing prominence in progressively more stringent translations, and products corresponding to subfamilies of more or less related sequences can be similarly identified by attention to relative intensities under different conditions. For example, from the data of Figure 6 we may conclude that clone 1453 encodes a class A chorion protein corresponding to band 3, that the closest homologies are those corresponding to bands 1, 2, and 5, and that components corresponding to band 4 are also homologous, but more distantly related; sequence homologies with class B chorion proteins are very limited or insignificant (Footnote). It is clear that this approach could be used whenever

Footnote: For very complex cases, such as the chorion multigene families, a possible danger is that a single electrophoretic band may contain polypeptides identical, closely homologous, distantly related or even unrelated to the polypeptide encoded by the clone. Depending on the relative abundance and the multiplicity of these components, misidentifications may occur. This danger is substantially reduced by increasing the resolution of the translated polypeptides, e.g. by two-dimensional electrophoresis. Another conceivable danger is that the "true" sequence corresponding to a clone may not be translatable by the particular cell-free system used, resulting in misidentification of a homologue as the "true" product. This danger may be overcome by using different cell-free translation systems[19]. On the other hand, hybrid-selected translation may be extremely useful for establishing that multiple, related polypeptides are encoded by distinct mRNAs, rather than resulting from different post-translational modifications: if distinct mRNAs are involved, multiple translation products should be detectable, following low-stringency hybridization and translation in a cell-free system which does not usually process translation products (e.g. the wheat germ system), while high-criterion hybridizations may yield fewer products, or products in different ratios, depending on the sequence homologies between the mRNAs.

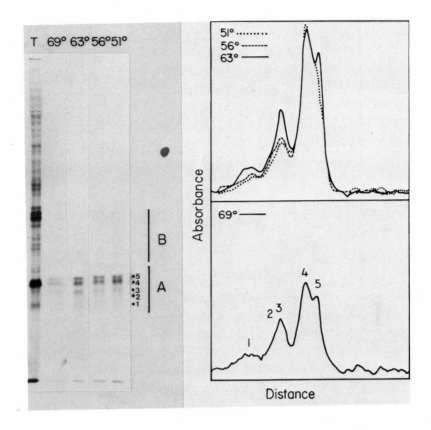

Figure 6: Characterization of a chorion gene family by hybrid-selected translation. Total RNA from B. mori follicles in early- to mid-choriogenesis (frac-
tional position 0.2 to 0.3, approximately stages IV and V; reference 20) was
hybridized with dots of DNA from the A family cDNA clone, ml453, and the hybridized RNA was eluted and translated in the wheat-germ system. Hybridizations
were performed at four temperatures, as shown; each used 4.2 µg DNA in three
dots, and 70 µl of hybridization mixture (see Experimental Protocols), containing
1.4 mg/ml total RNA. Washing was with 2X SET, 0.2 percent SDS (ten times at the
hybridization temperature) and with 10 mM Tris-HCl, pH 7.8, 2 mM EDTA (three
times at room temperature). As control, 6.8 µg of total RNA from total Gr^B mutant follicles was translated (T). Total and hybrid-selected translation products were analyzed on SDS-polyacrylamide gels. Autoradiograms are shown on the
left, with components of the A and B families indicated. Densitometric scans of
the A region are shown at the right, and reveal that the RNA most specifically
hybridizing with ml453 encodes band 3; bands 1, 2 and 5 are homologues, and band
4 is a more distant homologue.

some or all of the polypeptide products of a multigene family are distinguishable by electrophoresis (one- or two-dimensional), as in the globin, tubulin, actin or histone gene families. Since the approach depends on sequence homologies at the nucleotide, rather than the amino acid level, it yields information that extends rather than duplicates any immunochemical or protein sequence information that may be available. Furthermore, the approach can be further refined by using as immobilized DNA not the complete clone, but a fragment that corresponds to a rapidly evolving part of the gene (e.g. the 3' untranslated region; for resolution of close homologies), or to a conservative part of the gene (e.g. that encoding a functionally important polypeptide region; for detection of distant homologues).

III. Procedures

A. Dot Hybridization

Preparation of filters: All filters are washed in water for one hour. A plain nitrocellulose filter is mounted on top of two nitrocellulose filters with grids, on a sintered glass platform connected to a water aspirator. The filters are washed 3-4 times with 1M ammonium acetate before spotting the DNA.

Plasmid DNA is linearized by restriction endonuclease treatment and digested with Proteinase K (200 µg/ml in 50 mM Tris-HCl, pH 7.5; at 37°C, 30 minutes each with 0.2 percent and 2 percent SDS). After phenol extraction the DNA is denatured in 0.3N NaOH for ten minutes, chilled, and when needed is diluted with an equal volume of cold 2M ammonium acetate to a concentration of 1.4 µg/ml. It is then taken up in a capillary pipette attached to a micropipette filler (Clay Adams Suction Apparatus No. 4555), and is spotted on the filter under light vacuum; once the pipette touches the filter, contact is maintained continuously, while the DNA solution is delivered slowly with a combination of vacuum suction and positive pressure from the filler. For dot hybridizations, we routinely spot 0.7 µg DNA in 50 µl per dot, using a 100 µl micropipette. The dot is washed with a drop of 1M ammonium acetate, as is the area to be spotted next. After all the dots are made, the filter is washed again with 1M ammonium acetate under suction, air dried, treated with 2X Denhardt's solution[21] for one hour, drained, air dried, and baked at 80°C for 2 hrs.

Prehybridization: The dried filters are wetted evenly by slow immersion in 10X Denhardt's - 4X SET buffer (1X SET buffer is 0.15 M NaCl, 0.03 M Tris-HCl, pH 8, 1 mM EDTA), and are shaken in that solution for at least one hour. They

are transferred to sterile siliconized scintillation vials containing blank hybridization mixture, i.e. 50 percent deionized formamide, 2X Denhardt's solution,
4X SET, 0.1 percent SDS, 100 µg/ml yeast tRNA and 125 µg/ml poly(A), and are incubated for at least one hour at the hybridization temperature.

Hybridization and washing: The filters are hybridized for 16 to 48 hours
in hybridization mixture (prehybridization mixture supplemented with radioactive
probe). The temperature is selected to give the desired criterion of stringency[2,4]. Washes are performed at the same temperature, twice each with 4X, 3X,
2X, 1X, 0.5X and 0.2X SET, all with 0.1 percent SDS. Two final washes are performed at room temperature, with 0.1X SET without SDS. The filters are autoradiographed, either dry or moist and covered with Saran wrap (if melts are to be
undertaken).

B. Hybrid Selected Translation

Preparation of filters: The filters are prepared as for dot hybridization,
except that each spot receives 1.4 µg DNA in 100 µl solution, delivered with a
200 µl capillary pipette. After washing with 1M ammonium acetate, the filter
is air dried and baked, without treatment with Denhardt's solution. Small
squares, each containing one dot, are cut out with a razor blade.

Prehybridization, hybridization and washing: The filter squares are pre-
hybridized for one hour at incubation temperature, in 50 percent formamide, 0.6
M NaCl, 0.1M Pipes, pH 6.4, 100 µg/ml poly(A). They are then hybridized for
two hours in a siliconized test tube, using the same solution, minus poly(A)
but supplemented with the mRNA preparation. We usually use a volume of 30-100
µl (30 µl being sufficient for 3 dots), and a final concentration of total RNA
(of which approximately 2 percent is poly(A)[+] mRNA) of 1 to 2 mg/ml; these conditions easily permit recovery and translation of mRNA sequences representing
0.1 percent of the mRNA. The temperature is selected according the the criterion desired; for chorion sequences (approximately 57 percent G + C), we have
used successfully incubation temperatures as high as 69°C. Washing is usually
performed nine times at the same temperature in 1X SSC, 0.2 percent SDS, and
twice at room temperature in 10 mM Tris-HCl, pH 7.5, 2mM EDTA. Because of the
large amount of non-hybridized RNA, it is essential to wash thoroughly and in a
large volume; if non-specific RNA sticking is suspected, we recommend two initial washes with large volumes of prehybridization solution at the incubation
temperature, or transferring the squares to a clean tube after incubation, and
washing them in individual tubes for the last two (room temperature) washes.

<u>Recovery and translation of mRNA</u>: The RNA is eluted in 100 µl distilled water by boiling for one minute, quenched in ice, supplemented with 5 µg pure yeast tRNA, brought to 0.25M ammounium acetate, and precipitated with 2.5 volumes of 95 percent ethanol. The pellet is washed with 70 percent ethanol, dried, and redissolved in water for cell-free translation. We use the wheat germ cell free translation system, as described[22].

ACKNOWLEDGMENTS

Original work from F.C.K.'s laboratory was supported by grants from NSF, NIH and American Chemical Society. KI was supported by the Jane Coffin Childs Memorial Fund for Medical Research and the Leukemia Society of America.

REFERENCES

1. Kafatos, F.C., C.W. Jones and A. Efstratiadis, (1979) Nucleic Acids Res. 7, 1541-1552.
2. Sim, G.K., F.C. Kafatos, C.W. Jones, M.D. Koehler, A. Efstratiadis and T. Maniatis, (1979) Cell 18, 1303-1316.
3. Ricciardi, R.P., J.S. Miller and B.E. Roberts, (1979) Proc. Natl. Acad. Sci. USA 76, 4927-4931.
4. Iatrou, K., S.G. Tsitilou, M.R. Goldsmith and F.C. Kafatos, (1980) Cell 20, 659-669.
5. Dworkin, M.B. and I.B. Dawid, (1980) Devel. Biol. 76, 435-448.
6. Grunstein, M. and D.S. Hogness, (1975) Proc. Natl. Acad. Sci. USA 72, 3961-3965.
7. Thayer, R.E. (1979) Anal. Biochem. 98, 60-63.
8. Savakis, C. and P. Cherbas, (1981) in preparation.
9. Wolfner, M.F., D.J. Kemp, M.A.T. Muskavitch, G.M. Guild and D.S. Hogness (1981) in preparation.
10. Dworkin, M.B. and I.B. Dawid, (1980) Devel. Biol. 76, 449-464.
11. Williams, J.G. and J. Lloyd, (1979) J. Mol. Biol. 129, 19-38.
12. Lasky, L.A., Z. Lev, J-H. Xin, R.J. Britten and E.H. Davidson, (1980) Proc. Natl. Acad. Sci. USA 77, 5317-5321.
13. Weisrod, S. and Weintraub, H., (1981) Cell, in press.
14. Wahli, W., I.B. Dawid, T. Wyler, R.B. Jaggi, R. Weber and G.U. Ryffel, (1979) Cell 16, 535-549.
15. Howley, P.M., M.A. Israel, M-F. Law and M.A. Martin, (1979) J. Biol. Chem. 254, 4876-4883.
16. Jones, C.W., N. Rosenthal, G.C. Rodakis and F.C. Kafatos, (1979) Cell 18, 1317-1332.
17. Thireos, G. and F.C. Kafatos, (1980) Devel. Biol. 78, 36-46.
18. Moschonas, N.K., (1980) Ph.D. Thesis, University of Athens, Athens, Greece.
19. Bock, S.C. and M.R. Goldsmith, (1981) in preparation.
20. Nadel, M.R. and F.C. Kafatos, (1980) Devel. Biol. 75, 26-40.
21. Denhardt, D., (1966) Biochem. Biophys. Res. Comm. 23, 641-645.
22. Efstratiadis, A. and F.C. Kafatos, (1976) Methods in Molecular Biology 8, 1-124 (J.A. Last, ed.), Marcel Dekker, New York.

HYDROPHOBIC CHROMATOGRAPHY OF NUCLEIC ACIDS
AND PROTEINS ON TRITYLATED AGAROSE

PETER CASHION, ALI JAVED, DOLORES HARRISON,
JANE SEELEY, VICTOR LENTINI, and GANESH SATHE

Biology Department
University of New Brunswick
Fredericton, N.B. Canada E3B 5A3

I. INTRODUCTION

Trityl cellulose was developed in the early seventies by the senior author in collaboration with K. Agarwal in the laboratory of H. G. Khorana. It was used initially to simplify and speed up the purification of DNA fragments involved in the de novo chemical synthesis of a tRNA gene[1-3]. Later, it was observed[4,5] that trityl cellulose and more recently trityl agarose (TA) had a high binding affinity for poly A[+] mRNA, denatured DNA, and enzymes in general (see Figure 1).

II. Structure and Properties of Trityl-Agarose

A. Structure

Electron microscopic examination of beaded agarose shows regular bundles of polysaccharide fibers interspersed with large channels or voids, through which macromolecules diffuse when the resin is hydrated.[6] X-ray analysis[7] indicates that the basic polysaccharide structure is a left-handed duplex and that the fiber bundles seen with the electron microscope contain an average of about 500 duplexes each. Given the disaccharide repeat unit of agarose (Figure 1) and the stereochemical bulk of the trityl group and its electrophilic specificity, it can be predicted that, under average conditions, only the C_6-hydroxyl group of D-galactose can be tritylated. Furthermore, it may be calculated that high TA[a] (such as can be used to immobilize enzymes) has an overall trityl concentration of about 0.1 M, i.e., about 0.1 mmol of trityl groups per milliliter of bed volume. However, the fact that trityl groups are covalently anchored and hence confined within the volume domain of the agarose fibers (which themselves occupy only about 4 % of the column-bed volume) suggests that a more appropriate molarity designation for these trityl groups in a fiber's microenvironment would be at least 1 M. In turn, these heavily tritylated fibers would have a rather low dielectric constant, estimated to be close to that of ethanol (about 20), in sharp contrast with that of the "bulk solution" within the TA bead (about 80). Model-building indicates that one trityl group will cover or "umbrella" nearly two galactoses; hence, fully tritylated agarose fibers, with each alternating galactose monomer tritylated, will have a more or less continuous skin or veneer of trityl groups.

---

[a]"Low," "medium," and "high" TA-trityl groups at less than 20, 20-45, and 45-150 μmol/ml, respectively.

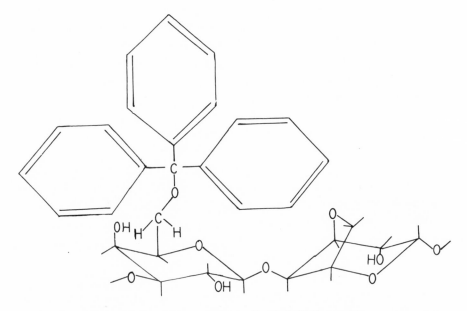

Fig. 1. Tritylated disaccharide repeat unit of trityl agarose.

B. Functions

Tritylated agarose, a novel hydrophobic resin, performs a wide variety of tasks for contemporary biochemists and molecular biologists, depending on the degree of tritylation:

- Isolation of poly A^+ mRNA (medium TA; Figure 4).
- Isolation of denatured DNA (medium TA; Figure 5).
- Isolation of individual RNA species, including mRNA, with tritylated complementary sequence synthetic ligands (low TA; Figure 6).
- Facile immobilization of enzymes and their stabilization in active form (medium-high TA; Table 2).

C. Binding/release mechanism

Polynucleotides or polypeptides that associate with TA are thought to do so exclusively through the formation of hydrophobic bonds between themselves and the trityl groups anchored to the resin. The surprising and unpredicted specificity for poly A tracts that lightly tritylated agarose (about one trityl group per 10 galactoses) exhibits is thought to be based on the more open, partially unstacked, and hence more accessible arrangement of this polypurine's bases.[4,5,8] A similar binding mechanism is postulated for denatured DNA.

554

Likewise, the generalized binding of proteins to more heavily tritylated agarose (about three to five trityl groups per 10 galactoses) is attributed to hydrophobic domains[9] on their surfaces, enriched in such residues as tryptophan, leucine, and phenylalanine. The acute sensitivity of these postulated hydrophobic bonds to the salt concentration in general (strengthened at higher salt, weakened at lower salt--see Figures 4-6), and to chaotropic, Hofmeister salts in particular, supports, with a variety of other experiments[4,5] the above binding mode. Finally, it has been suggested[4,5] that the pronounced "flypaper" stickiness of highly tritylated agarose for most enzymes could reflect merely the fact that, in their prepurification natural milieu, these enzymes were themselves membrane-associated and membrane-stabilized.[10,11] That is, it is conceivable that bundles of highly tritylated fibers resemble or mimic the amphipathic character of biomembranes enough for enzymes that were originally associated with membranes in vivo to be effectively induced or tricked into binding in vitro.

D. Functional resemblance to dT-cellulose and nitrocellulose

There is an obvious and uncanny resemblance between the behavior of poly A[+] mRNA on dT cellulose and on TA as a function of the salt concentration (Figure 4 and Table 1). This should not be interpreted, however, as indicating that similar binding/release mechanisms are operative. In the case of TA, elution of the mRNA at low salt[a] has been ascribed exclusively to the weakening of hydrophobic bonds. In the case of dT cellulose, it has been ascribed to the breakage of A=T hydrogen bonds, owing to interstrand phosphate repulsive forces-- although in this case, as well, hydrophobic forces are probably also involved. The functional homology with nitrocellulose is even more striking, in that it includes not only the moderate poly A specificity of nitrocellulose, but also the binding of denatured DNA, as well as the binding of most proteins at high salt (Figure 5 and Table 1). It appears difficult, at first glance, to account for these functional homologies between such apparently dissimilar materials as the hydrophilic nitrocellulose and the hydrophobic TA. However, as has been noted recently[12], the "nitrocellulose" designation is not very appropriate: in addition to the NO_2 groups, the cellulose is acetylated, which renders it sufficiently hydrophobic as to require the addition of some detergent (probably Triton X-100) to make it wettable and hence permeable to aqueous solutions.

---

[a]"Low" and "high" salt = 0.05 and 0.50 M NaCl, respectively, with 0.05 M Tris-HCl (pH, 7.5).

This more comprehensive description of the chemical makeup of nitrocellulose
clearly makes it easier to explain the functional similarity between it and TA.

TABLE I

COMPARISON OF BINDING SPECIFICITIES OF dT CELLULOSE, TRITYL AGAROSE,
AND NITROCELLULOSE[a]

	% Binding		
Polynucleotide	dT Cellulose	Trityl Agarose	Nitrocellulose
Poly A	100	100	100
Poly U	0	0	64
Poly G	0	0	9
Poly C	0	0	0
Native DNA	0	0	0
Heat-denatured DNA	0	100	100
Poly A[+] mRNA	100	100	--
3'-poly A-terminated -poly C	100	100	--

[a]Columns run as described in caption of Figure 5.

There are precedents for the binding of proteins and nucleic acids to hydrophobic polysaccharides. Lignins (polyaromatic natural products) contaminating
cellulose have been shown to bind poly A[+] mRNA.[13] Benzoylated celluloses have
been shown by Roberts[8] and others to bind poly A and usually poly U. Hofstee[14]
has shown that enzymes can be immobilized in active form by hydrophobic resins
containing quaternary nitrogens. Similarly, Butler has shown this with phenoxyacetyl cellulose[15] and Watanabe et al. with polyaromatic tannins.[16] In the
case of the poly A-binding resins just cited, high binding specificity is not
seen--poly U is often bound as well. Moreover, the benzoyl groups esterified
to cellulose are reported to have limited stability.[8] In the case of phenoxyacetyl cellulose,[15] the ester linkage is somewhat labile so that immobilized
enzymes have a relatively short column life. With respect to the other resins
cited,[14,16] ionic and hydrophobic forces appear to have an equivalent role in
the immobilization mechanism; hence, the number of enzymes that can be immobilized and the ease of immobilization might well be reduced by incompatible
isoelectric points, pH optimums, or salt requirements. Figure 2 shows the
general structures of some of the polysaccharide derivatives just mentioned.
The resemblance between these "natural" (lignin and tannin) and "unnatural"

(trityl, benzoyl, phenoxyacetyl, and Triton X) aromatic sugar adducts suggests a common mechanism for their binding of polynucleotides and proteins.

TABLE II

ACTIVITY OF ENZYMES IMMOBILIZED ON TRITYL AGAROSE COLUMNS

		% Activity Bound[a]
A.	Nucleases	
	1. Alkaline phosphatase (calf intestine)	83
	2. Alkaline phosphatase (E. coli)	100
	3. Acid phosphatase (wheat germ)	61
	4. Pancreatic DNase	17
	5. Venom phosphodiesterase (Crotalus adamanteus)	70
	6. Micrococcal nuclease	110
	7. ATPase	42
	8. S_1 nuclease	80
B.	Proteases	
	1. Trypsin	87
	2. Chymotrypsin	5[b]
	3. Carboxypeptidase A	5[b]
C.	Nucleotide bond synthesizing enzymes	
	1. Polynucleotide phosphorylase (M. luteus)	73
	2. DNA polymerase I (E. coli)	2
	3. RNA polymerase (E. coli)	30
	4. RNA ligase (T$_4$)	74
	5. Polynucleotide kinase (T$_4$)	359[c]
	6. AMV reverse transcriptase	13
	7. [γ-^{32}P] ATP preparation using coimmobilized glyceraldehyde-3-PO$_4$-dehydrogenase and 3-PO$_4$-glyceric acid kinase	83
D.	Restriction endonucleases	
	1. Eco RI	80
	2. Hind III	80
	3. Bam HI	20

[a]Column-immobilized activity compared with similar assay in free solution.
See Cashion et al.[5] for details.

[b]Given the specificity of these proteases for aromatic amino acids, we hypoth-
esize that the aromatic trityl groups themselves are inhibiting the enzymes.

[c]TA appears to stabilize the enzyme relative to the free solution assay during a 60-min incubation.

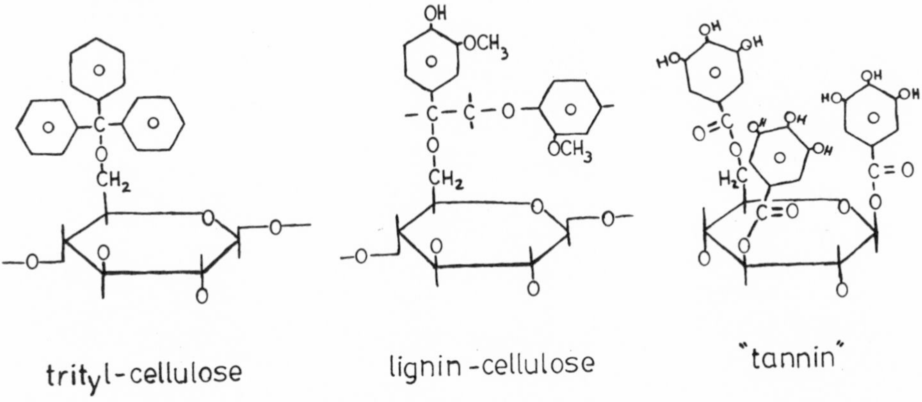

trityl-cellulose   lignin-cellulose   "tannin"

Fig. 2. Structural resemblances between synthetic and natural polyaromatic
sugar derivatives that hydrophobically bind polynucleotides and polypeptides.

III. EXPERIMENTAL PROCEDURES

A. Synthesis of Trityl Agarose

The approach followed involves suspending the polysaccharide in an anhydrous basic organic solvent, such as pyridine, and allowing the primary less-
hindered hydroxyl groups to react with the dissolved trityl chloride. Cellulose or beaded agarose may be used. Tritylated agarose is available from
Bethesda Research Laboratories (BRL) (Rockville, Md.). The procedure described below is similar to published ones.[1,2]

A 100-ml bed volume of beaded agarose (CL-4B, obtained from Sigma)[a] is
made anhydrous by suction filtration on a sintered-glass funnel with a few
bed volumes of 95% ethanol followed by a few bed volumes of pyridine. The
resin is then transferred to a 500-ml round-bottom flask and two or three
more bed volumes of pyridine are azeotropically flash evaporated. Finally,
to a total volume of about 200 ml, 8 g of trityl chloride is added. The reaction proceeds easily at room temperature with gentle agitation (see Figure 3).
Care must be taken during the drying process to ensure that the beads are
always wetted by the pyridine; otherwise, they could crack. Pyridine is
usually distilled over CaH_2. At the appropriate time, the reaction can be
stopped by pouring the suspension back into the sintered-glass funnel and

---

[a]It should be noted that cross linking is important.

washing with many bed volumes of ethanol or ethyl acetate to remove unbound trityl groups. Determination of the degree of derivatization has been described elsewhere.[4,5]

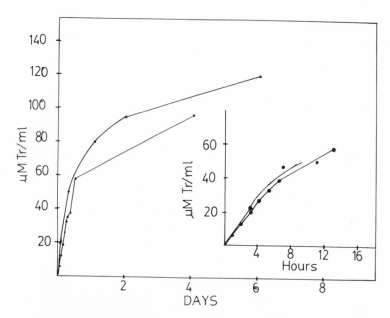

Fig. 3. Agarose tritylation kinetics. Two independent tritylation reactions are done as described in text. Medium TA is optimal for poly A$^+$ mRNA isolation; higher degrees of tritylation are satisfactory for most poly A$^+$ mRNA isolations, but the absolute specificity for poly A is not retained.[4,5] Good enzyme immobilization and stabilization are seen with high TA. For annealing experiments, as described in Figure 6 and accompanying text, low TA is used.

B. Isolation of poly A$^+$ mRNA or Denatured DNA

In general, the procedure for the TA isolation of either poly A, poly A$^+$ mRNA or denatured DNA is the same as for the well-known dT cellulose with respect to salt, pH, temperature, and bed volume. Figure 4 shows a typical high salt-low salt (or water) elution sequence of rabbit liver mRNA; comparison profiles with dT cellulose may be seen. To ensure quantitative and fast elution from TA, 0.1% sodium dodecyl sulfate (SDS) may be used in addition to 0.05 M Tris (pH, 7.5). Figure 5 shows the binding of native and heat-denatured DNA to TA; a comparison profile with nitrocellulose (Terochem Laboratories, Edmonton, Alberta, Canada) is also presented. Denatured DNA generally does not elute as readily as poly A$^+$ mRNA; the addition of 0.1% SDS facilitates the elution.

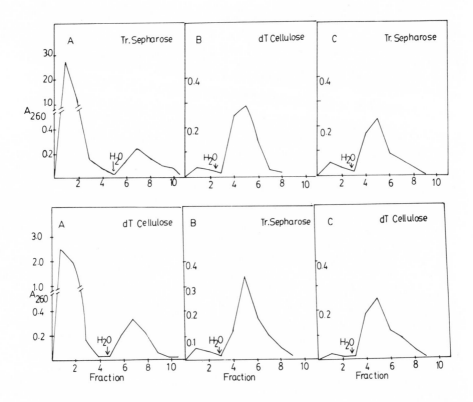

Fig. 4. Isolation of poly A⁺ mRNA on trityl agarose and dT cellulose columns. Each column (3-ml bed volume) was equilibrated with high salt. To each was added 40 A_{260} units of phenol-extracted (Darnell's method) rabbit liver RNA dissolved in high salt; after washing with high salt, the bound poly A⁺ mRNA was eluted with water (or low salt). For each column, the latter elute was made to high salt and reapplied in mutual crossover experiments. See Cashion et al.[4] for further details.

C. Isolation of Specific Polynucleotides With Complementary Tritylated Oligomers

These experiments are run as above for the isolation of poly A⁺ mRNA and denatured DNA, except that low TA may (but need not be) be used. Figure 6 shows (B) a typical elution sequence and (A) the requisite control patterns. This simple experiment can be seen as a prototype for the isolation of specific mRNA species, such as fibroin mRNA[4]; the use of a noncovalently anchored tritylated ligand allows annealing with the target mRNA molecules on the column proper or in free solution before applying to the TA column. The latter approach has some advantages.

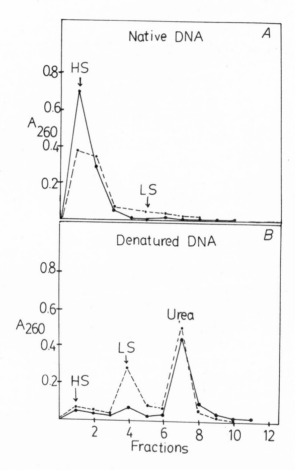

Fig. 5. Isolation of native and denatured DNA on trityl agarose and nitrocellulose columns. Columns run as described in Figure 4; 5 A_{260} units of native (calf thymus) or denatured (100°C, 4', then quick-cooled) DNA in high salt applied to each column. A, native DNA. B, denatured DNA. Solid line, TA column. Broken line, nitrocellulose column.

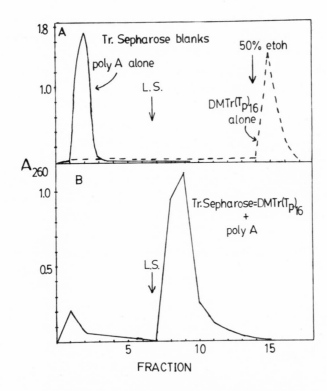

Fig. 6. Poly A isolation on low TA columns with hydrophobically anchored di-p-methoxytrityl $(Tp)_{16}$--DMTr$(Tp)_{16}$--as ligand. A, 5.6 A_{260} units of poly A in 2 ml of high salt was added to a low TA (13 μ mol/ml, 2-ml bed vol. equilibrated with high salt) column, eluted with high salt, and then with low salt. Later, 8 A_{260} units of 5'-DMTr$(Tp)_{16}$ in high salt was applied to the same column reequilibrated with high salt. Elution was continued with high salt, low salt, and finally 50% ethanol. Fractions were checked both for A_{260} and A_{500} (the latter being the A_{max} of the DMTr group in 35% $HClO_4$, with a molar extinction coefficient of about 45,000). B, after the ethanol was flash-evaporated from the DMTr$(Tp)_{16}$ solution, it was made to high salt and reapplied to the same column. Poly A (5.7 A_{260} units in high salt) was applied (as fraction 1) and washed in with high salt and then low salt as above. Fractions of about 3.3 ml were collected; recovery of poly A was quantitative. Later washing of the column with 50% ethanol gave full recovery of the tritylated ligand. Alternatively, the column may be reequilibrated with high salt and used for other poly A isolations without removing the ligand.

D. Immobilization of Enzymes

For enzyme immobilization, one simply adapts the normal free solution incubation to a TA-immobilized solid-phase context. This involves using medium or high-TA columns. The enzyme, in whatever buffered solution it is stable and active is applied to the TA column (bed volume, about 200 μl or as large as feasible, considering amounts of substrate and enzyme available), which has been equilibrated with the same buffer, and washed with a number of bed volumes of buffer to confirm that there has been no leakage off the column. The substrate (dissolved in the same buffer volume as that of the TA column) is then allowed to adsorb into the resin. Finally, the resin is incubated and the product is washed out with buffer solution, which is assayed for yield. High concentrations of glycerol or other organic enzyme stabilizers should be avoided, because they interfere with the hydrophobic adsorption to the TA fibers. In any case, they are usually found to be unnecessary in a TA resin, inasmuch as the trityl groups themselves seem to perform the same stabilizing function. The higher the salt concentration, the higher the enzyme binding capacity; likewise, higher trityl concentrations (over 100 μmol/ml) favor binding. Table 2 summarizes enzymes that have been checked. TA columns with immobilized enzyme are stored in buffer at 5°C; freezing must be avoided, lest the beads break. In our experience, the pH, salt, and co-factor requirements that are optimal for enzyme catalysis are not necessarily optimal for the long-term TA-adsorbed stability of the enzyme. For instance, calf intestinal alkaline phosphatase, which has a catalytic pH optimum between 9 and 10 with $(NH_4)_2 CO_3$ as buffer, is much more stable (half-life, over 6 mo) if stored in $ZnCl_2$ at a pH of about 6.

REFERENCES

1. Cashion, P., Fridkin, M., Agarwal, K., Jay, E. and Khorana, H. (1973) Biochemistry 12, 1985-1990.
2. Fridkin, M., Cashion, P., Jay, E. and Agarwal, K. (1974) in Methods in Enzymology, vol. 34, pp. 645-649.
3. Agarwal, K., Yamazaki, A., Cashion, P. and Khorana, H. (1972) Agnew. Chem. Internat. Edit. 11, 451-459.
4. Cashion, P., Sathe, G., Javed, A. and Kuster, J. (1980) Nucl. Acids Res. 8, 1167-1185.
5. Cashion, P., Javed, A., Sathe, G. and Ali, G. (1980) Nucl. Acids Res. Symp. series 7, 173-189.
6. Amsterdam, A., Er. El, Z. and Shaltiel, S. (1975) Arch. Biochem. Biophys. 171, 673-677.
7. Arnott, S., Fulmer, A. and Scott, W. E. (1974) J. Mol. Biol. 90, 269-284.
8. Roberts, W. K. (1974) Biochemistry 13, 3677-3682.
9. Lee, B. and Richards, F. M. (1971) J. Mol. Biol. 55, 379-400.
10. Schrock, H. L. and Gennis, R. B. (1977) J. Biol. Chem. 252, 5990-5995.
11. Lin, L.F.H. (1980) Biochemistry 19, 5135-5140.
12. Freifelder, D. (1976) in Physical Biochemistry, pp. 143-152, W. H. Free-man, San Francisco.
13. DeLarco, J. and Guroff, G. (1973) Biochem. Biophys. Res. Comm. 50, 486-492.

14. Hofstee, B.H.J. (1973) Biochem. Biophys. Res. Comm. 50, 751-757.
15. Butler, L. G. (1975) Arch. Biochem. Biophys. 171, 645-650.
16. Watanabe, T., Mori, T., Tosa, T. and Chibata, I. (1979) Biotech. Bioeng. 21, 477-486.

HYBRIDIZATION ANALYSIS OF SPECIFIC RNAs BY ELECTROPHORETIC SEPARATION IN AGAROSE GELS AND TRANSFER TO DIAZOBENZYLOXYMETHYL PAPER

J. Claiborne Alwine

Department of Microbiology, School of Medicine, University of Pennsylvania
Philadelphia, Pennsylvania, 19104

I. INTRODUCTION

The easy detection of specific nucleic acids in a heterogeneous population has become a necessity in molecular genetics. The technique developed by Southern[1] for the transfer of discrete fragments of DNA from gels to nitrocellulose has become one of the most widely used methods for detection of specific DNAs by hybridization analysis. However, RNA has a limited affinity for nitrocellulose, so the transfer of RNA to nitrocellulose has proved less reliable than the transfer of DNA.[*] Diazobenzyloxymethyl (DBM) paper was developed[2] to allow complete and consistent retention of RNA by covalent coupling to the paper as the RNA is transferred from the gel. This covalent interaction is primarily through the 2-position of guanosine or the 5-position of uridine residues[3]. Besides being used for the transfer of RNA[2], DBM paper has more recently been used for the transfer of DNA[3,4,6] and protein.[5] This paper deals with the transfer of RNA and is intended to be a bench reference for the transfer and the later hybridization analysis.

II. PREPARATION AND PROPERTIES OF DIAZOBENZYLOXYMETHYL PAPER

DBM paper is the final activated product of a series of reactions. Nitrobenzyloxymethyl (NBM) paper is reduced to aminobenzyloxymethyl (ABM) paper, which is diazotized to form the activated paper, DBM paper. The procedure for making NBM or ABM paper has been described elsewhere.[2,3] NBM and ABM paper are commercially available from Schleicher and Schuell, Inc., Keene, N. H. NBM paper is the more stable and can be stored at 4°C for at least a year. ABM paper has been kept for as long as a year at 4°C in a dry nitrogen atmosphere or a vacuum; it may turn yellow with time, especially if exposed to air; however, limited yellowing has not proven to hinder performance.

ABM paper is converted to DBM paper just before its use in a transfer experiment. A shallow pan containing 120 ml of 1.2 M HCl is placed in an icebath, and 3.2 ml of a freshly prepared 10 mg/ml solution of $NaNO_2$ (sodium nitrite) is added to the HCl. This forms nitrous acid for the diazotization; formation of nitrous acid can be checked by using starch iodide paper, which turns black in the presence of nitrous acid. A piece of ABM paper, cut to the size of the gel to be transferred, is placed in the cold nitrous acid solution and allowed to

---

[*]Addendum in proof: The problem of limited affinity of RNA for nitrocellulose appears to be resolvable using new denaturing gel systems. See this volume and Thomas, P. (1980). Proc. Natl. Acad. Sci. USA.

react for 30 min with occasional shaking. With these conditions, a piece of ABM paper of up to 200 cm^2 can be diazotized. If a larger surface area is used, a larger amount of nitrous acid solution should be prepared with HCl and NaNO$_2$ in the proportions stated above. The presence of nitrous acid should be checked with starch iodide paper periodically during diazotization. If the nitrous acid has been depleted, add more HCl and NaNO$_2$. At the end of the reaction, the paper is washed three times for 5 min each with 100 ml of ice-cold water and twice for 5 min each in 100 ml of ice-cold 25 mM potassium phosphate buffer (pH, 6.8). The paper will turn yellow as the pH is raised and should be used immediately for a transfer. If a transfer cannot be done immediately, the paper should not be washed, but should be left in the cold nitrous acid solution until the transfer can be done. DBM paper saturated in the nitrous acid solution can be stored at -20°C for several weeks.

III. TRANSFER PROCEDURES

A. Gel Electrophoresis Conditions

The conditions selected vary with the needs of individual experiments; however, RNA is most efficiently transferred from agarose gels (0.5-2.0%). Transfer from acrylamide gels has proved inefficient, except for small RNAs; however, composite agarose-acrylamide gels, with reversible cross-links, may be advantageous for some experiments.[3,6] The most widely used denaturing-gel systems for RNA transfers are agarose gels that contain 5-20 mM methylmercury hydroxide[7]* and agarose gels in which the samples have been treated with glyoxal[8]. Other denaturants (e.g., formaldehyde) can be used, but they must be soaked out of the gel before transfer, because they may interact with diazonium groups on the paper. Any amine will react with DBM paper. Therefore, amines in the gel (e.g., denaturants, Tris, EDTA, etc.) must be removed before transfer, lest the binding capacity of the DBM paper be substantially lowered.

Whichever gel system is chosen, some consideration must be given to the RNA samples to be subjected to electrophoresis. RNA must, of course, be prepared in intact form. More importantly, any RNAs that migrate near the ribosomal RNA will be difficult to detect after transfer, because the abundant rRNAs will

*Methylmercury is very toxic; use it only in a fume hood.

saturate the diazonium groups locally on the paper. Thus, for these species, selection of polyadenylated RNAs may be advantageous.

B. Preparation of the Gel for Transfer

For efficient and complete transfer of RNA from the gel to the DBM paper, the gel must be carefully prepared. During transfer, the first interaction between the RNA and the diazonium groups is an ionic interaction between a positive charge on the diazonium group and the negative charges of the RNA; this is followed quickly by covalent coupling. The primary charge interaction is very important; thus, transfers cannot take place under the high-salt conditions used for DNA transfers to nitrocellulose.[1] However, it has been found that RNAs larger than 19S do not transfer efficiently from the gel under low-salt conditions. Thus, a limited alkaline hydrolysis was introduced to hydro- lyze the RNA partially, in the gel, to produce smaller pieces. This allows complete transfer of large RNA species without substantial loss of smaller RNAs. The alkaline treatment is followed by neutralization and equilibration with transfer buffer. The following protocol describes the pretreatment procedures for an agarose gel (1-1.5%) 10 X 15 cm in area and 0.3-0.5 cm thick. Variations from these dimensions require adjustment in the volumes and lengths of treatment.

1. Alkaline Treatment. After electrophoresis, the gel is treated with two 150-ml portions of 50 mM NaOH for 15 min each at room temperature with gentle shaking. If the gel contained methylmercury hydroxide, the 50 mM NaOH solution should also contain 5-20 mM 2-mercaptoethanol to remove methylmercury from the RNA and gel. In the case of glyoxal gels, the alkali treatment alone will remove the glyoxal adducts from the RNA. If ethidium bromide staining of the RNAs is desired, the 50 mM NaOH solution can contain ethidium bromide at 1 µg/ml. The stained band pattern can be visualized with ultraviolet light after the gel has been neutralized (see below). Thus, the alkaline treatment is designed to remove denaturants, partially hydrolyze the RNA for efficient transfer, and stain the gel with ethidium bromide.

2. Neutralization and Equilibration of the Gel. Immediately after the alkaline treatment, the gel is neutralized by soaking it in two 150-ml portions of 0.2 M potassium phosphate buffer (pH, 6.8) for 15 min each at room temperature with gentle shaking. If 2-mercaptoethanol was included in the alkaline treatment, then the neutralization buffer should contain 10 mM iodoacetic acid; this removes excess 2-mercaptoethanol, which otherwise can react with diazonium groups.

After the first 15-min neutralization wash, the gel can be examined with ultraviolet light if it was stained with ethidium bromide during the alkaline treatment. Also at this point, the conversion of ABM paper to DBM paper should be started; this will allow the rest of the gel treatment and the preparation of the paper to conclude at the same time. After the second neutralization wash, the gel is equilibrated into the transfer buffer by being soaked in two 100-ml portions of 25 mM potassium phosphate buffer (pH, 6.8) for 7 min each at room temperature with gentle shaking. The gel is then ready for transfer with 25 mM potassium phosphate buffer (pH, 6.8).

Another transfer buffer that has been used is 0.2 M sodium acetate (pH, 4).[4] Although this has an advantage in DNA transfer (at a pH of 4, the denatured DNA is less likely to reassociate[4]), it appears not to be advantageous for RNA transfer. In the author's hands the acetate buffer causes broadening of the RNA bands.

C. Transfer.

Any transfer method used for the Southern transfer technique[1] can be used for RNA transfer to DBM paper. There should be a reservoir of transfer buffer (25 mM potassium phosphate buffer, pH 6.8) below the gel, DBM paper above the gel, and dry blotting paper with a weight on top. The transfer setup used in my laboratory consists of 10 pieces of Whatman 3MM or Schleicher and Schuell 597 paper soaked with transfer buffer to serve as the reservoir. The gel is laid on the soaked paper, and then the DBM paper is placed on the gel. Above the DBM paper are placed two pieces of 3MM or 597 paper soaked in transfer buffer and then a 3- to 4- in. stack of dry blotting material (paper towels). All the papers should be cut to the exact dimensions of the gel, with no overhang. The entire transfer setup should be wrapped in plastic wrap to prevent drying, with a weight on top (a 500-ml bottle of water works very well). Transfer at room temperature; wet blotting paper may be removed periodically, if desired, to hasten the transfer. However, the transfer can be left unattended and will be complete in 4-6 h, although it is often convenient to leave it overnight. The DBM paper turns a deep orange with time. At the end of the transfer procedure, care must be taken in the removal of the DBM paper from the gel, because fragments of the dehydrated gel may stick to the paper. This may be avoided by soaking the paper and gel in transfer buffer for a few minutes; the paper can then be blotted dry and stored in a sealed plastic boiling bag at 4°C for hybridization at a later time.

IV. HYBRIDIZATION

A. Treatment of the Transfer before Hybridization.

Before hybridization of labeled probe to a DBM paper transfer, the paper must be pretreated to remove unreacted diazonium groups and to block sites on the paper where the labeled probe would otherwise bind nonspecifically. Pretreatment is done in hybridization buffer (see below) containing 1% glycine. The glycine reacts with the diazo groups, and other components of the hybridization buffer (denatured calf thymus or salmon sperm DNA, Ficoll, and polyvinylpyrrolidone[9]) block nonspecific binding sites. Pretreatment is usually done in a sealable plastic boiling bag with 200 µl of pretreatment buffer per square centimeter of paper at 37-42°C for at least 5 h. After pretreatment incubation, the buffer can be squeezed out of the bag through a small hole cut in the corner. The paper is then ready for hybridization, but it can be stored, sealed in the bag, at 4°C. If stored for a long time, the paper should be washed once with hybridization buffer before the hybridization reaction starts.

B. Hybridization Conditions.

Either aqueous or formamide hybridization conditions can be used with DBM paper transfer; dextran sulfate can also be added to the hybridization buffer.[4] Hybridization conditions and incubation times vary with the experiment. Hybridization is done in plastic boiling bags; the same bag can be used for pretreatment and hybridization. In my laboratory the standard conditions for hybridization are: at least 24 h at 37-45°C in 50% formamide, 0.75 M NaCl, 75 mM sodium citrate, 25 mM sodium phosphate buffer (pH, 6.8), 0.2% w/v SDS, 0.02% w/v Ficoll, 0.02% w/v polyvinylpyrrolidone, and sonicated denatured calf thymus DNA at 1 mg/ml. For each square centimeter of paper, 50-100 µl of hybridization buffer are used. The labeled probe is denatured and added last. The entire mixture is put into the boiling bag, saturating the paper; bubbles are squeezed out, and the bag is sealed and submerged in a water bath. Occasional shaking is useful.

1. The Labeled Hybridization Probe

The ^{32}P-labeled probe most often used is DNA, labeled _in vitro_ by nick translation[10] or 5'-end labeling.[11] Whichever labeling method is used, the probe must be clean to maintain a low hybridization background. The easiest and best way to prepare a suitable probe is to stop the nick-translation reaction by making it 0.3 M in NaOH, boiling for 5 min, and separating incorporated from unincorporated deoxynucleotide triphosphates by chromatography on a column of Sephadex G-50 or G-75 equilibrated in 10 mM Tris (pH, 7.5), 10 mM NaCl, and 1 mM EDTA. After chromatography, the probe is ready for use; however, it should be denatured by boiling for 5 min just before it is added to the hybridization mixture. As a general rule, 100,000 to 500,000 cpm of nick-translated DNA probe (10^8 cpm/μg) is added to the hybridization reaction for each gel lane transferred to the paper.

C. Posthybridization Wash and Autoradiography

After hybridization, the paper is removed from the boiling bag, washed twice for 30 min at 37°C in 50-100 ml of 50% formamide, 0.75 M NaCl, and 75 mM sodium citrate, and then washed at least four times for 20 min each at 37°C in 50 to 100-ml portions of 0.15 M NaCl and 15 mM sodium citrate. More stringent washing conditions can be used, if necessary, to remove nonspecific hybridization (e.g., 15 mM NaCl and 1.5 mM sodium citrate at 45°C for 5 min). After the last wash, the paper is blotted dry, wrapped in plastic wrap, and autoradiographed.

V. REUSE

After autoradiography of DBM paper, the labeled probe can be removed by washing the paper several times in 25-ml portions of 99% formamide at 45°C. After this treatment, the paper is washed in 0.15 M NaCl and 15 mM sodium citrate, blotted dry, and stored in a sealed plastic boiling bag at 4°C or used in another hybridization reaction. Occasionally, some residual counts remain on the paper; if this is a problem for interpretation of later hybridizations, the phosphorus-32 may be left to decay. With care, a single DBM paper transfer can be reused many time. A second prehybridization treatment may be necessary after three to five reuses.

VI. CONCLUSIONS

The procedures presented here, if followed carefully, result in very sensitive detection of specific RNA species. It has been estimated[3] that 10-50 pg of specific RNA can be detected using a nick-translated probe and a reasonable period of autoradiography (5 d). In addition, relative quantitation can be calculated from autoradiographic band intensities, if hybridizations take place in probe excess.

572

REFERENCES

1. Southern, E. M. (1975) J. Mol. Biol. 98, 503.
2. Alwine, J. C., Kemp, D. J. and Stark, G. R. (1977) Proc. Natl. Acad. Sci. U.S.A. 74, 5350.
3. Alwine, J. C., Kemp, D. J., Parker, B. A., Reiser, J., Renart, J., Stark, G. R. and Wahl, G. M. (1980) Methods in Enzymology 68, 220.
4. Wahl, G. M., Stern, M. and Stark, G. R. (1979). Proc. Natl. Acad. Sci. U.S.A. 76, 3683.
5. Renart, J., Reiser, J. and Stark, G. R. (1979). Proc. Natl. Acad. Sci. U.S.A. 76, 3116.
6. Reiser, J., Renart, J. and Stark, G. R. (1978). Biochem. Biophys. Res. Commun. 85, 1104.
7. Bailey, J. M. and Davidson, N. (1976). Anal. Biochem. 70, 75.
8. McMaster, G. M. and Carmichael, G. G. (1977). Proc. Natl. Acad. Sci. U.S.A. 74, 4835.
9. Denhardt, D. (1966). Biochem. Biophys. Res. Commun. 23, 641.
10. Rigby, P. W. J., Dieckmann, M., Rhodes, C. and Berg, P. (1977). J. Mol. Biol. 113, 237.
11. Maxam, A. H., and Gilbert, W. (1977). Proc. Natl. Acad. Sci. U.S.A. 74, 560.

ELECTRON MICROSCOPY OF NUCLEIC ACIDS

CLAUDE F. GARON

Laboratory of Biology of Viruses, National Institute of Allergy
and Infectious Diseases, National Institutes of Health
Bethesda, Maryland   20205

I. INTRODUCTION

While electron microscopy has often played a significant role in the study
of various biological systems, that role was expanded appreciably with the description of a reliable procedure for visualizing nucleic acid molecules by

574

Kleinschmidt and Zahn[1]. Their basic protein monolayer technique, or modifications thereof, has been used successfully to ask both qualitative and quantitative questions about nucleic acids. And, since nucleic acid structure often
suggests function, the potential for defining features of replication, transcription and recombinational processes often exists and is frequently exploited. Recent advances in recombinant DNA technology and its widespread use
would promise to further expand the value of this tool. However, given the
intricacies of the various sample preparation procedures and the high failure
rate of many, the true potential of this rapid method of nucleic acid structural analysis probably has not been fully realized. Therefore, at the request
of the editors, an attempt will be made to describe in some detail, simple
protocols for preparing and mounting nucleic acid molecules for electron microscopy. These procedures have proven reliable under a wide variety of experimental conditions and have given consistant results in our laboratory for many
years. An attempt will be made to define those factors which, in our view,
contribute most often to the success or failure of this experimental approach.
And, finally, we will suggest ways in which these procedures might be more
fully exploited in what will undoubtedly be a rapidly expanding world of
nucleic acid analysis. This expansion is made possible, in large part, by new
techniques of molecular cloning and gene amplification now available.

The mounting of nucleic acid molecules for electron microscopic analysis
may be conveniently described in four steps: 1) the formation of a nucleic
acid-protein complex; 2) the spreading of this complex onto a fluid surface;
3) the transfer of the monolayer to a solid support medium; and 4) contrast
enhancement and viewing. Since prior sample preparation as well as subsequent
mounting protocols has a profound effect on the final image and its usefulness
for analysis, conditions must be carefully monitored throughout and, in most
cases, samples prepared with the requirements of electron microscopy in mind.

Although numerous procedural variations have been described for various
kinds of structural analyses, all may be placed into one of two categories:
aqueous or formamide. The aqueous procedure remains in our hands one of the
most useful and reliable methods for visualizing double-stranded DNA or RNA
molecules in the electron microscope. It appears less affected by minor variations in procedure. Stock reagents are more stable. Quantitative results are
more consistent and reproducible. Differentiation between single and double
stranded molecules or between single and duplex regions of an individual molecule is absolute and requires no interpretive skill. Although duplex DNA and

RNA are not visually distinguishable, the single important disadvantage of the aqueous procedure is that the procedure gives no quantitative information about single strands. Under the conditions employed single strands collapse into tight aggregates (bushes) due to random base to base interactions.

These random interactions may, however, be discouraged, and the single strands extended, by the addition of any of a number of chemical denaturants[2]. These denaturants form the basis for the second category of mounting procedures. The one most often used is formamide since[3,4] it, 1) is effective over a wide range of concentrations and, 2) enhances rather than reduces contrast over that range. At low concentrations of formamide single strands are extended to some degree and duplex molecules are not visually affected. At higher concentrations single strands may be well extended and partially homologous regions of an appropriate molecule (a heteroduplex, for example) may show areas of strand separation. However, native or duplex molecules under these conditions appear unaffected. At very high formamide concentrations A-T rich regions of native or duplex molecules may denature. Most duplex DNA molecules may be completely denatured in 98% formamide. While it is true that the addition of formamide to the basic protein monolayer spreading technique may provide valuable structural information about single strands, may allow the detection of partial homology, or the production of denaturation profiles, the price may be a considerable loss of reliability and consistancy. Image quality may be affected by all manner of procedural variation and, at certain formamide concentrations, the maintenance of reasonable differentiation between single and double strands may be a problem. Thus, formamide mounting protocols must always be carefully controlled and contour lenghts stringently calibrated by the inclusion of appropriate standards.

II. SAMPLE MOUNTING TECHNIQUES

Parlodion coated 200-400 mesh copper grids are freshly prepared each day using a 2.9% solution of oven dried (90°C for 24 hours) parlodion strips in reagent grade butyl acetate. The strips take several days at room temperature to dissolve completely, but the solution is usable for one month or more thereafter. New solutions are prepared when expanding pin holes appear in the film upon electron beam exposure. Rapid tearing of the film generally denotes an excessively thin coating rather than a faulty parlodion solution. Grids are rinsed individually in acetone prior to placing them on the stainless steel screen beneath the water surface of the trough shown in Fig. 1. The initial drop of parlodion solution is allowed to dry on the surface of the fluid for at least one minute and is then discarded. The second drop is dried for a similar time before the water is drained slowly from the bottom of the trough allowing the film to settle gently upon the grids. The screen is removed, covered with a glass petri dish and allowed to dry at 55°C for 35 minutes. After cooling to room temperature the grids are used without further treatment.

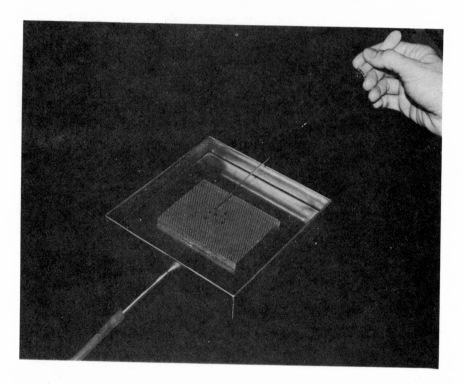

Fig. 1. Procedure for parlodion coating microscope grids.

In both the aqueous and formamide procedures, two solutions must be prepared from stocks immediately prior to use. The spreading solution contains the nucleic acid, cytochrome C and appropriate buffers and salt, while the second solution, the hypophase, serves as the support fluid upon which the protein monolayer is formed.

A. Aqueous Procedure

The hypophase solution for aqueous spreading consists of 120 ml of 0.25 M

ammonium acetate. This volume fills a 100x15 mm square plastic petri dish
(Lab Tek No. 4021). Teflon bars provide a non-wettable surface in the front of
the dish and a convenient microscope slide support in the rear. Plain glass
slides, which are stored in sulfuric acid-dichromate cleaning solution, are
washed exhaustively in deionized water and allowed to air dry immediately prior
to use. A typical aqueous spreading solution (50 μl total volume) containing
35 μl of 1 mM EDTA, 5 μl of 5 M ammonium acetate, 5 μl of nucleic acid sample
(concentration 1-5 μg/ml) and 5 μl of cytochrome C (1 mg/ml) is drawn up in a
capillary pipette and gently extruded onto the glass ramp approximately 1 cm
above the hypophase solution. The monolayer is aged for 50-60 sec before a parlodion coated grid is touched briefly to the surface of the hypophase solution
at a point approximately one grid's diameter from the glass-fluid interface (Fig.
2B). The grid is immediately transferred to a 10^{-5} M uranyl formate[6] staining
solution freshly prepared each day in 90% ethanol. The grid is stained for 30
seconds, transferred to a 90% ethanol wash for 10 seconds, blotted with filter
paper and allowed to air dry. All solutions and stains should be filtered (pore
size 0.2 μm) being careful to exclude any detergents or wetting agents that might
have been incorporated into the filter membrane. Grids may be viewed either before or after further contrast enhancement by rotary shadowing.

B. Formamide Procedure

While the formation of the nucleic acid-protein monolayer in the formamide
procedure is similar to that of the aqueous procedure, the composition of both
spreading solution and hypophase differ markedly. These differences relate
to 1) the adequate buffering of the formamide, which tends to become rapidly
acidic, and 2) the proper balance of salt, temperature and formamide concentration necessary to achieve the desired level of denaturation during speading.
The formamide procedure that follows has, in our hands, the advantages of
simplicity, reliability and produces excellent differentiation between single
and double strands. A stock spreading solution (1 ml total volume) is prepared
each day and consists of 700 μl of formamide (99% Matheson, Coleman and Bell,
No. FX420), 100 μl of 1 M Tricine buffer, pH 8.0, 50 μl of 5 M NaCl, 20 μl of
0.5 M EDTA and 130 μl of distilled water. Immediately prior to spreading 8 μl
of nucleic acid sample is added to 40 μl of the above stock solution and mixed
gently but thoroughly with 2 μl of cytochrome C (1 mg/ml). The sample is taken
up in a capillary pipette and extruded down a glass slide ramp onto the surface
of a deionized water hypophase. Since components of the spreading solution
diffuse away from the monolayer and into the water hypophase with a significant

Fig. 2. The apparatus for spreading the protein monolayer upon the hypophase (A), and for transferring the monolayer to grids.

loss in contrast, the monolayer must be transferred to a parlodion coated grid within 20 seconds. Grids are stained with uranyl formate as previously described. Nucleic acid strands are clearly visible following staining although with less contrast than is achieved with the aqueous procedure.

C. Contrast Enhancement

Rotary shadowing of grids with heavy metals remains the method of choice for achieving both contrast enhancement and greater monolayer stability during electron microscopy of nucleic acids. Approximately 3 cm of platinum-palladium (80:20, 8 mil diameter) is wrapped tightly around the tip of a 25 mil tugsten V-filament, pumped to a vacuum of 2.5×10^{-5} Torr before vaporizing for 15 sec. at a "distance from the grids" to "height above the grids" ratio of 8:1. See Fig. 3.

Fig. 3. An apparatus for rotary shadowing several microscope grids with platinum-palladium for contrast enhancement.
The grids are rotated at approximately 40 rpm during the procedure.

As shown several grids may be rotary shadowed simultaneously provided they are clustered near the center of the rotating table. Additional levels of contrast may be achieved by 1) viewing grids at the lowest (40kV or less) accelerating voltages available, 2) by the use of small (50 μm or less) objective apetures and 3) by the use of a pointed electron gun filament. Dark field illumination provides added contrast as well. Finally, photographic contrast enhancement methods may be employed during the initial development of microscope negatives or during the production of final prints.

Generally, however, since contrast is largely determined by the size of the column of protein surrounding the nucleic acid molecule and the thickness of the protein monolayer, efforts are better spent at that level of the procedure. While grids may be successfully stored in open dishes or grid storage boxes for several weeks, some loss in contrast is to be expected over that time period. For that reason more immediate photographic recording is to be encouraged.

D. Contour Length Measurement and Calibration

Devices available for determining the contour lengths of nucleic acid molecules range from simple mechanical "map measurers" to elaborate image analyzing systems which can be interfaced to computers for data storage and/or statistical analysis. However, as important as the measurement instrumentation is the understanding that the absolute length of both single and double stranded nucleic acid molecules depend to a great degree upon the conditions under which the molecules were mounted. Although single strands are clearly more sensitive to alterations in ionic strength, temperature, pH and other characteristics of the spreading mixture, duplex molecules may also be affected[7,8]. Because of these effects, meaningful linear densities can only be attained by direct comparison with carefully chosen calibration standards. The best standards are nucleic acid molecules which are 1) structurally related to the sample in size and form; 2) of acurately known molecular weight, and 3) clearly distinguishable from the sample when mounted and viewed together. Circular calibration standards (SV-40, polyoma or ØX174 RF II DNA for duplex; ØX174 DNA for single strands) are now commercially available and offer by their circular structure a guarantee of intactness. Covalently closed molecules must, however, be relaxed prior to use. A convenient method involves the incubation of supercoiled molecules with pancreatic DNAase I (Worthington Biochemical Corp. Cat. No. DPFF) at a concentration of 0.02 units/ml for 25 min at 25°C. These conditions produce approximately one single-strand break per DNA molecule[9].

E. Trouble Shooting

Difficulties encountered with mounting molecules for electron microscopy can often be traced to one of several problems with the nucleic acid sample itself. Purification procedures which result in inadequate removal of either bound protein or certain chemical agents from the final sample may serve in one degree or another to disrupt mounting protocols. Effective spreading is difficult in the presence of even residual amounts of such agents as ethanol, phenol, cesium chloride, SDS or other detergents, high salt or sucrose.

Another major category of difficulty involves structural damage to the nucleic acid molecules. Often molecules which appear suitably intact for analysis by gel electrophoresis or other techniques will prove unsuitable for electron microscopic analysis, particularly those that involve denaturation and reannealing. Nucleic acid molecule damage may result from either mechanical shear forces or from the action of contaminating nucleases. The solution to the former category is extraordinary care in handling at all stages of the procedure. Large molecules are invariably more susceptible to the shear damage produced by small bore pipetting, stirring and shaking. Sonication and freeze-thaw procedures should also be avoided. Often the introduction of a viscous agent such as sucrose (20%) serves to protect large molecules from shear damage during handling. The agent must, however, be removed prior to mounting for microscopy. The damage of nucleic acid molecules by contaminating nucleases may be minimized by the use of sterile reagents, by the maintenance of low temperature and alkaline pH (8.0-8.5) where possible and by the inclusion of at least millimolar levels of chelating agents, such as EDTA, into all purification, spreading and storage reagents.

F. An EM Nicking Assay

Since intact nucleic acid molecules are the sine qua non of many kinds of structural analyses, particularly those involving denatured and reannealed molecules, a procedure is described whereby the source of troublesome nuclease damage can be rapidly and sensitively detected in the electron microscope. The assay takes advantage of the easily observable structural conversion of a covalently closed, supercoiled molecule into a relaxed open circular form as a consequence of one single strand break[10]. Relaxed circular molecules may accumulate, depending on size, a number of additional single strand breaks before they are converted to full length linear duplexes. The production of smaller linear fragments would indicate considerable nuclease activity. The entire purification procedure or any part of it may be tested for nuclease contamination by this method. It is critical that the cumulative nuclease activity throughout the isolation, purification and mounting procedures results in a total of less than a single break per molecule if successful structural analysis such as heteroduplexing or R-looping are to be attempted.

Restriction endonuclease digestion protocols have often been identified as a source of considerable random nicking activity, particularly when excessive enzyme is added or when long incubation times are used to insure limit digests. While nuclease contamination may be assayed using the procedures described above, a supercoiled molecule which is not specifically cleaved by the particular restriction endonuclease being tested is required. Single stranded, circular molecules such as ØX174 DNA may be used to assay certain enzymes although care should be exercised since specific cleavage of single stranded molecules has been described.

III. APPLICATIONS - DNA ANALYSIS

The technology for visualizing single strands and the introduction of the heteroduplex analysis method by Davis and Davidson[11] and by Westmoreland, Szybalski and Ris[12] have permitted the growth of electron microscopic DNA analysis much beyond the simple definitions of size and conformation. Using appropriately denatured and reannealed samples, direct genetic comparisons have been between members of related viral groups based on the identification of characteristic structures representing heterology, deletion or substitution.

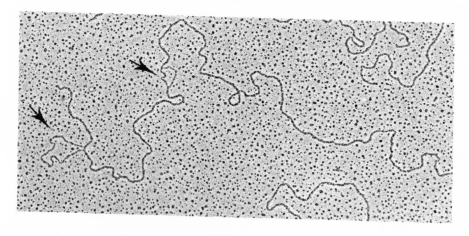

Fig. 4. Electron micrograph of heteroduplex molecules formed between DNA isolated from adenovirus serotypes 12 and 31. The arrows indicate the positions of a characteristic deletion loop and a heterology bubble. The molecules were mounted as described in the text.

Further improvements, mostly in the titration of formamide concentrations used in spreading solution and hypophase, have also allowed a more precise look at regions of partial homology or genetic drift.

A. Heteroduplex Formation

This method requires that DNA from two different sources be mixed, denatured and renatured under appropriate conditions. The resulting heteroduplex molecules (i.e., duplexes composed of one strand from one source and one strand from the other) can then be identified and characterized if they contain a sufficiently large region of noncomplementarity between strands. The detection limit of these regions is generally agreed to fall in the range of 50-100 nucleotide pairs. In certain heteroduplex molecules, however, the boundaries of homology and non-homology are not sharp and the observation that the fraction of heteroduplex observed as double-stranded decreases with an increase in denaturing conditions (Fig. 5) is a common and useful indicator of partial homology[13]. A convenient protocol involves mixing an equal volume of the purified DNA and 0.2 N NaOH. Following an incubation period of 10 minutes at room temperature (all strands are demonstrably single stranded by aqueous spreading) the pH is adjusted to 8.5 by the addition of 2 M Tris-hydrochloride buffer. Formamide is added to give a 50% final concentration. The moleucles are mounted for microscopy using the formamide procedure previously described.

Although the conditions affecting the reassociation of nucleic acids in solution are well defined[14,15], optimum incubation conditions for the formation and subsequent mounting of specific heteroduplex molecules may require careful consideration. For example, extremely heterologous molecules, or molecules which differ markedly in length (Fig. 6), do not form as efficiently or as stably as homoduplexes in the same solution. Thus, these molecules may be seen only transiently during the incubation period. These structures must be protected from subsequent reassortment into homoduplexes by underannealing.

Heteroduplex structures with many branches, bubbles or loops are invariably more difficult to spread unambiguously and are seen more often as tangled aggregates. Therefore, these structures may be more easily viewed following longer incubations at lower concentrations of DNA. Although one would predict a heteroduplex population approaching 50% of the reannealed duplexes, the actual number of heteroduplex molecules present in a typical field may be considerably less.

584

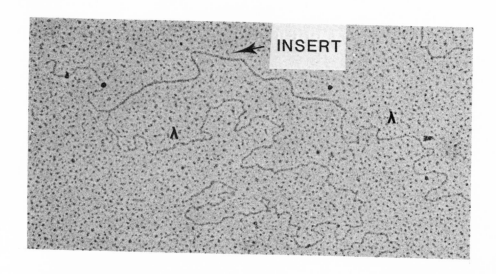

Fig. 6. Electron micrograph showing the hybridization of a viral sequence to a bacteriophage lamda amplified recombinant DNA molecule. Lambda arms extend far beyond the edge of the photograph.

While the sensitivity of electron microscopic heteroduplexing methods does not approach that of restriction endonuclease analysis in detecting small regions of mismatch, its utility lies in the rapid detection of nucleotide sequence arrangements or reassortments. An example of that type of analysis is illustrated by the terminal sequence arrangements that were deduced, in part, based on the appearance of DNA molecules in the electron microscope following various treatments. A schematic summary of 4 classes of molecules are presented in Figure 7 and represent the categories of terminal redundancy found in various viral DNA molecules. The circular molecules formed from bacteriophage lambda simply by annealing are a consequence of sequence arrangement which features exposed, complementary termini[16]. Category II molecules, represented by T3, T7 and P22, circularize following (and only following) sequential removal of a number of 3' nucleotides by exonuclease digestion[17]. Cyclically permuted[18] populations of DNA molecules, such as the genomes of T2, T4 and others, will readily form duplex circles following denaturation and reannnealing. Category IV molecules, represented by members of several animal virus

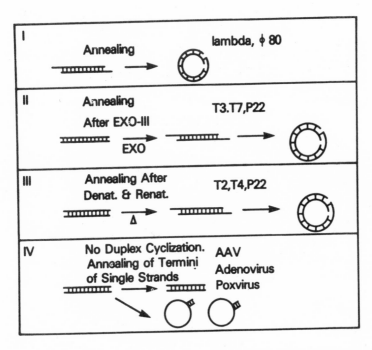

Fig. 7. A schematic representation of behavior of molecules having different terminal sequence arrangements.

groups, posses an inverted terminal repetition which does not permit the formation of duplex circles, but rather circularizes only in single stranded form - a consequence of terminal, intrastrand interaction [19-23]. The formation and visualization of circular molecules from eukaryotic DNA fragments as a result of tandomly repeated DNA sequences [24,25], together with the repeated identification of inverted repeats in a number of procaryotic transposable elements [26], represent major areas of DNA structural analysis where electron microscopy may contribute. Doubtless our recently acquired ability to produce assayable quantities of relatively rare nucleotide sequences, such as integrated viral genomes, by recombinant DNA techniques, will expand our use of this technique even further.

IV. APPLICATIONS - RNA ANALYSIS

A. R-Looping

The ability to accurately map individual viral RNA transcripts in the electron microscope was made possible by a description of conditions[27-29] which would allow RNA to hybridize with double stranded DNA, causing the displacement of the non-complementary DNA strand. The resulting structure, called an R-loop, is comprised of a double stranded DNA-RNA hybrid arm and a single stranded DNA arm (Fig. 8) and can be maintained during appropriate mounting procedures. Rapid and efficient formation of R-loops occurs under conditions in which the DNA-RNA hybrids are more stable than the DNA duplex. Such conditions are achieved by using a combination of temperature and chemical denaturation (usually formamide) which nearly result in DNA-DNA strand separation. A convenient reaction mixture (100 μl total volume) consists of 70 μl of formamide (99% Matheson, Coleman and Bell), 10 μl of 1 M Tricine buffer, pH 8.0, 5 μl of 5 M NaCl, 2 μl of 0.5 M EDTA and 13 μl of the DNA-RNA mixture. Since the rate of R-loop formation is directly proportional to the RNA concentration incubation times vary. Incubation temperature is critical, the maximum rate of R-loop formation being within one degree of the denaturation temperature for that DNA molecule. That temperature may be imperically determined by raising the temperature of the reaction mixture containing the specific duplex DNA incrementally until the conversion to single strands is observed in the electron microscope[27].

Although poly (A) tails are occasionally visible directly in R-loops, the detection limits have been greatly increased by the use of a visual marker[30]. Furthermore, spreading and measurement of R-loop structures may be simplified

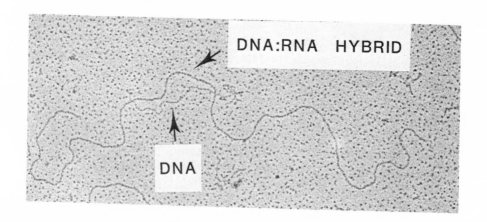

Fig. 8. Electron micrograph of a typical R-loop structure showing the DNA strand being displaced by a hybridized RNA molecule. A poly (A) tail is visible at one end of the R-loop, an unhybridized region of RNA at the other.

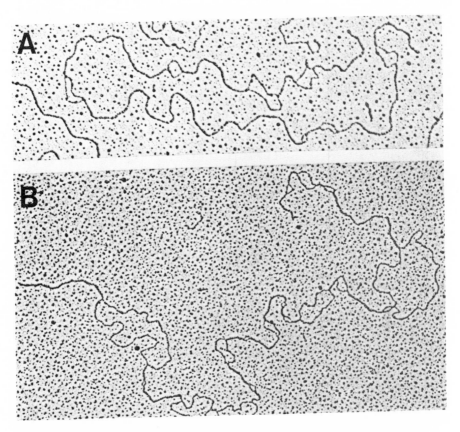

Fig. 5. Electron micrographs of heteroduplex molecules formed with DNA from members of the highly oncogenic group of adenoviruses. Panel A was mounted in 60% formamide, panel B in 70%. Note the greatly increased areas of heterology at the higher formamide concentration.

by removal of unhybridized RNA molecules[31]. Enormous potential still exists for further identifying and accurately mapping both viral[32-34] and cellular[35] sources of RNA along duplex DNA molecules.

ACKNOWLEDGMENTS

I wish to thank P.A. Sharp for conducting a 2 day training session on structural analysis of nucleic acid molecules, at which time many of these procedures and an appreciation of their use was conveyed.

REFERENCES

1. Kleinshchmidt, A.K. and Zahn, R.K. (1959) Naturforsch. B 14, 770-779.
2. Davis, R.W., Simon, M. and Davidson, N. (1971) Methods in Enzymology D 21, 413-428.
3. Bonner, J., Kung, G. and Bekhor, I. (1973) Biochemistry 6, 3650-3653.
4. McConaughy, B.L., Laird, C.D. and McCarthy, B.J. (1969) Biochemistry 8, 3289-3295.
5. Inman, R.B. and Schnos, M. (1974) In: Principles and Techniques of Electron Microscopy. Ed. M.A. Hayat 4, 64-80.
6. Leberman, R. (1965) J. Mol. Biol. 13, 606.
7. Lang, D., Bujard, H., Wolff, B. and Russell, D. (1967) J. Mol. Biol. 23, 163-181.
8. Bujard, H. (1970) J. Mol. Biol. 49, 125-137.
9. Martin, M.A., Howley, P.M., Byrne, J.C. and Garon, C.F. (1976) Virology Virology 71, 28-40.
10. Vinograd, J., Lebowitz, J., Radloff, R., Watson, R., and Laipis, P. (1965) Proc. Natl. Acad. Sci. USA 53, 1104-1110.
11. Davis, R.W. and Davidson, N. (1968) Proc. Natl. Acad. Sci. USA 60, 243-250.
12. Westmoreland, B.C., Szybalski, W. and Ris, H. (1969) Science 163, 1343-1348.
13. Davis, R.W. and Hyman, R.W. (1971) J. Mol. Biol. 62, 287-301.
14. Wetmur, J.G. and Davidson, N. (1968) J. Mol. Biol. 31, 349-370.
15. Bluthmann, H., Bruck, D., Hubner, L. and Schoffski, A. (1973) Biochem. Biophys. Res. Comm. 50, 91-97.
16. MacHattie, L.A. and Thomas, C.A., Jr. (1964) Science 144, 1142-1144.
17. Richardson, C.C. and Kornberg, A. (1964) J. Biol. Chem. 239, 242-250.
18. Streisinger, G., Edgar, R.S. and Denhardt, G.H. (1964) Proc. Natl. Acad. Sci. USA 51, 775-779.
19. Garon, C.F., Berry, K.W. and Rose, J.A. (1972) Proc. Natl. Acad. Sci. USA 69, 2391-2395.
20. Wolfson, J. and Dressler, D. (1972) Proc. Natl. Acad. Sci. USA 69, 3054-3057.
21. Garon, C.F., Berry, K.W. and Rose, J.A. (1975) Proc. Natl. Acad. Sci. USA 72, 3039-3043.
22. Koczot, F.J., Carter, B.J., Garon, C.F. and Rose, J.A. (1973) Proc. Natl. Acad. Sci. USA 70, 215-219.
23. Garon, C.F., Barbosa, E. and Moss, B. (1978) Proc. Natl. Acad. Sci. USA 75, 4863-4867.
24. Pyeritz, R.E. and Thomas, C.A., Jr. (1973) J. Mol. Biol. 77, 57-73.
25. Lee, C.S. and Thomas, C.A., Jr. J. Mol. Biol. 77, 25-42.
26. Kleckner, N. (1977) Cell 11, 11-23.

27. Thomas, M., White, R.L. and Davis, R.W. (1976) Proc. Natl. Acad. Sci. USA 73, 2294-2298.
28. White R.L. and Hogness, D.S. (1977) Cell 10, 167-176.
29. Kaback, D.B., Angerer, L.M. and Davidson, N. (1979) Nucleic Acid Res. 6, 2499-2517.
30. Bender, W. and Davidson, N. (1976) Cell 7, 595-607.
31. Chan, H.W., Garon, C.F., Chang, E.M., Lowy, D.R., Hager, G.L., Scolnick, E.M., Repaske, R. and Martin, M.M. (1980) J. Virol. 33, 845-855.
32. Meyer, J., Neuwald, P.D., Lai, S.P., Maizel, J.V., Jr., and Westphal, H. (1977) J. Virol. 21, 1010-1018.
33. Chow, L.T., Gelinas, R.E., Broker, T.R., and Roberts, R.J. (1977) Cell 12, 1-8.
34. Wittek, R., Barbosa, E., Cooper, J.A., Garon, C.F., Chan, H. and Moss, B. (1980) Nature 285, 21-25.
35. Tilghman, S.M., Tremeier, D.C., Seidman, J.G., Peterlin, B.M.M. Sullivan, M.M., Maizel, J.V. and Leder, P. (1978) Proc. Natl. Acad. Sci. USA 75, 725-729.

Subject Index